The Art of the Botanist

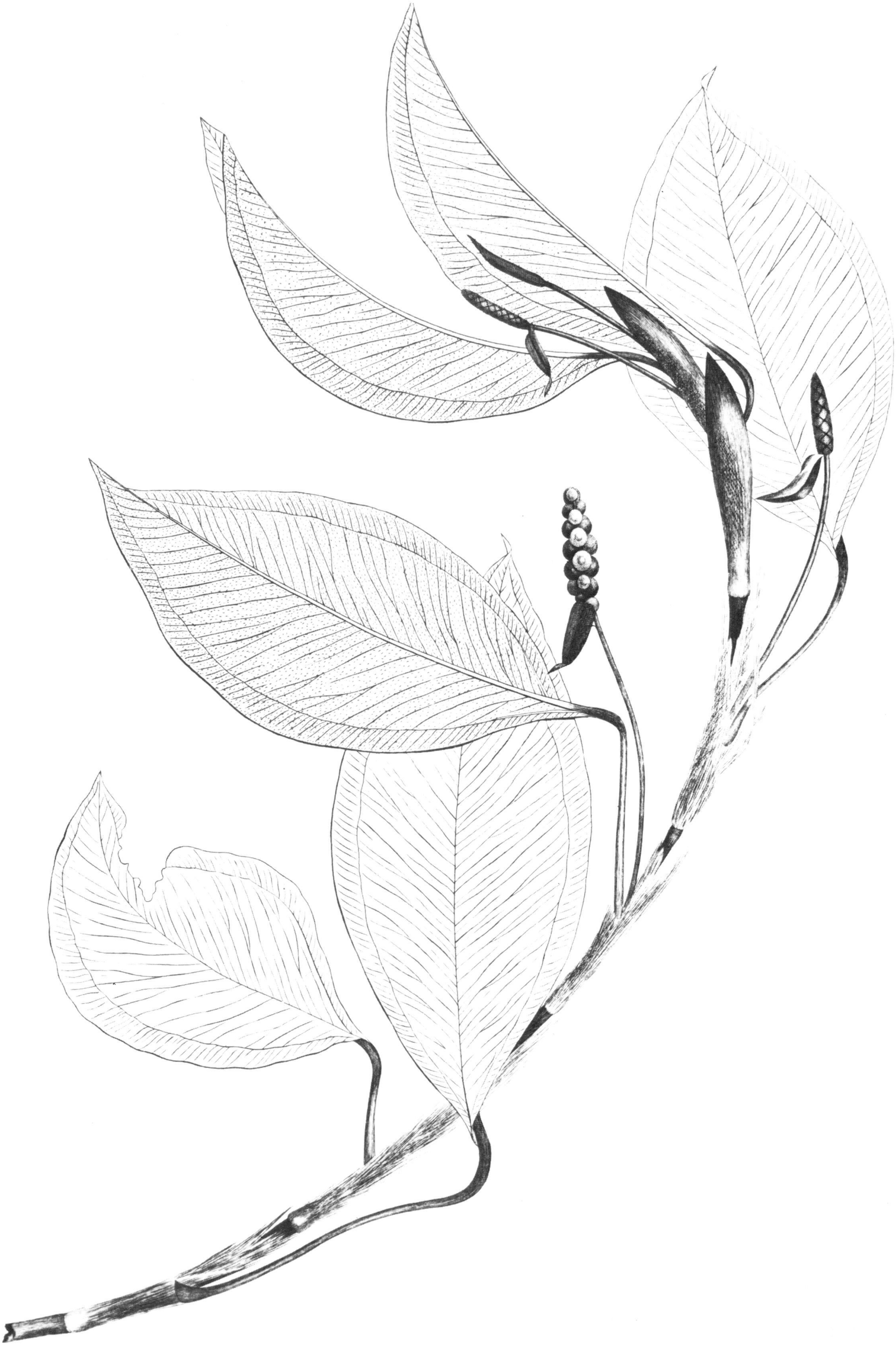

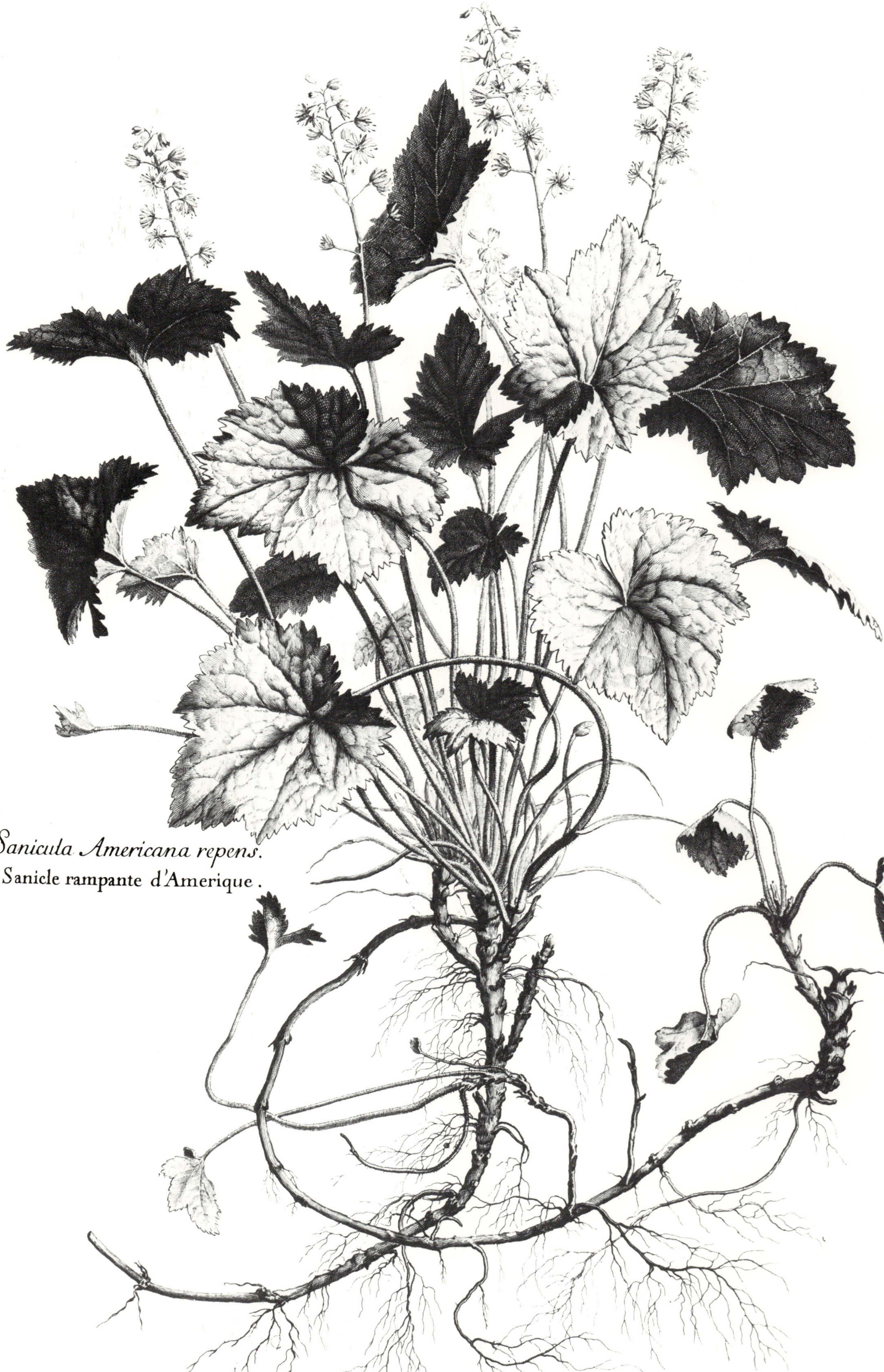

Sanicula Americana repens.
Sanicle rampante d'Amerique.

N. Robert del. de Chastillon Sculp.

The Art of the Botanist

MARTYN RIX

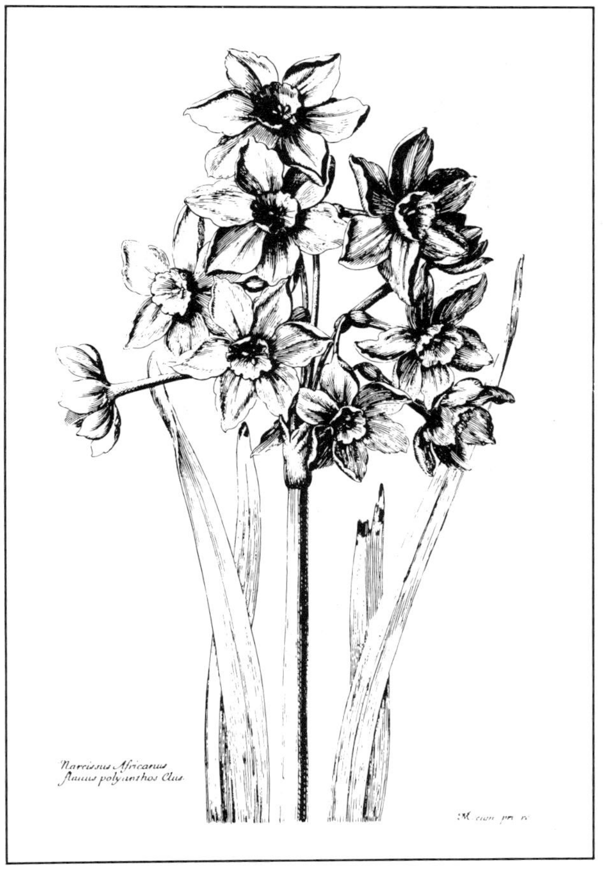

LUTTERWORTH PRESS
Guildford & London

Produced by Cameron & Tayleur (Books) Limited, 25 Lloyd Baker Street, London WC1X 9AT.
Published by Lutterworth Press, Luke House, Farnham Road, Guildford, Surrey.

Filmset by SX Composing Limited, 22 Sirdar Road, Weir Industrial Estate, Rayleigh, Essex.
Monochrome reproduction by Tenreck Limited, 172a Brent Crescent, London NW10.
Colour reproduction, printing and binding by de Lange/van Leer b.v., Deventer, Holland.

Printed in Holland

ISBN 0-7188-2482-2 - 1

Published simultaneously in the United States as *The Art of the Plant World*.

Edited and designed by Ian Cameron.
Photographed by Ian Cameron and Donna Thynne.

The great majority of the illustrations in this book are from the library of the Linnean Society of London. We wish to express our gratitude to the Society and in particular to its Librarian, Gavin Bridson, for his generous assistance. The remaining illustrations from printed sources were photographed in the Lindley Library of the Royal Horticultural Society, London, to whose Librarian, Peter Stageman, we also extend our thanks. Illustrations from original manuscripts and paintings are from the collections indicated in the captions. The two pictures on page 23 are reproduced by gracious permission of Her Majesty The Queen.

Additional photographic credits:
Ray Gardner 8
Bodleian Library, Oxford 13a
British Library 18, 19
Scala 20, 29, 38
Cooper Bridgeman 21, 24b, 25, 80
Mansell Collection 22
Phaidon Press 23a, 23b, 26, 27

To PETER STAGEMAN

Note on the illustrations
Plant names in quotation marks are those used by the authors of the books from which the illustrations are reproduced and are not those used by botanists today. Wherever possible, a modern botanical name has also been given.

Page 1:
'*Pothos violaceus*'. Engraving after a drawing by Pierre-Antoine Poiteau from *Nova Genera et Species Plantarum* (1815–25) by Alexander von Humboldt, Aimé Bonpland and C. S. Kunth. The plant, a climber on trees that is native to the West Indies, Central America and Brazil, is correctly called *Anthurium scandens* var. *violaceum*. Plants of the genus *Pothos* are natives of Ceylon.

Frontispiece:
'*Sanicula Americana repens*' – *Tiarella cordifolia*, a member of the saxifrage family found in rich woods in eastern North America. Engraving by Louis de Châtillon after a drawing by Nicolas Robert from *Mémoires pour servir à l'Histoire des Plantes* by Dionys Dodart, 1788 edition.

Title page:
Narcissus tazetta, a native of the Mediterranean region that has been cultivated for many centuries both in Europe and in China. Engraving by Nicolas Robert after his own drawing in the undated *Variae ac multiformes florum species appressae ad vivum et aeneis tabulis incisae*, which was printed around 1660.

Page 6:
The Garden of Eden, filled with exotic flowers and trees. Woodcut, the frontispiece of *Paradisi in Sole, Paradisus Terrestris* (1629) by John Parkinson.

Contents

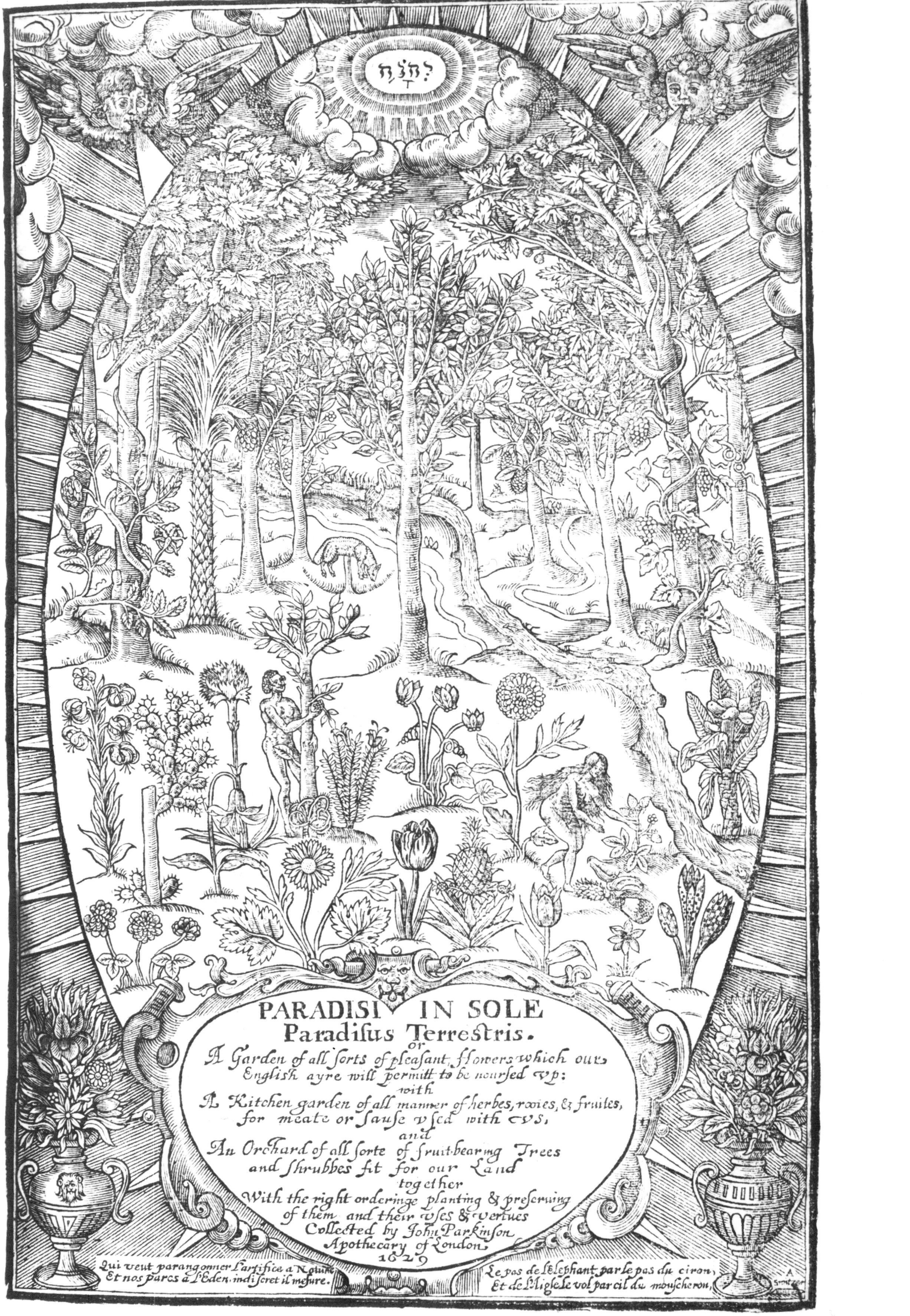

PARADISI IN SOLE
Paradisus Terrestris.
or
A Garden of all sorts of pleasant flowers which our English ayre will permitt to be noursed vp:
with
A Kitchen garden of all manner of herbes, rootes, & fruites, for meate or sause vsed with vs,
and
An Orchard of all sorte of fruitbearing Trees and shrubbes fit for our Land
together
With the right orderinge planting & preseruing of them and their vses & vertues
Collected by John Parkinson Apothecary of London
1629
Qui veut parangonner l'artifice a Nature, Et nos parcs a l'Eden, indiscret il mesure.
Le pas de l'elephant par le pas du ciron, Et de l'Aigle le vol par cil du mouscheron,

PART ONE

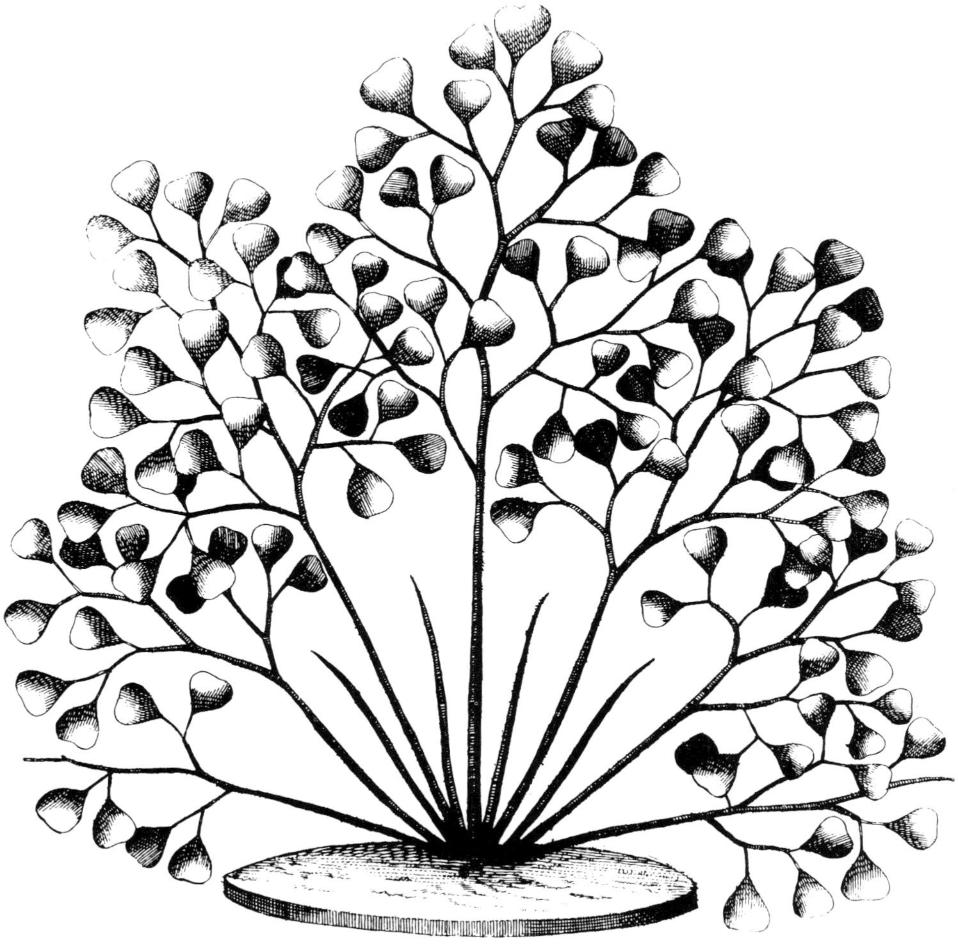

CHAPTER I

Early Botany

Previous page:
Maidenhair fern (*Adiantum capillus-veneris*). Engraving made from the *Codex Vindobonensis* under the direction of Nikolaus von Jacquin during the reign of Empress Maria Theresa, who died in 1780.

Botany became a recognised branch of knowledge in the fourth century BC. The first major figure in the systematic study of plant biology was the Greek philosopher Aristotle, although his surviving works, which include four major treatises on animals, have little to say about plants. However, the study of botany was pursued by Theophrastus, his pupil and successor as head of the school he founded, the Lyceum. Two botanical works by Theophrastus have come down to us: *Inquiry into Plants* and *The Growth of Plants*. In the *Inquiry into Plants* can be found a discussion of plant anatomy, a proposed classification and sections on the plants of different lands. Most of it, though, is concerned with the flora of the Mediterranean area and concentrates, like all botanical books up to the Renaissance, on medicinal uses. Much of the information that is included on plants from further east would probably have come from geographers attached to the army of Alexander the Great.

No traces of any illustrations have survived from Theophrastus's works, nor indeed are there any good examples remaining of botanical art from the great eras of Greek or Roman civilisation. A few representations of plants survive on frescoes and in mosaics, and the one form in which Greek plant portraits are commonly found is on coins, where the range depicted is quite considerable. Greek coins show vines, papyrus, ears of wheat and barley, leaves of oak, ivy and olive, as well as one or two medicinal herbs such as the giant fennel or silphion (*Ferula chiliantha*), from which the drug asafoetida was obtained. Silphion was associated on coins with Cyrene, just as rose flowers usually were with the island of Rhodes.

In spite of the lack of direct evidence, there can be no doubt that Greek painting reached very high levels of realism and accuracy – after all, the finest of the Greek sculpture that does survive is quite outstanding. There is a story that, in the fifth century BC, a competition was held between two famous painters, Zeuxis, a native of Heracleia in southern Italy, and Parrhasius from Ephesus. Zeuxis painted grapes so cleverly that birds flew down to peck at them. Thinking that he was sure to win, he asked Parrhasius to pull aside the curtains which hid the rival picture. Parrhasius, though, had painted the curtain, and Zeuxis had to admit defeat: 'I deceived the birds, but you have deceived me.'

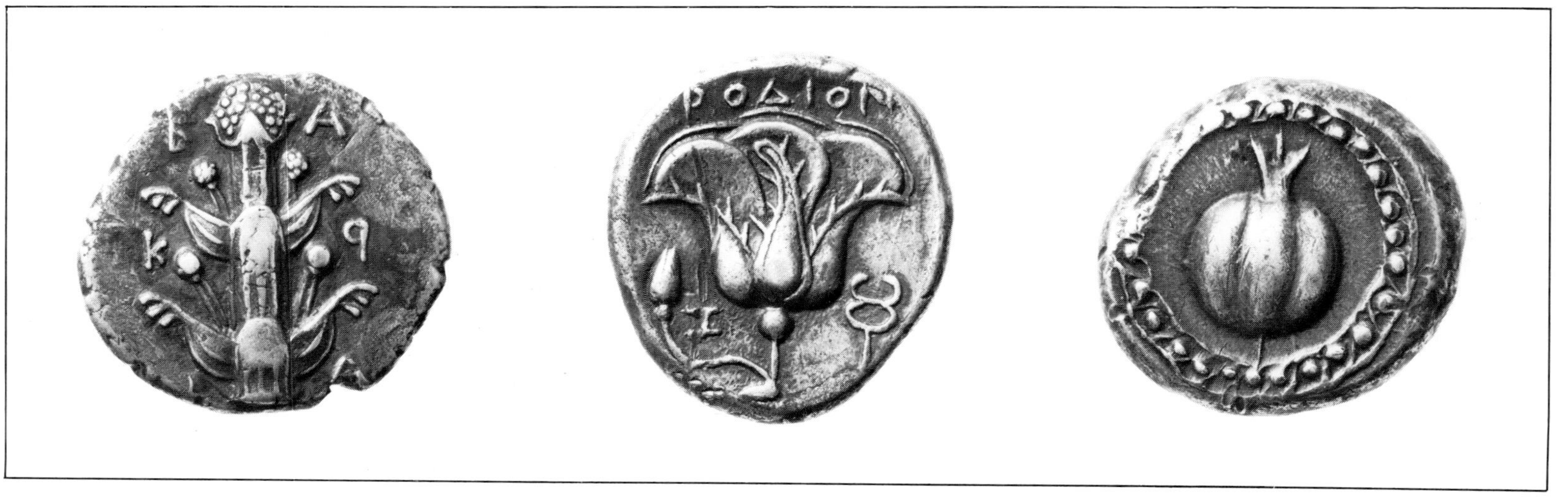

Plants on Greek coins. *Left to right:* giant fennel or silphion (*Ferula chiliantha*) on a silver tetradrachm, Cyrenaica-Barce, from the period 435–308 BC; rose, also on a silver tetradrachm, Rhodes, from the period 400–333 BC; pomegranate on a silver stater, Pamphylia-Side, fifth century BC.

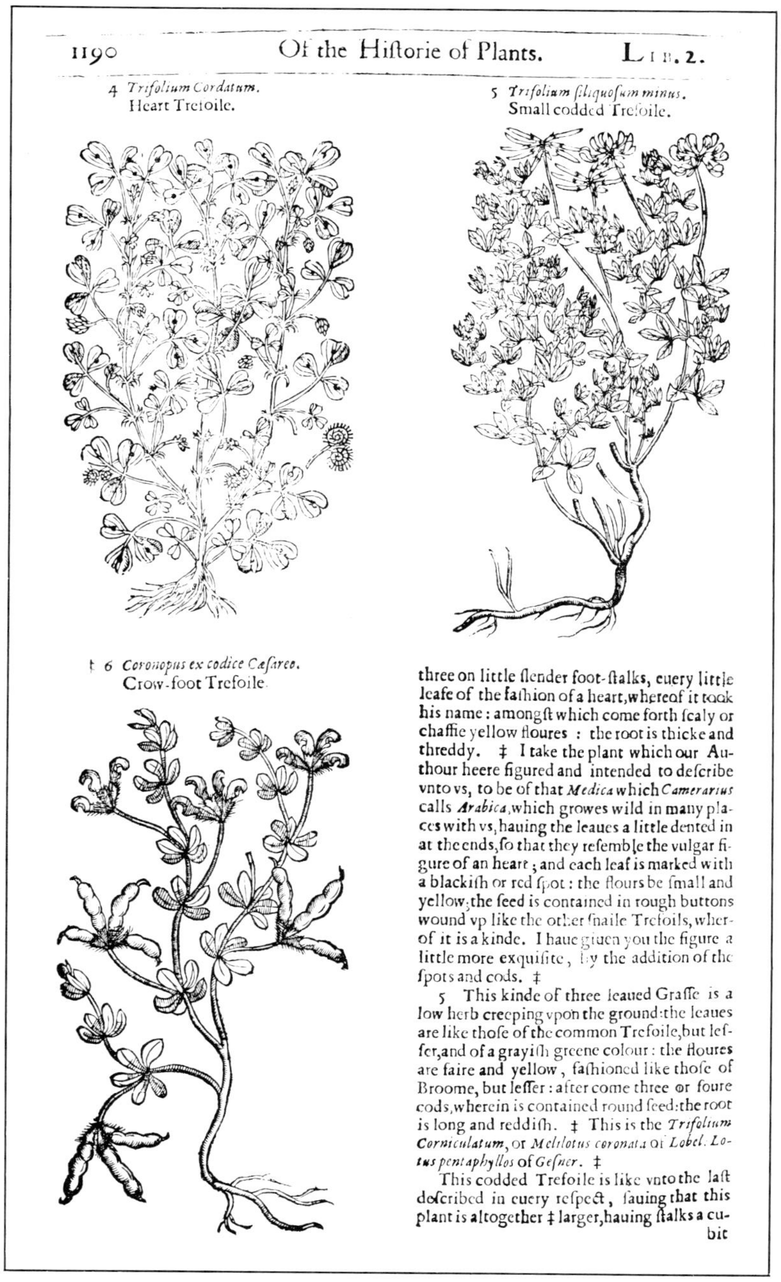

1190 Of the Hiſtorie of Plants. Lib. 2.

4 *Trifolium Cordatum.*
Heart Trefoile.

5 *Trifolium ſiliquoſum minus.*
Small codded Trefoile.

† 6 *Coronopus ex codice Cæſareo.*
Crow-foot Trefoile.

three on little ſlender foot-ſtalks, euery little leafe of the faſhion of a heart, whereof it took his name: amongſt which come forth ſcaly or chaffie yellow floures: the root is thicke and threddy. ‡ I take the plant which our Authour heere figured and intended to deſcribe vnto vs, to be of that *Medica* which *Camerarius* calls *Arabica*, which growes wild in many places with vs, hauing the leaues a little dented in at the ends, ſo that they reſemble the vulgar figure of an heart; and each leaf is marked with a blackiſh or red ſpot: the flours be ſmall and yellow; the ſeed is contained in rough buttons wound vp like the other ſnaile Trefoils, whereof it is a kinde. I haue giuen you the figure a little more exquiſite, by the addition of the ſpots and cods. ‡

5 This kinde of three leaued Graſſe is a low herb creeping vpon the ground: the leaues are like thoſe of the common Trefoile, but leſſer, and of a grayiſh greene colour: the floures are faire and yellow, faſhioned like thoſe of Broome, but leſſer: after come three or foure cods, wherein is contained round ſeed: the root is long and reddiſh. ‡ This is the *Trifolium Corniculatum*, or *Melilotus coronata* of *Lobel. Lotus pentaphyllos* of *Geſner*. ‡

This codded Trefoile is like vnto the laſt deſcribed in euery reſpect, ſauing that this plant is altogether ‡ larger, hauing ſtalks a cubit

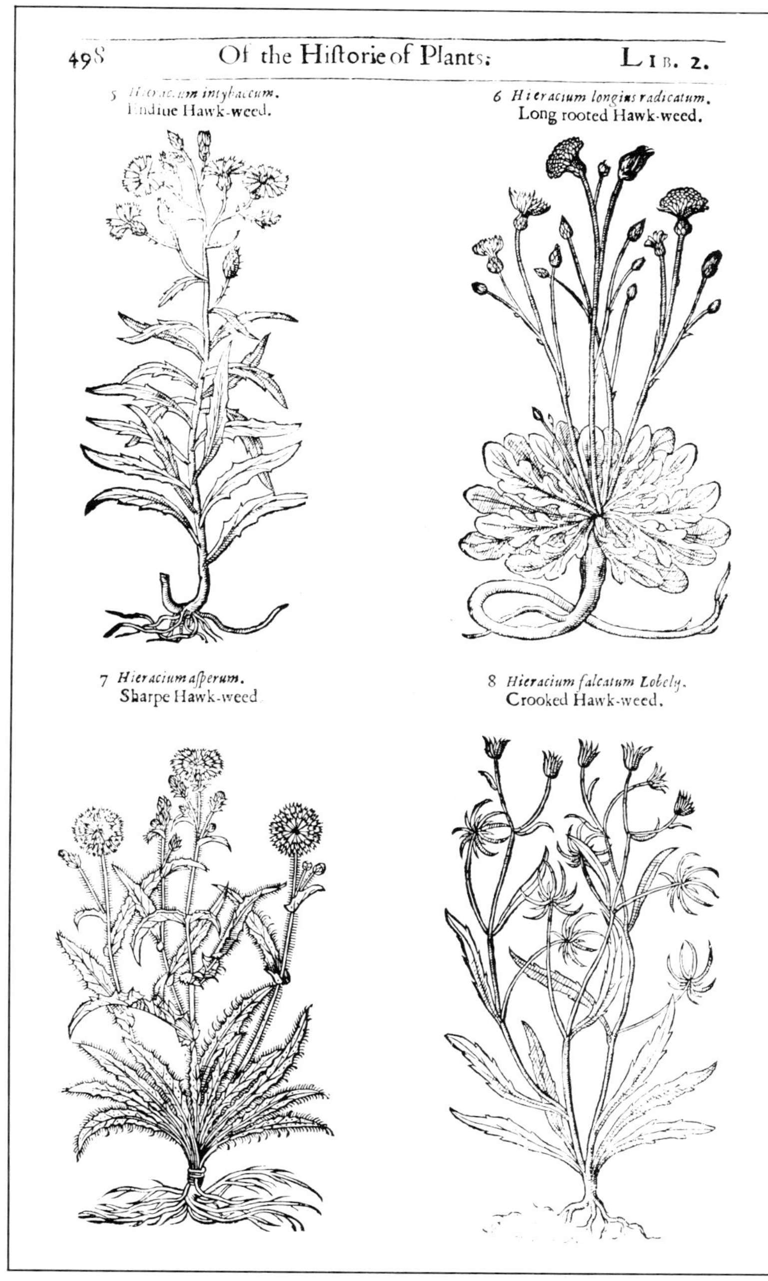

498 Of the Hiſtorie of Plants. Lib. 2.

5 *Hieracium intybaceum.*
Endiue Hawk-weed.

6 *Hieracium longius radicatum.*
Long rooted Hawk-weed.

7 *Hieracium aſperum.*
Sharpe Hawk-weed.

8 *Hieracium falcatum Lobelij.*
Crooked Hawk-weed.

Two pages, showing vetches and hawkweeds from *The Herball* by John Gerard in the 1636 edition, 'very much Enlarged and Amended by Thomas Johnson, Citizen and Apothecarye of London.' The top two vetches are spotted medick (*Medicago arabica*) and birdsfoot trefoil (*Lotus corniculatus*); the lower illustration, which is described as '*Coronopus ex codice Caesareo*' and is possibly of another species of *Lotus*, is derived from the *Codex Vindobonensis*, a manuscript of Dioscorides that dates from AD 512. The illustrations in the *Codex Vindobonensis* are in turn considered to be copies of Hellenistic originals.

One report of Hellenistic plant illustration was given by the Roman naturalist Pliny the Elder, who was killed by the fumes when he tried to get close to the great eruption of Vesuvius which destroyed Pompeii in AD 79. He left behind 37 books of his *Natural History*, an encyclopedic work that is dull in many ways and totally unselective, combining interesting information with wild hearsay. In it, he mentions some early botanical writers on the medicinal properties of plants. His opinion of their illustrations is none too high: '. . . the subject has been treated by Greek writers, whom we have mentioned in their proper places; of these, Cratevas, Dionysius and Metrodorus adopted a most attractive method, though one which makes clear little else except the difficulty of employing it. For they painted likenesses of the plants, and then wrote under them their properties. But not only is a picture misleading when the colours are so many, particularly as the aim is to copy nature, but besides this, much imperfection arises from the manifold hazards in the accuracy of copyists . . . For this reason other writers have given verbal accounts only.' But whatever the limitations of the pictures, they were clearly intended as aids to identification like those in modern illustrated floras.

Cratevas was the most famous botanist between Theophrastus and Pliny. He was physician to Mithridates IV Eupator, King of Pontus in Asia Minor, who by 88 BC had conquered much of present-day Turkey and northern Greece. Because his illustrations became attached to the writings of Dioscorides, his influence persisted in Europe until the eighteenth century. As late as 1633, plates originally by Cratevas were being printed in Thomas Johnson's edition of Gerard's *Herball*; the Roman origin of one picture of a vetch is indicated in its title, '*Coronopus ex codice Caesareo*'.

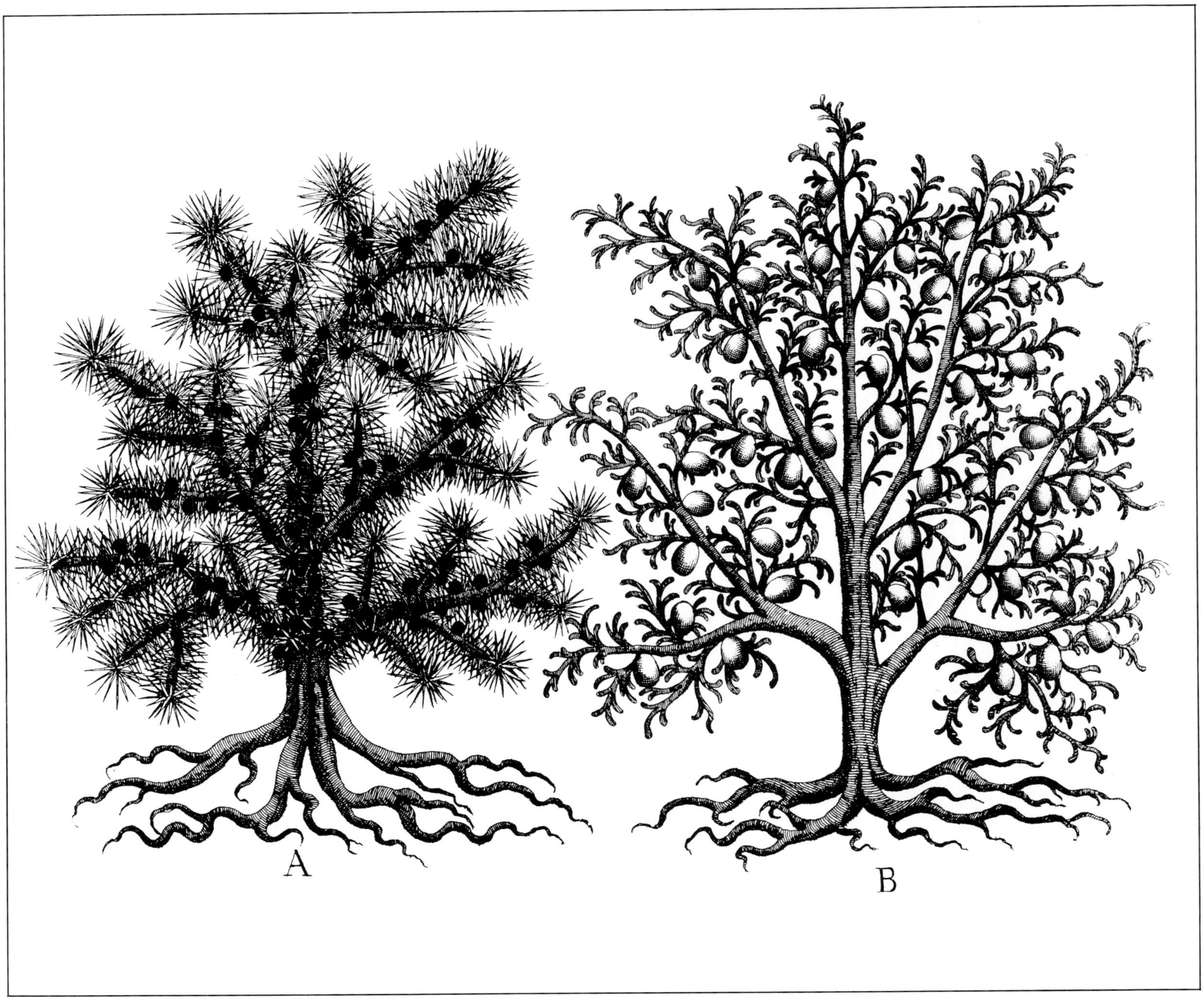

Dioscorides, the most influential of all early botanical writers, was born in Anazarba, which is now in Turkey. He travelled widely as a doctor in the Roman army during the 1st century AD. His work, which is usually known by its Latin title, *De Materia Medica*, was written in Greek, and was based both on his own experience and on the writings of others, notably Cratevas.

Dioscorides covers about five hundred plants, but his text gives little more than the name and healing properties of each herb. His reputation survived throughout the Dark Ages, and, at the beginning of the Renaissance, his work became the starting point of all botanical knowledge, as well as the chief source for pharmacy. Thomas Johnson wrote that it 'is as it were the foundation and grounde-worke of all that hath been since delivered in this nature.'

In 1785, before setting out for a plant-collecting expedition to the eastern Mediterranean, John Sibthorp, Professor of Botany at Oxford, visited Vienna to study the ancient manuscripts of Dioscorides. Here he met Ferdinand Bauer, a young flower painter, and the two combined to produce the *Flora Graeca*, perhaps the greatest illustrated flora ever written. The preface, written after Sibthorp's early death, mentions that their aim was to rediscover 'the plants which were found in Greece from the most ancient times until the present day, and mentioned by the most capable writers, Homer, of course, Theophrastus, Dioscorides etc . . .' The list of synonyms gives the names used by Dioscorides; 'the synonyms of Dioscorides were taken from Sibthorp's notes, mainly [written] in Vienna, where he saw the old and very famous *codex* with painted illustrations.' This was the *Codex Aniciae Julianae*, often called the *Codex Vindobonensis*.

Proofs of engravings made from the *Codex Vindobonensis* under the direction of Nikolaus von Jacquin during the reign of Empress Maria Theresa. Only two copies are said to have been taken from the plates, as the work was never completed. The copy from which these pictures are taken contains 142 plates in oblong folio format; it was sent to Linnaeus with notes by Jacquin and is now in the collection of the Linnean Society in London. The other copy was lent by Jacquin to John Sibthorp to use during his work on the *Flora Graeca* and is now in the library of the Botanic Garden in Oxford. *Above:* two shrubs. *Opposite page, top:* wormwood (*Artemisia*). *Bottom: Onosma* and a thistle.

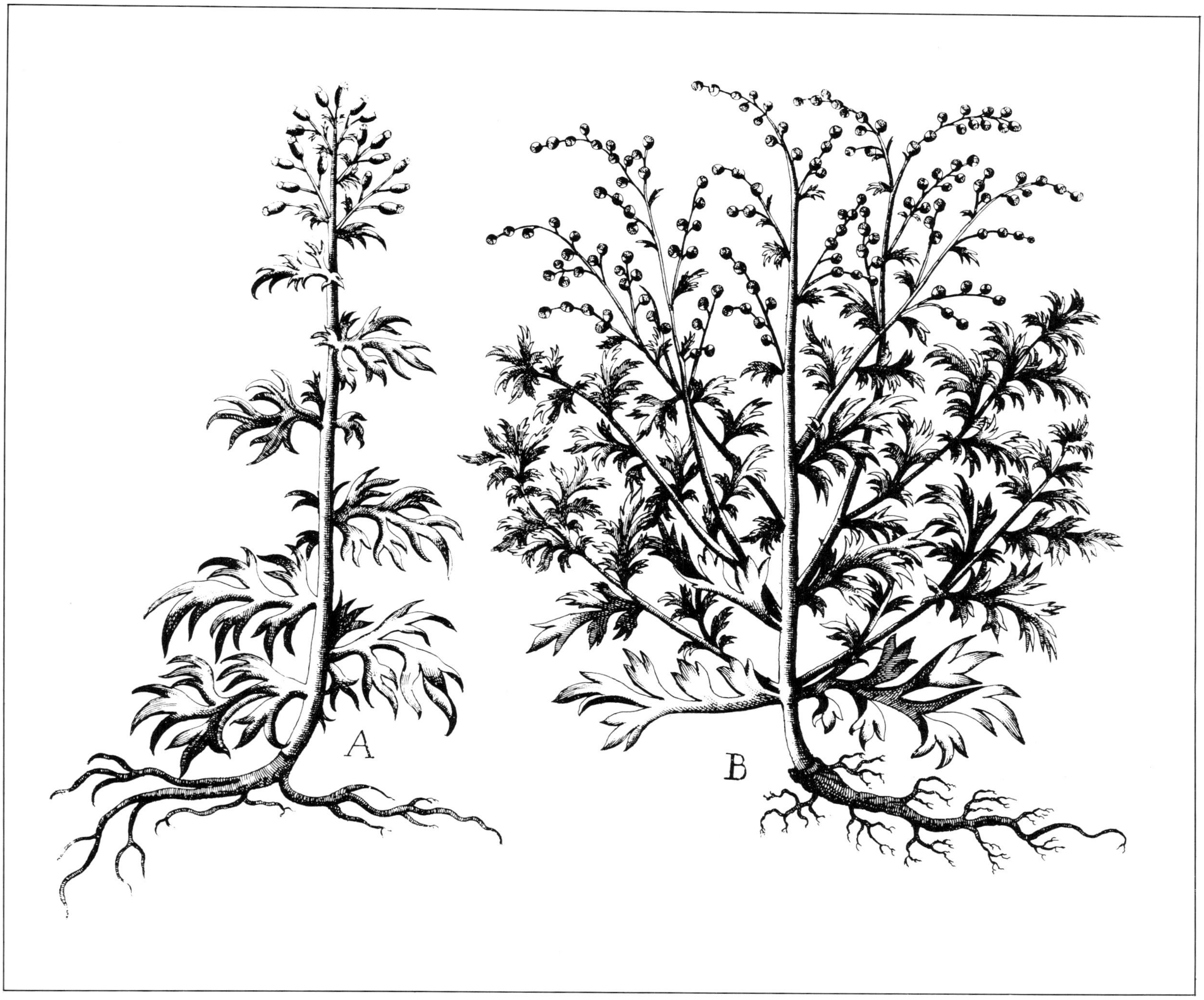
A
B

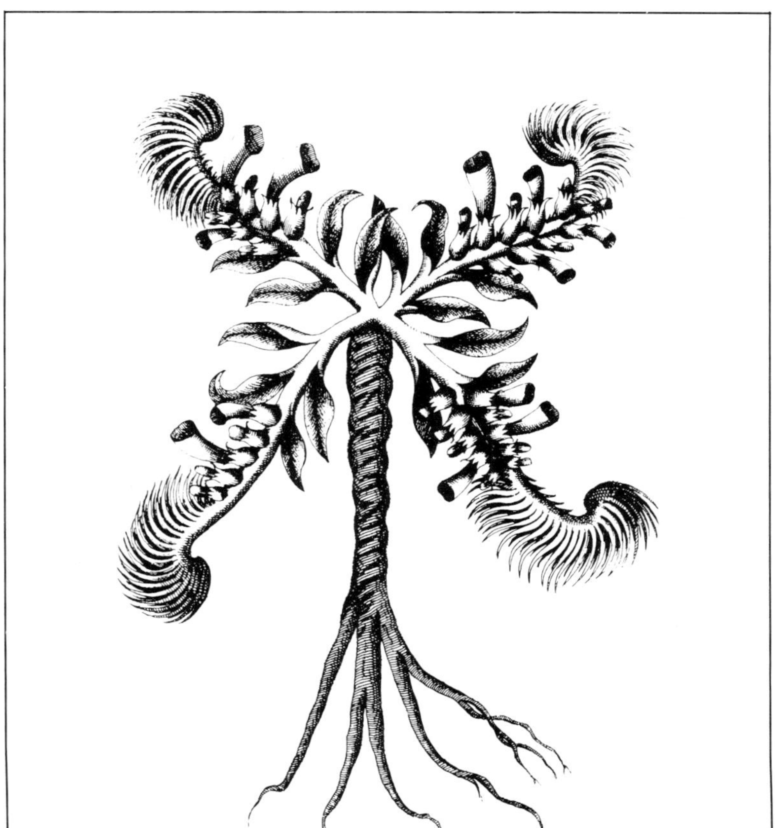

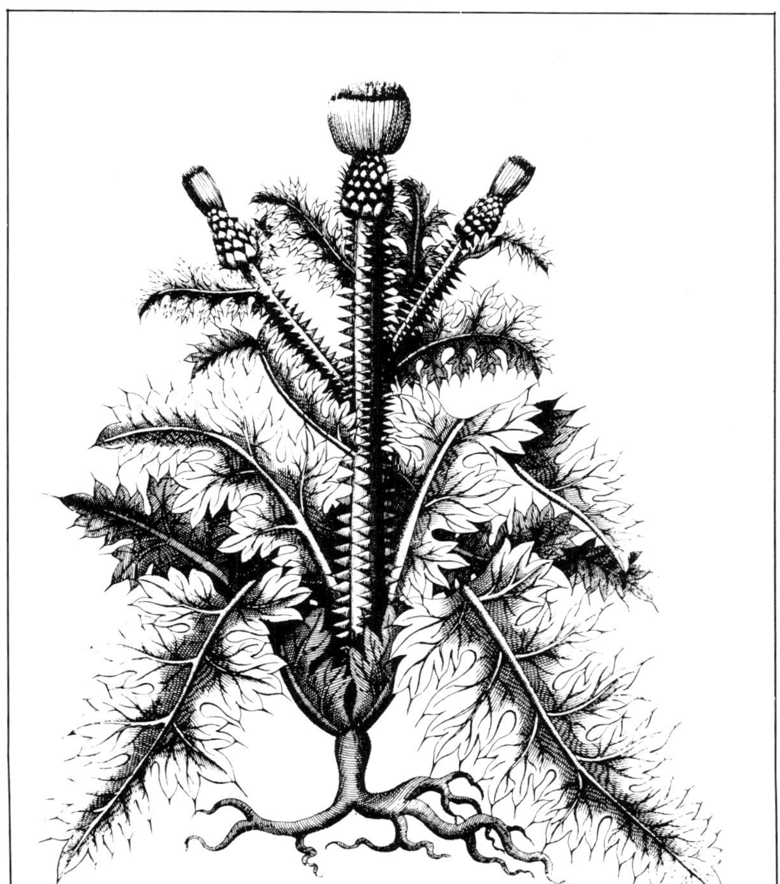

Just as the writings of Dioscorides formed the link between classical and modern botany, so the paintings of the *Codex Vindobonensis* form the link between ancient Greek and modern European botanical illustration. The actual manuscript was copied in AD 512 for Anicia Juliana, a lady renowned for her Christian faith and daughter of Flavius Anicius Olybrius, who was Emperor of the West in 472, in the last years of the Western Roman Empire. The *Codex* was probably written in Constantinople and remained there until the 16th century. In 1562, the ambassador of the Holy Roman Empire at Constantinople, Ogier Ghiselin de Busbecq, described in a letter how he saw it and tried to add it to the vast quantity of Greek manuscripts he took back to Vienna. 'One treasure I left behind in Constantinople, a manuscript of Dioscorides, extremely ancient and written in majuscules, with drawings of the plants and containing also, if I am not mistaken, some fragments of Cratevas . . . It belongs to a Jew, the son of Hamon, who, while he was still alive, was physician to Suleiman [the Magnificent]. I should like to have bought it, but the price frightened me; for a hundred ducats was named, a sum which would suit the Emperor's purse better than mine. I shall not cease to urge the Emperor to ransom so noble an author . . . the manuscript, owing to its age, is in a bad state, being externally so worm eaten that scarcely anyone, if he saw it lying in the road, would bother to pick it up.' Seven years later, the *Codex* found its way into the Imperial Library in Vienna.

It consists of nearly five hundred parchment sheets, roughly 30cm square, with about four hundred full page paintings of plants and some smaller ones of birds. Most of the plant species are common in Greece and the eastern Mediterranean, and are either native or grown as vegetables. At the beginning of the *Codex* are several pages of miniatures, including one showing Dioscorides at work, while Intelligence holds up a mandrake for Cratevas to draw. The original of this is clearly ancient – a very similar subject was found as a wall-painting at Pompeii. The style of the plant paintings is described by Wilfrid Blunt as having a 'naturalism alien to Byzantine art of that period,' and he concludes that they are obviously derived from originals of a much earlier date. A more typically Byzantine style can be seen in the formalised faces of the dedicatory miniature to Anicia Juliana. Not all the paintings are in the same style. Some show greater skill than others in handling awkward details such as the way that the leaf-bases clasp the stem in sow-thistles (*Sonchus* spp.); the illustrations of delicate-leaved plants such as fennel are especially successful. Also beautifully shown are cyclamen, various species of wormwood (*Artemesia*), *Delphinium staphisagria*, scarlet pimpernel and asphodel. Even these must have lost a great deal of detail through repeated copying before they appeared in the *Codex*. Nonetheless, no better plant illustrations were painted until the fifteenth century.

While the *Codex Vindobonensis* is the most beautiful and oldest of the existing manuscripts of Dioscorides – indeed the oldest of all illustrated botanical works – it is not the only one. The *Codex Neapolitanus* at Naples probably dates from the eighth century and appears to be derived from the same source, or even copied from, the Vienna *Codex*, and there is another probable copy in the Pierpoint Morgan Library in New York. There is also a ninth-century codex in Paris which is probably derived from a separate source, and there are several extant manuscripts in Arabic, some with reasonable illustrations.

Sir Arthur Hill, who was Director of the Royal Botanic Gardens at Kew in the 1930s, discovered that the influence of Dioscorides was not entirely dead even in the twentieth century. He describes a visit he made to Mount Athos in 1934: 'The official botanist monk . . . was a remarkable old man with an extensive knowledge of plants and their properties. Though fully gowned in a long black cassock he travelled very quickly, usually on foot, and sometimes on a mule, carrying his "flora" with him in a large black bulky bag. Such a bag was necessary as his "flora" was nothing less than four manuscript volumes of Dioscorides, which apparently he himself had copied out. This "flora" he invariably used for determining any plant which he could not name at sight, and he could find his way in his books and identify his plants – to his own satisfaction – with remarkable rapidity.'

The second major herbal of antiquity dates from the fourth or fifth century AD. Nothing is known of the life of its writer, Apuleius Platonicus, who is often called Pseudo-Apuleius or Apuleius Barbarus to distinguish him from the second century author of *The Golden Ass*. His herbal is a prescription book with about 130 illustrations of medicinal herbs, written in Latin and derived mainly from Dioscorides and Pliny. The illustrations of the many surviving manuscripts are much inferior to those of the *Codex Vindobonensis*. The oldest of these, the *Codex Vossianus* in the University Library at Leiden, Holland, was probably written in southern France in the seventh century. One of the several Anglo-Saxon manuscripts of Pseudo-Apuleius is particularly interesting. It was written at Bury St Edmunds around 1120 and is now in the Bodleian Library in Oxford. What makes it remarkable is that the paintings appear to be new instead of the usual inferior copies of pictures from an earlier manuscript. It mainly illustrates native English flowers, which do not necessarily match the adjacent text.

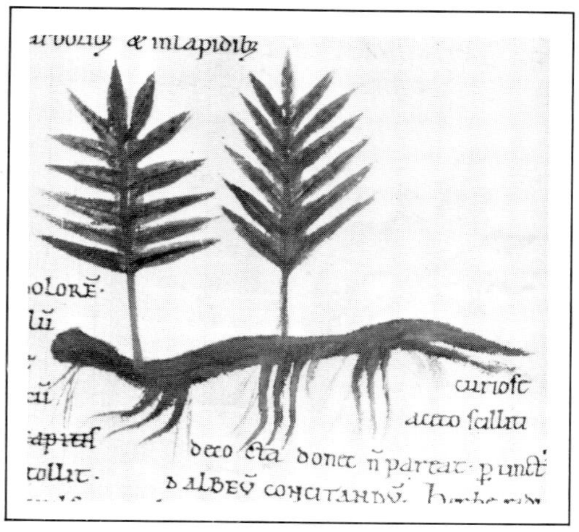

Polypody fern (*Polypodium*), watercolour on vellum, from the Bury St Edmunds manuscript of the herbal of Pseudo-Apuleius, dating from about 1120. Bodleian Library, Oxford (Bodley MS 130).

The first printed version of Pseudo-Apuleius was published by Johannes Philippus de Lignamine in Rome in 1481. The dedicatory epistle to Lignamine's first edition states that it was taken from a manuscript he had recently found at Monte Cassino. Facsimiles of a ninth-century manuscript from the monastic library at Monte Cassino and of the printed version were published side by side by F. W. T. Hunger in 1935, and the similarities between the two are striking enough to indicate that the manuscript was probably used in the making of Lignamine's edition. The drawing is simple and stylised but nevertheless quite decorative.

The last quarter of the fifteenth century saw the appearance in printed form of the herbals that had been popular during the Middle Ages. This was quite early in the history of printing – the first European printed books had been issued by Johann Gutenberg of Mainz in the 1450s – and the illustrations were crude woodcuts derived from the repeatedly copied paintings in the manuscript herbals. When a Latin translation of Dioscorides was first printed in 1478 in Germany and when it was published in the original Greek eleven years later in Venice, the *Codex Vindobonensis*, with its wonderful illustrations, was still in the hands of the Turks in Constantinople. Three of the early printed herbals are worthy of mention, but more for their influence on contemporary thought than for their artistic or scientific excellence.

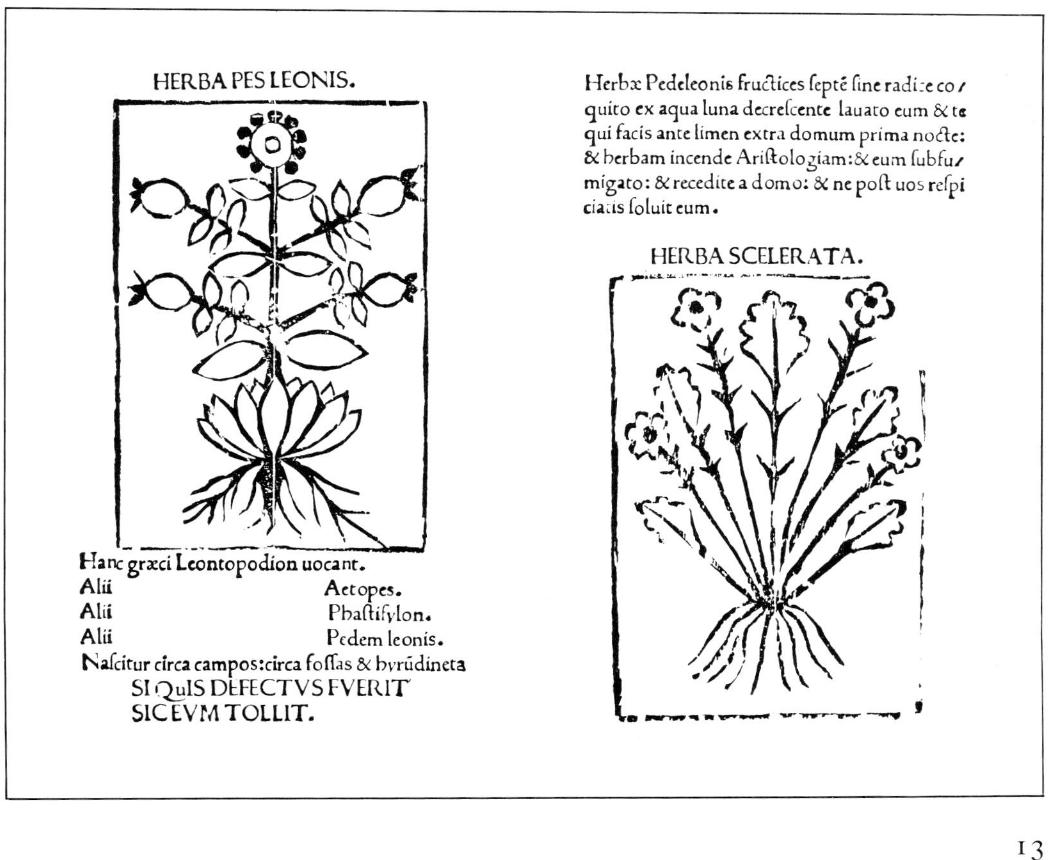

A double-page spread from the first printed herbal, the *Herbarium Apulei*, published by Johannes Philippus de Lignamine in 1481.

xlij

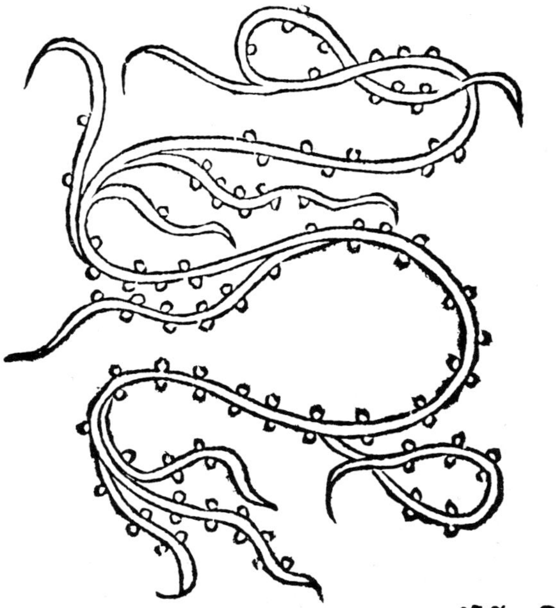

Cuſtuta **Syde uff flaß**

Cuſtuta eſt res inuoluta ſup linū· eſt calidum in p̄mo et ſiccū citra ſcōm· Eſt mūdificatiuū et pur gatiuū melancolie principaliter ſecūdario fleg‑ ma· Iſto modo· Rx· cuſtute ſtolopendrie añ·m·j· polipodij· radic· eſule· foliorum ſene añ·ʒ·ſ· florum violarum boraginis añ·m·ſ· omnia buliantur in vino et aqua añ·lb·j· aceti parum ad cōſumpcō‑ nem tercie ptis colatū dulcoretur cū zucro et fiat potus ſumendus ut ſup̄· poſtea ſumātur pillule

Liij

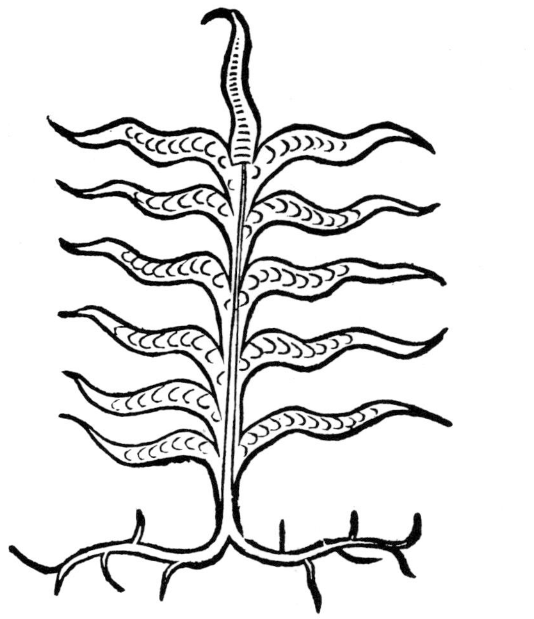

Polipodiū **engelſuß**

Polipodiū calidū eſt in ſcdo ſiccū in tercio g̃du· Virtus eſt in radice q̄ in ſapore dulcis eſt et no‑ doſa· et qd naſcitur ſup radices quercus eſt effi‑ cacius· Eſt reſolutiuū ventoſitatū et humidita‑ tum et in decoctione polipodij deb3 poni aliqd excluſiuū ventoſitatis ut eſt aniſū et ſemen fe‑ niculi et cimini· qꝛ polipodiū reſoluit humores in ventoſitates· Virtutē etiā habet polipodiū diſſoluendi attrahēdi et purgandi p̄ncipaliter

xxviij

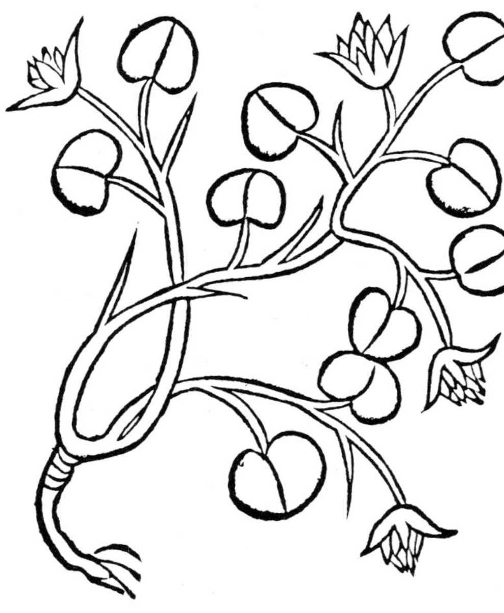

Nenufar **Seeblomen**

Nenufar frigid· et hu· in ſcdo gradu· Et eſt du‑ plex quedā defert albū florem que eſt melior· et eſt croceū florem ꝓducens qd non eſt adeo bo‑ num· Et flos eius vſui medicine cōpetit· Et ex floribus fit ſiropus cōtra febrē acutā et epatis calorē· Flores nenufaris ⁊ flores violarū et ſemē endiuie et ſemē lactuce et portulace et ſemia q̃tu‑ or frigida q̄ oia decoquant̄ in aq̈ ⁊ parū aceti· et

lx

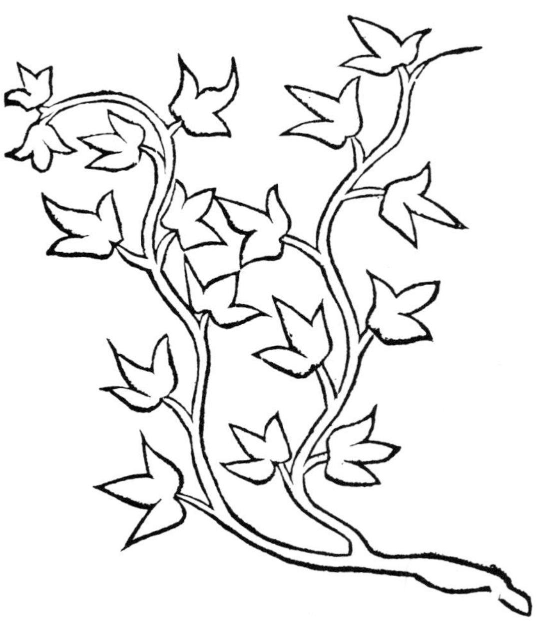

Edera arborea **ebich**

Edera arborea ab herendo eſt dicta eo qꝛ arbo‑ ribus adhereat· vel dicitur ab edo qꝛ a capris editur et ipis lac multiplicat· Radix eius frigi‑ de eſt nature et· frigidam inducit eſſe terrā vbi creſcit· virorē ſeu viriditatē diu ſeruāt folia eius Et ē amari et ſtiptici ſaporis· Edere autē due ſūt ſpēs ſcȝ alba et nigra ſiue maſculus ⁊ femia· Ede‑ ra alba fructū defert albū· et nigra nigrū· et edera

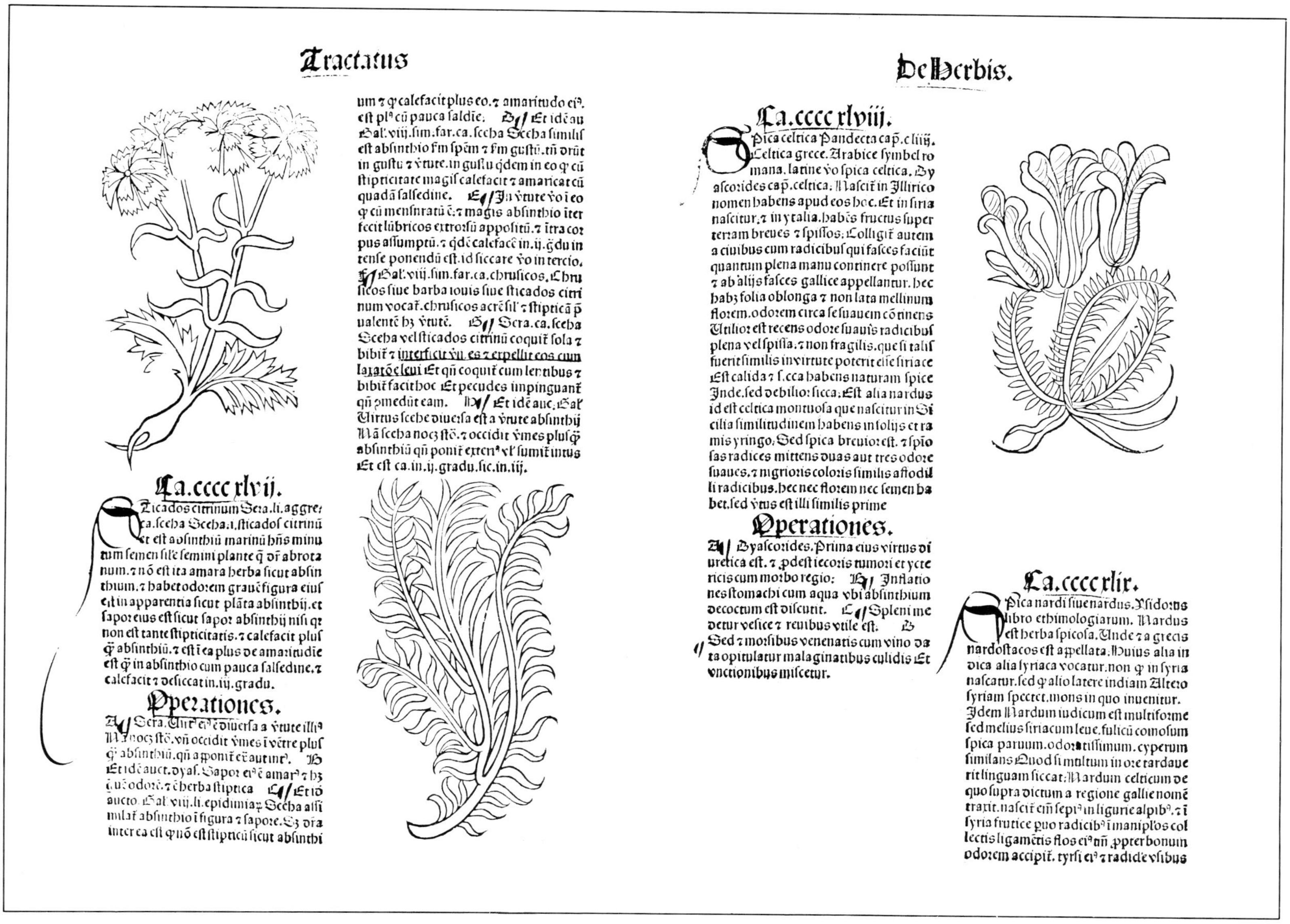

A double-page spread from the *Hortus Sanitatis* (1491), with woodcuts showing a pink (*Dianthus*), wormwood (*Artemisia*) and possibly a milk-vetch (*Astragalus*).

The Latin *Herbarius* is a small quarto volume first issued in 1484 by Peter Schoeffer, one of Gutenberg's successors in Mainz. The woodcuts are so formalised as to be in many cases totally unrecognisable. At best, they have a bold sense of design, as in the pictures of '*Serpentaria*' (perhaps a plantain) – so called because it was thought to be effective for snake bites – and '*Brionia*'; neither is identifiable with any particular species.

A larger volume printed by Schoeffer and published a year later than the Latin *Herbarius* is the German *Herbarius*, which is variously known as the

'*Eruca*', which does not look much like rocket, the salad plant that now has the botanical name *Eruca sativa*, and opium poppy (*Papaver somniferum*). Two woodcut illustrations in a double-page spread from the *Macer Floridus*, published around 1510.

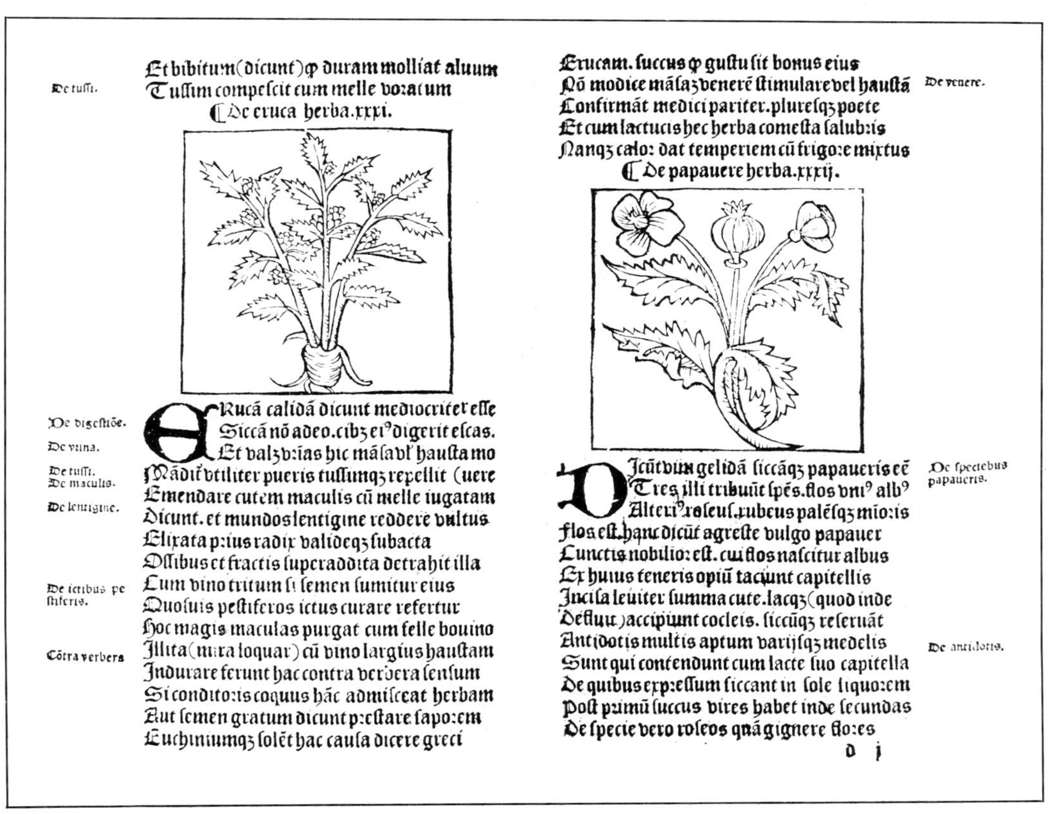

Opposite page: Four pages with woodcut illustrations from the Latin *Herbarius* of 1484, printed by Peter Schoeffer in Mainz. *Top left:* dodder (*Cuscuta*), a scrambling parasite with thread-like stems and small spherical clusters of flowers. *Top right:* polypody fern (*Polypodium*). *Bottom left:* '*Nenufar*', a waterlily. *Bottom right:* '*Edera arborea*', ivy (*Hedera helix*).

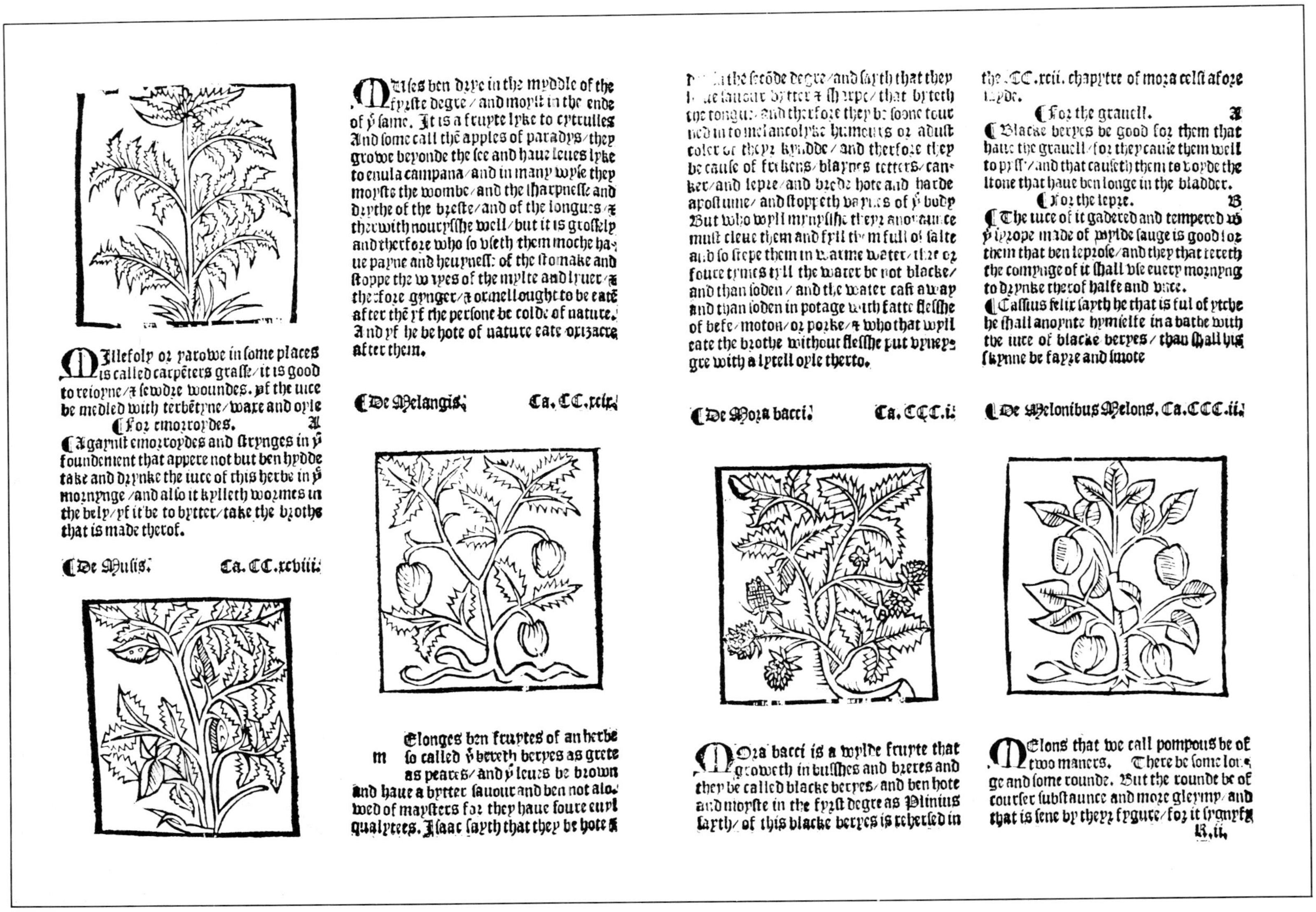

Millefoly or yarowe in some places is called carpēters grasse / it is good to ioyne & sowdre woundes. yf the iuce be medled with terbētyne / ware and oyle
¶ For emoroydes.
¶ Agaynst emoroydes and strynges in ý foundement that appere not but ben hydde take and drynke the iuce of this herbe in ý mornynge / and also it kylleth wormes in the bely / yf it be to bytter / take the brothe that is made therof.

¶ De Musis. Ca. CC.xcviii.

Muses ben drye in the myddle of the fyrste degre / and moyst in the ende of ý same. It is a frute lyke to cytrulles And some call the apples of paradys / they growe beyonde the see and haue leues lyke to enula campana / and in many wyse they moyste the wombe / and the sharpnesse and drythe of the breste / and of the longues & therwith nourysshe well / but it is grossely and therfore who so useth them moche haue payne and heuynesse of the stomake and stoppe the wayes of the mylte and lyuer & therfore gynger / & otmell ought to be eatē after thē yf the persone be colde of nature. And yf he be hote of nature eate oxizacra after them.

¶ De Melangis. Ca. CC.xcix.

Melonges ben frutes of an herbe so called ý bereth beryes as grete as peares / and ý leues be browne and haue a bytter sauour and ben not allowed of maysters for they haue foure euyl qualytees. Isaac sayth that they be hote & […] the secōde degre / and sayth that they […] bytter & sharpe / that byteth the tonge: and therfore they be soone tourned into melancolyke humours or adust coler of theyr kynde / and therfore they be cause of frakens / blaynes tetters / canker / and lepre / and brede hote and harde apostume / and stoppeth the pores of ý body But who wyll mynysshe theyr anoyaunce must cleue them and fyll them full of salte and so kepe them in warme water / thre or foure tymes tyll the water be not blacke / and than soden / and the water cast away and than soden in potage with fatte flesshe of befe / moton / or porke / & who that wyll eate the brothe without flesshe put vynegre with a lytell oyle therto.

¶ De Mora bacci. Ca. CCC.i.

Mora bacci is a wylde frute that groweth in busshes and breres and they be called blacke beryes / and ben hote and moyste in the fyrst degre as Plinius sayth / of this blacke beryes is reherced in the .CC.xcii. chapytre of mora celsi afore […]de.
¶ For the grauell. A
¶ Blacke beryes be good for them that haue the grauell / for they cause them well to pysse / and that causeth them to voyde the stone that haue ben longe in the bladder.
¶ For the lepre. B
¶ The iuce of it gadered and tempered wt ý syrope made of wylde sauge is good for them that ben leprose / and they that receyue the compuge of it shall vse euery mornyng to drynke therof halfe and more.
¶ Cassius felix sayth he that is ful of ytche he shall anoynte hym selfe in a bathe with the iuce of blacke beryes / than shall his skynne be fayre and smote.

¶ De Melonibus Melons. Ca. CCC.ii.

Melons that we call pompous be of two maners. There be some longe and some rounde. But the rounde be of courser substaunce and more gleymy and that is sene by theyr fygure / for it sygnyfyeth
B.ii.

Various herbs and fruits including yarrow (*Achillea millefolium*), aubergine (*Solanum melongena*), blackberry and melon. Woodcuts derived from the illustrations in the German *Herbarius* of 1485 in a double-page spread from *The Grete Herball* (1526), 'which giveth parfyt knowlege and understandyng of all maner of herbes & there gracyous vertues whiche god hath ordeyned for our prosperous welfare and helth / for they hele & cure all maner of dyseases and sekenesses that fall or misfortune to all maner of creatoures of god created / practysed by many expert and wyse maysters / as Avicenna & other. &c.'

Herbarius zu Teutsch, the German *Hortus Sanitatis* or Cube's *Herbal*, after Dr Johann von Cube, the physician who may have contributed the medical information in the book. Like Sibthorp two and a half centuries later, the author travelled to the East to find the plants mentioned by Dioscorides that were not known in Germany. Although the illustrations are again simple woodcuts, their subjects are on the whole much more recognisable than those in the Latin *Herbarius*. These pictures were the source for most other herbals for the next fifty years, including the *Hortus Sanitatis* proper. Popular mythology, however, was still in evidence in the portrait of the mandrake, perhaps because the writer of the *Herbarius* did not manage to see one during his travels. He clearly had no such problem with the portraits of ordinary German plants like herb Robert and the common polypody, which are accurate enough to be immediately identifiable, as is the daisy in spite of its being surprisingly called '*Primula veris*', the modern botanical name for the cowslip.

The other influential herbal of the late fifteenth century was the *Hortus Sanitatis*, which also emerged from Mainz, where it was first printed by Jacob Meydenbach in 1491. The botanical section is in part a German adaptation of the German *Herbarius*, from which the more realistic illustrations are also taken. About a third of the figures of herbs are new, and there are sections on various animals and on rocks. The *Hortus Sanitatis* was certainly intended to be a popular book, and its Latin text would have given it an international readership. Some of the witty little woodcuts that illustrate it are meant to illustrate the effects of the herbs as well as – or instead of – their appearance. Little figures pop up out of the flowers in the illustration of a '*Narcissus*', which has more to do with the youth of classical mythology who fell in love with his own image than with the flower that bears his name. The *Hortus Sanitatis* appeared only forty years before the *Herbarum Vivae Icones* of Otto Brunfels; the difference between them is the difference between medieval herbalism and modern science.

Opposite page :
Botanical illustration in the period between the German *Herbarius* of 1485 and Brunfels's *Herbarum Vivae Icones* of 1530. Handcoloured woodcuts of various herbs in the *Liber de Arte Distillandi de Simplicibus* (1500) by Hieronymus Brunschwyg.

Mandrake, a grape vine, an unidentified herb and a clover. Woodcuts on a double-page spread from *Le Grand Herbier*, which was printed in Paris before 1522.

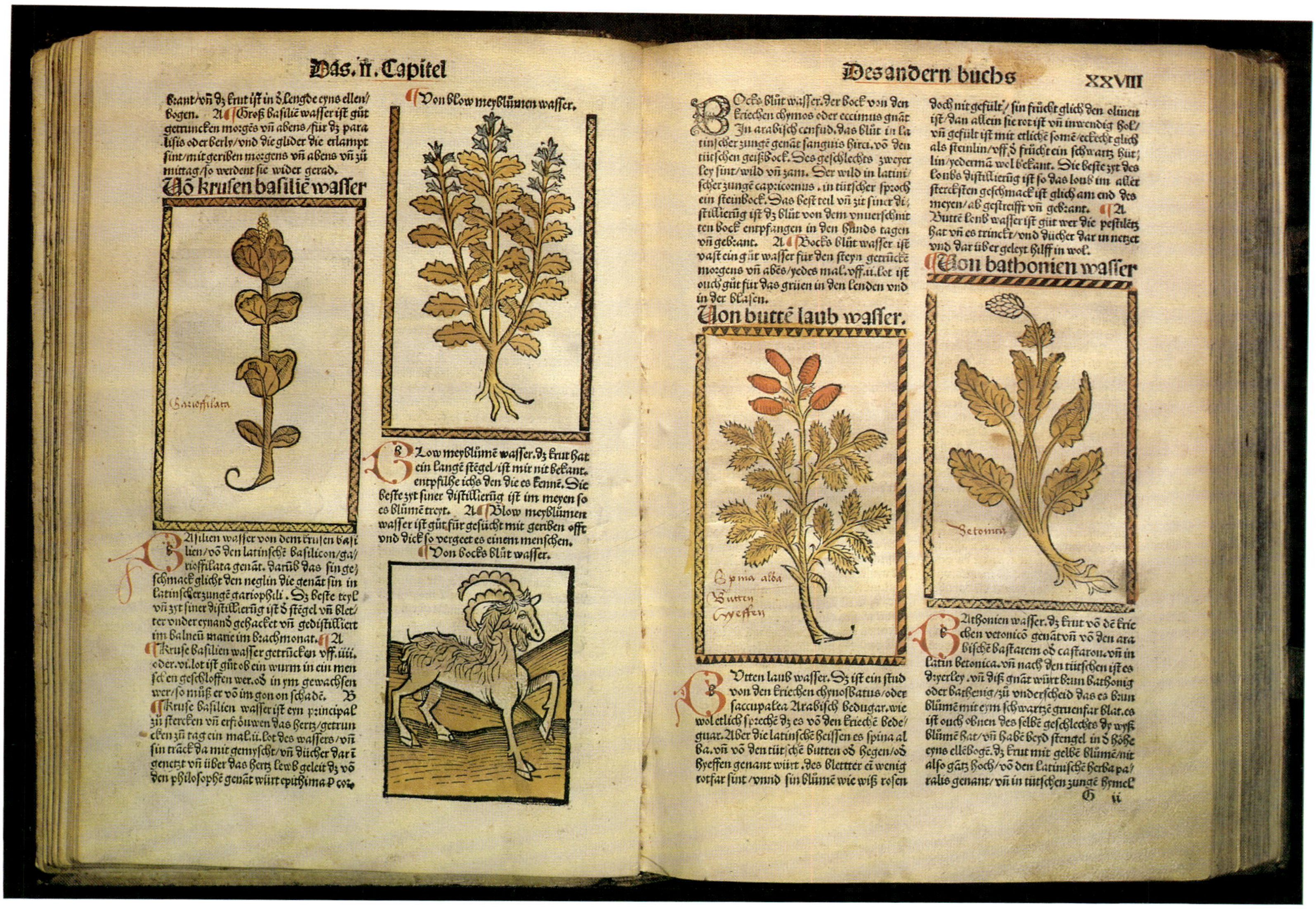

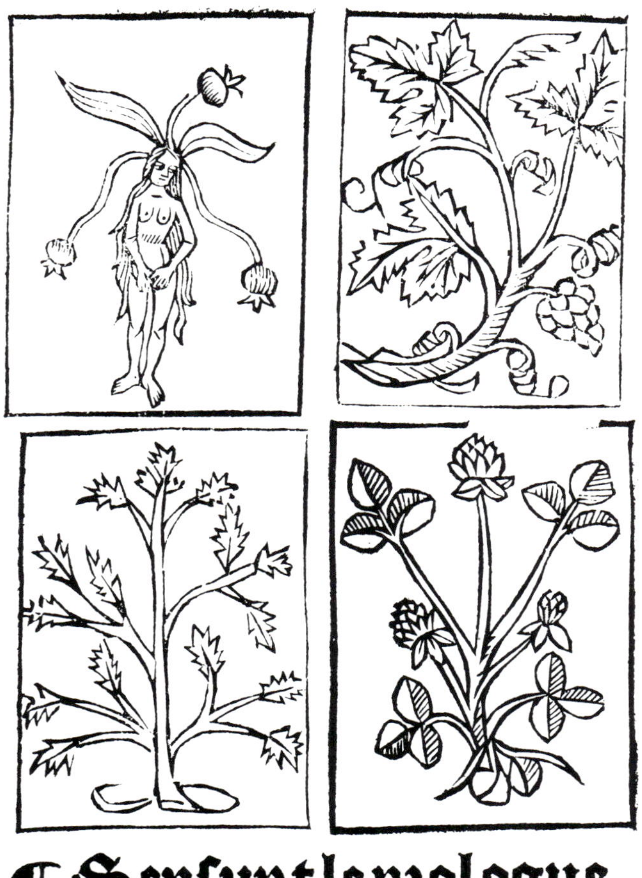

Le prologue
Fueillet.i.

Prologue

En ceste presente besogne est nostre ppos ⁊ intēcion de traicter des siples medecines. Et est assavoir que la medecine est dicte simple pource quelle est telle que nature la produycte et formee ainsi comme giroffle ⁊ noix muguettes. Ou iacoit ce quelle soit preparee par artifice / toutesfoys elle nest point meslee avec autre medecine sicomme les tamarins / le sālz par artifice sōt causez ⁊ en sōt les escoces ostees Et aussi le aloen q̄ par artifice est faict du ius dune herbe cuytte. Mais on pourroit faire question assez raisonnable. Po[2] quoy on trouua les medecines cōposees / puis q̄ toute vertu q̄ est trouuees composees et trouuee aux simples / Car toute medecine qui est contre maladie est pour la superhabōdance des humeurs ou diminution diceulx / ou pour arrester les courans / ou elle est contre la feblesse des vertus po[2] alteration des calitez ou solutiō des cōtinuitez es playes ou aultre ētameure. Or est ainsi quō treuue medecine simple la patiue apetissant et diminuāt la superhabūdāce des humeurs. Et aussi medecine simple deffluās et decoutans / et aussi alteratiue et cōsolidatiue. Solution Plusieurs causes ya pourquoy on trouua les medecines composees esquelles les simples ne suffisoyēt point. Cestassavoir violence de maladie / contrarietez de mal.adies / contraire dispositiōs des membres du corps / la noblesse du membre malade / et la violence de medecine. Violence de maladie / car il est des maladies si fortes et si enracinees que iamais par simples medecines ne seroyent guaries si comme epilēce et lepre. Et pource fault il messer les simples pour acroistre la vertu delles. Contrariete de maladie aussi comme q̄ ensēble auroit fiebure et ydropisie appellee / leucofleumance / et la endroit auroit maladies causees / Lune de chault et lautre de froit / et pource il est necessaire quilz soyēt medecines composees de contraires / affin que par contraires proprietez ilz resistēt et obuient a contraires maladies. Et aussi ne treuue on point medecines simples sans cōposition q̄ puissent suruenir a membres de contraire disposition blessez / comme a lestomac froit ⁊ a foye chault Si fust necessaire y trouuer medecine composees qui par ses contraires qualitez peust obuenir a contraire qualite de membres. La noblesse du membre malade / comme se le foye auoit aulcune durte pour ressouldre et espartir / et aussi stiptique pour le membre conforter / car la chaulde le affoiblíroit trop se la stiptique nestoit a

a.i.

Qui se comencia a tratar segondo lordene de le medesine le quale e calde e seche in lo primo grado.

Tracta primo de le medesine calde e seche in lo primo grado.

Capitolo primo. Del riso. xiii.

El riso dice Galieno che el so sauore e stiptico e per
strençe el uentre temperamentre perche questa stip
ticita e pucha. E in la scorça del riso e un pucho de
roseça pui che noe in lo riso. e lo riso e de menore
nutrimento cha el formento. E alguna fia li sani ha
besogno de magnarne coto in forma de sugoli per
le puncion del uentre le quale uenisse per fluxo de
mulla humori colerici o per altra caxon simele a que
sta. E questo riso segundo como e dito se de cuser
che deuente a muo de sugolo e po darlo a magnar a lo infermo. E la ho
ra quando e pui conuegnente da usarli e quando el tempo he humido
e quando el stomego e forte a paidir. Et la uertu soa e dita in parte e ol
tra ço e de pucho nutrimento. e acresse la sperma e desmenuisse le fece e
la urina e la uentosita. Anchora e bono ale puncion e a le ulceracion deli
intestini quando el se fa crestiero deso. Ma lo riso rosso ha maore e pui
forte uertu a strençer.

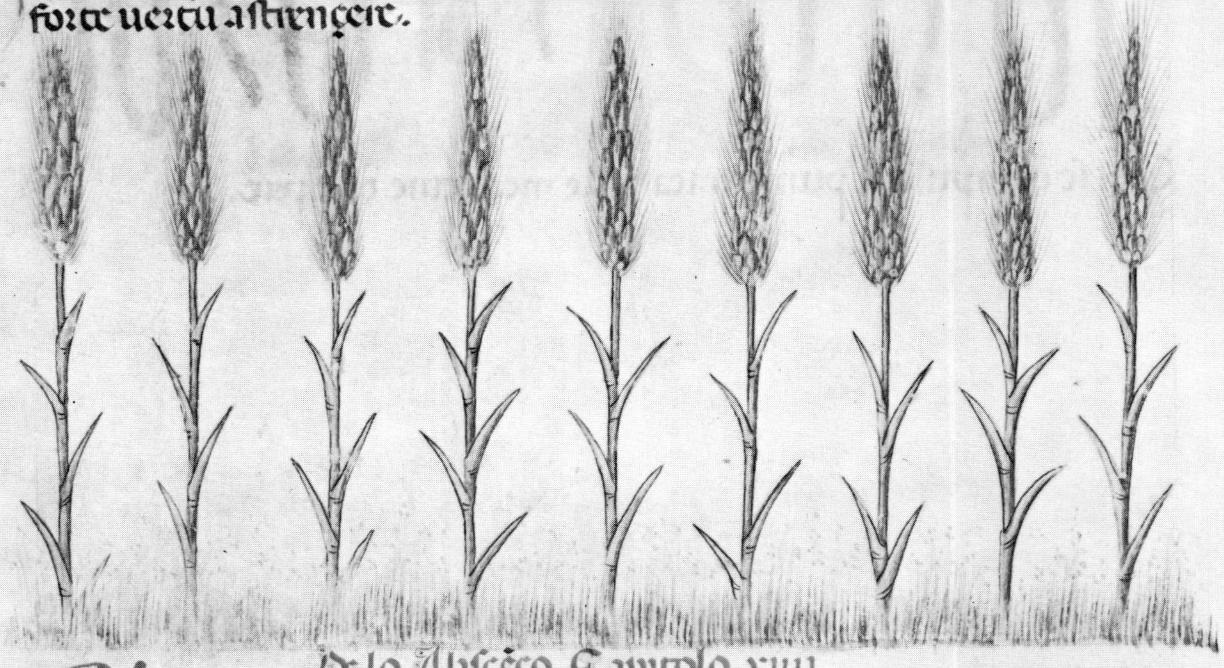

De lo Absenço. Capitolo xiiii.

Lo absenço ha tre specie. Una specia se chiama generalmentre absen
ço uero. o uer absenço roman e questo e el meiore. Galieno
dice che queste tre specie çoe maniere de absenço. no ha do qua
litade do uertu segundo uno modo. ço e a dire che luno noe egualmen

CHAPTER II

Nature Observed

Nature, as we have seen, did not intrude very much into the herbals of the Middle Ages; for the monastic copyists, it was of little significance compared to the orthodoxy established, albeit at many removes, by authors who had worked a thousand years before. Nevertheless, a movement towards naturalism was in progress in the more directly religious works that predominated in medieval art. In the borders of prayer books and the foregrounds of religious paintings, there began to appear flowers depicted

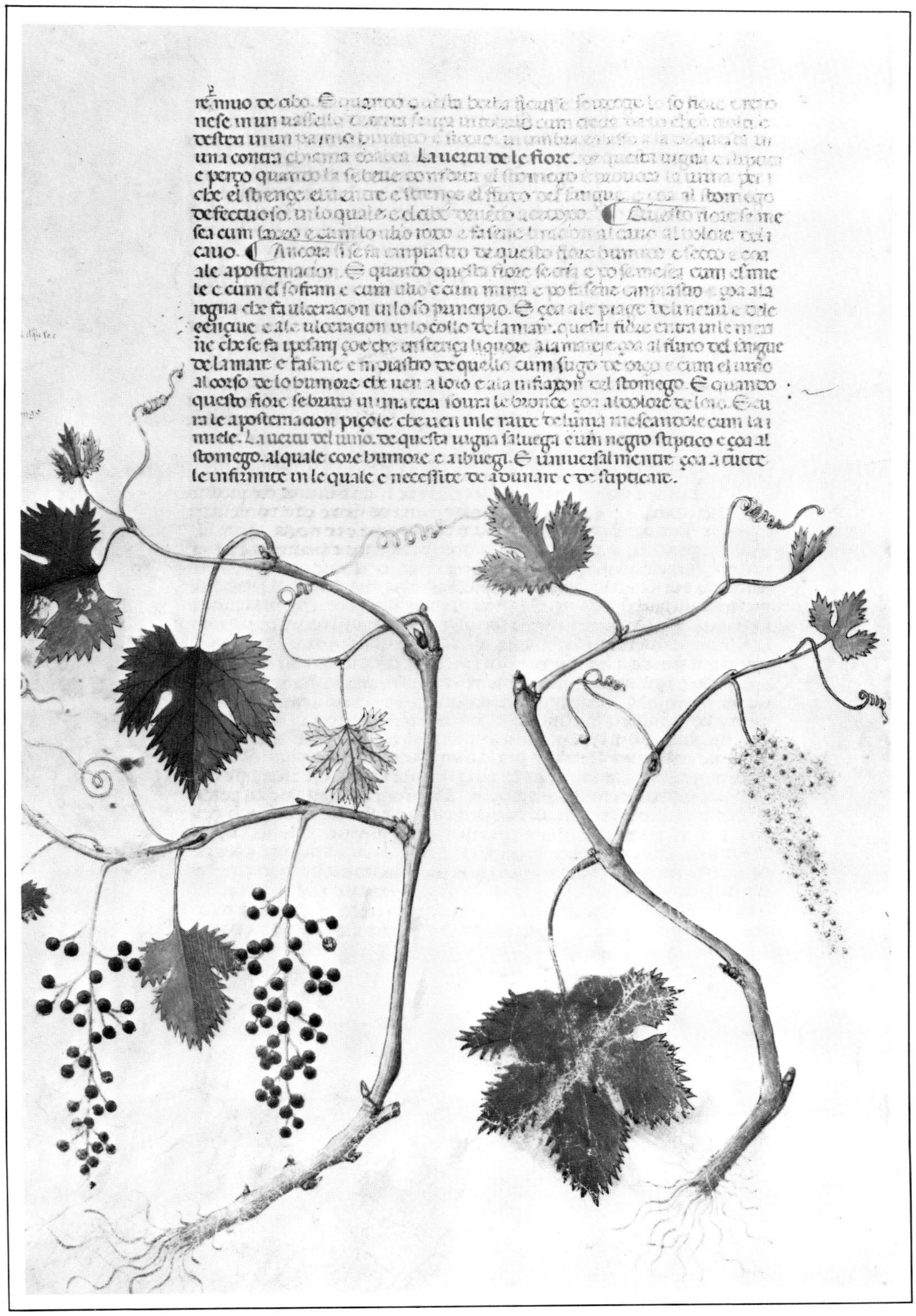

Ears of wheat, and vines with flowers and unripe grapes. Two pages from the Carrara herbal, painted at the end of the fourteenth century. British Library (Egerton MS 2020).

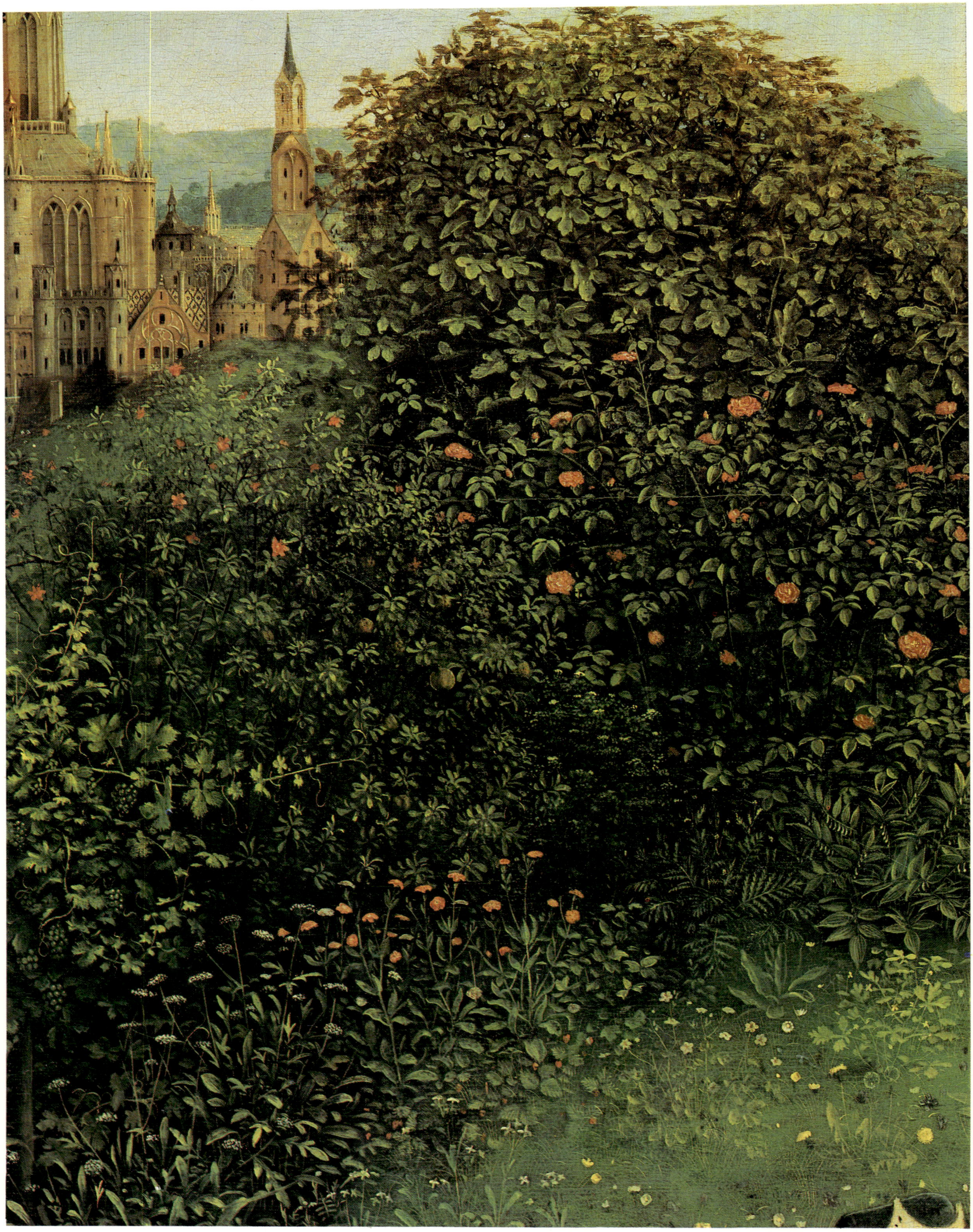

in such a fresh and lively way that they must have been taken from nature rather than from the musty pages of a herbal.

The twelfth-century Bury St Edmunds manuscript of Pseudo-Apuleius, with its new – though still formalised – pictures of English flowers, was very much the exception, and nothing like it in approach seems to have been produced in Europe for another two hundred years. We have to look to the

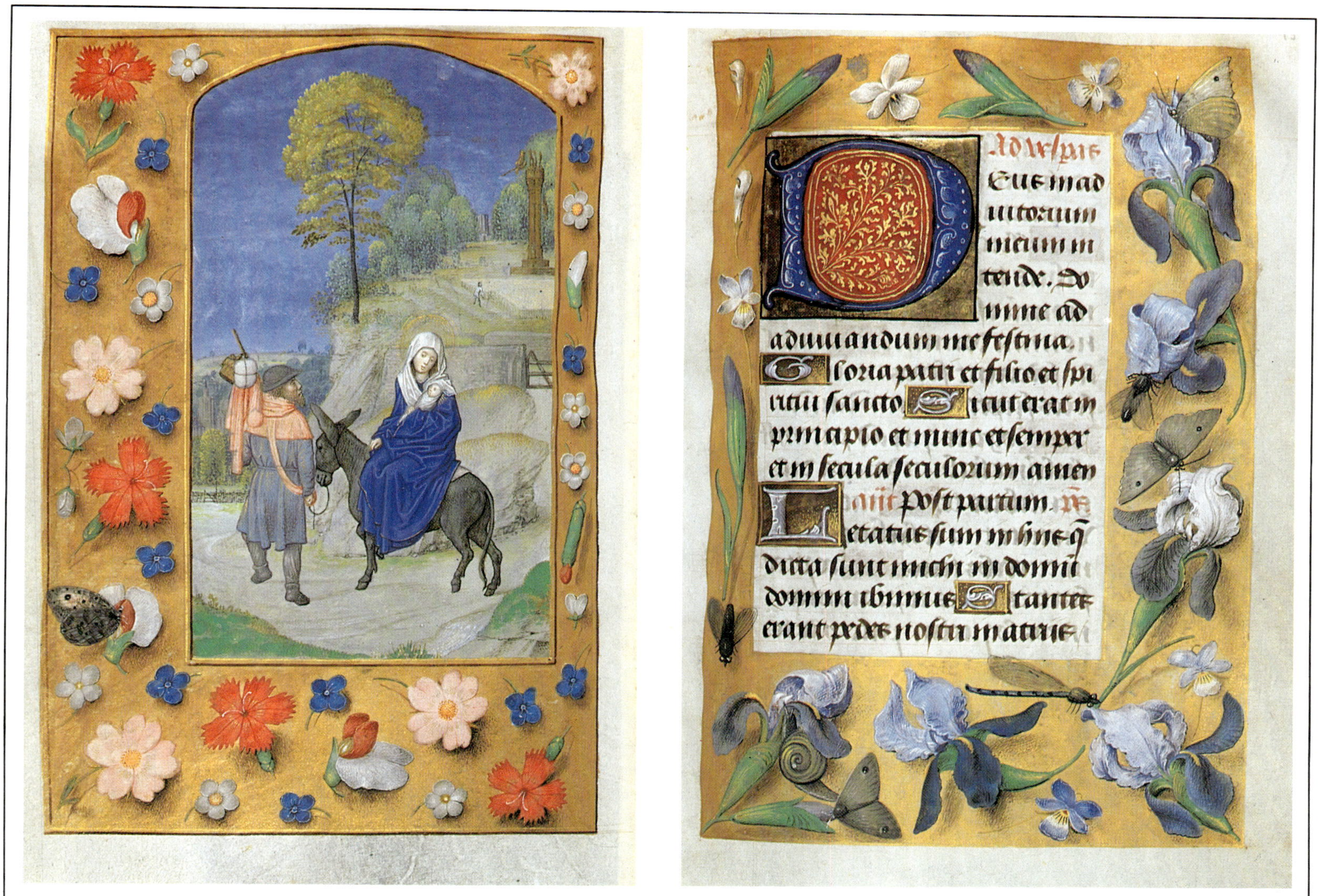

The Flight into Egypt in the *Hastings Hours* from the School of the Master of Mary of Burgundy, dating from around 1480, painted in opaque watercolour on vellum with a background of gold leaf. The borders show pinks, peas, dog roses, irises, speedwell, heartsease, lousewort and other flowers, with butterflies, moths, flies and a damsel fly. British Library.

Detail from *The Adoration of the Lamb*, the central panel of the Ghent Altarpiece by Hubert and Jan van Eyck, completed around 1430. In the medieval garden depicted in this small section of the polyptych can be recognised vines, pomegranates, rue and rose of Provins, with figs in the background and various flowers in the foreground including valerian, rose campion, wild strawberry, tansy, Solomon's seal, greater celandine, dandelion and daisy. Cathedral of St Bavon, Ghent.

paintings of Giotto (*c* 1266–1337) to find the first stirrings in art of the love of nature that had been kindled by St Francis of Assisi at the beginning of the thirteenth century. Without this love of nature, no tradition of flower painting could become established.

The few isolated examples of herbal illustration that are outside the usual tradition appeared in Italy. An early fourteenth-century herbal, now in the British Library, has simple but naturalistic illustrations delicately executed in watercolour and showing such common flowers as love-in-a-mist, field poppy and parsley. This herbal was probably produced at Salerno in southern Italy, where Islamic influence was strong and had indeed resulted in the establishment of a medical school as early as the eleventh century.

A hundred years after the Salerno herbal, a remarkable series of flower paintings was begun in Padua. It was undertaken by an unknown artist working for Francesco Carrara, the last Lord of Padua, who died in 1403. The text is an Italian translation of an Arabic medical treatise by Serapion the Younger who lived around AD 800. Although spaces have been left for many paintings, only about fifty have been completed. These are outstanding not just for their accuracy but also for their design: instead of being squeezed into a small rectangle, each dominates the page on which it has been painted. There is no attempt to portray the whole plant, and a stem, usually without roots, is shown as if it has been picked and laid across the page.

Neither of these Italian herbals seems to have had a great influence in its time, although one descendant of the Carrara herbal can be traced. This is the herbal of Benedetto Rinio, which was painted around 1415 and is now in the library of St Mark's Cathedral in Venice. The artist, Andrea Amadio, of whom nothing else is known, produced nearly five hundred full page illustrations, which in general are delicately realistic and finely coloured. According to the established herbal tradition, the plants are shown complete with their roots (which, after all, were often the valuable part). Wilfrid Blunt records that 'the purpose of the book was to assist herbalists

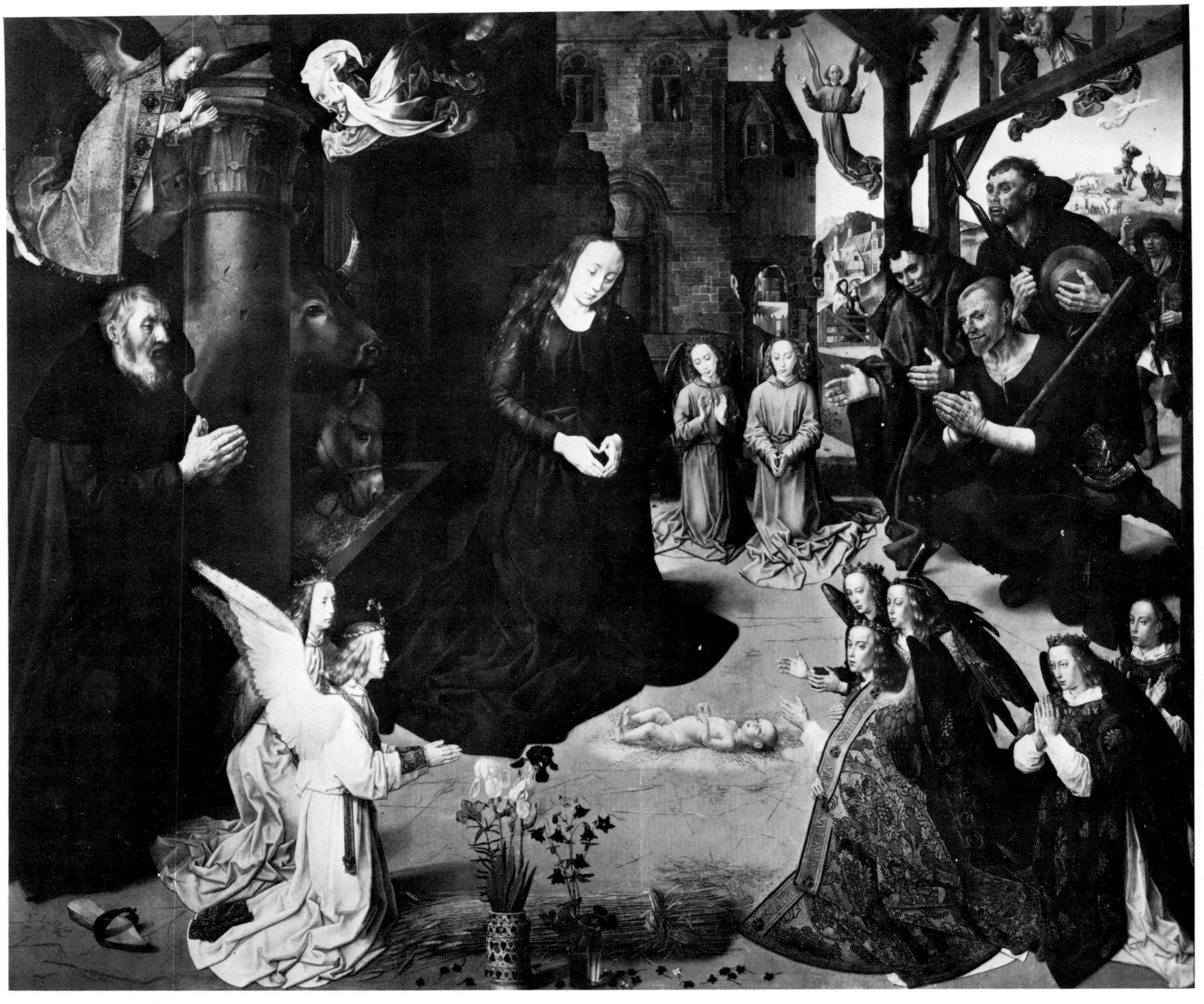

The central panel of the Portinari altarpiece by Hugo van der Goes, commissioned by the agent of the Medicis in Bruges around 1475. The vases in the foreground contain bearded irises, a lily (*Lilium bulbiferum*), a columbine and three carnations; sweet violets are strewn on the ground around them. Each flower has a particular religious significance. The blue iris is a royal flower, appropriate to the Madonna as Queen of Heaven, while the orange lily also signifies royalty. The columbine, so called because the petals were supposed to be in the shape of a dove, symbolises the Holy Ghost. Carnations represent the Incarnation, while violets stand for humility. Uffizi Gallery, Florence.

in gathering the correct plants,' a purpose for which most earlier herbals would have been of little use. Among Amadio's paintings, some twenty are direct copies from the Carrara herbal; although small details, like an extra tendril on a vine, have been left out, the leaves are in the same position. John Ruskin, who had presumably not seen the Carrara herbal, particularly admired this one, which he held to contain 'the earliest botanical drawings of approximate accuracy'.

Meticulously painted flowers can be found as details in Italian paintings from the middle of the fifteenth century. A hedge of roses, oranges and pomegranates adorns the background of the panel of Paolo Uccello's *The Rout of San Romano* (*c*1456) that is in the National Gallery, London. In another painting in the gallery, *The Martyrdom of St Sebastian* by Antonio and Piero del Pollaiuolo, dating from around 1470, the spiny butcher's broom and St Benedict's thistle by the stake to which Sebastian is bound make a visual allusion to the sharp arrows fired into his body. Many flowers had a specific religious significance, and it would have been for this that Hugo van der Goes, a Flemish painter working in Italy, included in the altarpiece commissioned by the Portinari family in 1475 two vases, one containing columbines and the other irises and an orange lily, *Lilium bulbiferum*, which grows wild in the hills above Florence. Although many individual flower studies must have been done at this time, almost the only one to survive is the famous watercolour of an iris painted by Jacopo Bellini around 1450 and now in the Louvre.

With one exception, flowers do not make such an early appearance in the painting of northern Europe. The one picture in which they do appear in

Job's tears (*Coix lacryma-jobi*), drawing by Leonardo da Vinci. The 'fruits' of this tropical grass are shiny, grey and hard when ripe and are often used in India as beads. Royal collection, Windsor Castle.

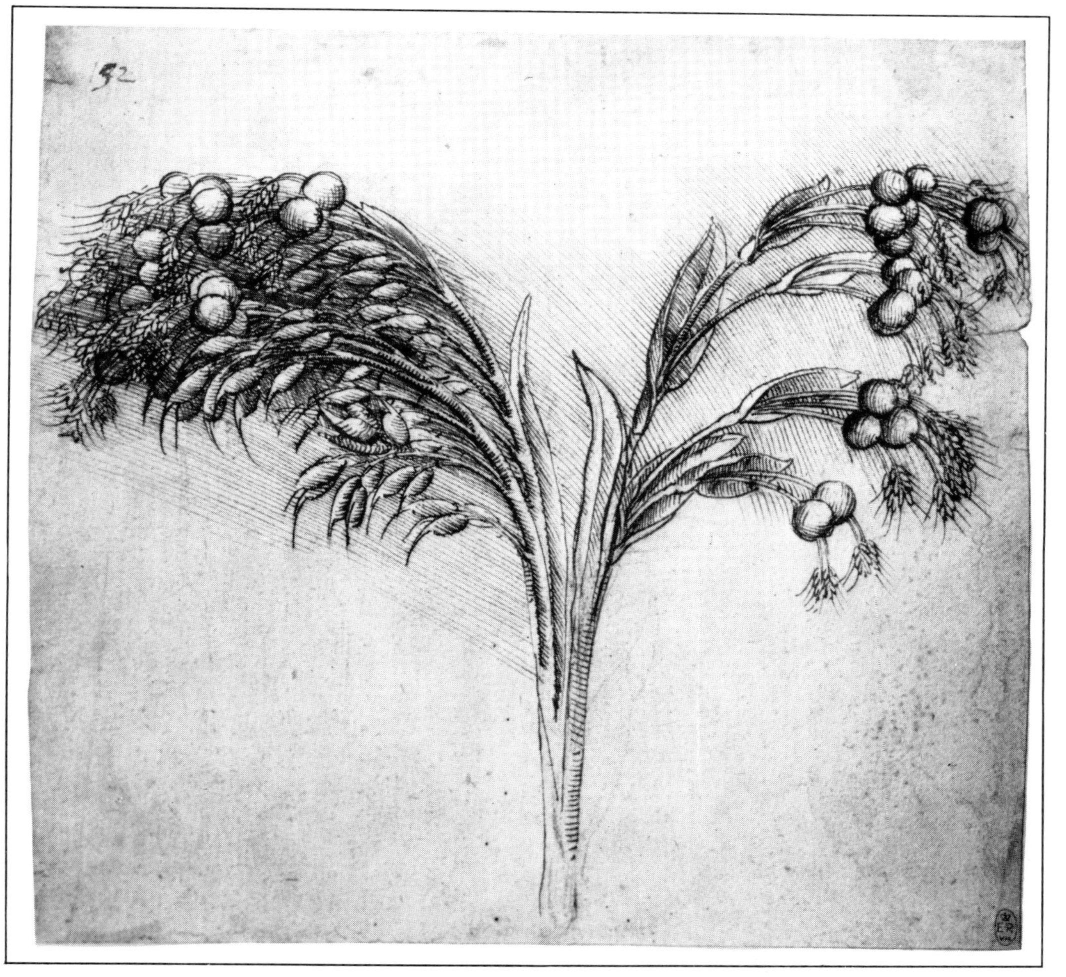

profusion is the central panel of the altarpiece painted around 1430 by the brothers Hubert and Jan van Eyck for Ghent Cathedral. The flowers in the meadow that forms the foreground are shown in great, if not impeccably accurate, detail – among those that can easily be identified are dandelions, lilies-of-the-valley, lady's smock, irises, madonna lilies and Solomon's seal. Many shrubs and trees, including a date palm, are also depicted. No similar examples have survived from the next fifty years of Flemish painting.

One expression of the increasing sophistication of painting from the late fourteenth century onwards was the International Gothic style, a development of French Gothic art characterised by its elegance of line, brilliance of colour and naturalism of detail. This was the style of some of the finest manuscript illumination, of which the most famous practitioners were the brothers Limbourg, whose *Très Riches Heures* was painted in the second decade of the fifteenth century for Jean, Duc de Berry. The International Gothic style, as its name suggests, spread across Europe; it lasted for the rest of the fifteenth century and through the Ghent-Bruges school of

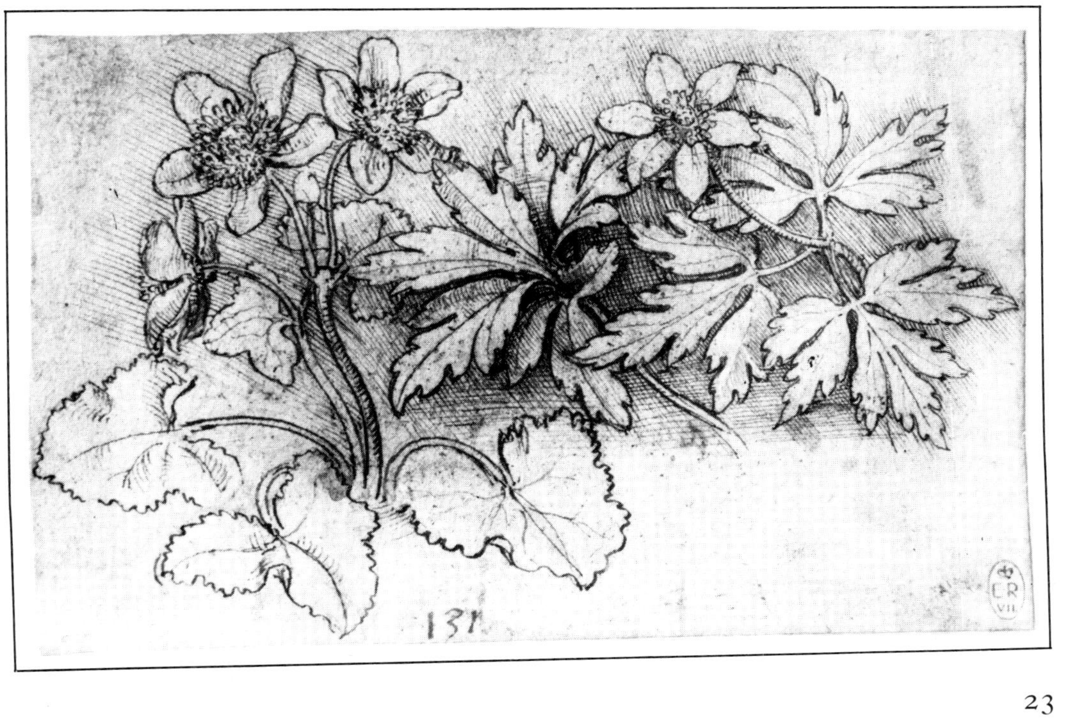

Marsh marigold (*Caltha palustris*) and wood anemone (*Anemone nemorosa*), ink and black chalk, by Leonardo da Vinci, around 1505. Royal collection, Windsor Castle.

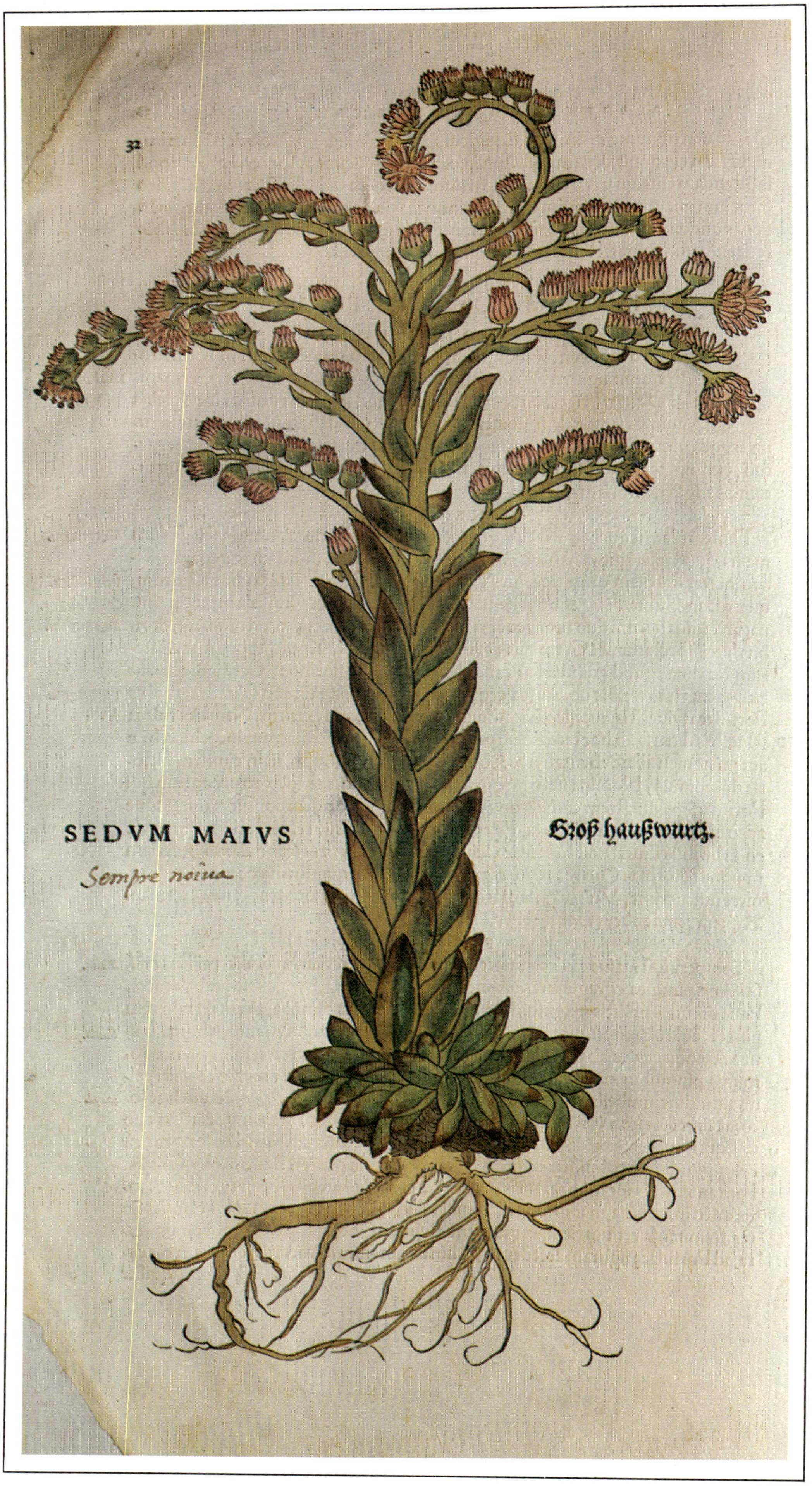

'*Sedum maius*' or '*Gross Hauswurtz*', a houseleek (*Sempervivum tectorum*). Woodcut with contemporary colouring from *De Historia Stirpium* (1542) by Leonhart Fuchs.

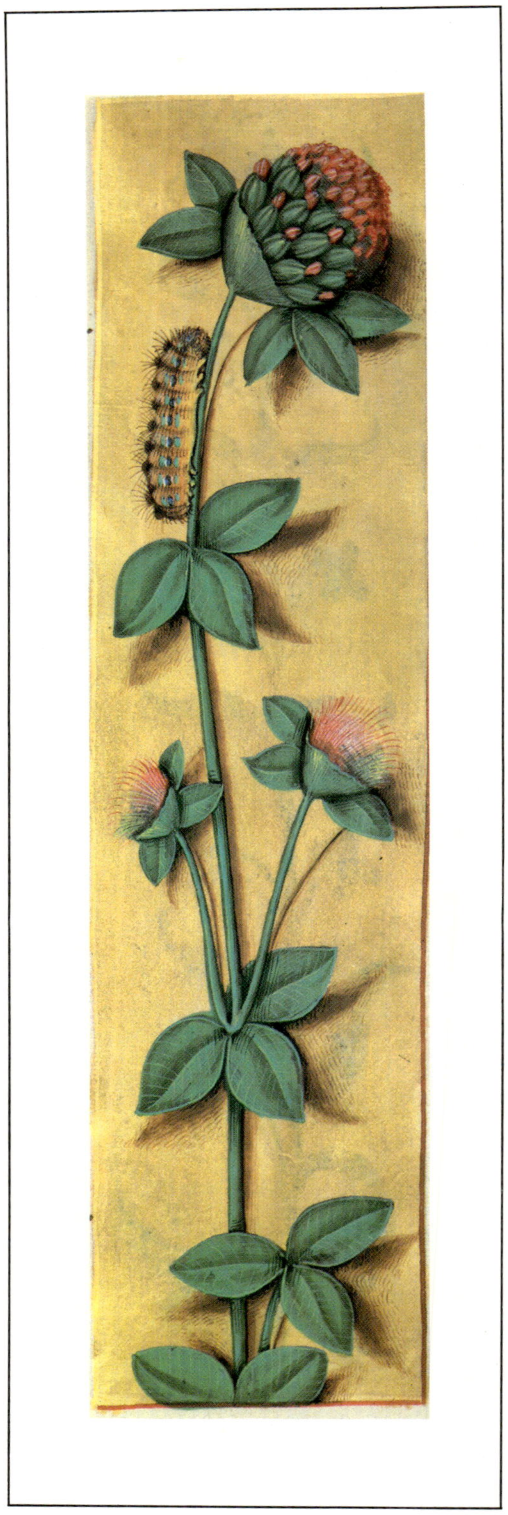

Clover (*Trifolium pratense*) with caterpillar. Detail from a Book of Hours, School of Jean Bourdichon, around 1510. British Library.

illumination, produced exquisite breviaries and books of hours with the decorative borders around the text strewn with flowers and insects. This style is particularly associated with illuminator known as the Master of Mary of Burgundy, who was working in Bruges in the 1480s. In his *Hastings Hours*, now in the British Library, each flower is meticulously painted as if lying on the paper, with careful shadows to produce a *trompe l'oeil* effect. From about the same date, but from the other side of Europe, the Marches of Italy, comes a painting, now in the National Gallery, London, by Carlo Crivelli in which the foreground is strewn with rosebuds,

Right :
Das Grosse Rasenstück (*The Large Piece of Turf*), watercolour on paper, by Albrecht Dürer, painted around 1503. The plants shown are salad burnet (*Poterium sanguisorba*), dandelion (*Taraxacum sp.*), meadow grass (*Poa trivialis*), great plantain (*Plantago major*), daisy (*Bellis perennis*) and yarrow (*Achillea millefolium*). Albertina Collection, Vienna.

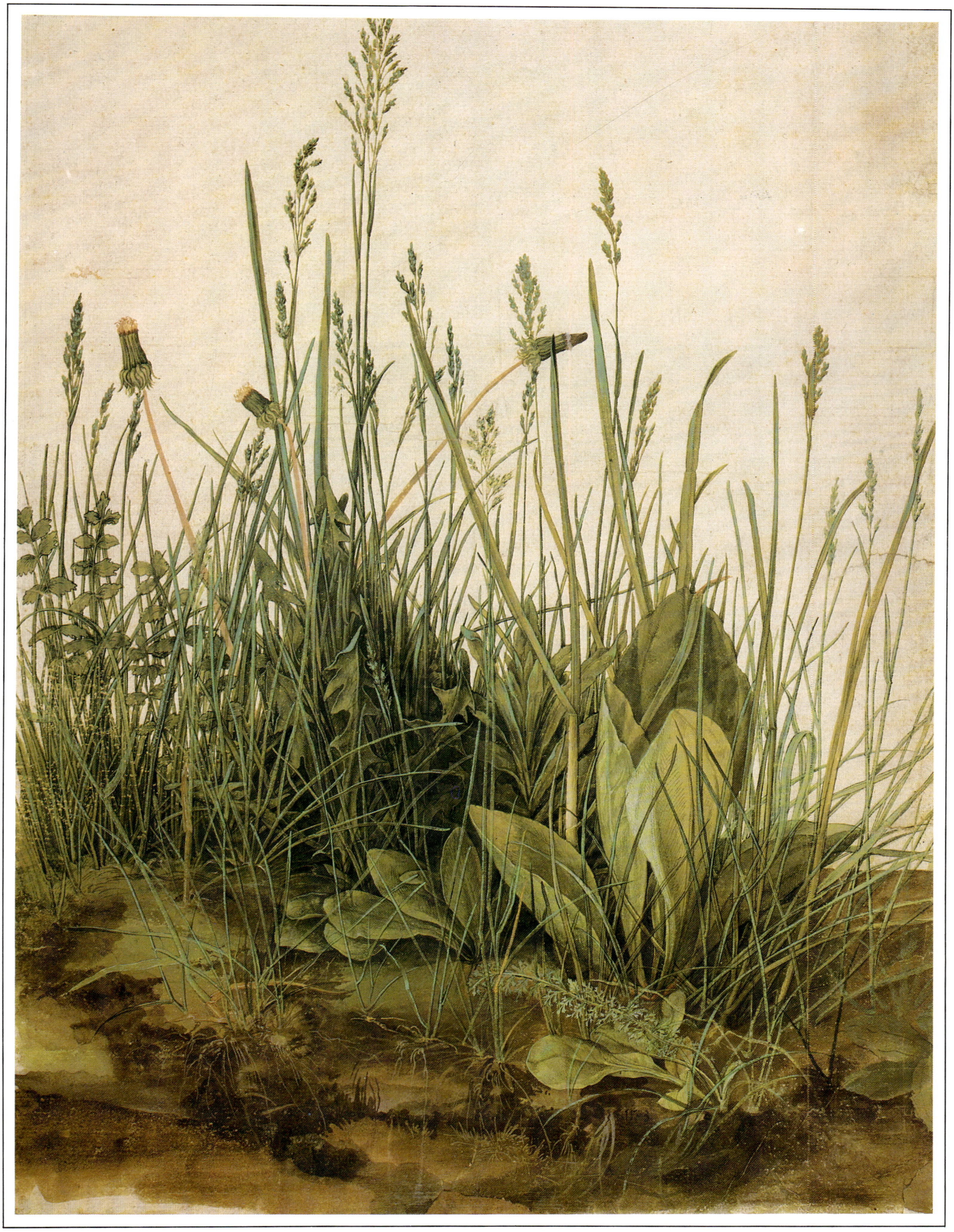

marigolds and daisy and poppy heads in a manner that clearly shows the International Gothic influence.

The style culminates in the illuminations of Jean Bourdichon and his school. Bourdichon himself was born in Tours around 1457 and became court painter to the kings of France. In the rectangular borders of his *Book*

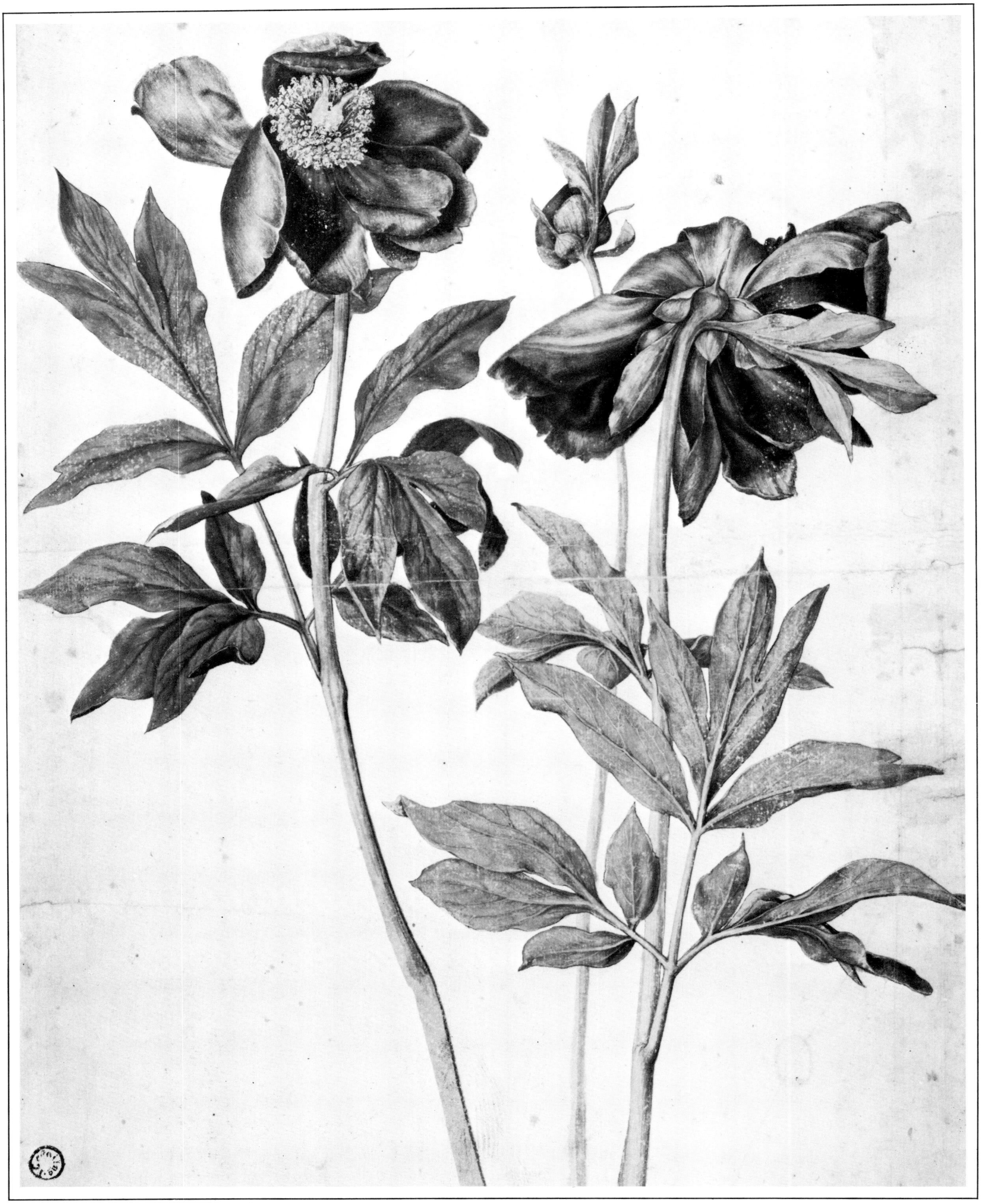

Peonies, watercolour by Albrecht Dürer, painted around 1503. Kunsthalle, Bremen.

of Hours of Anne of Brittany, painted between 1500 and 1508, whole plant stalks, rather than individual flowers, are vigorously rendered in opaque colour, with strong shading throwing them into high relief; sometimes the background is gilded, and caterpillars, butterflies and other insects are included. More than three hundred plants appear, with the artist indicating his particular love of nature by adding the name of each in French or Latin.

Nevertheless, the impulse behind Bourdichon's illuminations in devotional books for the French court is very different to the quest for scientific accuracy that is evident in Albrecht Dürer's flower paintings from the same decade. The antecedents to these include a single surviving sheet of plant studies by Pisanello, who died around 1450, and the flower drawings of Leonardo Da Vinci, some of which were done twenty years before the Dürer pictures. Although there is a splendid rhythm to them, they are still no more than preparatory sketches, and we have no paintings by him in which flowers play more than a very minor part. However, columbine, cyclamen, Jacob's ladder, jasmine, St John's wort and an iris have been identified in the Louvre version of *The Virgin of the Rocks*. The bulrush and bur-reed that appear in his cartoon of Leda are carefully observed, and there is an especially fine drawing by Leonardo of an *Ornithogalum*.

Dürer's watercolours, on the other hand, are the first botanical masterpieces, clearly intended as paintings in their own right and outstanding for their delicacy and detail. Dürer was the first artist to take a piece of nature, draw it as faithfully as he could and produce a perfect work of art. He stated his own philosophy with some force: '. . . study nature diligently. Be guided by nature and do not depart from it, thinking that you can do better yourself. You will be misguided, for truly art is hidden in nature and he who can draw it out possesses it.'

Most of Dürer's flower studies which survive, about ten in number, are believed to have been painted between 1503 and 1505 at the time of his second visit to Venice. Nearly all are in either the Albertina in Vienna or the Kunsthalle in Bremen. The most famous of Dürer's botanical pictures is *Das Grosse Rasenstück*, which is, as its name indicates, a large piece of turf – a detailed study of meadow grasses and dandelions on a dull day, growing up out of brown earth. The dandelion flowers are closed, the

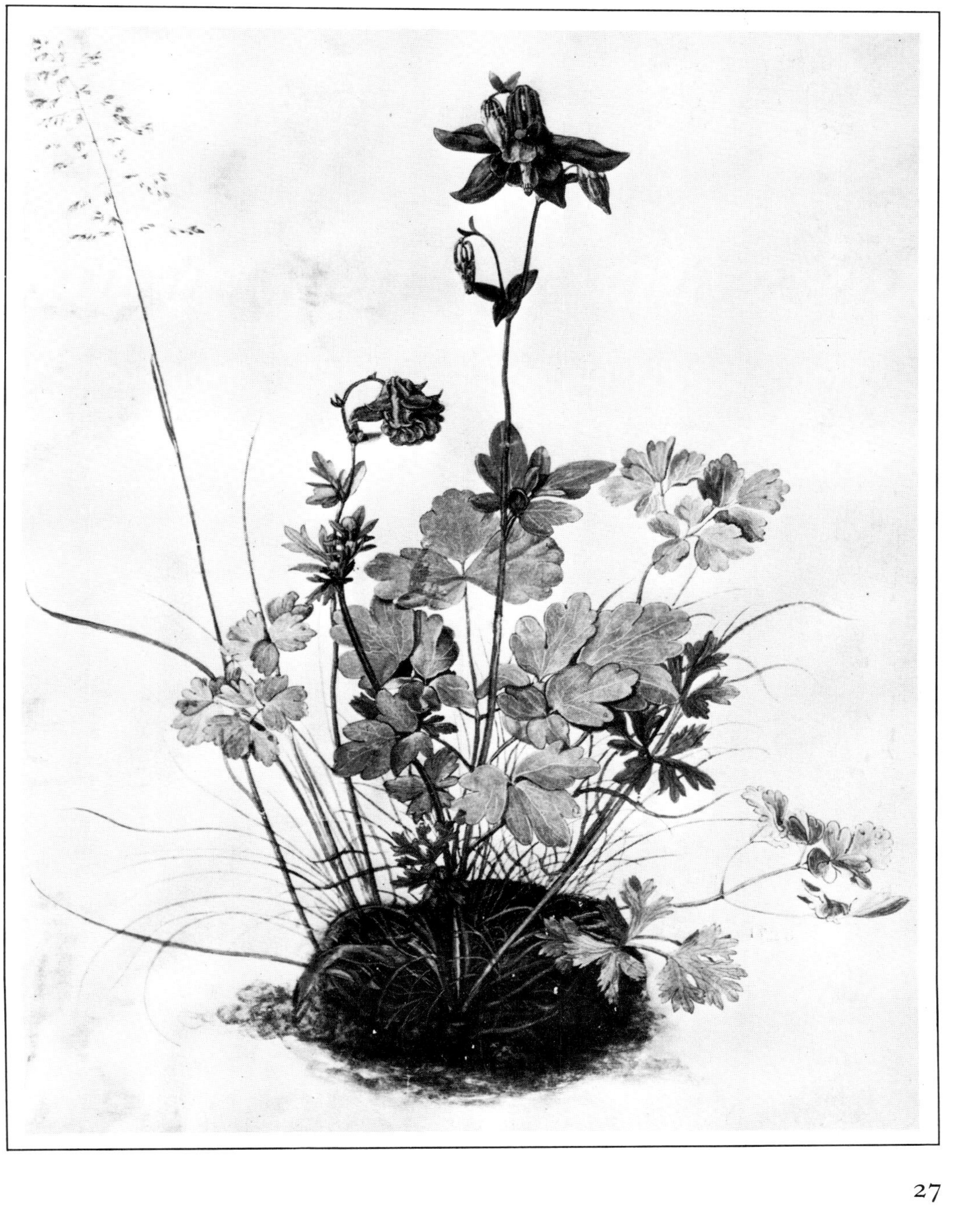

A columbine (*Aquilegia vulgaris*) and a meadow grass (*Poa trivialis*), with leaves of the creeping buttercup (*Ranunculus repens*). Watercolour by Albrecht Dürer, painted around 1503. Albertina Collection, Vienna.

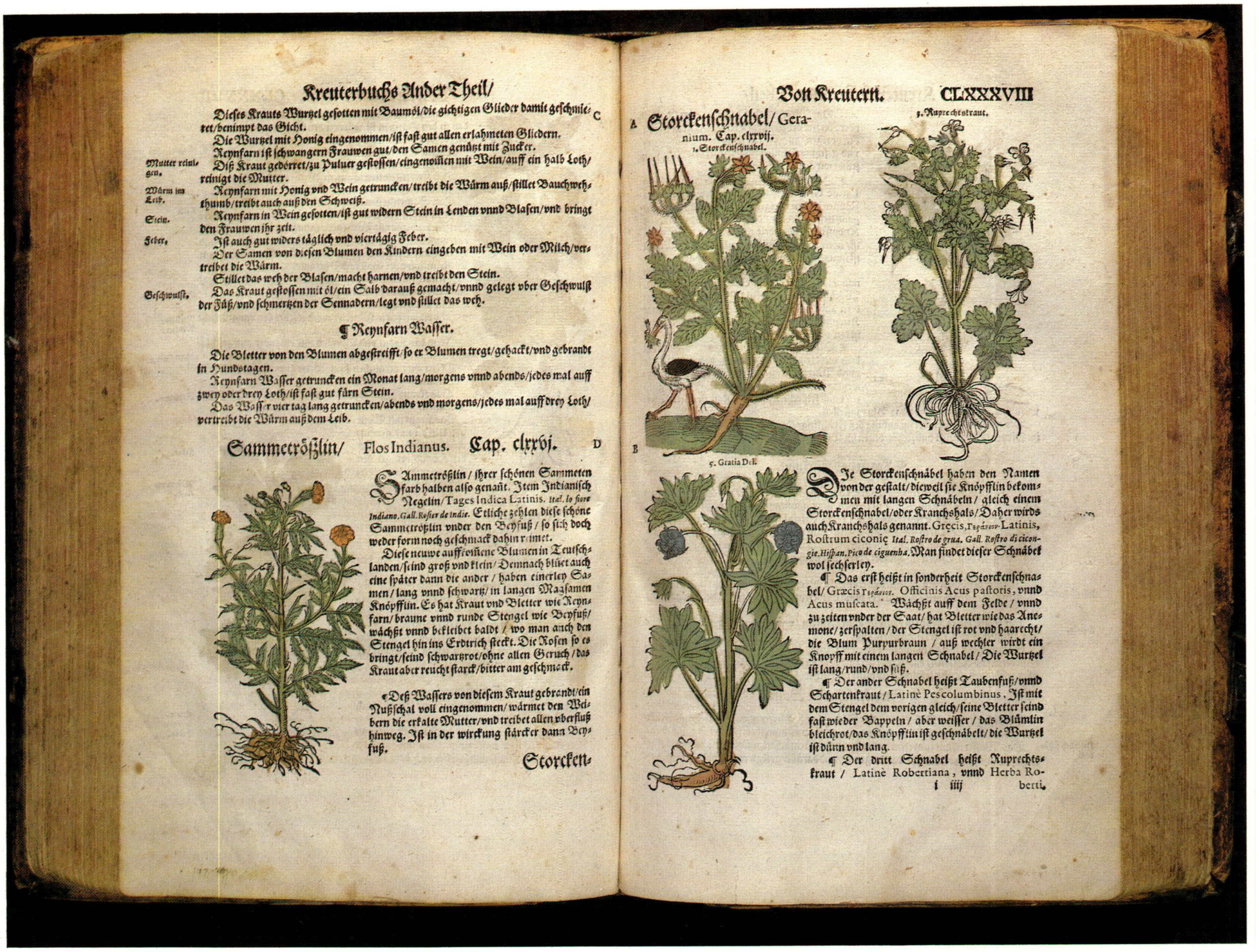

Late sixteenth century illustration: woodcuts with contemporary colouring in the *Kreuterbuch* (1582) of Adam Lonicerus. The plants shown are '*Sammetrösslin*' or '*Flos Indianus*', which is the French marigold (*Tagetes patula*), a native of Mexico, and three members of the geranium family: '*Storckenschnabel*', storksbill (*Erodium cicutarium*), '*Ruprechtskraut*', Herb Robert (*Geranium robertianum*), and perhaps, though the flower colour is completely wrong, bloody cranesbill (*Geranium sanguineum*).

grasses not yet in full flower; every detail is true. In a slightly more colourful painting, a common buttercup is the central subject, with grasses, clover and a plantain on similar brown earth. Two more pictures show flowers growing, a bugloss, *Anchusa hybrida*, called *Liebangel*, and a group dominated by a columbine, with buttercup leaves and meadow grass. These paintings are all similar in style as are the studies of individual flowers: a painting of peonies, a Martagon lily and an iris (probably *Iris trojana*). The iris was used for the painting *The Madonna with the Iris*, now in the National Gallery, London, but is in itself a fully finished work of art, drawn almost life size. There is another less spectacular iris painting in the Escorial in Madrid, and three more smaller paintings which are in the Winkler catalogue of Dürer's work, but possibly not by him: a small study of a bunch of violets, a small patch of turf much more crudely drawn and with more flowers than the others, and a watercolour of a single stem of wild lettuce which in the Musée Bonnat in Bayonne. With its shrivelled lower leaves and the lack of the precision in the drawing, it bears a very close resemblance to many of the plant studies of Hans Weiditz, and may have been painted by him rather than by Dürer.

Weiditz was a near-contemporary of Dürer and almost certainly his pupil. For many years his work was known only through the woodcuts in Brunfels's *Herbarum Vivae Eicones*, but in 1930 some of the original paintings by Weiditz were found in Berne in the herbarium of Felix Platter (1536–1614), a physician from Basle. They are very literal drawings in watercolour on paper. Weiditz seems to have adhered, if anything too strictly, to the letter of Dürer's advice. Where a leaf was eaten, dead or shrivelled, Weiditz painted it so, and this rejection of any interpretation shows up in the plates of the *Herbarum Vivae Eicones*, in which some of the plant specimens are rather feeble and atypical.

Mandrake (*Mandragora autumnalis*), opaque watercolour on paper, by Jacopo (also known as Giacomo) Ligozzi, painted around 1480. In its style and botanical accuracy, this painting is closest to those of Ehret and the Bauers two hundred years later. The accuracy of this particular painting is all the more remarkable because anthropomorphic mandrakes continued to be shown in reputable botanical books as late as 1700. Gabinetto Disegni, Uffizi Gallery, Florence.

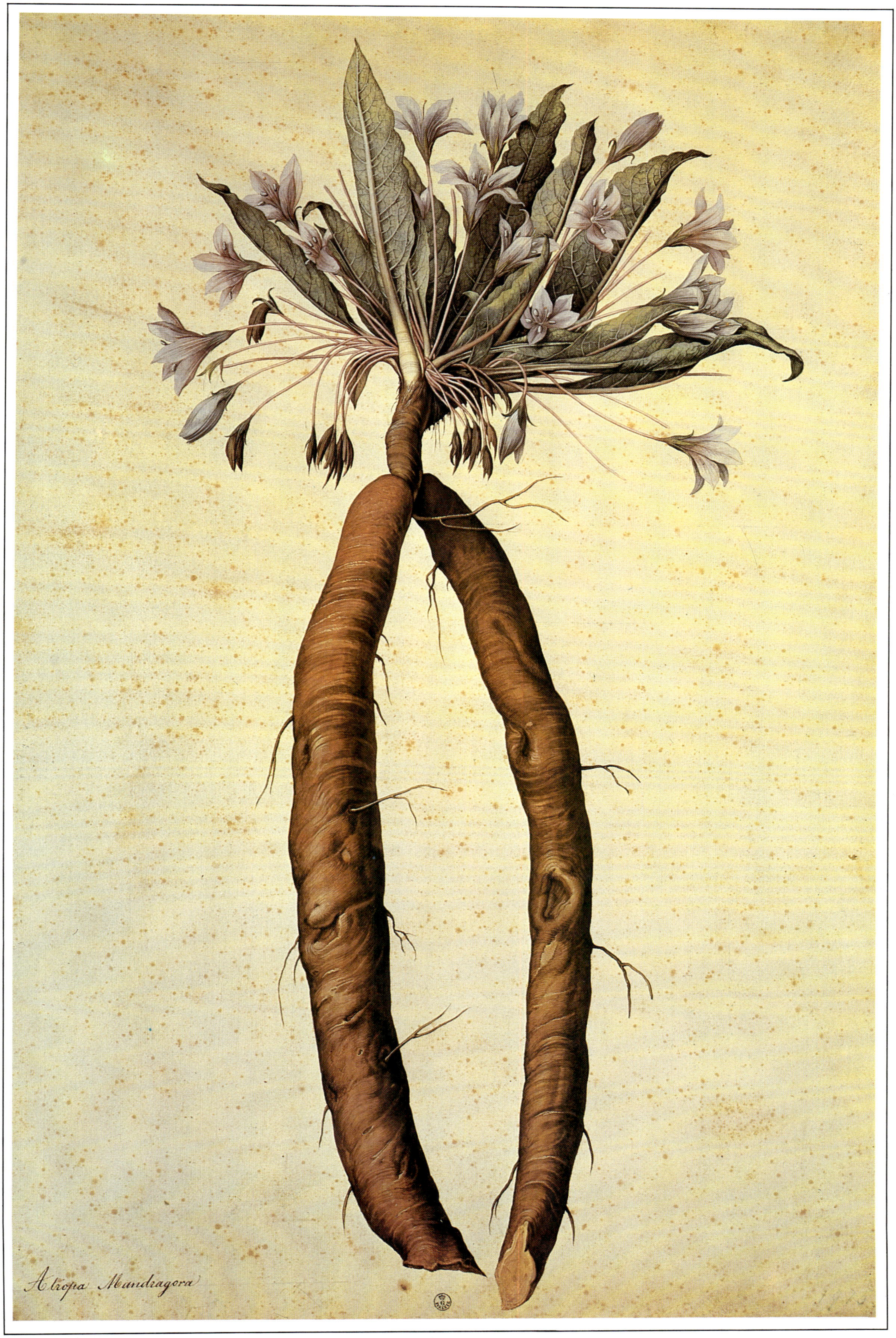
Atropa Mandragora

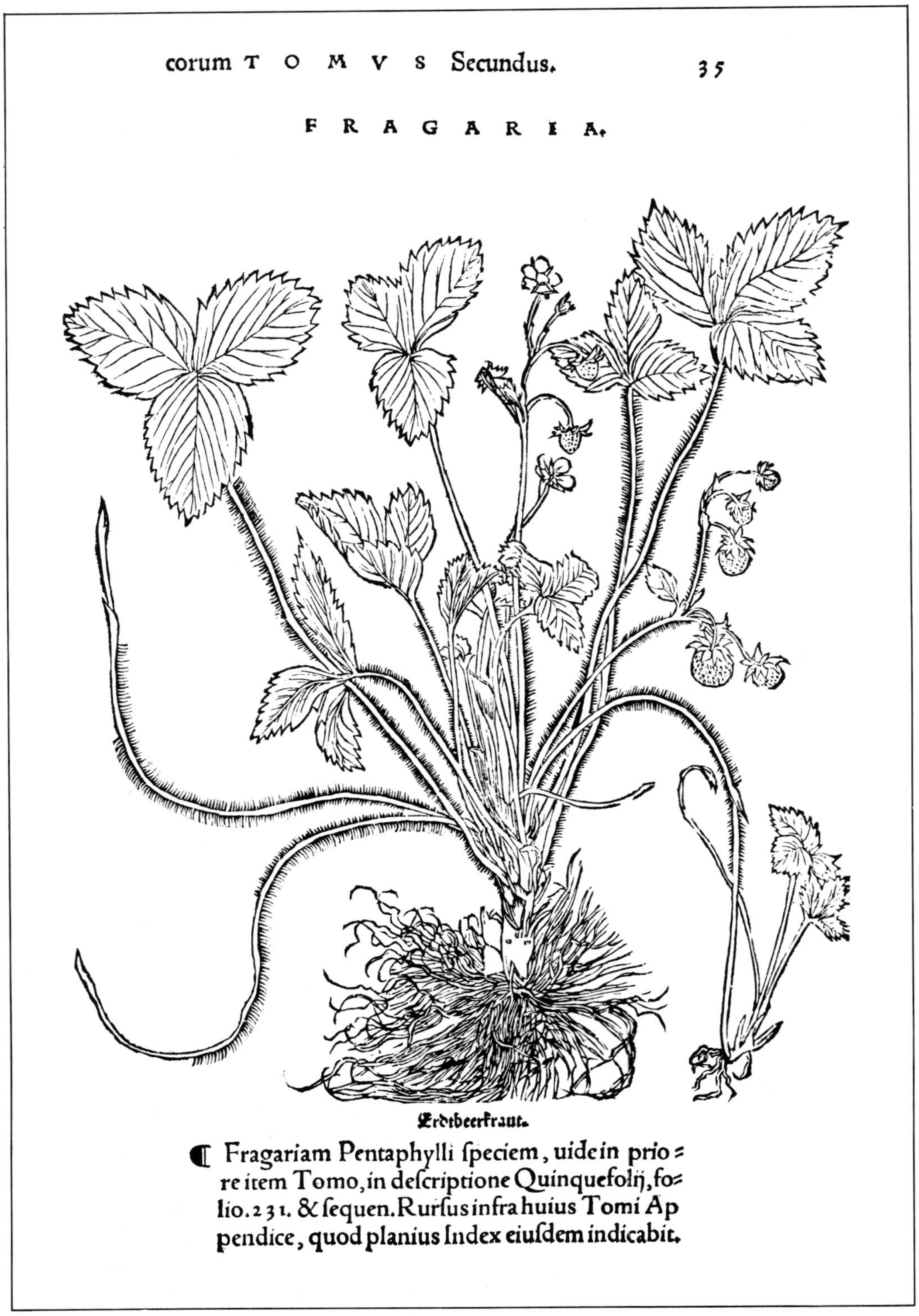

Right: Three woodcuts on consecutive pages of *Herbarum Vivae Eicones* (1530) by Otto Brunfels: stinging nettle (*Urtica dioica*), white dead-nettle (*Lamium album*) and spotted dead-nettle (*Lamium maculatum*).

Right: Pasque flower (*Anemone pulsatilla*), woodcut from *Herbarum Vivae Eicones* (1530) by Otto Brunfels.

Wild strawberry (*Fragaria vesca*), woodcut from *Herbarum Vivae Eicones* (which means 'Living Pictures of Plants') by Otto Brunfels, 1530.

Far right:
Lime, or linden, tree (*Tilia*) with dancers beneath. Hand-coloured woodcut from *De Stirpium* (1552 edition) by Hieronymus Bock, known as Tragus. Bock's illustrations, though jolly, are less precise than those of Brunfels, but his text marks a considerable advance, giving good descriptions of the plants and their manner of occurrence.

The publication of Brunfels's *Herbarum Vivae Eicones* in 1530 and that of Fuchs's *De Historia Stirpium* in 1542 were the two main landmarks of botanical illustration and scholarship in the first half of the 16th century. These were the first biological books which relied on scientifically accurate illustrations. Weiditz, the illustrator of Brunfels's herbal, did not cut the woodblocks himself, but their standard is nevertheless very high, and the figures of plants with leaves of unusual shapes, among them the common lady's-mantle, herb Robert (*Geranium robertianum*) and hart's-tongue fern, are particularly successful.

Brunfels himself first entered a Carthusian monastery but later became a Lutheran and finally, shortly before he died in 1534, had become town physician of Berne. His text, based primarily on Dioscorides and Pliny and consisting of descriptions of plants with a list of the uses of the herbs, does not approach the high standard of the figures.

Leonhart Fuchs also became a physician as well as a Lutheran in later life. Born in Wemding, Bavaria, in 1501, he practised for a time as a physician in Munich before returning in 1526 to the University of Ingolstadt, where he had once studied, as Professor of Medicine. Through his successful treatment of the terrible plague that spread across Germany in 1529, his reputation spread abroad, even as far as England. In 1535, he

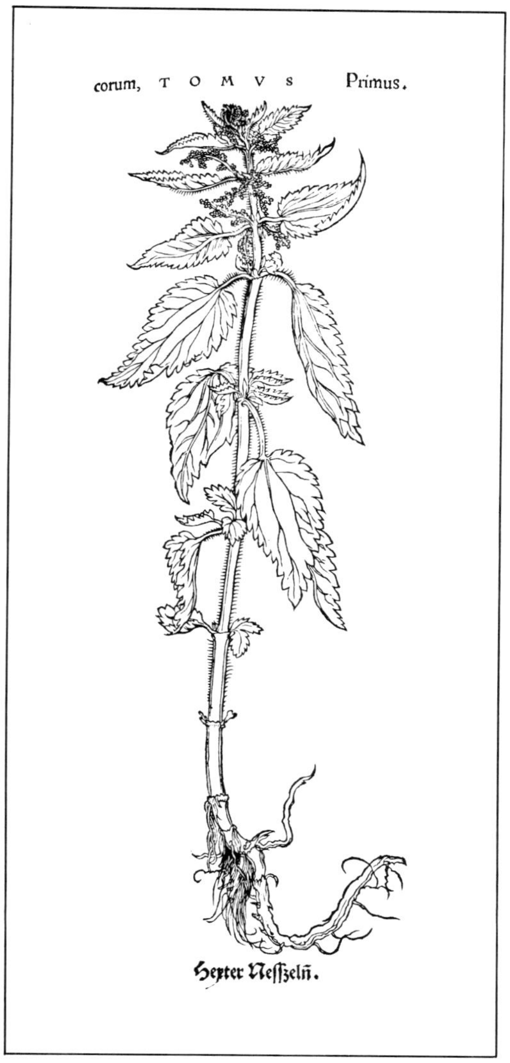
corum, TOMVS Primus.

Hepter Nesselñ.

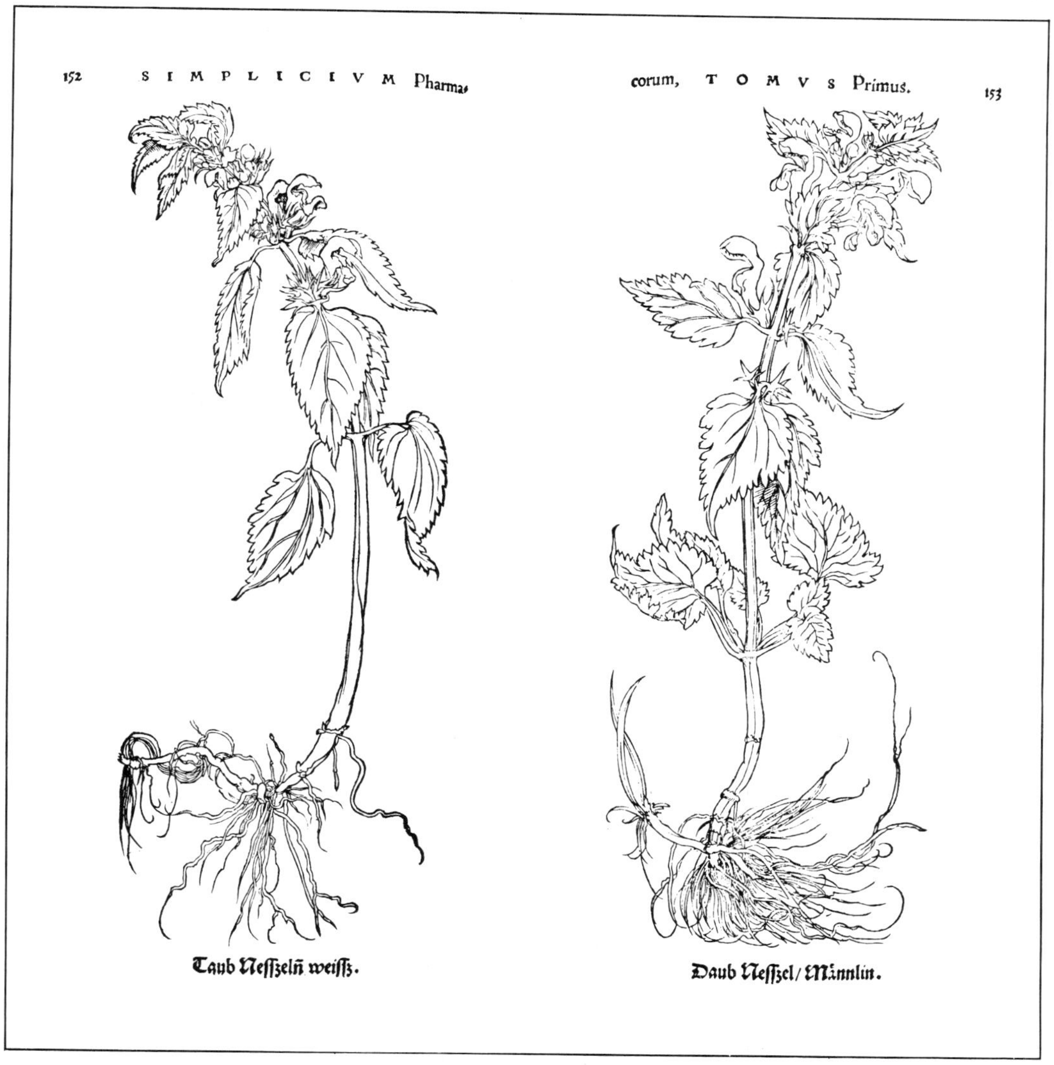
152 SIMPLICIVM Pharma-

Taub Nesselñ weissz.

corum, TOMVS Primus. 153

Daub Nesszel / Männlin.

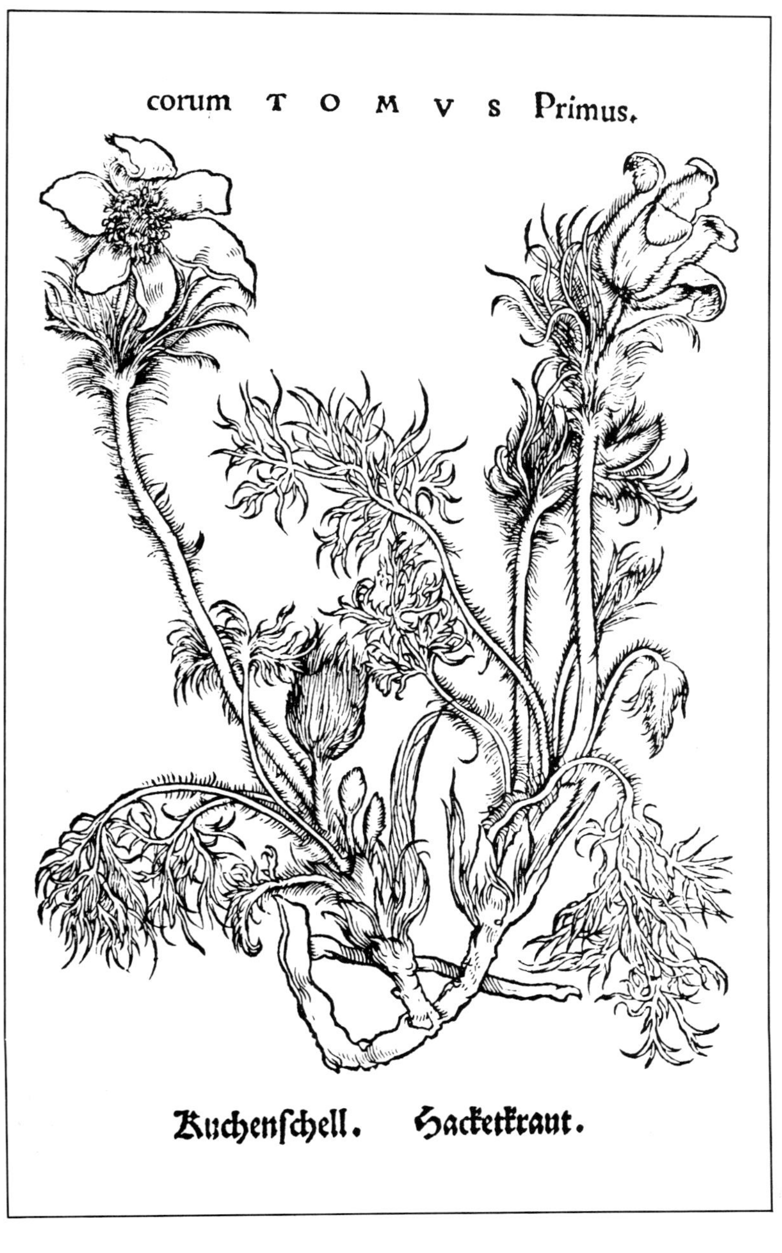
corum TOMVS Primus.

Kuchenschell. Hacketkraut.

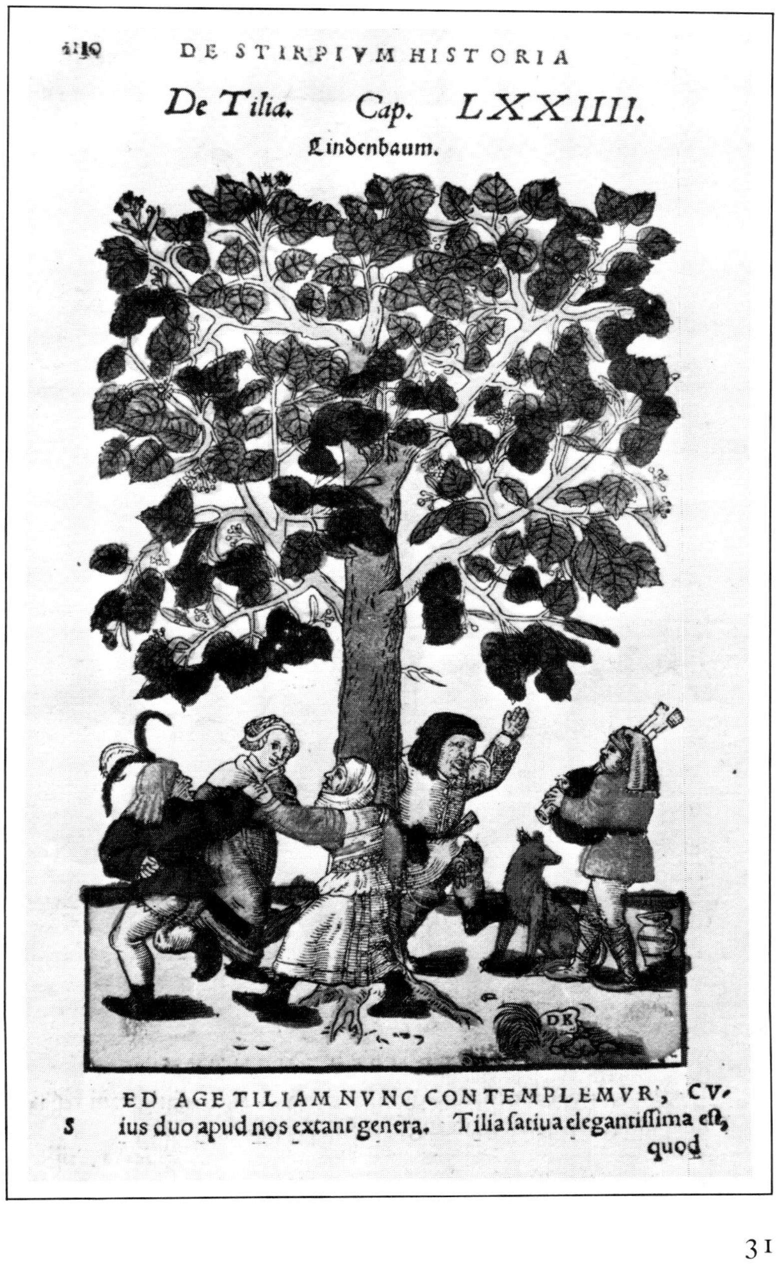
DE STIRPIVM HISTORIA

De Tilia. Cap. LXXIIII.

Lindenbaum.

ED AGE TILIAM NVNC CONTEMPLEMVR, CV-
S ius duo apud nos extant genera. Tilia satiua elegantissima est,
quod

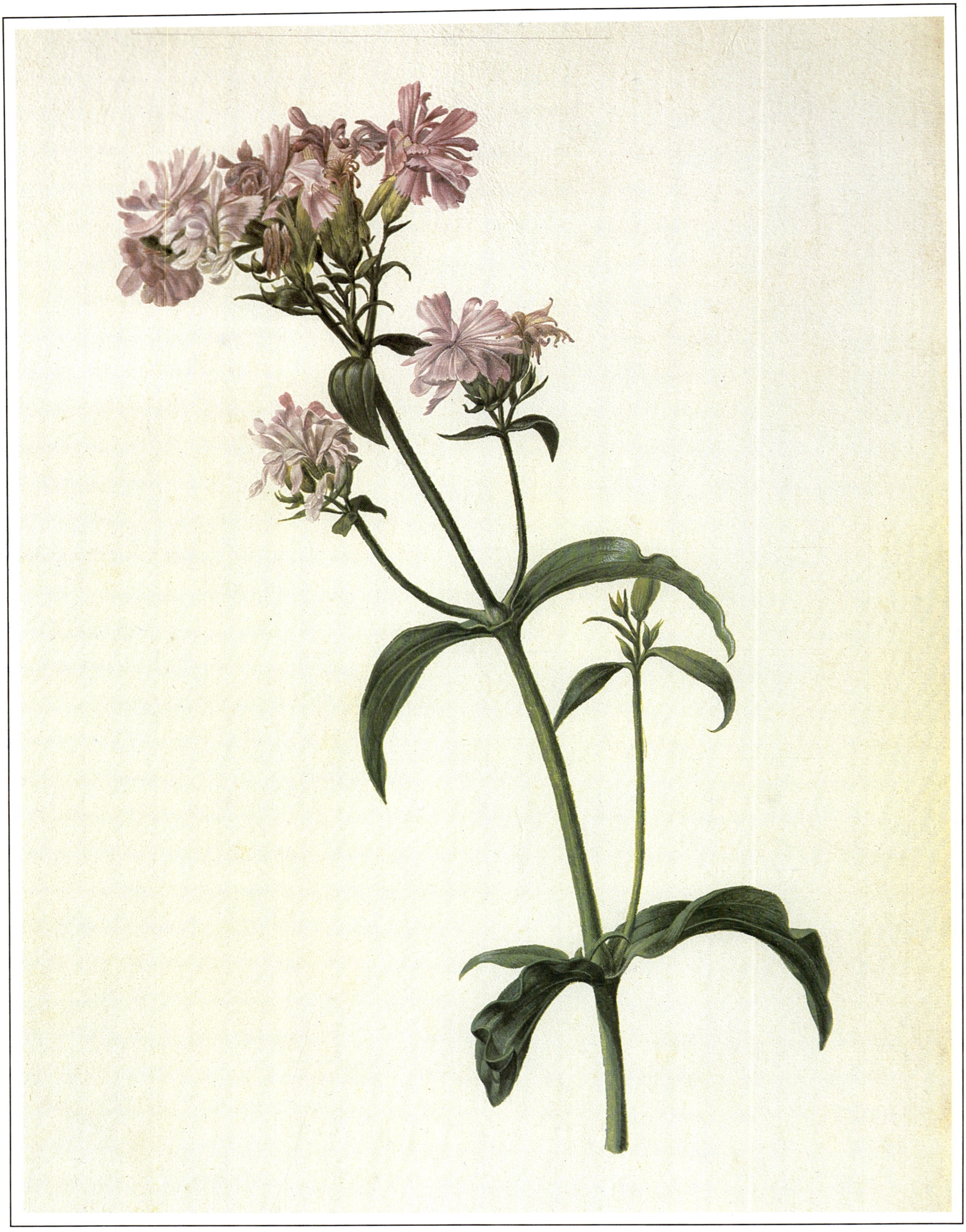

moved to Tübingen. His great work, *De Historia Stirpium*, was printed in Basle in 1542. Five hundred plants are illustrated in folio format, carefully printed in outline. They were designed to be coloured. Unlike Brunfels, he shows a large, well-developed and somewhat idealised specimen of each plant. He is most successful with some of the smaller herbs; in some of the plates of trees and climbers, the page seems rather over-crowded. A full

Double soapwort (*Saponaria officinalis*), opaque watercolour on vellum, by Peter van Kouwenhoorn, around 1630. Soapwort is an ancient, but still common, garden flower of the campion family, Caryophyllaceae. Lindley Library, Royal Horticultural Society, London.

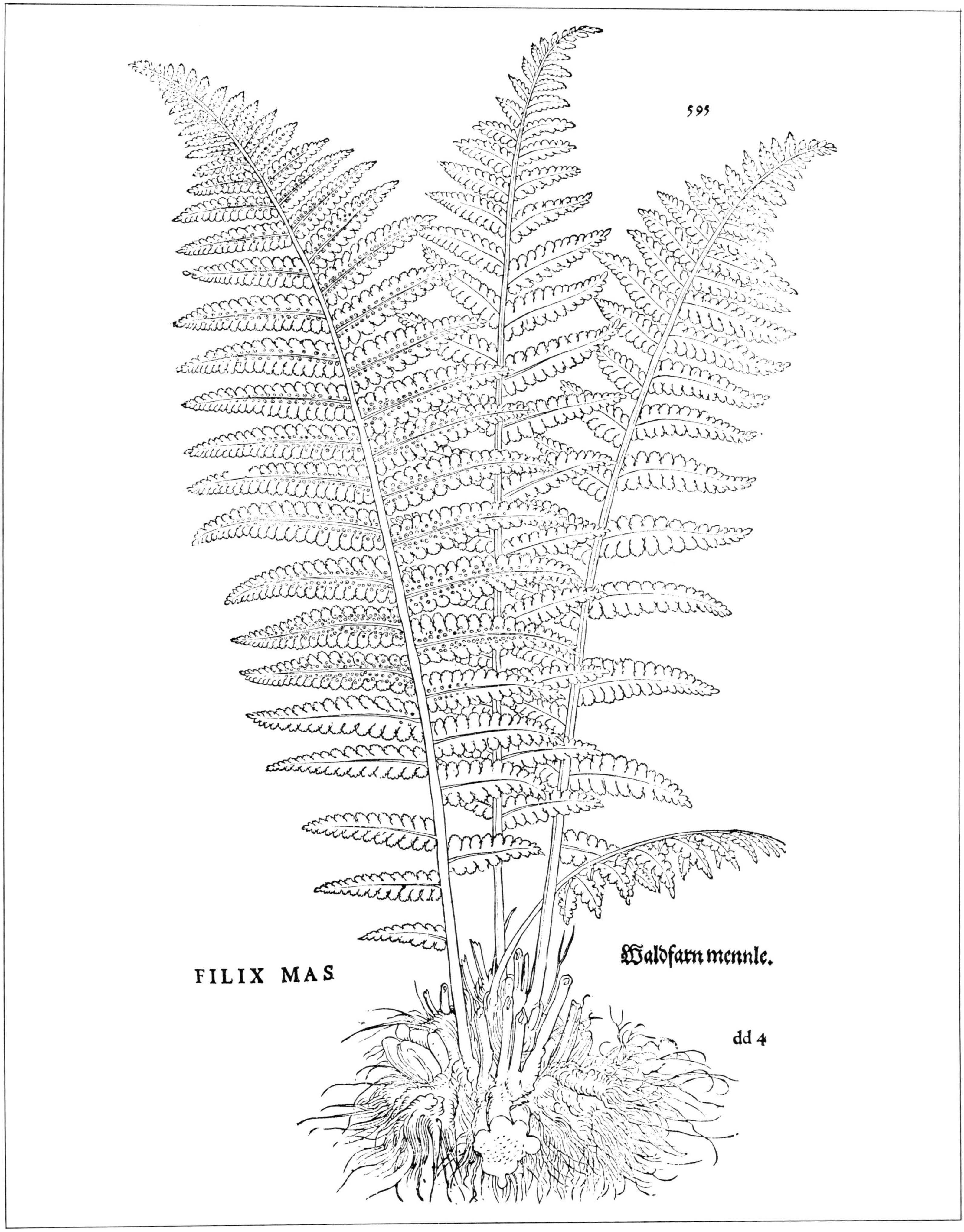

Male fern (*Dryopteris filix-mas*). Woodcut from *De Historia Stirpium* (1542) by Leonhart Fuchs.

page plate shows Albrecht Mayer, who drew the plants from nature, and Heinrich Füllmaurer and Veit Rudolf Speckle, who prepared the woodblocks. The text is a great improvement on Brunfels: it gives the names, habits, and localities, time of flowering, and tempers of the plants, together with their uses according to the ancient authorities Dioscorides, Pliny and the second-century physician Galen.

CHAPTER III
Plants from an Expanding World

The sixteenth century saw the spread of European exploration and trade around the world. Until Turkish power began to decline after the failure of the seige of Malta in 1565, the land routes to India and China were still dominated by Islam. As a result, European nations sought maritime trade routes to the Orient. In 1499, the Portuguese explorer Vasco da Gama sailed around the Cape of Good Hope and landed near Cochin in southern India. Christopher Columbus, a Genoese navigator in the service of Spain, also seeking a trade route to the East, had first reached the West Indies in 1492, and he later explored some of the coast of South America. The explorers and the travellers who followed them brought back new plants as well as spices and gold, and with the result that where plants had previously been cultivated for their medical properties they now came to be grown as much for their ornamental value.

Ogier Busbecq, who was instrumental in bringing the great *Codex Vindobonensis* from Constantinople to Vienna, also introduced to Europe several new ornamental plants which had been grown for centuries by the Turks; he sent back tulip seeds in 1554, at about the same time as he introduced the Crown Imperial (*Fritillaria imperialis*) and the hyacinth. Busbecq probably brought them as far as Vienna, Clusius spread them around Europe. Busbecq is also credited with introducing horse chestnut,

Aubergine or eggplant (*Solanum melongena*) and a sedge (*Cyperus esculentus*). Woodcuts on two pages of *Reuse in die Morgenländer* (1583) by Leonhart Rauwolff.

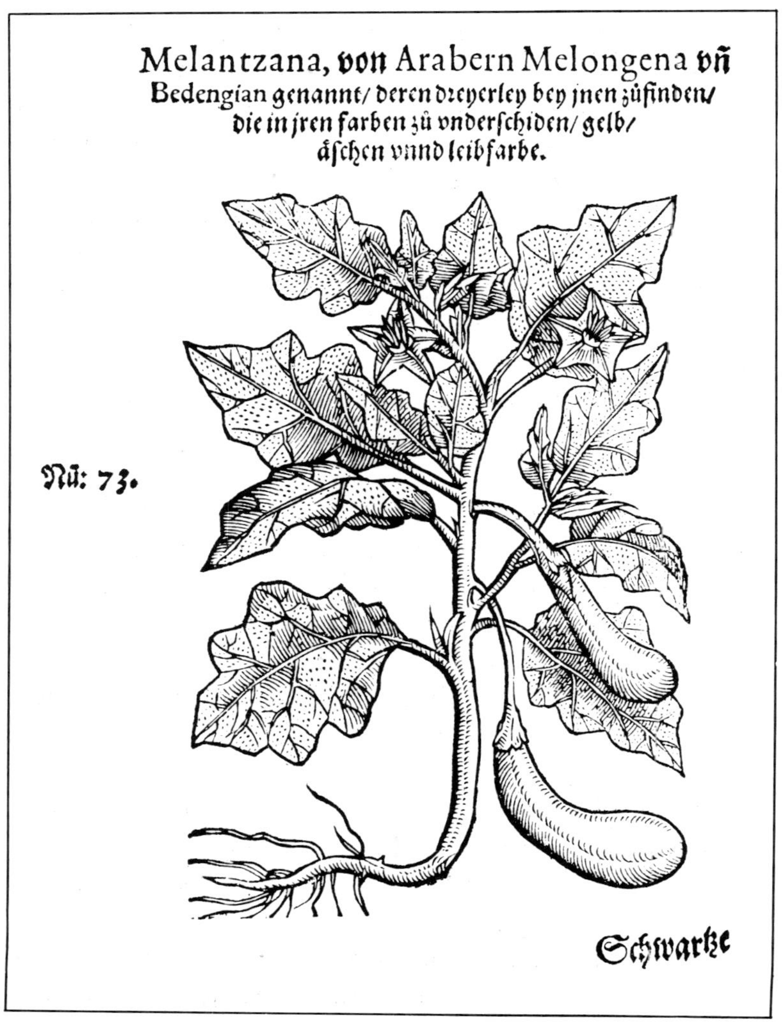

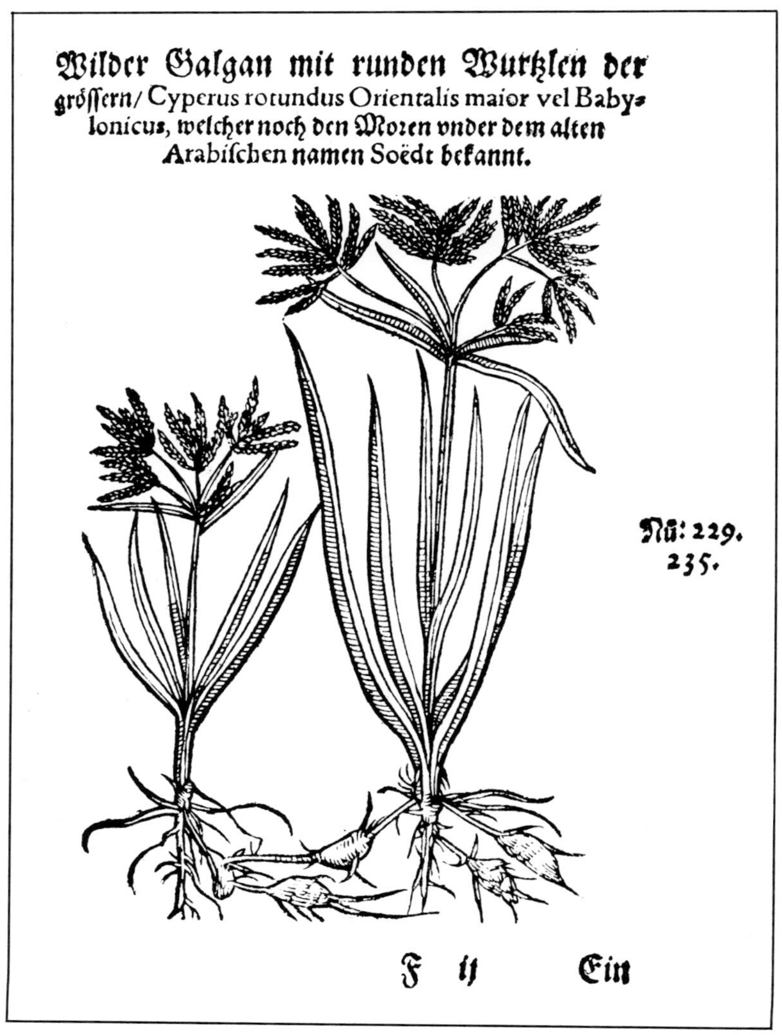

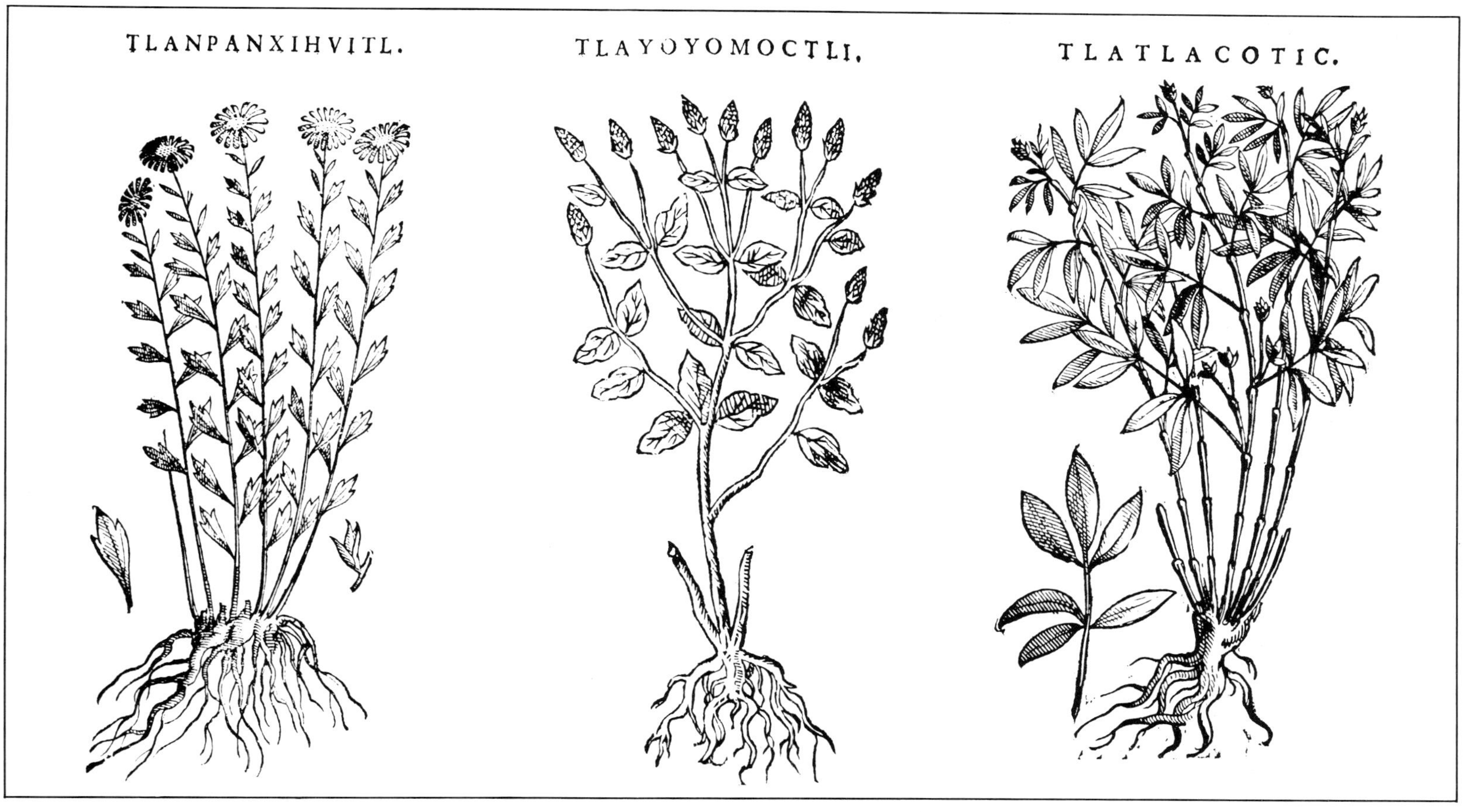

Three Mexican plants. Woodcuts from *Plantarum, Animalium et Mineralium Mexicanorum Historia* by Francisco Hernández, published in Rome in 1651.

mock orange and lilac to western European gardens. He gave the seeds to the leading botanists of the day, some to Clusius in Vienna, some to Matthioli in Prague.

Busbecq was not the only European visitor to the Levant who had an interest in botany. Two other travellers in the area returned to write travel books which included illustrations of plants. The Frenchman Pierre Belon (1517–64) was a contemporary of Gesner and probably his equal as a zoologist. He visited Greece and Turkey from 1546 to 1549 and published an account of his journey in 1553; a Latin translation by Clusius was published in Antwerp in 1605.

A slightly later journey to the Middle East was made by Leonhart Rauwolff, a physician from Augsburg. In 1583, some six years after his return, he published an account of his travels containing 42 illustrations of plants. Some of the specimens in his herbarium, which is still preserved in the University of Leiden, were used a century later by Robert Morison for his *Plantarum Historia Universalis Oxoniensis* of 1680. Rauwolff climbed Mount Lebanon and visited the aged Patriarch of the Maronites whom he treated for gout. They drank much of the Patriarch's wine, which Rauwolff found excellent, and, after climbing all day, reached the monastery, a small building protected by an overhanging rock. There, Rauwolff could count only 24 of the famous cedars and found no young plants. He did, however, collect an everlasting (*Helichrysum sanguineum*) and a spiny burnet (*Poterium spinosum*).

Very little has survived of the drawings made by the earliest explorers to America, and the *Navigationi et Viaggi* of G. B. Ramusio, published in Venice in 1556, has some of the first published illustrations of American plants, including maize and the first picture of a cactus, a prickly pear. Francisco Hernández, physician to Philip II of Spain, made coloured drawings of plants he found during his exploration of Mexico in the 1570s, but the originals were destroyed by a fire at the Escorial in 1671. A woodcut of a dahlia, one of the American plants that were first brought back from this expedition, was finally published in 1651, after Hernández's death, in his *Thesaurus*. Little else remains showing Mexican plants found by the early explorers, apart from a copy at Windsor Castle of an Aztec herbal of 1552 in which only one or two of the plants are actually recognisable.

We are not much better off for early illustrations from North America. Jacques Le Moyne de Morgues was a Frenchman, a Huguenot born in

Dieppe, which was then a centre of cartography. He was recruited for Laudonnière's expedition to Florida in 1564, as its recording artist. None of his paintings of North American flowers (if he did any) have survived, but he left behind a graphic description of the troubles which beset the expedition, of the Frenchmen's stupidity in their bad treatment of the Indians, and of Spanish brutality.

Laudonnière landed in Florida on 25th June and built Fort Caroline. To begin with, relations with the Indians were good; Le Moyne drew scenes of Indian life and mapped the rivers. But later in the year, the Indians refused to give the expedition food. By the Spring of 1565, they were starving, and some of the party died. Survivors were saved by the timely arrival of the English privateer John Hawkins, who sold them provisions and a ship. In September they were attacked by the Spaniards and only Le Moyne and Laudonnière, with about fourteen others, escaped. Meanwhile, a French expedition under Ribault, sent to relieve them, was shipwrecked nearby and persuaded to surrender to the Spaniards occupying Fort Caroline. Ribault and the officers were stabbed, and his men were tied together in fours and clubbed to death. Le Moyne and the other fifteen managed to reach their own small ships, which had sheltered in a creek. They struggled back across the Atlantic, landing in Bristol late in 1565. But Le Moyne's troubles were not over. In 1572, he had to flee to England to escape the massacre of the Huguenots and lived for some years in Blackfriars, working for Sir Walter Raleigh. He died there in 1588.

Most of Le Moyne's flower paintings are of plants common in European gardens at the time. His style owes a lot to French miniature painting of the school of Jean Bourdichon: in his earlier work, now in the Pierpoint Morgan library, flowers, fish, and insects are strewn over the page. The slightly later examples in the Victoria and Albert Museum are small paintings of single plants. Later work still, probably done after Le Moyne fled to England and now in the British Museum, is stiffer and more stylised. Some of the paintings were published in 1586 as simple woodcuts, designed to be used as models for painting, embroidery or tapestry, in Le Moyne's *La Clef des Champs*, which was dedicated to Lady Mary Sidney, mother of Sir Philip.

Another flower painter to visit North America was John White, who was official artist on Sir Walter Raleigh's abortive attempt to found an English colony in Virginia in 1585 and 1586. He subsequently made several other visits to North America, and was in charge of the disastrous second colonisation attempt in 1587. Like Le Moyne, he drew maps and scenes of Indian life; a few simple flower paintings by him still exist. One of milkweed (*Asclepias*) in fruit was used in Gerard's *Herball*.

The early explorations and the plants that were brought back proved a great stimulus to botanical studies in Europe.

Pierandrea Matthioli or Matthiolus, one of the friends to whom Ogier Busbecq sent seeds from Turkey, was born in Siena in 1501 and brought up in Venice where his father was a doctor, but he spent much of his later life in Prague as physician to the Grand Dukes. His herbal, *Commentarii in Sex Libros Pedacii Dioscoridis*, was published in Venice in 1544, first in Italian and later translated into Latin and other languages. More than forty editions were published; the early ones which had small woodcuts are said to have sold 32,000 copies. Later editions, published from 1562 to 1585, had a new set of large illustrations by Giorgio Liberale of Udine and Wolfgang Meyerpeck, a German, which include a small number of animal pictures that are the equal of the best zoological illustration of the time. The text contains plants of all types but is difficult to follow as no attempt is made to classify them beyond arranging them alphabetically by their common names. The illustrations, although many are new, are very much in the style of Fuchs.

Another friend of Matthioli was the scientist Ulisse Aldrovandi or Aldrovandus (1522–1605), who was the founder and director of one of the first natural history museums, in Bologna, and one of the earliest botanic gardens. Though he published little on botany himself, he had great

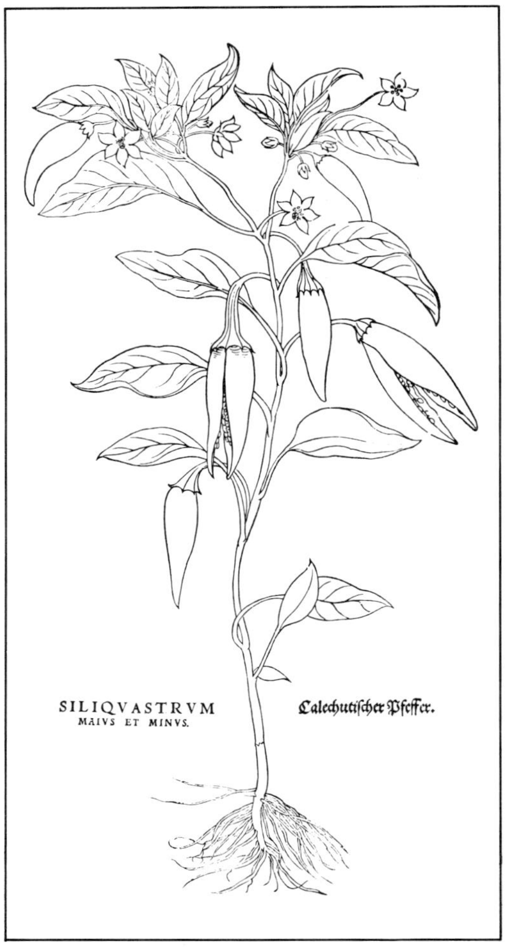

A pepper (*Capsicum*) – the genus *Capsicum* is native to the West Indies and tropical America, and was an early introduction to Europe. Woodcut from *De Historia Stirpium* (1542) by Leonhart Fuchs.

Right:
Four woodcuts from *Commentarii in Sex Libros Pedacii Dioscoridis* (1565 edition) by Pierandrea Matthioli: *'Rhus'*, possibly the stag's horn sumach (*Rhus typhina*); ivy (*Hedera helix*); *'Cyanus maior'*, perennial cornflower (*Centaurea montana*); monkshood (*Aconitum*).

RHVS.

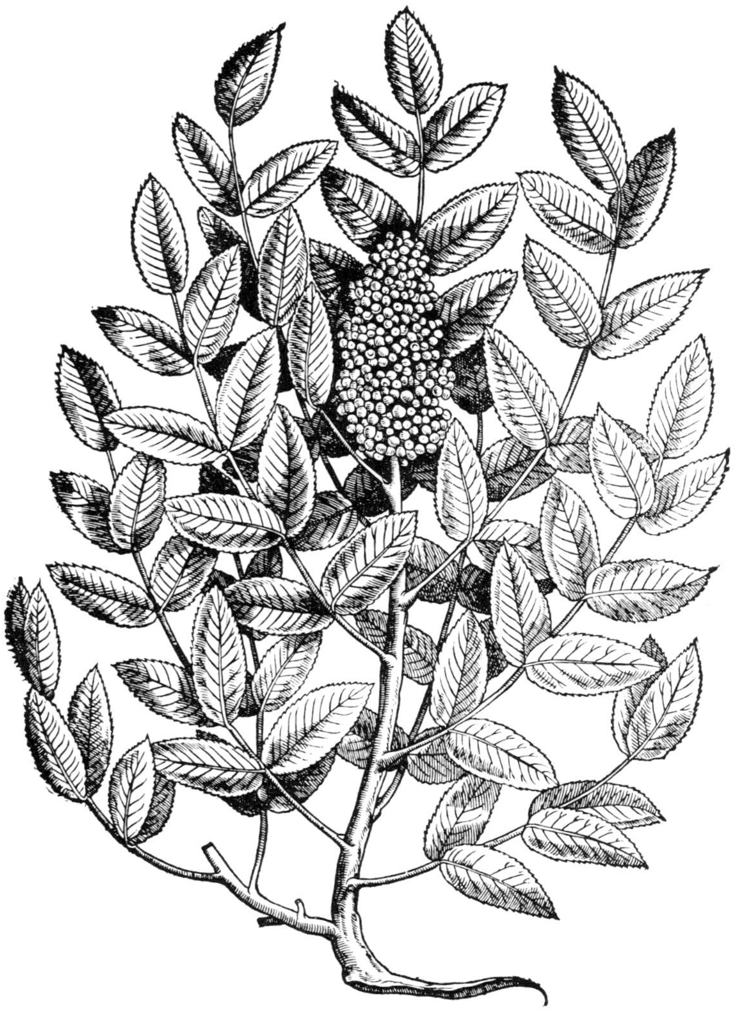

HEDERA HELIX.

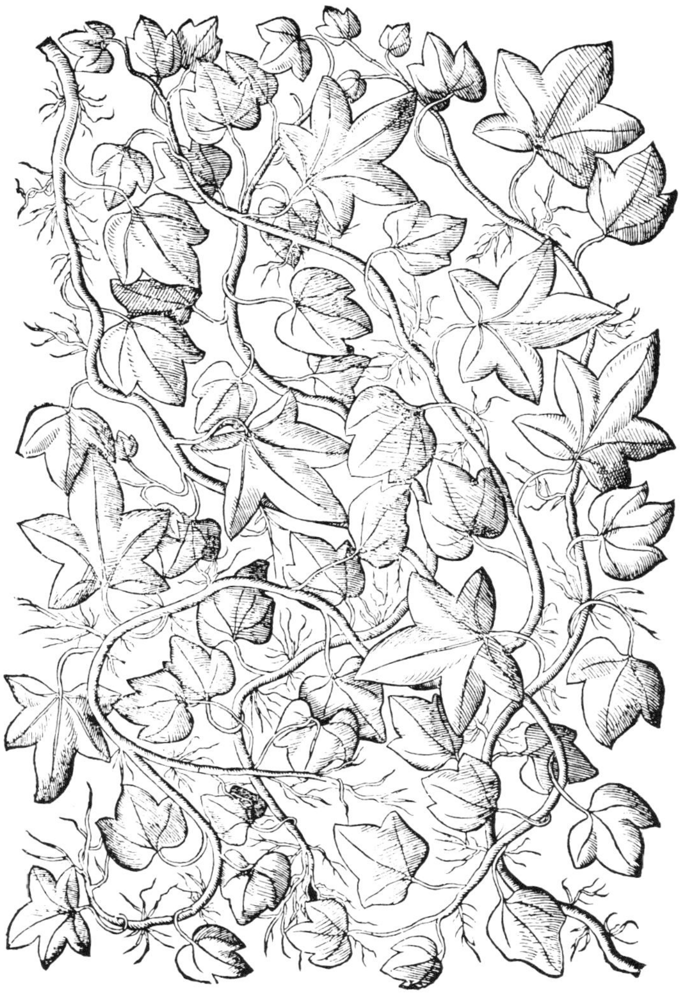

CYANVS MAIOR.

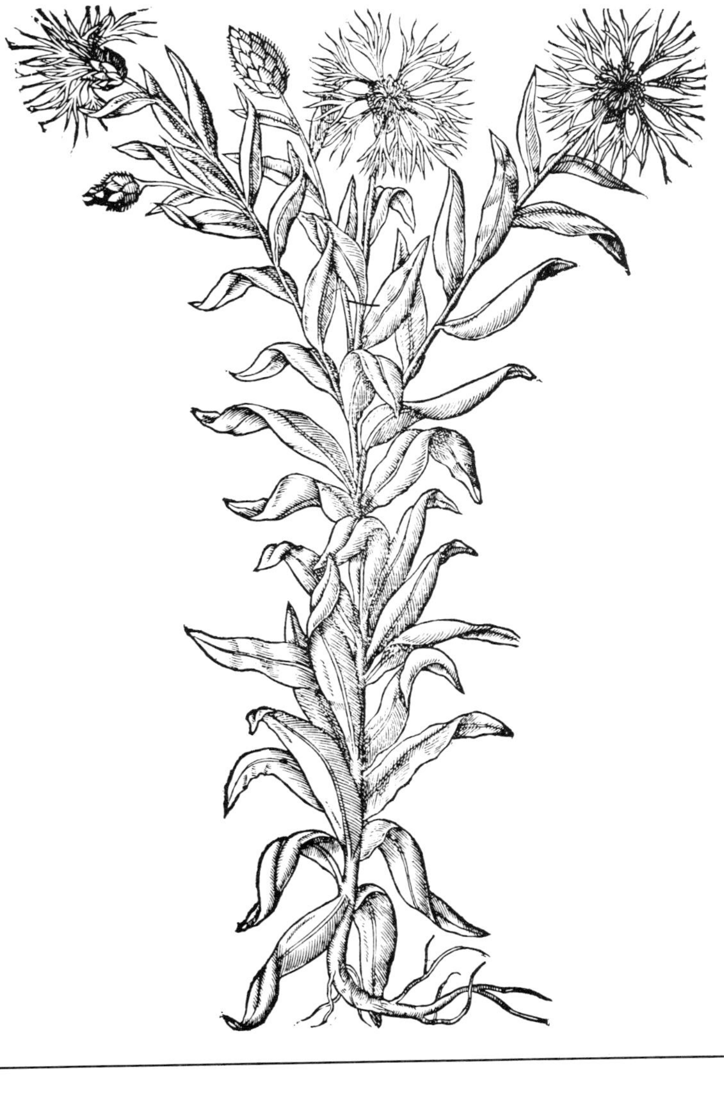

ACONITVM VIII.

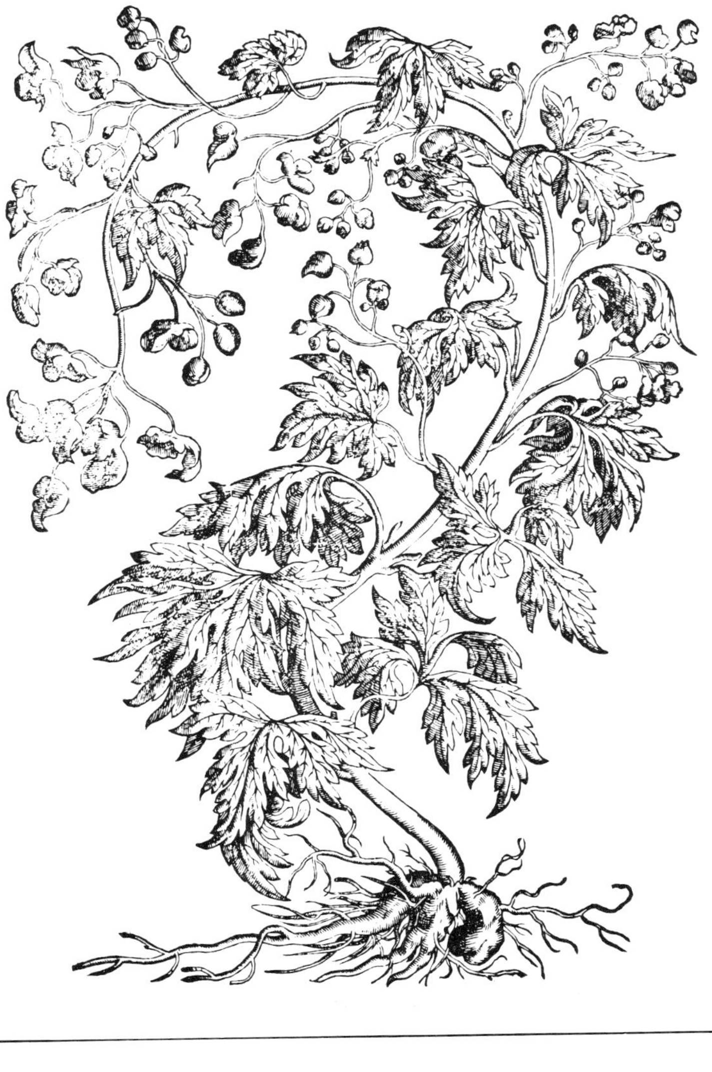

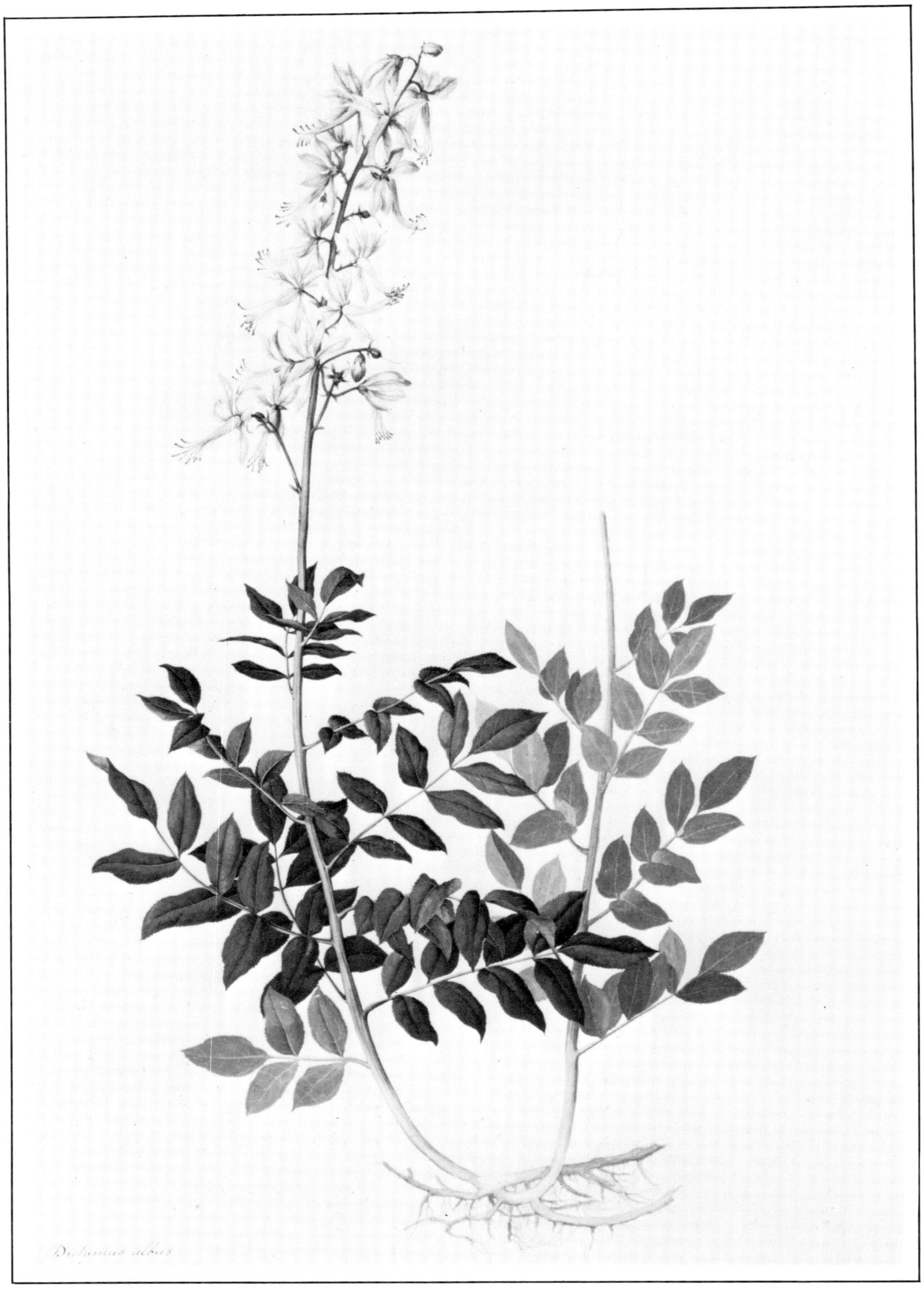

Burning bush or dittany (*Dictamnus albus*), watercolour on paper, by Jacopo Ligozzi, painted around 1580. Gabinetto Disegni, Uffizi Gallery, Florence.

Below:
An evergreen oak. Woodcut from Ulisse Aldrovandi's *Dendrologiae*, published in Bologna in 1668.

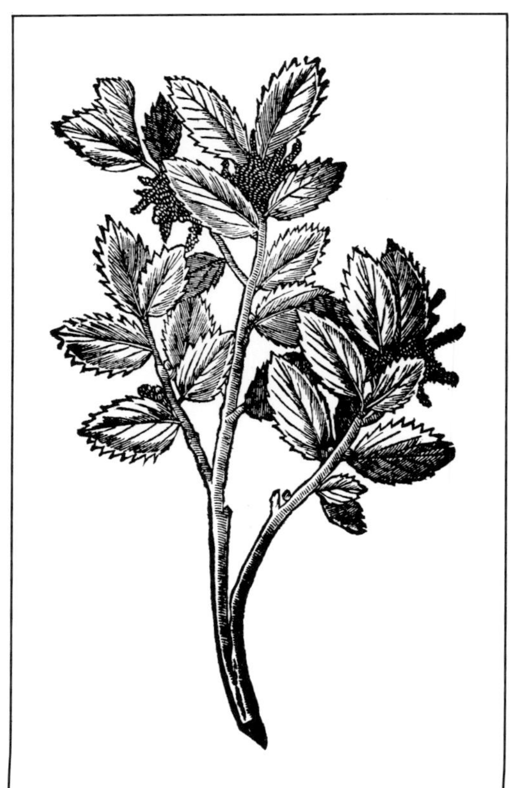

influence through his teaching. It was for him, too, that probably the finest botanical illustrations of the 16th century were painted. They are the work of Jacopo Ligozzi, who was born in Verona in 1547. He arrived in Florence around 1577 and became court painter to the Grand Duke of Tuscany and later Superintendent of the Uffizi Gallery. His work can be found both in the Uffizi and at the University of Bologna. Most of Ligozzi's paintings are in watercolour and gouache. They are large, almost life-size, and reminiscent in style of the best work produced at the zenith of botanical illustration, two hundred years later, with particular refinement in such details as the veins on the leaves of a leopard's bane (*Doronicum*). When most of the plants were painted in the 1580s, many were recently introduced: *Agave americana*, a morning glory (*Ipomoea quamoclit*) and marvel of Peru (*Mirabilis jalapa*) from Mexico, mourning iris (*Iris susiana*), a spurge (*Euphorbia dendroides*) and a grape hyacinth (*Muscari macrocarpa*) from the Levant and the cashew tree from Africa. Ligozzi also painted native Italian plants, as well as a number of animals (probably from Aldrovandi's museum) and classical scenes such as *The Rape of the Sabine Women*.

Aldrovandi's teacher had been the Swiss Conrad Gesner, who is known primarily as a zoologist for his huge *Historia Animalium*, but was also a botanist of distinction and was preparing a companion volume *Historia Plantarum* when he died of the plague in 1565. Had he lived longer, he would be remembered as one of the foremost botanists of the 16th century. His interests were universal: he also wrote on bibliography, linguistics, medicine and mineralogy. He was born in Zurich in 1516, and studied in Paris and Basle where he worked on a Greek and Latin lexicon. He was Professor of Greek at Lausanne for a time and finally returned to Zurich as Professor of Philosophy and Natural History. His botanical work was never published as a whole, but his carefully annotated watercolours of plants, many painted around 1550, are scientifically excellent. Among the many paintings, primarily of alpine plants, preserved in the university library at Erlangen is a study of *Nicotiana rustica*, which was probably the first species of tobacco to be introduced into Europe for smoking.

Apart from Aldrovandi, Gesner's students included the brothers Jean and Caspar Bauhin, protestant refugees from France who lived in Switzerland. Caspar Bauhin (1560–1624) was the younger of the two by nineteen years, but is the more famous, primarily because of his *Pinax Theatri Botanici*, which was published in Basle in 1623. It was a comprehensive concordance or register of all plant names, a very important work at a time when no standardised system of nomenclature had been established. It followed a similar list published by Gesner in 1542 and continued in use until it was superseded by the *Species Plantarum* of Linnaeus in 1753.

The centre of botanical learning in northern Europe in the late sixteenth and early seventeenth centuries was Flanders, especially the city of

Spinach. Woodcut from *Commentarii in Sex Libros Pedacii Dioscoridis* (1565 edition).

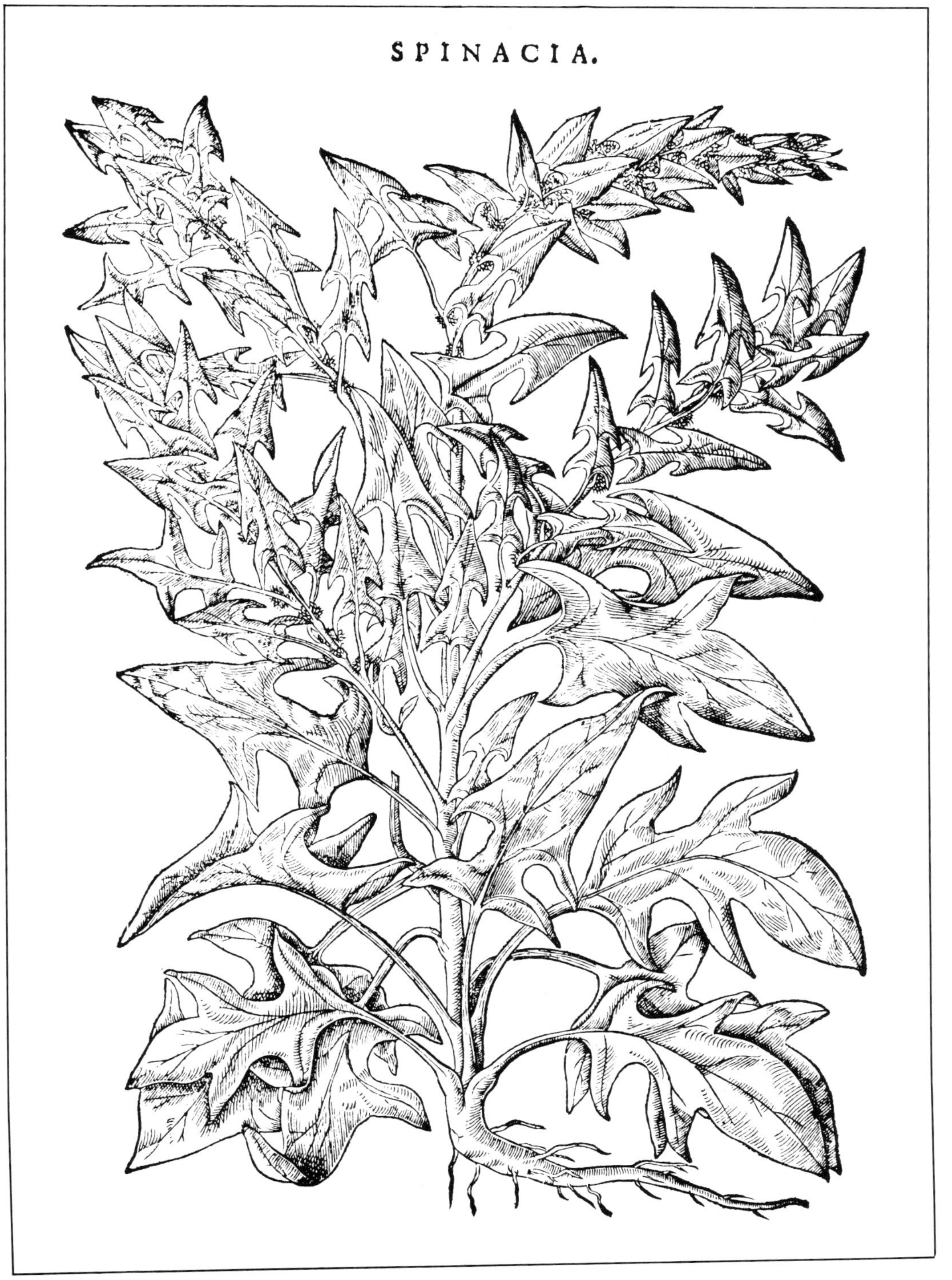

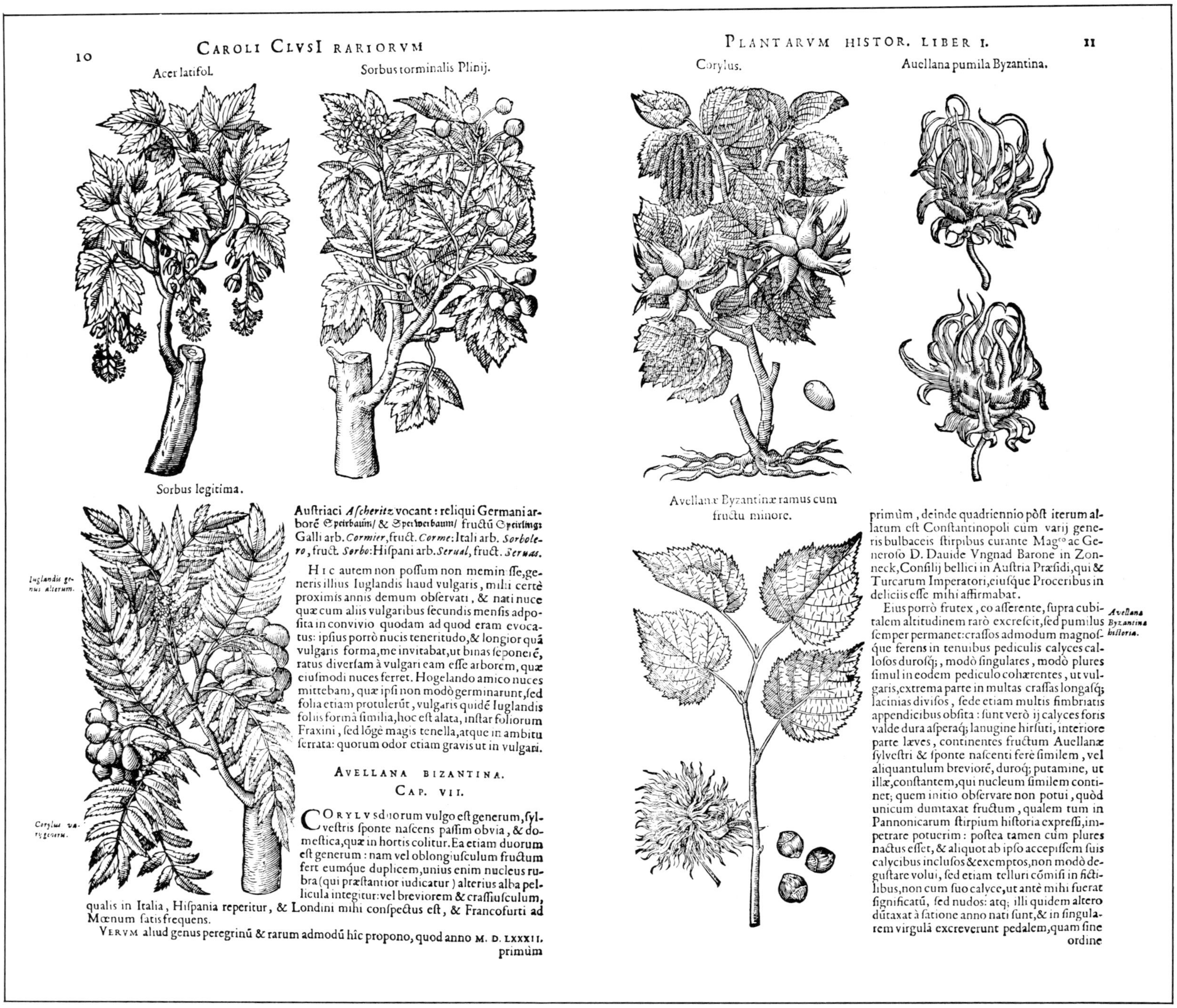

10 CAROLI CLVSI RARIORVM

Acer latifol.

Sorbus torminalis Plinij.

Sorbus legitima.

Iuglandis genus alterum.

Austriaci *Ascheritz* vocant: reliqui Germani arborē Speirbaum/ & Speirwerbaum/ fructū Speirling: Galli arb. *Cormier*, fruct. *Corme*: Itali arb. *Sorbolero*, fruct. *Sorbo*: Hispani arb. *Serual*, fruct. *Seruas*.

Hic autem non possum non meminisse, generis illius Iuglandis haud vulgaris, mihi certè proximis annis demum observati, & nati nuce quæ cum aliis vulgaribus secundis mensis adposita in convivio quodam ad quod eram evocatus: ipsius porrò nucis teneritudo, & longior quā vulgaris forma, me invitabat, ut binas seponerē, ratus diversam à vulgari eam esse arborem, quæ eiusmodi nuces ferret. Hogelando amico nuces mittebam, quæ ipsi non modò germinarunt, sed folia etiam protulerūt, vulgaris quidē Iuglandis foliis formâ similia, hoc est alata, instar foliorum Fraxini, sed lōgè magis tenella, atque in ambitu serrata: quorum odor etiam gravis ut in vulgari.

AVELLANA BIZANTINA.
CAP. VII.

Corylus varij generis.

CORYLVS duorum vulgo est generum, sylvestris sponte nascens passim obvia, & domestica, quæ in hortis colitur. Ea etiam duorum est generum: nam vel oblongiusculum fructum fert eumque duplicem, unius enim nucleus rubra (qui præstantior iudicatur) alterius alba pellicula integitur: vel breviorem & crassiusculum, qualis in Italia, Hispania reperitur, & Londini mihi conspectus est, & Francofurti ad Mœnum satis frequens.

VERVM aliud genus peregrinū & rarum admodū hîc propono, quod anno M. D. LXXXII. primùm

PLANTARVM HISTOR. LIBER I. 11

Corylus.

Auellana pumila Byzantina.

Avellanæ Byzantinæ ramus cum fructu minore.

primùm, deinde quadriennio pòst iterum allatum est Constantinopoli cùm varij generis bulbaceis stirpibus curante Mag^co ac Generoso D. Dauide Vngnad Barone in Zonneck, Consilij bellici in Austria Præsidi, qui & Turcarum Imperatori, eiusque Proceribus in deliciis esse mihi affirmabat.

Eius porrò frutex, eo asserente, supra cubitalem altitudinem rarò excrescit, sed pumilus semper permanet: crassos admodum magnosque ferens in tenuibus pediculis calyces callosos durosque, modò singulares, modò plures simul in eodem pediculo cohærentes, ut vulgaris, extrema parte in multas crassas longasque lacinias divisos, sede etiam multis fimbriatis appendicibus obsita: sunt verò ij calyces foris valde dura asperaque lanugine hirsuti, interiore parte læves, continentes fructum Auellanæ sylvestri & sponte nascenti ferè similem, vel aliquantulum breviorē, duroque putamine, ut illæ, constantem, qui nucleum similem continet; quem initio observare non potui, quòd unicum dumtaxat fructum, qualem tum in Pannonicarum stirpium historia expressi, impetrare potuerim: postea tamen cùm plures nactus esset, & aliquot ab ipso accepissem suis calycibus inclusos & exemptos, non modò degustare volui, sed etiam telluri cōmisi in fictilibus, non cum suo calyce, ut antè mihi fuerat significatū, sed nudos: atque illi quidem altero dūtaxat à satione anno nati sunt, & in singularem virgulā excreverunt pedalem, quam sine ordine

Avellana Byzantina historia.

Antwerp and the University of Leiden, and its foremost botanist was a Frenchman, Charles de L'Ecluse, who is usually known by his Latin name, Clusius. He had been born at Arras in 1526 and had studied at many universities, notably Montpellier, travelled widely and corresponded with other botanists and travellers from all over Europe. His first publication, in 1557, was a French translation of the *Cruydeboek* of Rembert Dodoens, which had first been published in Flemish in Antwerp in 1554. Like the original edition and various other herbals from the second half of the sixteenth century, this was largely illustrated with woodblocks that had originally been prepared for the octavo edition of Fuchs's *De Historia Stirpium* (first published in 1545).

In 1564, Clusius made a collecting expedition to Spain and brought back about two hundred new species of plants, which were described in his *Rariorum aliquot Stirpium per Hispanias Observatorum Historia* of 1576. This contained the first picture of the dragon tree (*Dracaena draco*), a native of the Canary Islands which today is sometimes grown in more northerly climates as an indoor pot plant. In 1573, Clusius went to Vienna at the invitation of the Emperor; he stayed there for about fourteen years and was involved in growing the collections of seeds and living plants that had been sent back from Constantinople. Descriptions of these were published in 1583 in his *Rariorum aliquot Stirpium per Pannoniam, Austriam et Vicinas Historia*. New plants from further afield were included in Clusius's collected works, *Rariorum Plantarum Historiae*, published in Antwerp in 1601, the most notable being the Jacobean lily from Central

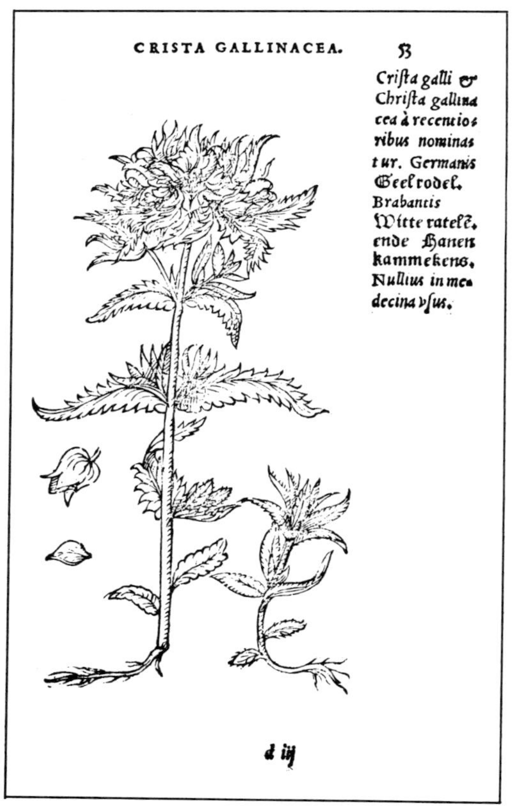

Yellow rattle (*Rhinanthus*). Woodcut from *Stirpium Historia* (1554), the Latin version of the *Cruydeboek* of Rembert Dodoens.

America, *Sprekelia formosissima*, of which Clusius reports, 'I had only one plant . . . sent by the learned doctor Simon D. Tovar, Spanish physician, which flowered in June 1594 . . . called in Mexico *azcal xochitl*.' Appended to some editions is the *Fungorum Historia*, the first monograph on mushrooms and toadstools.

Clusius's connections with both Austria and the Netherlands led to the introduction into gardens of the turban Ranunculus (*Ranunculus asiaticus*), the garden anemone (*Anemone coronaria*) and *Narcissus tazetta*, as well as many of the bulbs such as tulip and hyacinth which became such a feature of seventeenth-century gardening and flower painting in the Low Countries. He visited England several times and obtained flowers from America through Sir Francis Drake. When he died in 1609, he was Professor of Botany at Leiden.

Clusius was not himself a flower painter, but he collected together a large number of illustrations, some of which were used several times by his friends, and by others whose work was published by the house of Christophe Plantin in Antwerp, the leading printer of the time. Of Clusius's friends, the most eminent botanist was a pupil of Aldrovandi, Mathias de l'Obel, whose

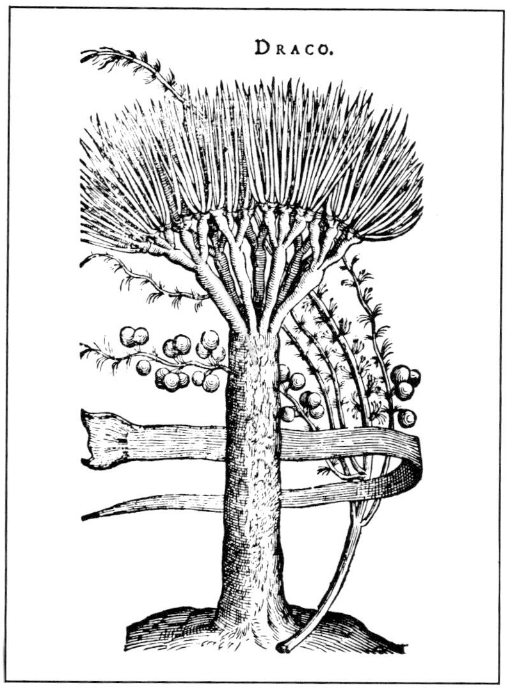

Top :
Dragon tree (*Dracaena draco*) from the Canary Islands. Woodcut from *Rariorum Plantarum Historiae* (1601), the collected works of Clusius (Charles de l'Ecluse).

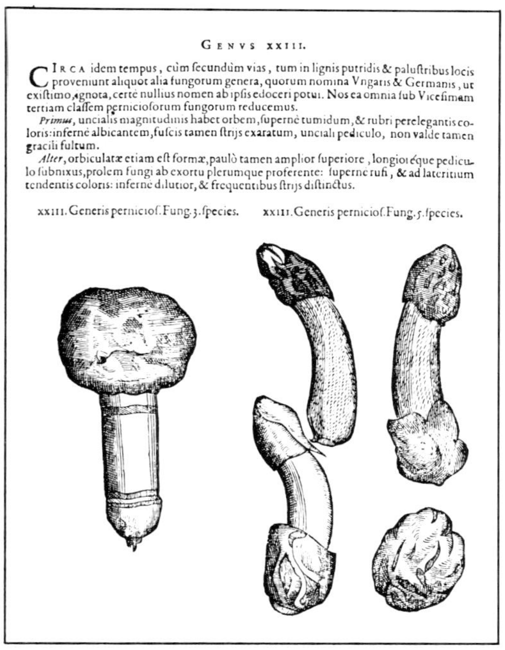

GENVS XXIII.

CIRCA idem tempus, cùm secundùm vias, tum in lignis putridis & palustribus locis proueniunt aliquot alia fungorum genera, quorum nomina Vngaris & Germanis, ut existimo, ignota, certè nullius nomen ab ipsis edoceri potui. Nos ea omnia sub Vicesimam tertiam classem perniciosorum fungorum reducemus.

Primus, uncialis magnitudinis habet orbem, supernè tumidum, & rubri perelegantis coloris: infernè albicantem, fuscis tamen strijs exaratum, unciali pediculo, non valde tamen gracili fultum.

Alter, orbiculatæ etiam est formæ, paulò tamen amplior superiore, longioréque pediculo subnixus, prolem fungi ab exortu plerumque proferente: supernè rufi, & ad lateritium tendentis coloris: infernè dilutior, & frequentibus strijs distinctus.

XXIII. Generis pernicios. Fung. 3. species. XXIII. Generis pernicios. Fung. 5. species.

Above :
Stinkhorn fungi (*Phallus*). Woodcut from *Rariorum Plantarum Historiae*.

Opposite page, top :
Double-page spread from Clusius's *Rariorum Plantarum Historiae* with woodcuts of sycamore (*Acer pseudoplatanus*), wild service tree (*Sorbus torminalis*), service tree (*Sorbus domestica*), filbert (*Corylus maxima*) and a tree hazel (*Corylus colurna*).

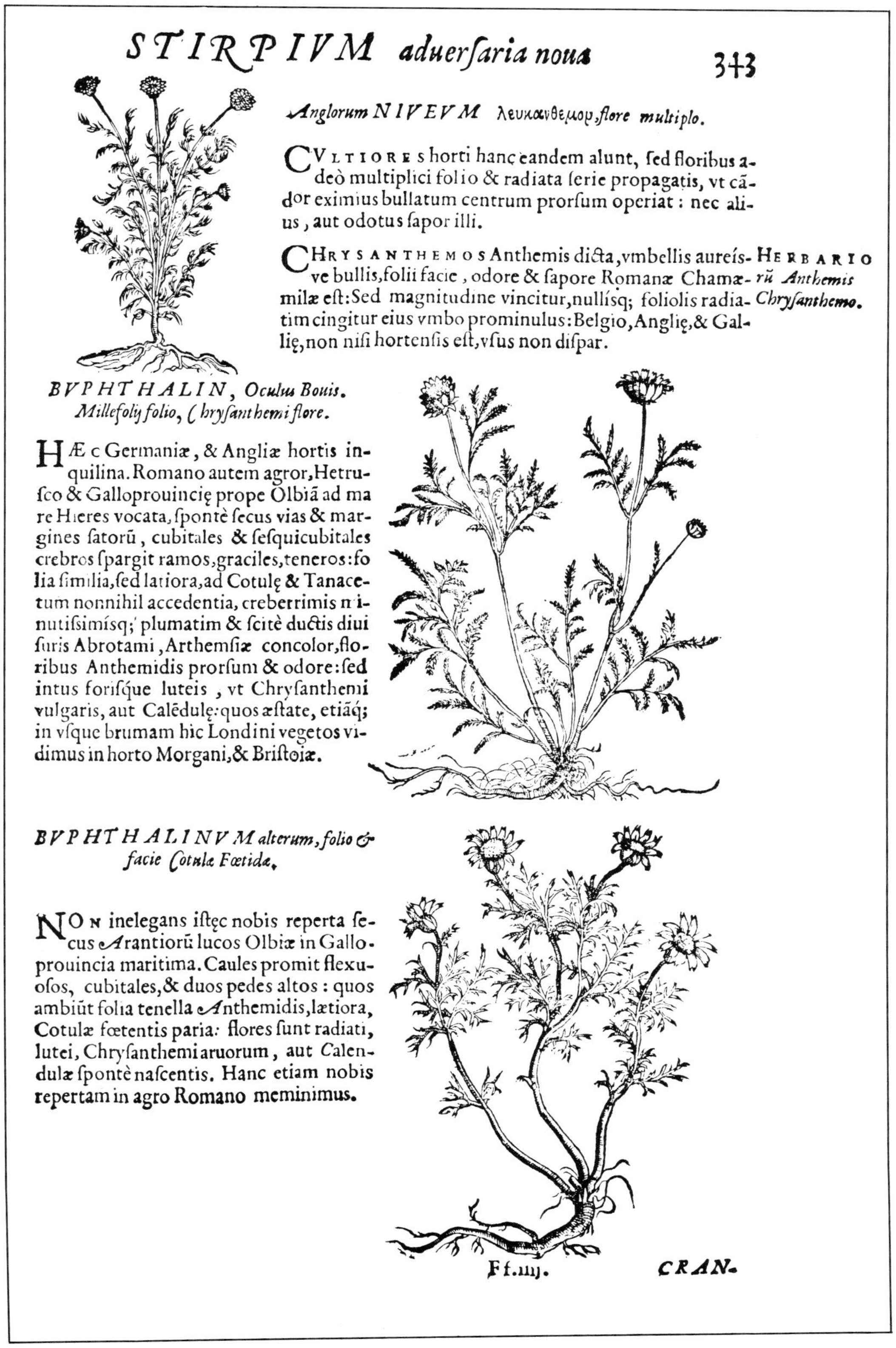

STIRPIVM aduersaria noua 343

Anglorum NIVEVM λευκανθεμον, *flore multiplo.*

CVLTIORES horti hanc eandem alunt, sed floribus adeò multiplici folio & radiata serie propagatis, vt cãdor eximius bullatum centrum prorsum operiat: nec alius, aut odotus sapor illi.

CHRYSANTHEMOS Anthemis dicta, vmbellis aureísve bullis, folii facie, odore & sapore Romanæ Chamæmilæ est: Sed magnitudine vincitur, nullísq; foliolis radiatim cingitur eius vmbo prominulus: Belgio, Angliȩ, & Galliȩ, non nisi hortensis est, vsus non dispar. HERBARIOrũ *Anthemis Chrysanthemo.*

BVPHTHALIN, Oculus Bouis. Millefolij folio, Chrysanthemi flore.

HÆc Germaniæ, & Angliæ hortis inquilina. Romano autem agror, Hetrusco & Galloprouinciȩ prope Olbiã ad mare Hieres vocata, spontè secus vias & margines satorũ, cubitales & sesquicubitales crebros spargit ramos, graciles, teneros: folia similia, sed latiora, ad Cotulȩ & Tanacetum nonnihil accedentia, creberrimis minutissimísq; plumatim & scitè ductis diuisuris Abrotami, Arthemsiæ concolor, floribus Anthemidis prorsum & odore: sed intus forisque luteis, vt Chrysanthemi vulgaris, aut Calẽdulȩ: quos æstate, etiãq; in vsque brumam hic Londini vegetos vidimus in horto Morgani, & Bristoiæ.

BVPHTHALINVM alterum, folio & facie Cotulæ Fœtidæ.

NON inelegans istȩc nobis reperta secus Arantiorũ lucos Olbiæ in Galloprouincia maritima. Caules promit flexuosos, cubitales, & duos pedes altos: quos ambiũt folia tenella Anthemidis, lætiora, Cotulæ fœtentis paria: flores sunt radiati, lutei, Chrysanthemi aruorum, aut Calendulæ spontè nascentis. Hanc etiam nobis repertam in agro Romano meminimus.

Ff.iij. CRAN-

Three species of chamomile illustrated in woodcuts on a page from *Stirpium Adversaria Nova* (1570) by Mathias de l'Obel.

writings were also published by the house of Plantin. De L'Obel's chief contribution to botany was a system of classification based on leaf form which brought out many natural affinities that had previously been overlooked or ignored, but, at the same time, provided some strange juxtapositions. This classification was first published in *Stirpium Adversaria Nova* in 1570 in London, where de l'Obel had arrived in 1559. He eventually obtained the title of botanist to James I and died in Highgate in 1616.

The early herbals in England were not particularly significant either for their botany or their illustration. William Turner's *A New Herball* was published in instalments, starting in 1551, and then as a whole in 1568. Turner himself was an unconventional character, one of the first of the botanical clerics. He was banned under Henry VIII, made Dean of Wells under Edward VI, exiled by Mary Tudor, but restored by Elizabeth I, only to be suspended for non-conformity. He studied medicine in Italy, travelled throughout Europe and corresponded with Gesner and Fuchs. The plates in *A New Herball* were derived primarily from the octavo Fuchs, as were those for Henry Lyte's *A Niewe Herball* of 1578, which was printed in Antwerp but published in London. This was largely a translation of Clusius's French version of Dodoens's *Cruydeboek*, but included many additions and corrections.

Without doubt, the most famous English herbal is *The Herball or General Historie of Plants* by John Gerard, first published in 1597, but it is by no means the most reliable. Gerard was primarily a gardener, and his herbal was in part a compilation and again in part a translation from Dodoens. In 1599, he published a list of the plants he grew in his garden in Holborn, and it contains such interesting species as two cacti – a *Cereus*, which was

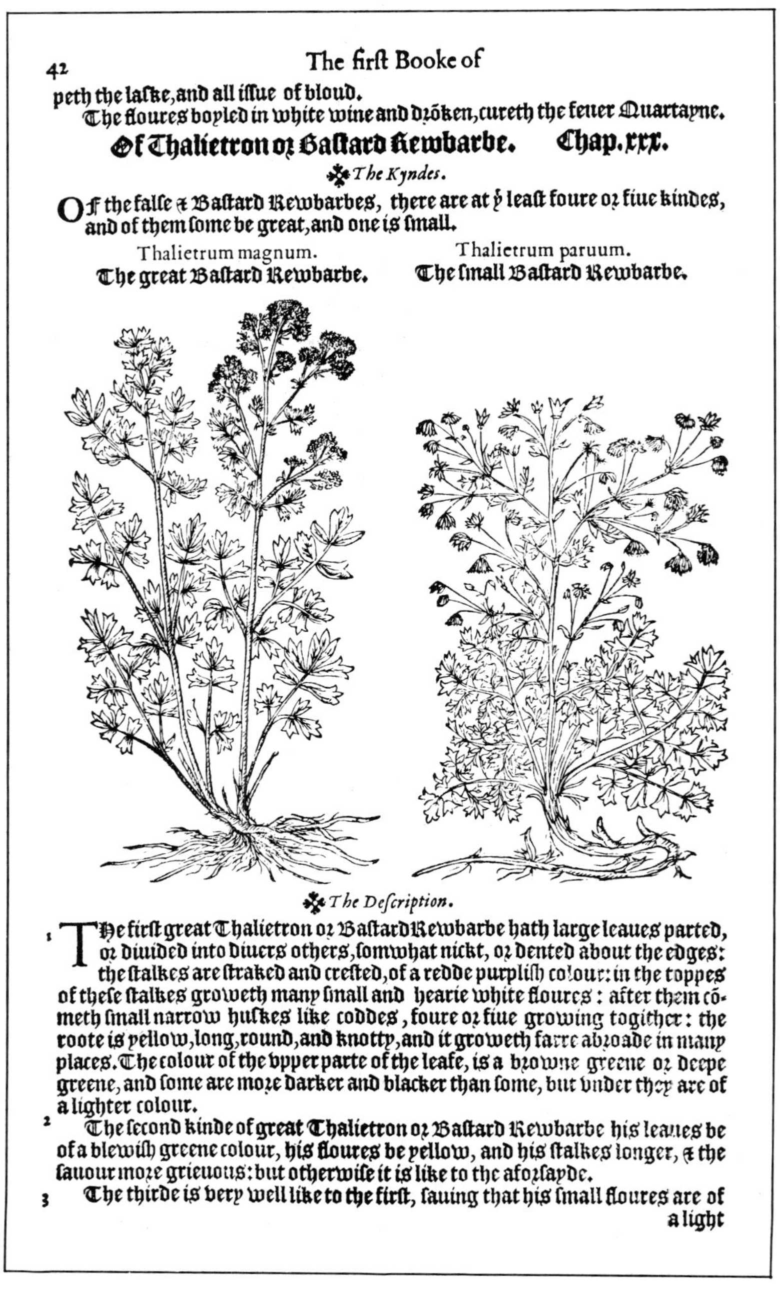

42 The first Booke of

peth the laske, and all issue of bloud.

The floures boyled in white wine and drōken, cureth the feuer Quartayne.

Of Thalietron or Bastard Rewbarbe. Chap. xxx.

The Kyndes.

Of the false & Bastard Rewbarbes, there are at y least foure or fiue kindes, and of them some be great, and one is small.

Thalietrum magnum. The great Bastard Rewbarbe. Thalietrum paruum. The small Bastard Rewbarbe.

The Description.

1 The first great Thalietron or Bastard Rewbarbe hath large leaues parted, or diuided into diuers others, somwhat nickt, or dented about the edges: the stalkes are straked and crested, of a redde purplish colour: in the toppes of these stalkes groweth many small and hearie white floures: after them cōmeth small narrow huskes like coddes, foure or fiue growing togither: the roote is yellow, long, round, and knotty, and it groweth farre abroade in many places. The colour of the vpper parte of the leafe, is a browne greene or deepe greene, and some are more darker and blacker than some, but vnder they are of a lighter colour.

2 The second kinde of great Thalietron or Bastard Rewbarbe his leaues be of a blewish greene colour, his floures be yellow, and his stalkes longer, & the sauour more grieuous: but otherwise it is like to the aforsayde.

3 The thirde is very well like to the first, sauing that his small floures are of a light

Left:
Two species of meadow-rue (*Thalictrum*), referred to as 'bastard rewbarbe', in woodcuts on a page from Henry Lyte's *Niewe Herball* (1578).

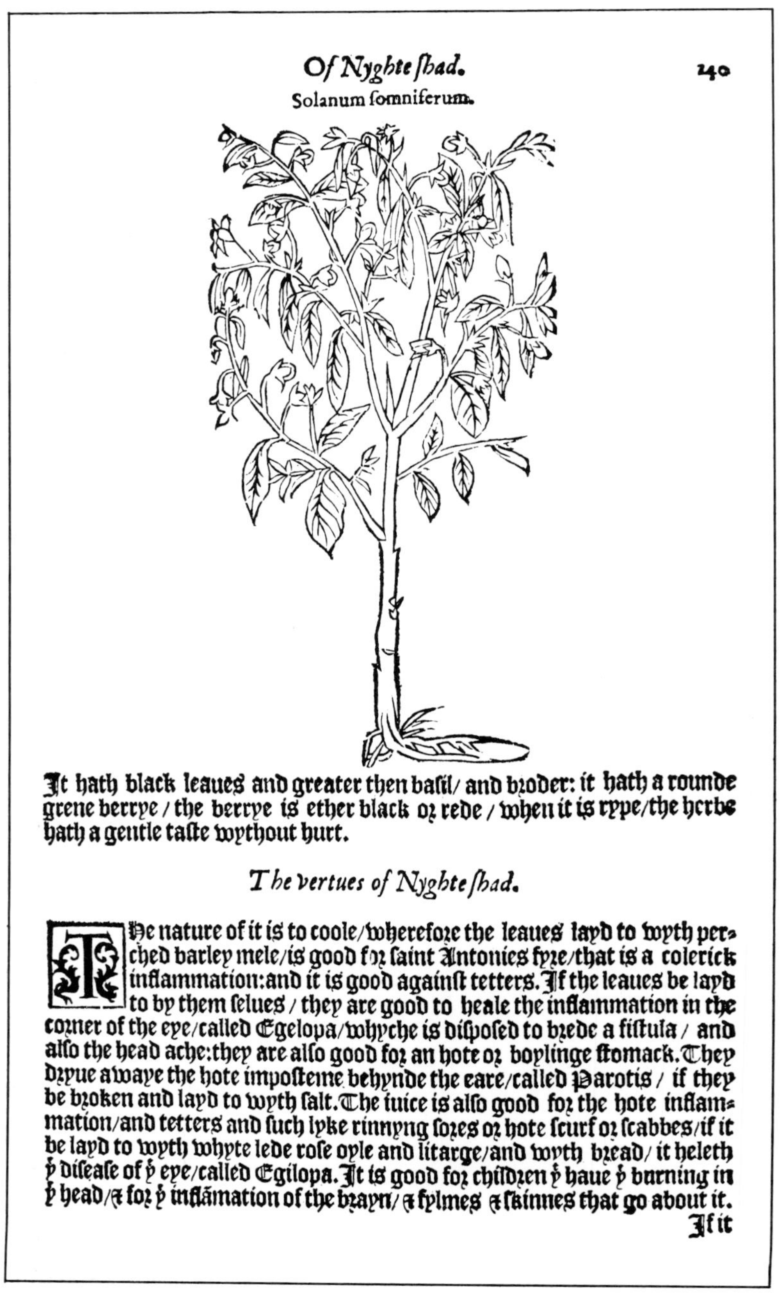

Of Nyghte shad. 240

Solanum somniferum.

It hath black leaues and greater then basil/ and broder: it hath a rounde grene berrye/ the berrye is ether black or rede/ when it is rype/ the herbe hath a gentle taste wythout hurt.

The vertues of Nyghte shad.

The nature of it is to coole/wherefore the leaues layd to wyth perched barley mele/is good for saint Antonies fyre/that is a colerick inflammation: and it is good against tetters. If the leaues be layd to by them selues/ they are good to heale the inflammation in the corner of the eye/called Egelopa/whyche is disposed to brede a fistula/ and also the head ache: they are also good for an hote or boylinge stomack. They dryue awaye the hote imposteme behynde the eare/called Parotis/ if they be broken and layd to wyth salt. The iuice is also good for the hote inflammation/and tetters and such lyke rinnyng sores or hote scurf or scabbes/if it be layd to wyth whyte lede rose oyle and litarge/and wyth bread/ it heleth y disease of y eye/called Egilopa. It is good for children y haue y burning in y head/& for y inflāmation of the brayn/& fylmes & skinnes that go about it. If it

Below:
Deadly nightshade (*Atropa belladonna*). Woodcut illustration on a page from *A New Herball* (1568) by William Turner.

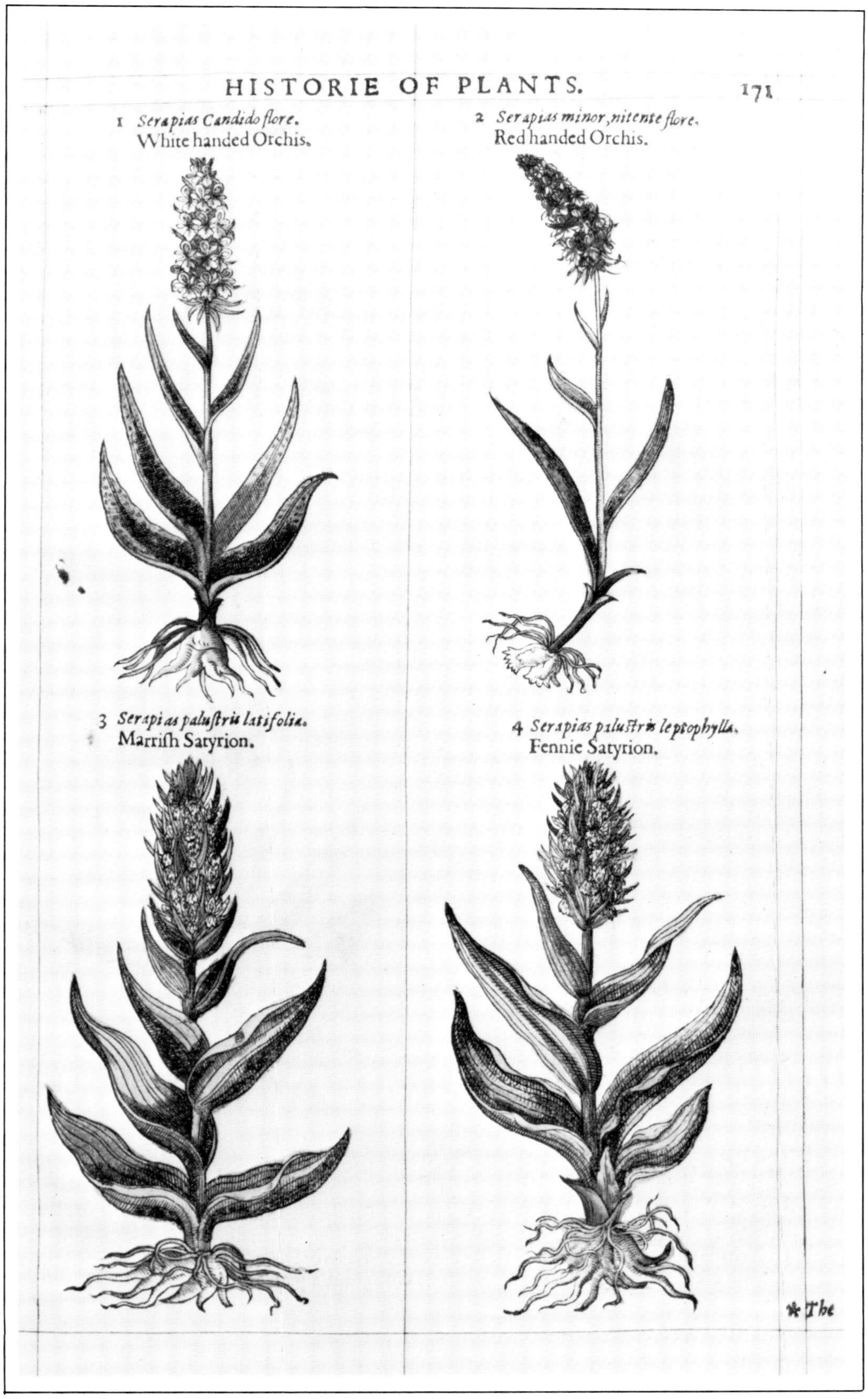

Various orchids, including the southern marsh orchid (*Dactylorhiza praetermissa*), in hand-coloured woodcuts on a page of *The Herball* by John Gerard, first edition (1597).

'destroyed by cold weather', and a prickly pear (*Opuntia vulgaris*), on which 'I have bestowed great pains and cost in keeping it from the innuri of our cold climate.' He mentions receiving several plants from Jean Robin of Paris, including *Crocus vernus flore luteo* (probably the cloth-of-gold crocus, *Crocus susianus*) and the snake's-head fritillary (*Fritillaria meleagris*). Gerard's *Herball* was a great financial success, and a revision by Thomas Johnson was published in 1632. Plantin's blocks, by this time nearly 2,800 in number, were used for the illustrations. Johnson himself was a well-known London apothecary who published one of the first local floras, an account of flowers found growing in parts of Kent and on Hampstead Heath: *Descriptio Itineris Plantarum Investigations ergo Suscepti in Agrum Cantianum AD 1632, et Enumeratio Plantarum in Ericeto Hampstediano Locisque Vicinis Crescentium.* Johnson died during the Civil War from the effects of a shot wound received when he was fighting for the Royalist cause in the defence of Basing House.

LIB. 2. Of the Hiſtorie of Plants. 393

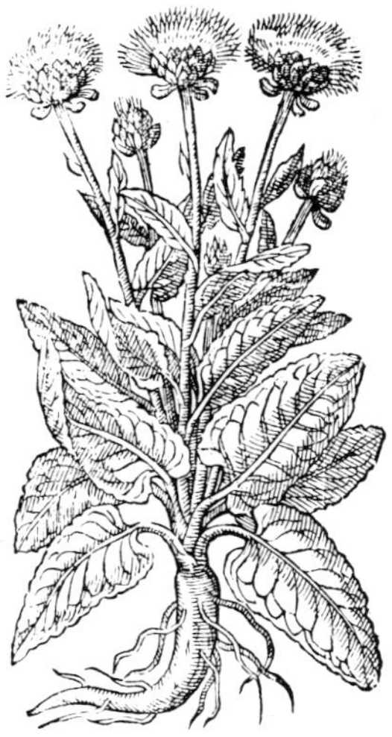

1 *Rha Capitatum Lobelÿ.*
Turky Rubarb.

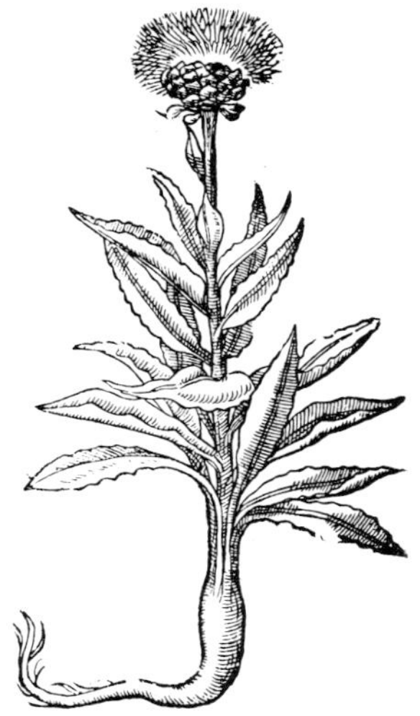

2 *Rha Capitatum anguſtifolium.*
The other baſtard Rubarb.

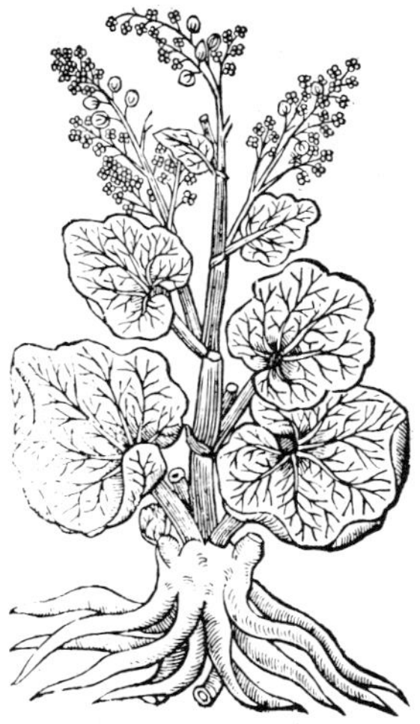

‡ 3 *Rha verum antiquorum.*
The true Rubarb of the Antients.

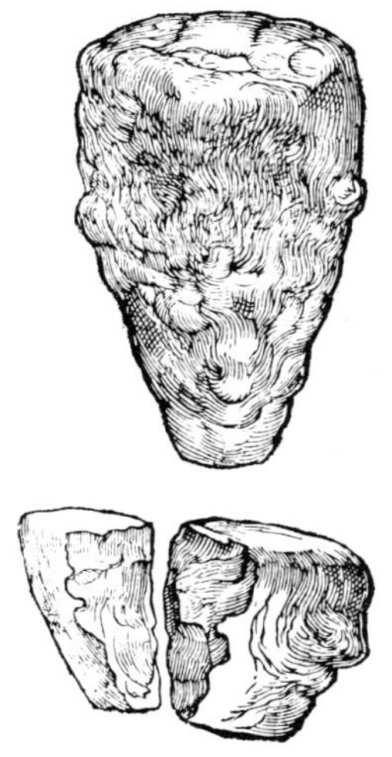

Rhabarbarum ſiccatum.
The dry roots of Rubarb.

Kk 3

LIB. 3. Of the Hiſtorie of Plants. 1515

with a thin rinde like that of the Fig, of a yellow colour when they be ripe: the pulpe or ſubſtance of the meate is like that of the Pompion, without either ſeeds, ſtones or kernels, in taſte not greatly perceiued at the firſt, but preſently after it pleaſeth, and entiſeth a man to eate liberally thereof, by a certaine entiſing ſweetneſſe it yeelds: in which fruit, if it be cut according to the length (ſaith mine Author) oblique, tranſuerſe, or any other way whatſoeuer, may be ſeene the ſhape and forme of a croſſe, with a man faſtned thereto. My ſelfe haue ſeene the fruit, and cut it in pieces, which was brought me from Aleppo in pickle; the croſſe I might perceiue, as the forme of a ſpred-Egle, in the root of Ferne; but the man I leaue to be ſought for by thoſe which haue better eies and judgement than my ſelfe.

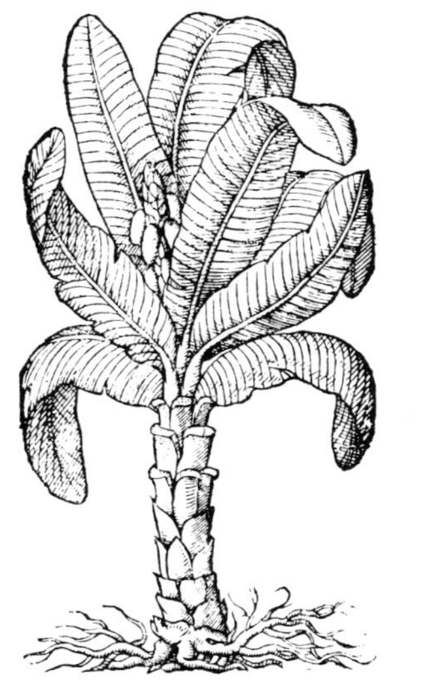

Muſa Serapionis.
Adams Apple-tree.

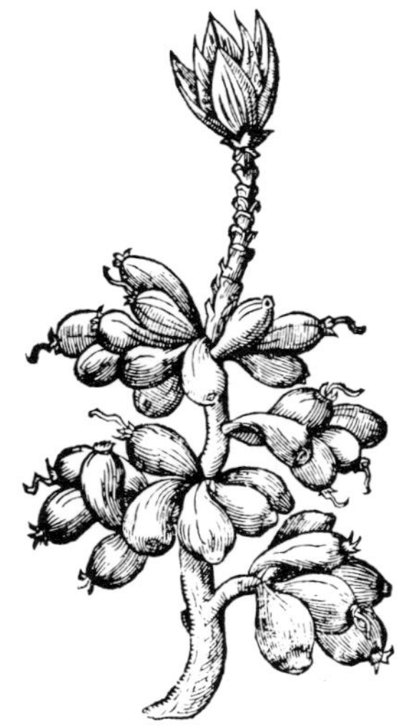

Muſæ Fructus.
Adams Apple.

‡ Aprill 10. 1633. my much honoured friend D. Argent (now Preſident of the Colledge of Phyſitions of London) gaue me a plant hee receiued from the Bermuda's: the length of the ſtalke was ſome two foot; the thickeneſſe thereof ſome ſeuen Inches about, being creſted, and full of a ſoft pith, ſo that one might eaſily with a knife cut it aſunder. It was crooked a little or indented, ſo that at each two or three inches ſpace it put forth a knot of ſome halfe inch thickneſſe, and ſome inch in length, which encompaſſed it more than halfe about, and vpon each of theſe joints or knots, in two rankes one aboue another, grew the fruit, ſome twentie, nineteene, eighteene, &c. more or leſſe, at each knot: for the branch I had, conteined nine knots or diuiſions, and vpon the loweſt knot grew twenty, and vpon the vppermoſt fifteene. The fruit which I receiued was not ripe, but greene, each of them was about the bigneſſe of a large Beane; the length of them ſome fiue inches, and the bredth ſome inch and halfe: they all hang their heads downewards, haue rough or vneuen ends, and are fiue cornered; and if you turne the vpper ſide downeward, they ſomewhat reſemble a boat, as you may ſee by one of them expreſt by it ſelfe: the huske is as thicke as a Beane, and will eaſily ſhell off it: the pulpe is white and ſoft: the ſtalke whereby it is faſtned to the knot is very ſhort and almoſt as thicke as ones little finger. Tſte ſtalke with the fruit thereon I hanged vp in my ſhop, where it became ripe about the beginning of May, and laſted vntill Iune: the pulp or meat was very ſoft and tender, and it did eate ſomewhat like a Muske-Melon. I haue giuen you the figure of the whole branch, with the fruit thereon, which I drew as ſoone as I receiued it, and it is marked with this figure 1. The figure 2 ſheweth the ſhape of one particular fruit, with the lower ſide

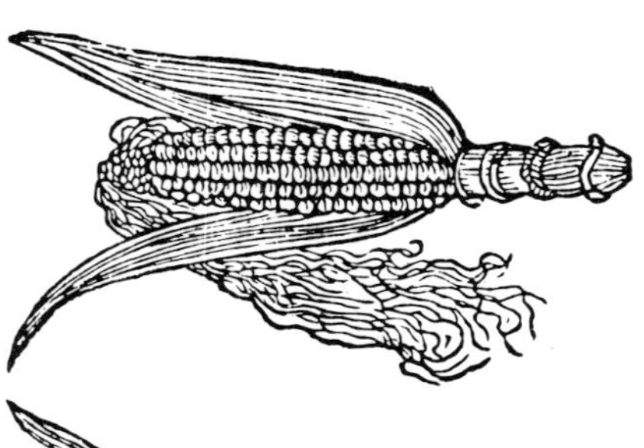

5 *Frumentum Indicum rubrum.*
Red Turky Wheat.

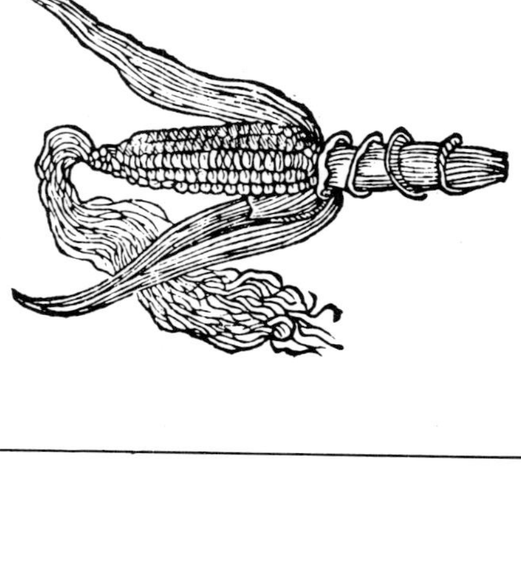

4 *Frumentum Indicum luteum.*
Yellow Turky Wheat.

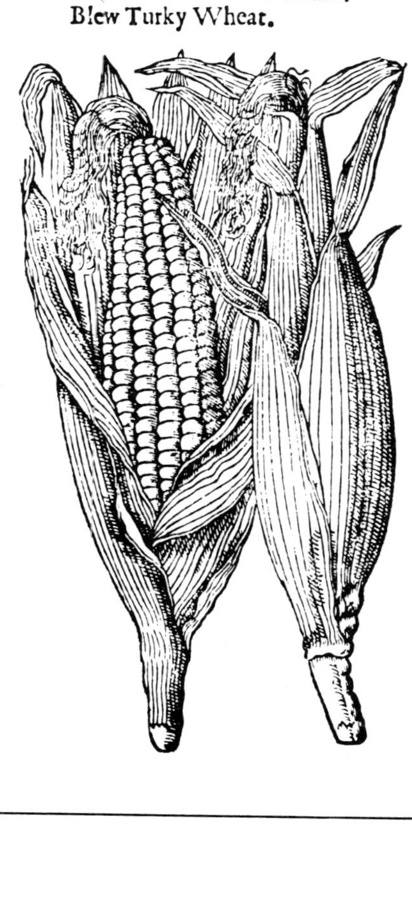

6 *Frumentum Indicum cæruleum.*
Blew Turky Wheat.

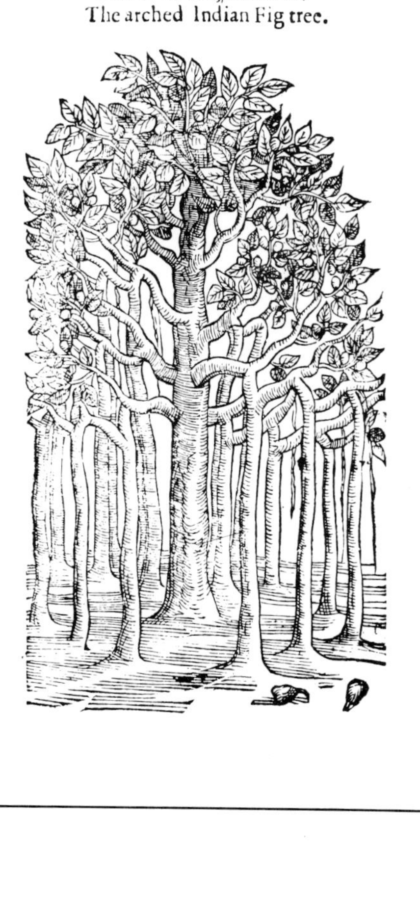

Arbor ex Goa, ſiue Indica.
The arched Indian Fig tree.

CHAPTER IV
Gardens and Travellers

Two pages and illustrations from Thomas Johnson's 1636 edition of *The Herball* by John Gerard. Page 393: various plants described as 'rubarb', none of them the species of *Thalictrum* that were called 'rewbarbe' by Lyte (see page 42); 'Turky Rubarb' and 'The other bastard Rubarb' are species of *Centaurea*, while 'The true Rubarb of the Antients' is indeed rhubarb (*Rheum rhaponticum*), which was introduced to England from Siberia before 1573 – at first, only the roots were valued for their medicinal properties, and the petioles were not used for eating until the eighteenth century. Page 1515: 'Adam's Apple', the banana. 'Turky Wheat', which is also referred to as 'Corne of Asia', is maize (*Zea mays*) and is American in origin. The Indian sacred fig (*Ficus religiosa*) starts life as an epiphyte, putting down aerial roots and soon strangling its host. It can then continue to grow, making many trunks and finally covering several acres.

The seventeenth century was a period of development in European botany, rather than of active exploration and collection. The discoveries from America and the tropics were cultivated and studied, while a few more species filtered in from the Ottoman Empire. In the gardens of Europe, especially in the Low Countries, cultivation of ornamental plants as opposed to vegetables and medicinal plants became fashionable. Some of the hundreds of new varieties collected and grown are represented in the flower pieces that are such a characteristic form of seventeenth-century Dutch and Flemish art. The floral miscellanies that appeared in vases in these paintings were also gathered together in books as florilegia, which are no more than unsystematic anthologies of flower portraits usually with little, if any, supporting text. The crucial difference between these florilegia and the herbals of the previous century is that they are essentially concerned with flowers for their decorative qualities rather than with herbs for their usefulness – a difference that had its origin in the new emphasis in gardening.

Some of this reached England through the agency of the Tradescants, father and son, who were the leading English gardeners of the century. In 1609 John Tradescant the Elder became gardener to Robert Cecil, First Earl of Salisbury at Hatfield House. He set about creating the finest garden in the country, filled with rare plants and exotic fruit trees, for which he travelled to Holland and France visiting nurserymen as well as famous gardeners such as Jean Robin in Paris. An album of the fruit at Hatfield, now in the Bodleian Library in Oxford, was painted by Alexander Marshall, who called it *Tradescant's Orchard.* It included fruit varieties from Europe, among them a black, pear-shaped cherry which Tradescant had introduced into England and which was named after him.

After Tradescant left the Cecils' employment he became gardener to Sir Dudley Digges, near Canterbury, and accompanied him to Russia on an embassy to arrange help for the Tsar in a war against Poland. The mission ended in failure, but gave Tradescant a chance to see and collect new plants. A further chance came in 1620, when Tradescant joined a naval expedition against pirates in Algiers.

Tradescant was given the commission to set up new gardens for Charles I and his bride Henrietta Maria, sister of the French king Louis XIII, at Oatlands. This Tudor palace between Walton and Weybridge in Surrey had been built by Henry VIII, but was destroyed by Cromwell and no trace of it now remains. At Oatlands, Tradescant planted orchards, vineyards and flower gardens, but at the same time he continued to add to his garden at Lambeth and acquire a collection of oddities and curios from all over the world that later became the basis of the Ashmolean Museum in Oxford. He had a particular interest in American plants, listing over a hundred species in his garden catalogue of 1634.

Tradescant's son, also called John, made three expeditions to Virginia, even acquiring an interest in land in York County. It was in 1638, while he was on the first of these visits 'to gather all varieties of flowers, plants and shells,' that his father died. The younger Tradescant continued his father's work; his *Museum Tradescanteanum*, published in 1656, contained a catalogue of the contents of the museum as well as a revised garden list.

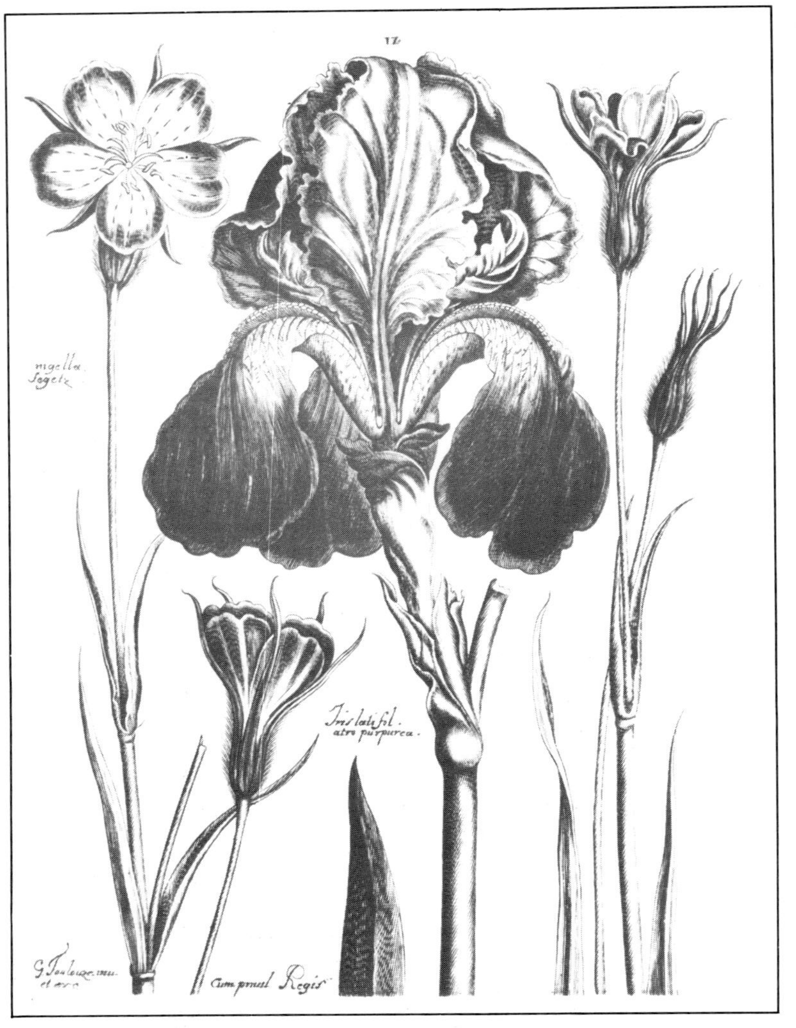

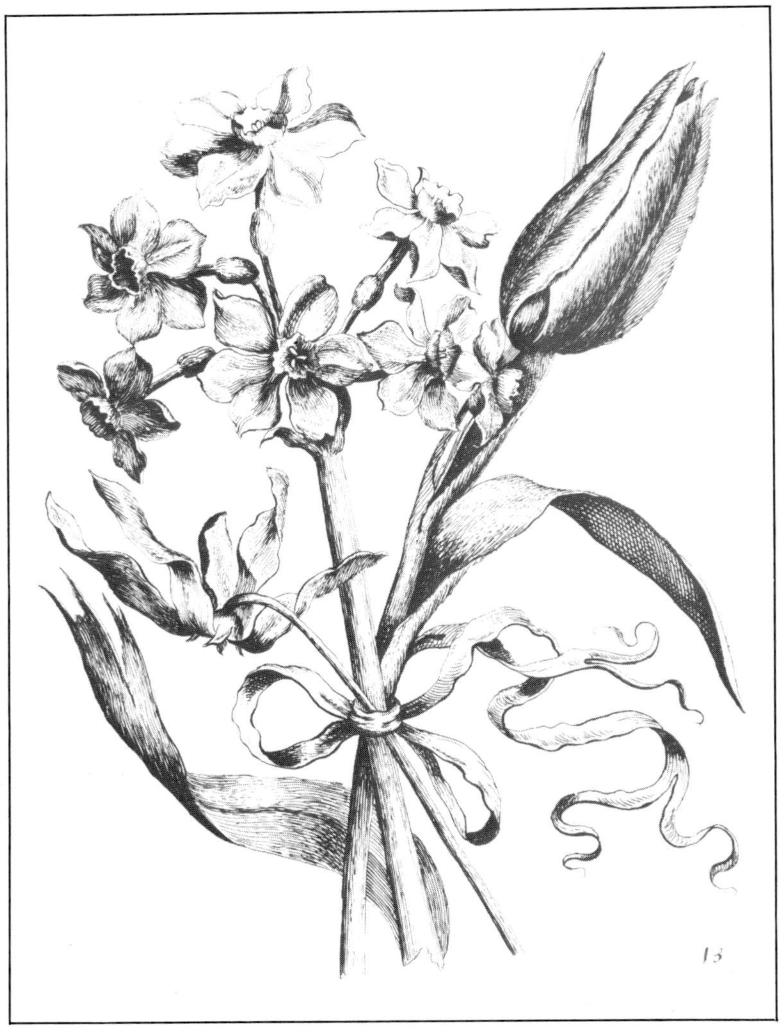

The only English florilegium of the period dates from this time and is by the same Alexander Marshall who had painted the fruit at Hatfield. It is now in the royal library at Windsor and contains nearly 2,000 flowers boldly painted in opaque water colour. It is very likely that Marshall painted it for the Tradescants or at least obtained many of his models from their garden, as almost all the plants are also listed in the *Museum Tradescantianum*. Marshall's book contains one of the earliest pictures of a *Nerine* ('this flower was sent to me by Generall Lambert August 29th 1659 from Wimbleton'), and another particularly interesting flower is *Sutherlandia frutescens*, a striking red-flowered broom from South Africa. Some of the fritillaries painted by Marshall are not in Tradescant's list, for example the Greek *Fritillaria obliqua* from Attica and *Fritillaria pontica* from the southern shore of the Black Sea, but Tradescant may nevertheless have grown them, as he records having received various bulbs from Sir Peter Wyche, English ambassador at Constantinople.

Nearly all the plants grown in northern Europe at this time were relatively hardy and able to survive frost in winter, although a few were cultivated that had to be brought inside in winter and protected in orangeries where a fire could be lit on particularly cold nights. Heated greenhouses had not yet been invented and were not to make a significant contribution to horticulture until the early nineteenth century. Books on tropical plants, of which several appeared in the seventeenth century, were thus either of interest purely to botanists or mainly of use to the increasing number of Europeans who already lived in the tropics, like the famous *Hortus Indicus Malabaricus* of 1678–1703, compiled by H. A. van Rheede tot Draakestein.

Among the flowers grown in Holland, the tulip became outstandingly popular in the 1620s and 1630s, as Wilfrid Blunt has described so well in his book *Tulipomania*. The first tulips had been brought to Europe through the agency of Busbecq, who probably sent seed to Clusius in 1554, and the first illustrations appeared in Gesner's book on the German garden in 1561. Some merchants in Antwerp imported a cargo of bulbs in 1562, and these, together with some further imports of seeds, formed the basis of Dutch tulip cultivation.

Below:
Caraway (*Carum carvi*) and three tubers of American origin: 'Spanish [sweet] Potatoes' (*Ipomoea batatas*), 'Virginia Potatoes' (*Solanum tuberosum*) and 'Potatoes of Canada, or Artichokes of Jerusalem' (*Helianthus tuberosus*). Woodcut from *Paradisi in Sole, Paradisus Terrestris* (1629) by John Parkinson.

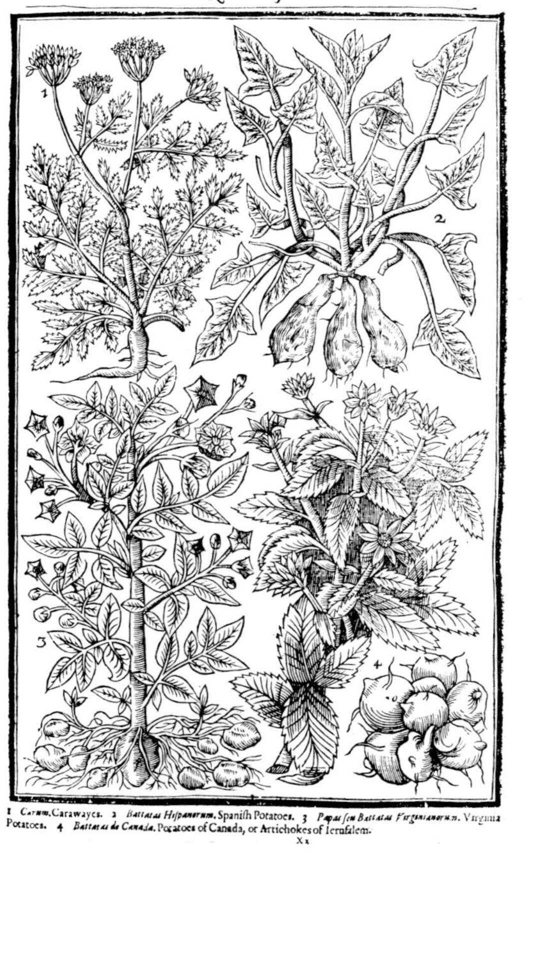

The Kitchen Garden. 517

1 *Carum.* Carawayes. 2 *Battatas Hispanorum.* Spanish Potatoes. 3 *Papas seu Battatas Virginianorum.* Virginia Potatoes. 4 *Battatas de Canada.* Potatoes of Canada, or Artichokes of Ierusalem.

Xx

Far left:
Corn cockle (*Agrostemma githago*) and a bearded iris (*Iris germanica*). Engraving from *Livre de Fleurs, Feuilles, et Oyzeaus dessiné après le naturel* by Guillaume Toulouse, published in Montpellier in 1656.

Left:
A cyclamen, a narcissus and a tulip. Engraving by Nicolaus Guillelmus, published in Rome in 1638.

The Turks had cultivated tulips for many centuries and had already produced several different colours. Clusius mentioned having obtained yellow, red, white and purple self-coloured flowers, as well as 'broken' forms from a batch of seed he received from Busbecq in 1573. Breaking, the presence of stripes and patches of a paler colour, is the sign of a virus infection that is spread from plant to plant by aphids. The amount of breaking may therefore vary from year to year, and this uncertainty was one of the fascinations of tulip growing. In the early 1630s, at the height of tulipomania, up to 5,000 florins were paid for a single bulb.

Tulips are always well represented in the florilegia of the period, and nearly every seventeenth century Dutch flower painting has at least one broken tulip, together with predominantly the common garden flowers of the period. After tulips, the next most conspicuous flowers are roses, particularly *alba* and *foetida* roses and the *centifolia* roses that were being developed in Holland throughout the century. Popular flowers from the Orient were the Crown Imperial, hyacinths, the purple and gold *Iris variegata*, and many forms of the turban Ranunculus, which had also been very popular with the Turks. From Greece came the madonna lily (*Lilium candidum*) and the scarlet *Lilium chalcedonicum*. The south of France and Italy produced *Narcissus tazetta*, yellow fritillaries, and many forms of anemone. From Spain came members of the genus *Narcissus* such as the jonquil and the double *Narcissus odorus*. American plants were often cultivated but rarely painted and appear infrequently in flower pieces, although *Canna indica* and *Tropaeolum minus* from Peru do appear in a picture by Georg Flegel

An epiphytic orchid of the genus *Bulbiphyllum*. Engraving from *Hortus Indicus Malabaricus* by Henrik Adriaan van Rheede tot Draakestein, which was published in Amsterdam between 1678 and 1703.

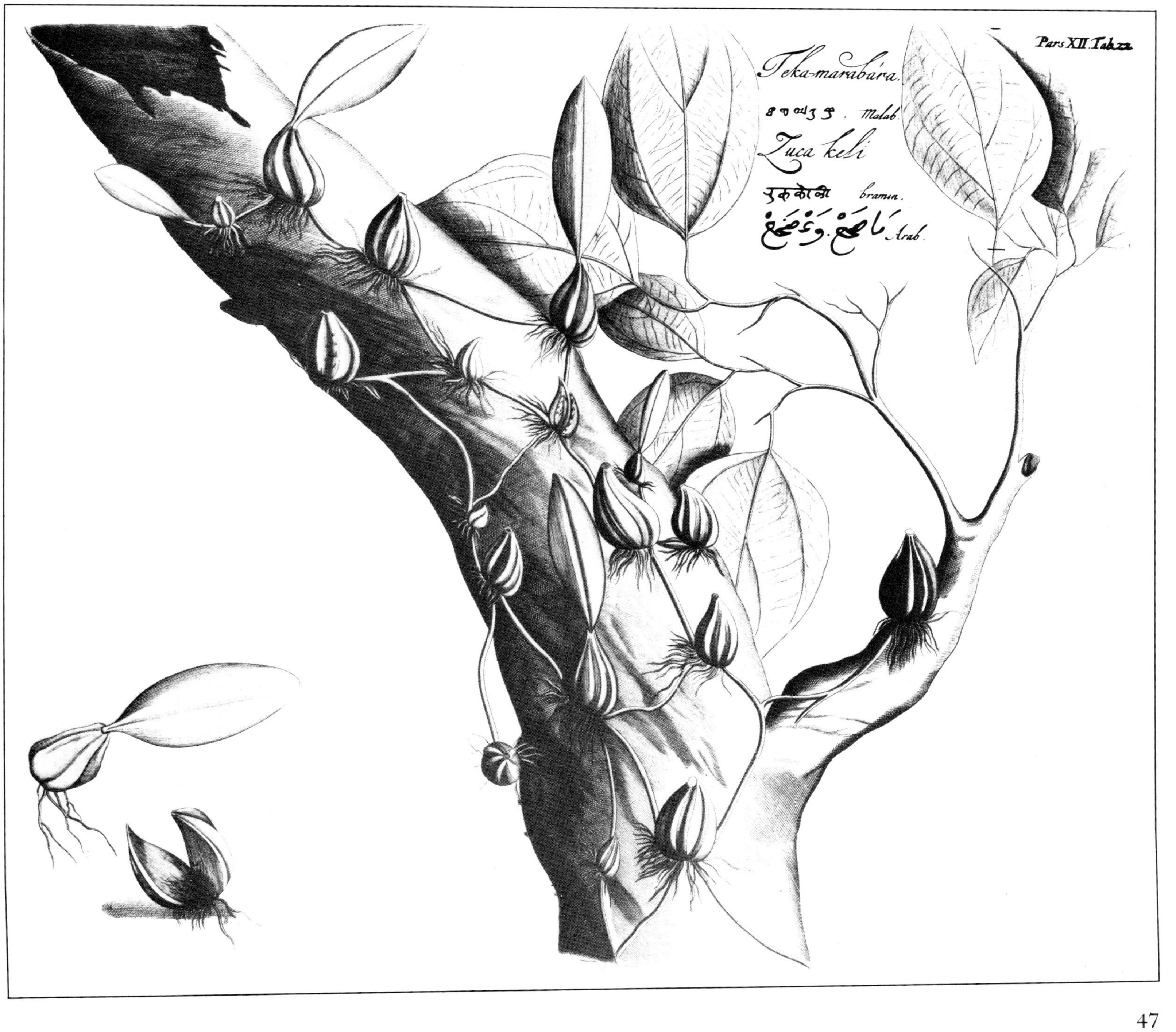

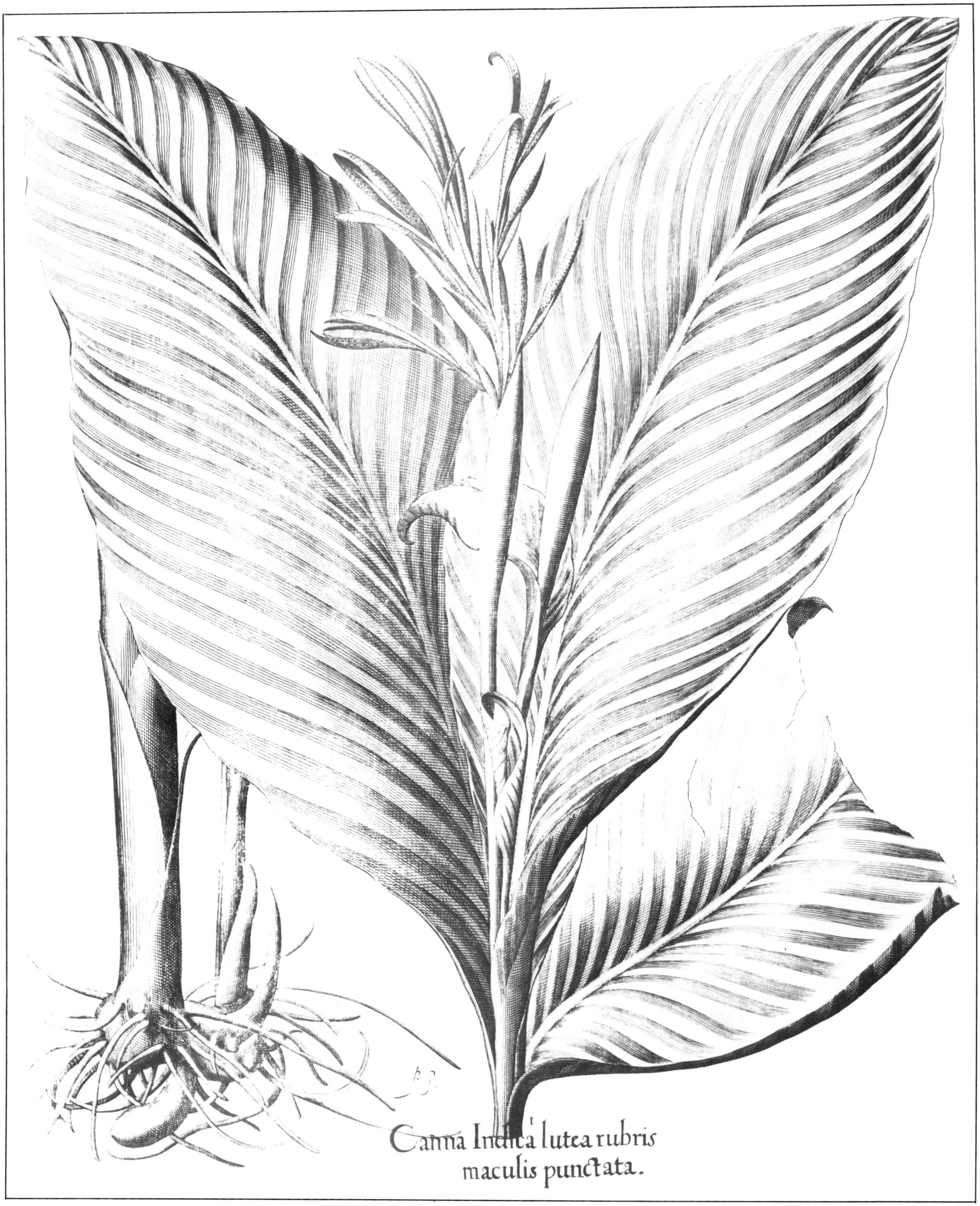

Canna indica. Engraving from *Hortus Eystettensis* (1613) by Basil Besler.

dated 1638, and probably painted in Frankfurt-am-Main. The belladonna lily, which was not grown in England until 1712, was painted in 1670 by the Italian Paolo Porporra in Rome. The scented tuberose (*Polianthes tuberosa*) from Mexico was also popular in the latter part of the 17th century.

The foremost painters of flower pictures were the families of Ambrosius Bosschaert the elder (1573–1621), his brother-in-law, Balthasar van der Ast, and of Jan Breughel the Elder (1568–1625). As the larger flower pieces bring together blooms from all seasons of the year, the painters must have

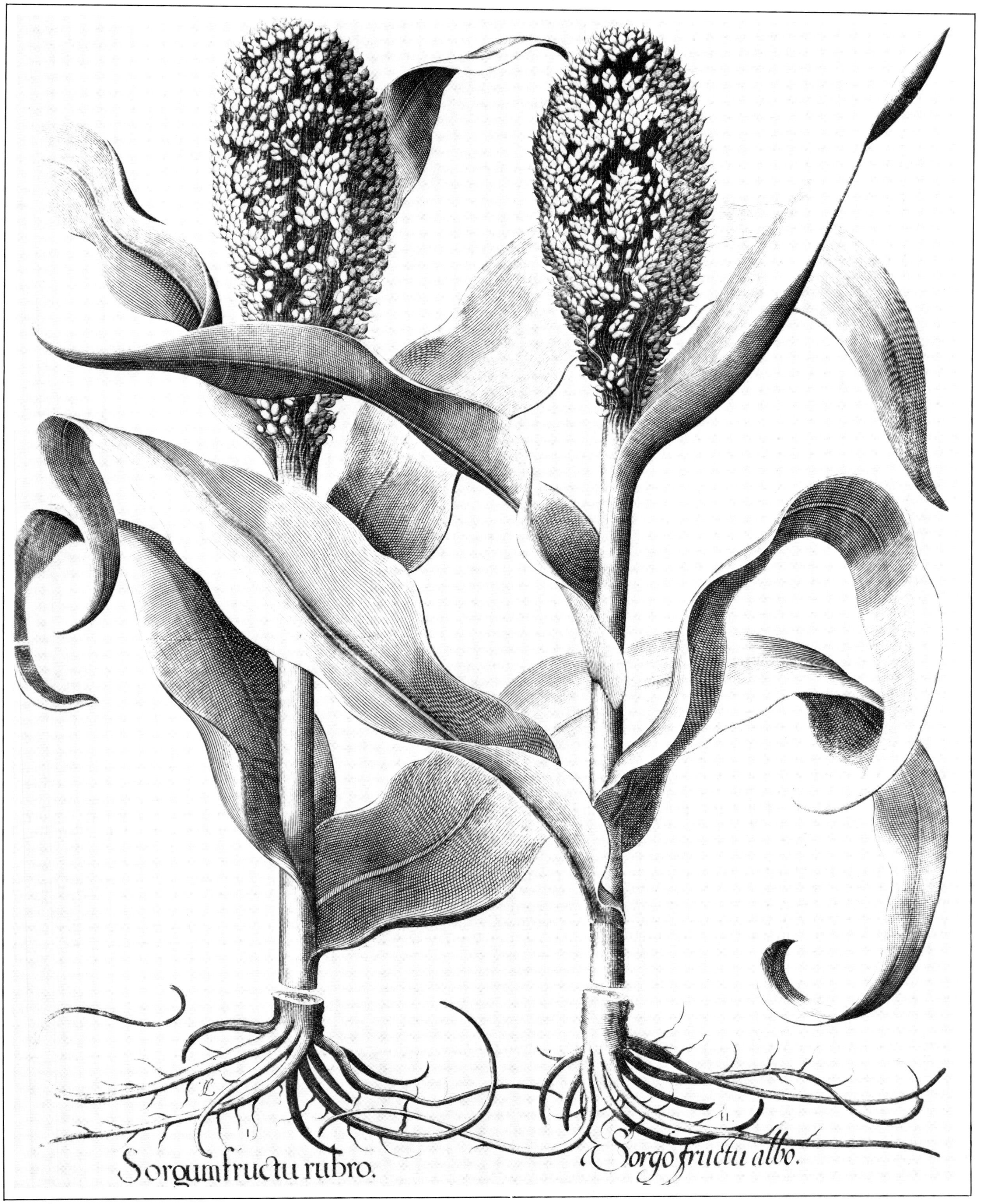

Two varieties of sorghum millet. Engraving from *Hortus Eystettensis* (1613) by Basil Besler.

had sketch books rather like florilegia, but none of these has survived.

One early manuscript florilegium that still exists is by Peter van Kouwenhoorn and is in the Lindley Library of the Royal Horticultural Society in London. It was probably painted in Holland in the 1630s, but nothing is known of van Kouwenhoorn, except that he was probably a glass painter. His paintings, boldly executed in opaque watercolour, are of the usual garden plants of the period, including a masterful crown imperial fritillary and a number of fine irises. The double flowers are particularly well handled.

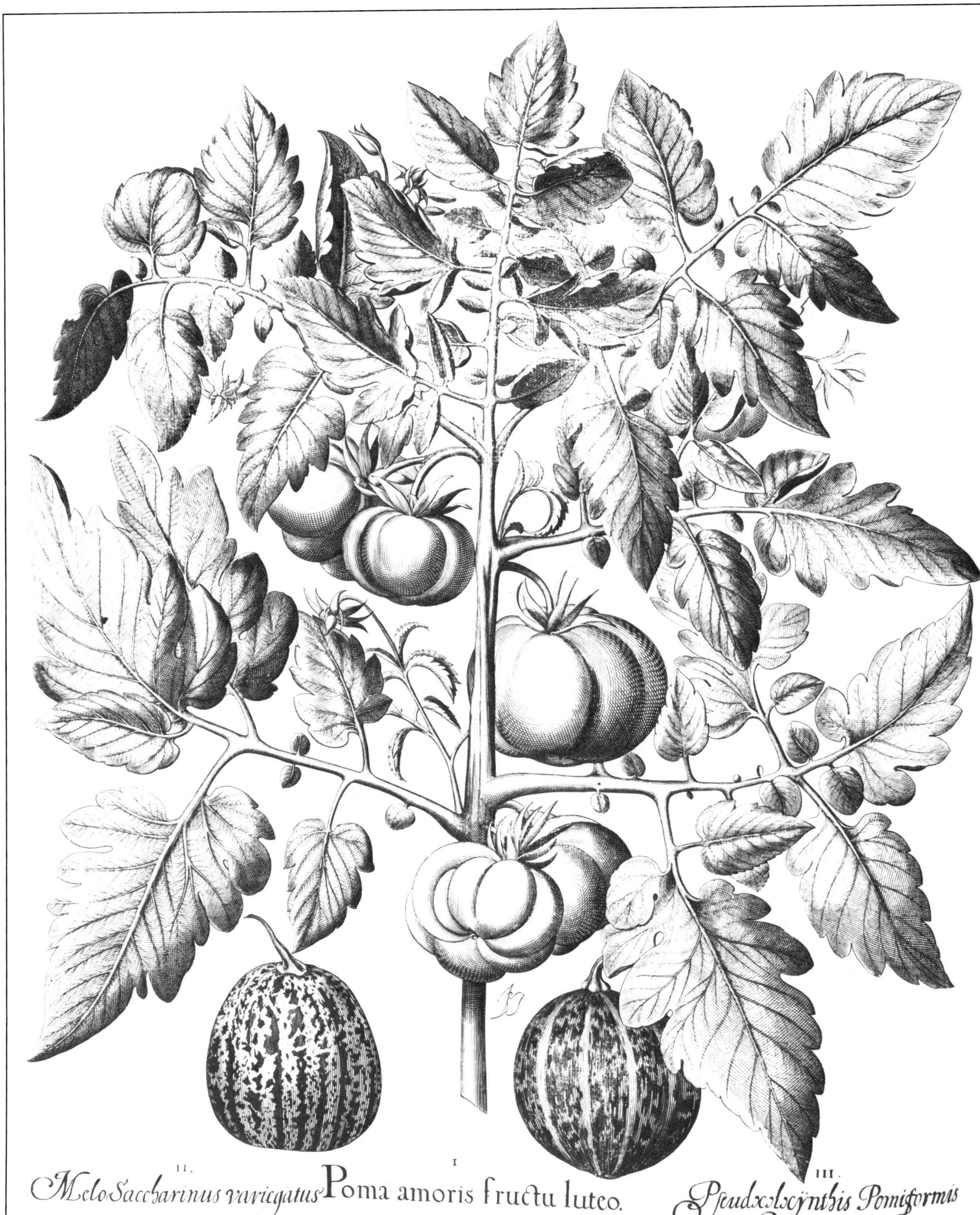

Yellow-fruited tomato with two melons. Engraving from *Hortus Eystettensis* (1613) by Basil Besler.

Some other florilegia were printed and published, usually with engravings or etchings, which by this time were superseding the woodcuts that had commonly been used in the 16th century. The most famous ones were published in Germany, like that of Emmanuel Sweert, a Dutchman employed by the Emperor Rudolph II. His florilegium was published in 1612 in Frankfurt-am-Main. It has no text, but a catalogue in Latin, German, English and French. The first part contains 67 plates of bulbous plants including some exotic items, such as South African species of blood lily

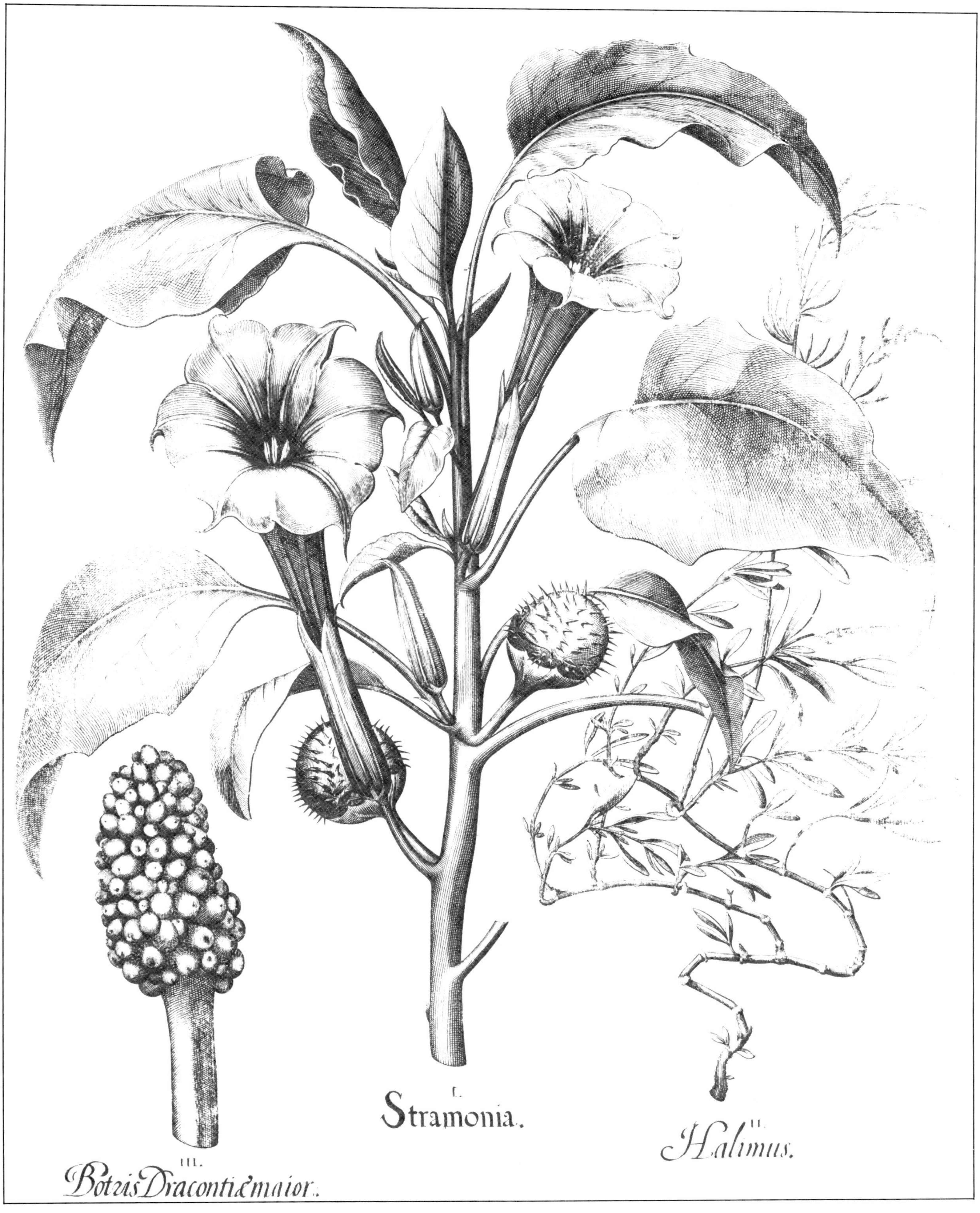

Thorn-apple (*Datura stramonium*) with the fruit of the dragon arum (*Dracunculus vulgaris*) and a shrubby orache (*Atriplex halimus*). Engraving from *Hortus Eystettensis* (1613) by Basil Besler.

(*Haemanthus*) and *Gladiolus* and the giant bulbed *Boophone disticha*, which is found throughout southern Africa. Its leaves are extremely poisonous and have been used to tip poisoned arrows; the huge bulb may be up to a foot in diameter. Some of Sweert's plates are copied directly from another florilegium, the *Florilegium Novum* of Johann Theodor de Bry, published the previous year, which in turn had some plates copied from *Le Jardin du très Chrestien Henry IV* (1608), which was the work of a Frenchman, Pierre Vallet.

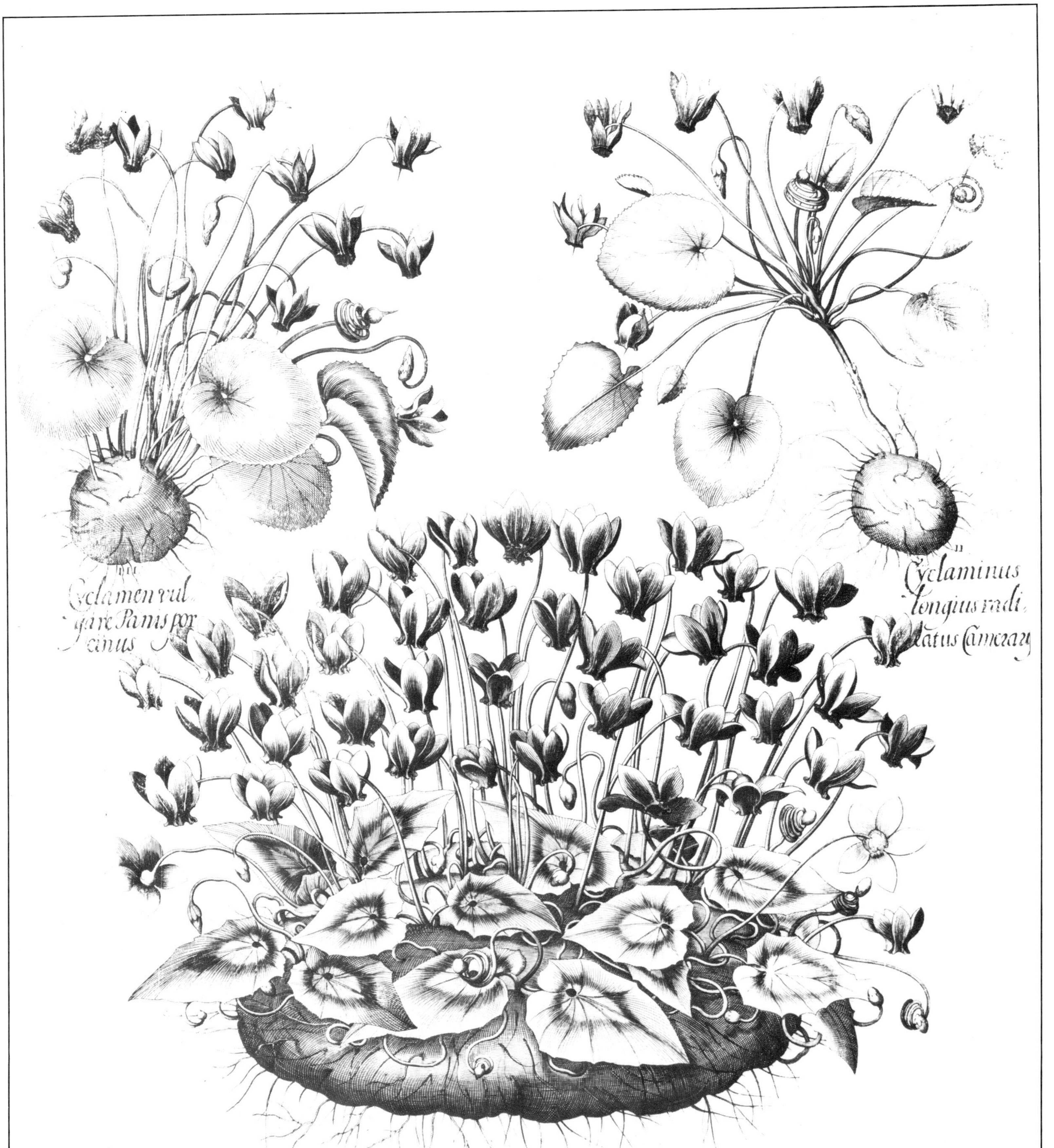

The most impressive of the early seventeenth-century florilegia was the *Hortus Eystettensis*, published at Eichstätt near Nuremberg in 1613 under the patronage of Johann Conrad von Gemmingen, Bishop of Eichstatt, in whose garden most of the plants were grown. This was a massive enterprise; its author, the Nuremberg apothecary Basil Besler (1561–1629) worked on and off for sixteen years on the drawings and six or more engravers produced the plates. Its two large folio volumes with 374 plates show each plant at life size, starting with spring flowers and progressing through the year. The

Two cyclamens: *Cyclamen purpurascens* (above) and *Cyclamen hederifolium*. Engraving from *Hortus Eystettensis* (1613) by Basil Besler.

Christmas rose (*Helleborus niger*). Engraving from *Hortus Floridus* (1614) by Crispin de Passe.

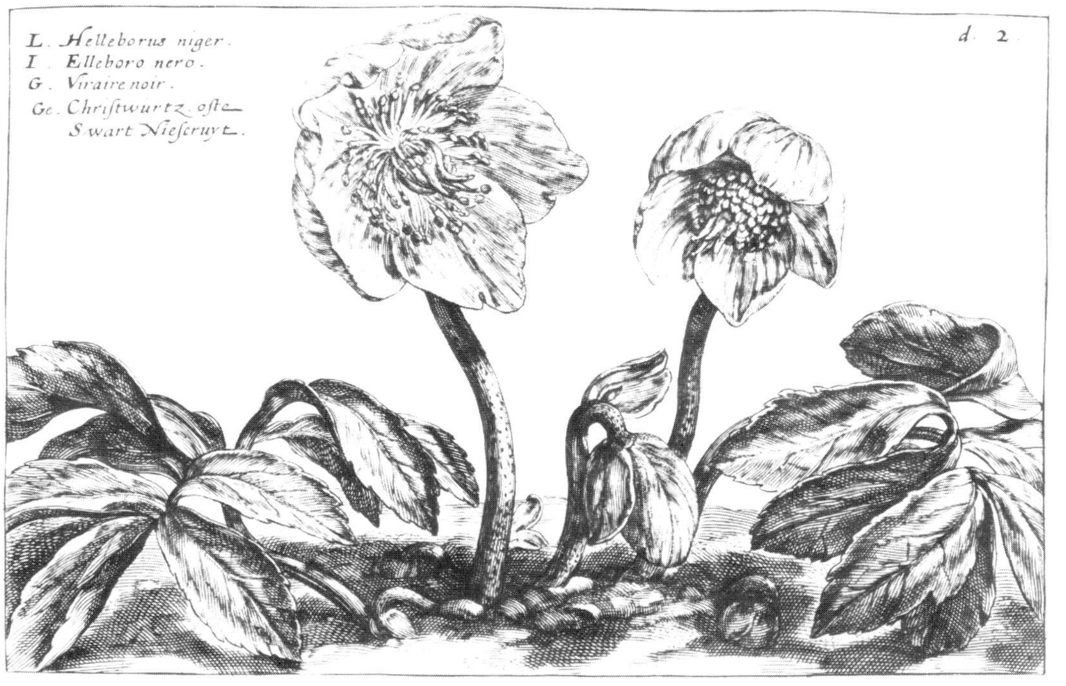

copper-plate engravings are rather stiff and coarse but the layout of the plates makes many of them highly decorative. Weeds, wild plants and familiar garden flowers are often portrayed together. Again, bulbs are particularly well represented, with eleven pages of tulips. There is a brief description of each plant figured and a list of references to previous authors. A section on vegetables includes aubergines, green and red peppers of many shapes and sizes, and a tomato ('*Poma amoris fructu luteo*') with yellow fruit the size of oranges. Among the exotic plants are several tobaccos, the highly poisonous thorn-apple, sorghum (a tropical relation of millet) and potatoes, which had not yet gained acceptance as a vegetable.

The *Hortus Floridus* in Latin, French, Dutch and English, had, as the publishers doubtless hoped, a wide sale throughout Europe and lasting influence on seventeenth-century gardening. The author, Crispin de Passe, was a Dutchman who lived first in Cologne, then in Utrecht and later in Paris, where he became Professor of Drawing. The *Hortus Floridus* is an oblong quarto with a brief and largely descriptive text. Although small in comparison with other florilegia, it is unusually charming; many of the plants are shown growing, with bees and other insects flying around or feeding from the flowers. The emphasis is on the common garden flowers of the period, with a great preponderance of spring bulbs, especially tulips; the arrangement is by season, with an appendix of fruit. A frontispiece shows a Dutch formal walled garden, with small hedges and with tulips, irises, fritillaries and primulas among the many other flowers in the beds.

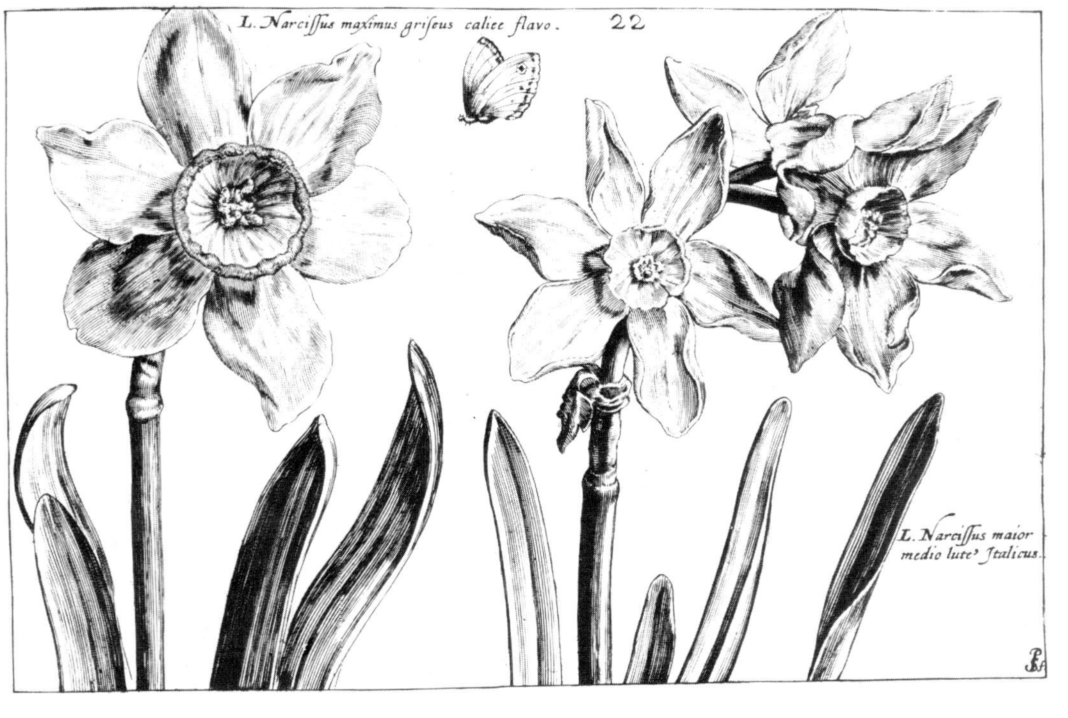

Two hybrid narcissi. Engraving from *Hortus Floridus* (1614) by Crispin de Passe.

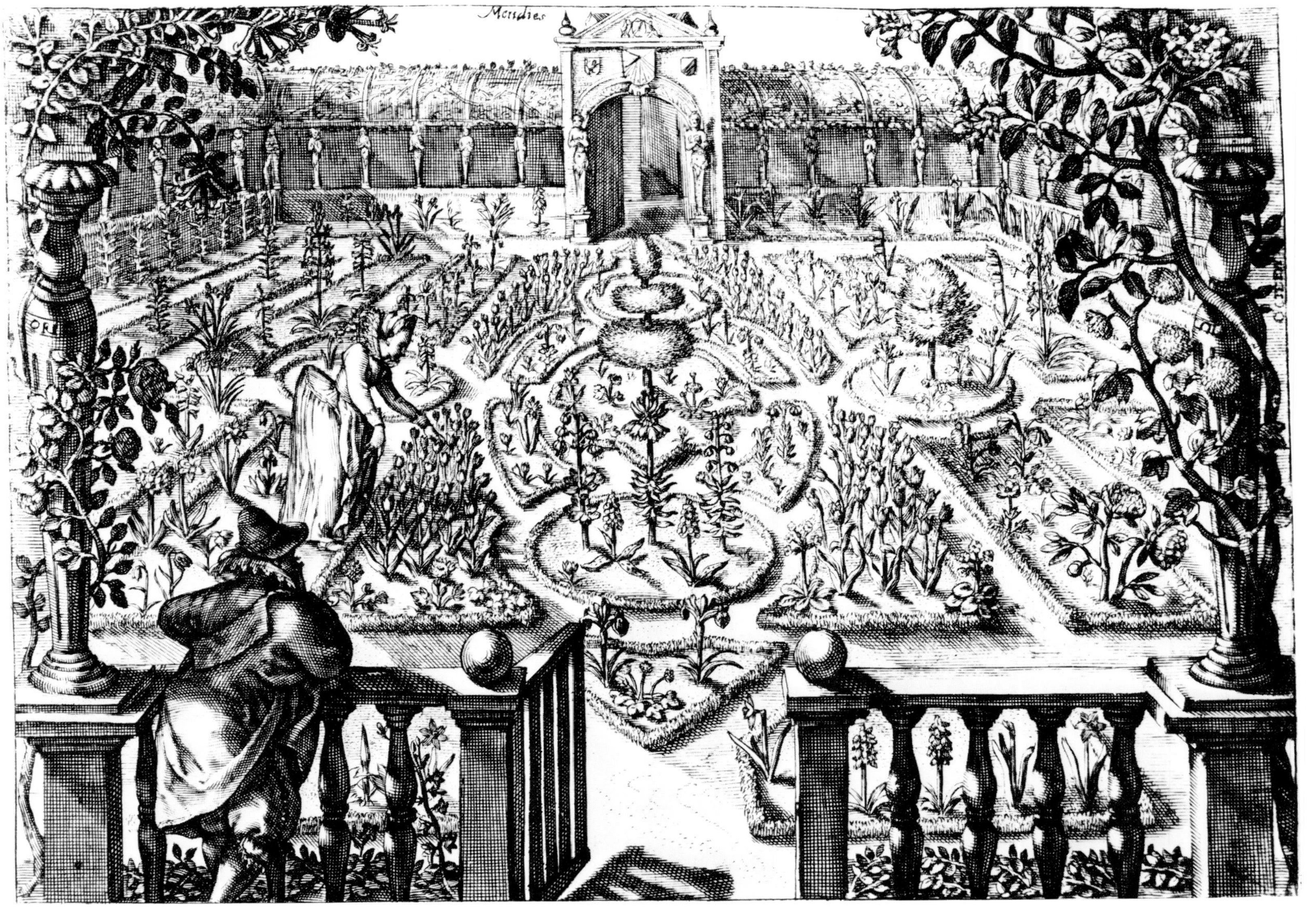

A lady tending her tulips in a Dutch formal garden. Among the many flowers that can be identified, tulips are predominant, although there are also many fritillaries, lilies, daffodils, irises and crocuses, all plants that had become garden favourites and have remained popular ever since. Engraving from *Hortus Floridus* by Crispin de Passe, published in Utrecht in 1614.

The florilegium retained its popularity from the beginning of the seventeenth century to the early nineteenth. Indeed, its spirit continues to the present day in the *Botanical Magazine*. One of the foremost late seventeenth-century examples is the laboriously titled *Horti Medici Amstelodamensis Rariorum Plantarum Descriptio et Icones* (1697–1701) by Jan Commelin, who was director of the Amsterdam Physic Gardens and a senator of the city. He died in 1698, and his nephew Caspar finished the second volume. I have not seen the original illustrations, which were painted by Johan and Maria Moninckx, but Wilfrid Blunt says of them: 'the engravers, who have taken considerable liberties with the drawings, have not succeeded in giving

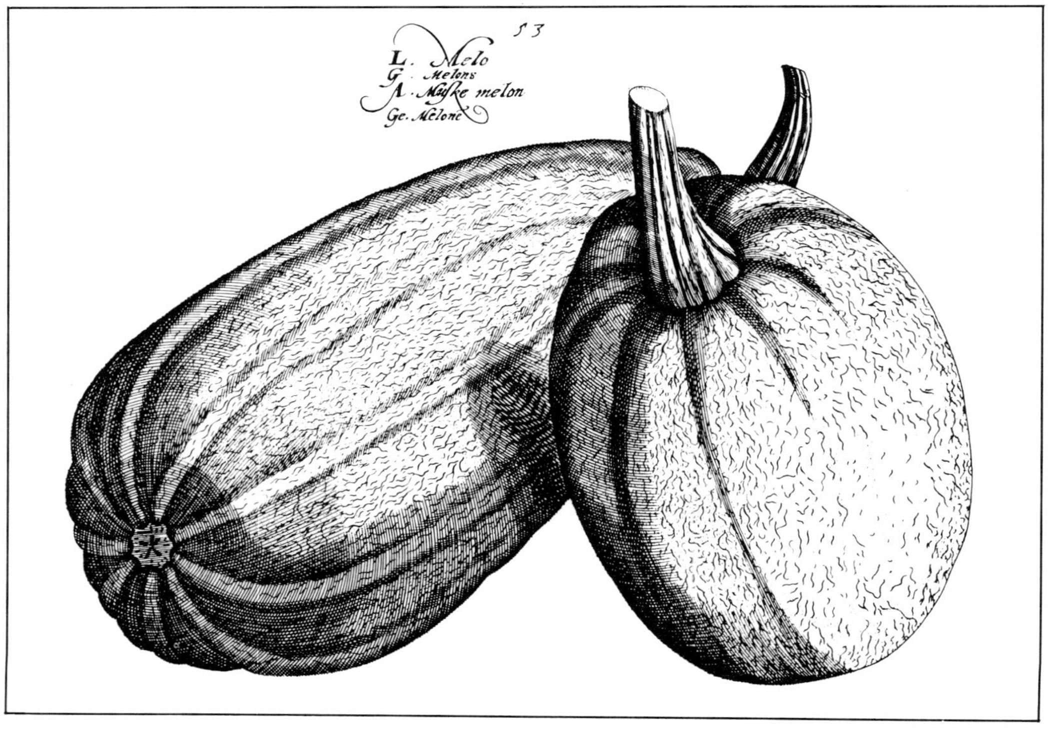

Melons. Engraving from *Hortus Floridus* (1614) by Crispin de Passe.

any idea of the quality of the originals' – nor, indeed, is it easy to identify the species in some of them with any confidence. The plants shown are mostly exotic rarities grown in the Amsterdam Physic Gardens and include some introduced from South Africa or Ceylon by Paul Hermann. One of the most spectacular is the lily, *Gloriosa superba*, which is described as '*Lilium zeylanicum superbum*' and was doubtless seen by Hermann if not actually introduced by him. An interesting new arrival is okra, from tropical Africa, which is said to have been introduced in 1686.

Although a few plants from South Africa, notably the familiar blue *Agapanthus africanus* and the belladonna lily, were cultivated in Europe before 1630, Hermann *en route* for the East Indies in 1672, was the first botanist to visit the Cape region of South Africa. A garden had been set up at Cape Town in 1652, but primarily to grow vegetables for the Dutch East India Company's ships. Hermann made collections of plants on both his outward and his homeward journeys. Slightly later, two other collectors sent plants back from South Africa to Holland: Hendrick Oldenland, who was superintendent of the Cape Town garden and Jan Hartell, its master gardener, who made collecting journeys up country in 1699, 1705 and 1707.

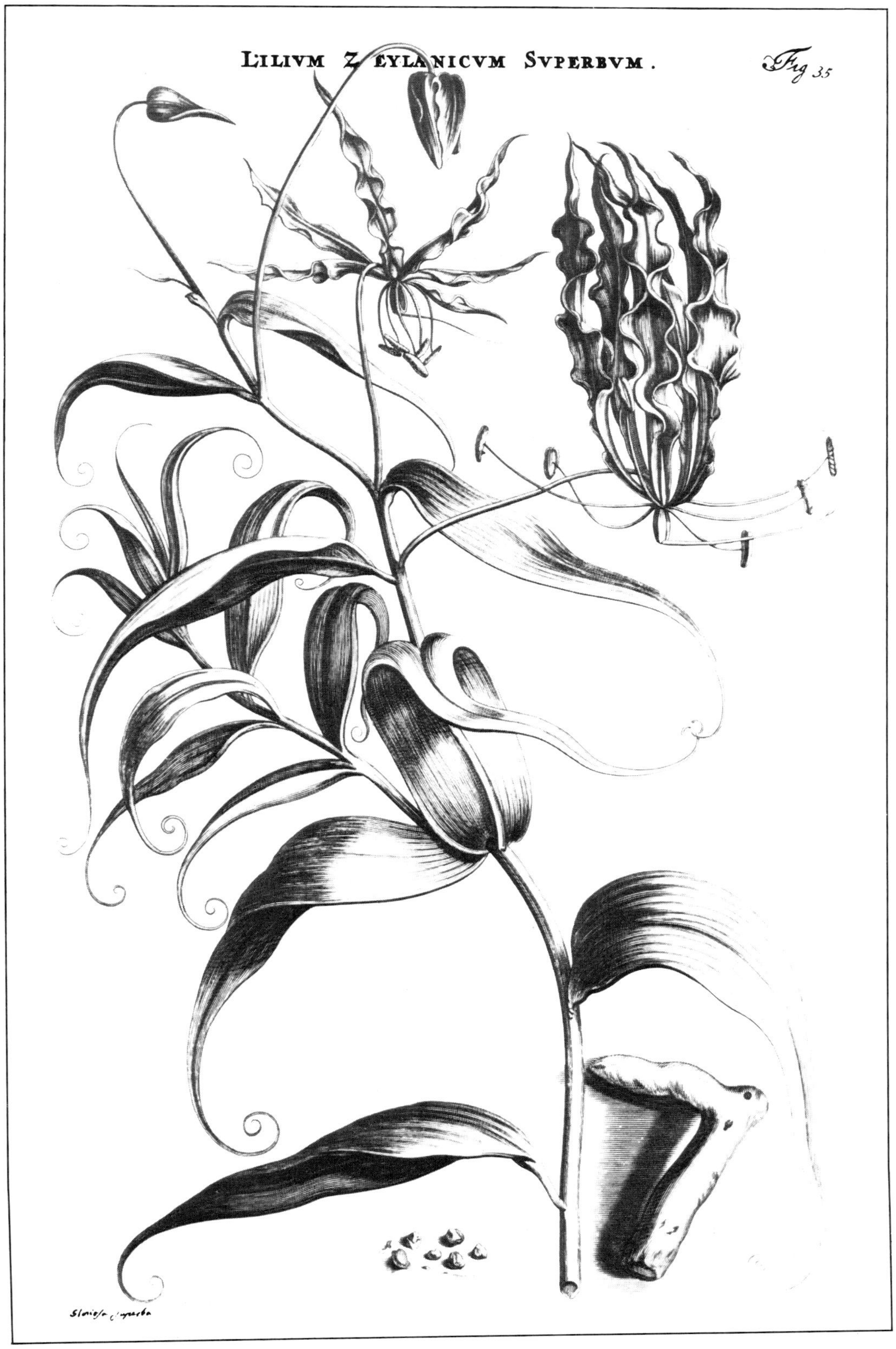

'*Lilium zeylanicum superbum*', the glory lily (*Gloriosa superba*), an African plant that climbs up through grass and scrub; its striking flowers are orange, red or yellow. Engraving from *Horti Medici Amstelodamensis Rariorum Plantarum Descriptio et Icones* (Part I, 1697) by Jan Commelin.

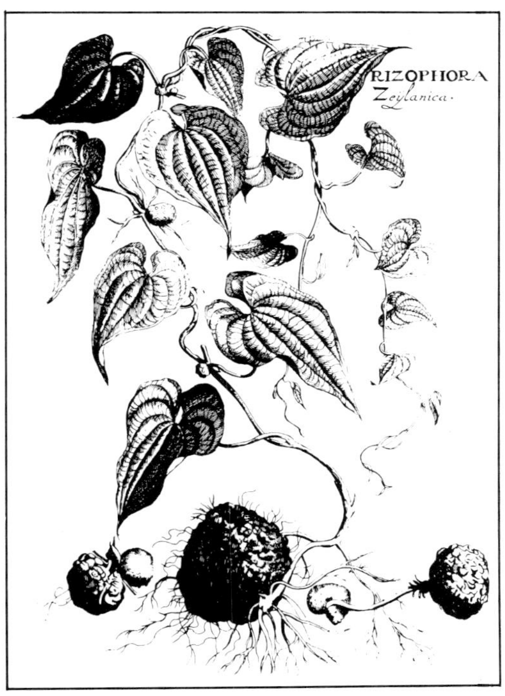

A yam (*Dioscorea bulbifera*) which produces aerial bulbils that are edible. Engraving from *Paradisus Batavus* (1698) by Paul Hermann.

Some plants from Hermann's collection are shown in his *Paradisus Botanicus* (1698) and *Museum Zeylanicum* (1698). Among these are a giant edible arum, the elephant's foot yam (*Amorphophallus*), brought back from Ceylon, and a tuberous *Pelargonium* from the Cape.

During the seventeenth century, the Dutch were very active in the Orient through the Dutch East India Company, which was founded in 1602 and given a government monopoly of trade over the whole area between the Cape of Good Hope in the west and the Straits of Magellan in the east. One Dutch colonial administrator with an interest in plants was the Governor of Malabar on the southwestern coast of India, H. A. van Rheede tot Draakestein. The twelve large folio volumes of engraved plates that make up his *Hortus Indicus Malabaricus* were published between 1678 and 1703 in Amsterdam. They show cultivated plants and wild ones, including grasses, ferns and mosses, as well as many now familiar tropical fruits such as coconuts, bananas and paw paws, several of them very beautifully engraved. The names are given in Latin, Arabic, Sanskrit and Malayalam. In the *Hortus Indicus Malabaricus* can be found what must be the first illustrations of a large number of tropical orchids. Another plate shows a characteristic

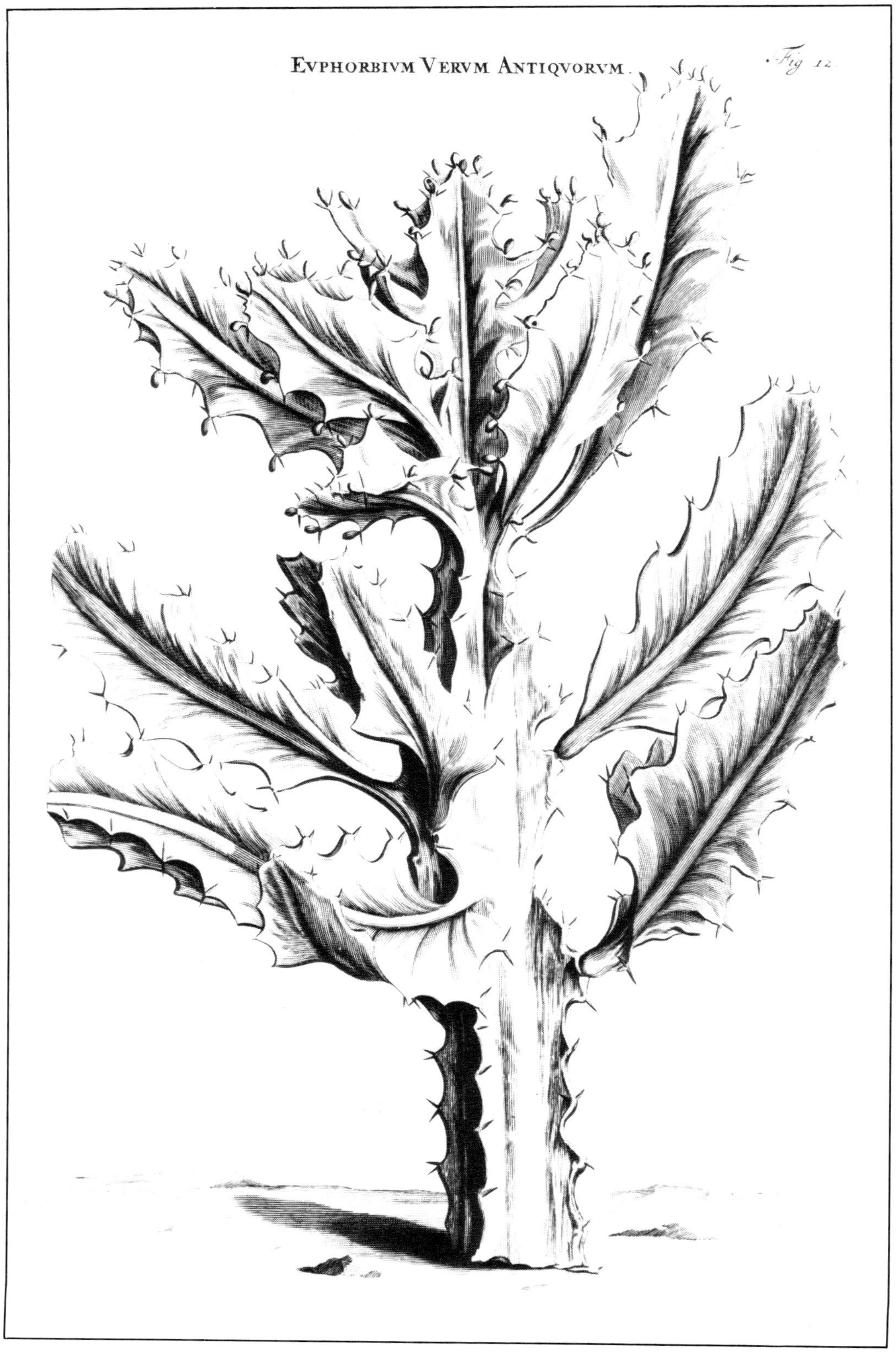

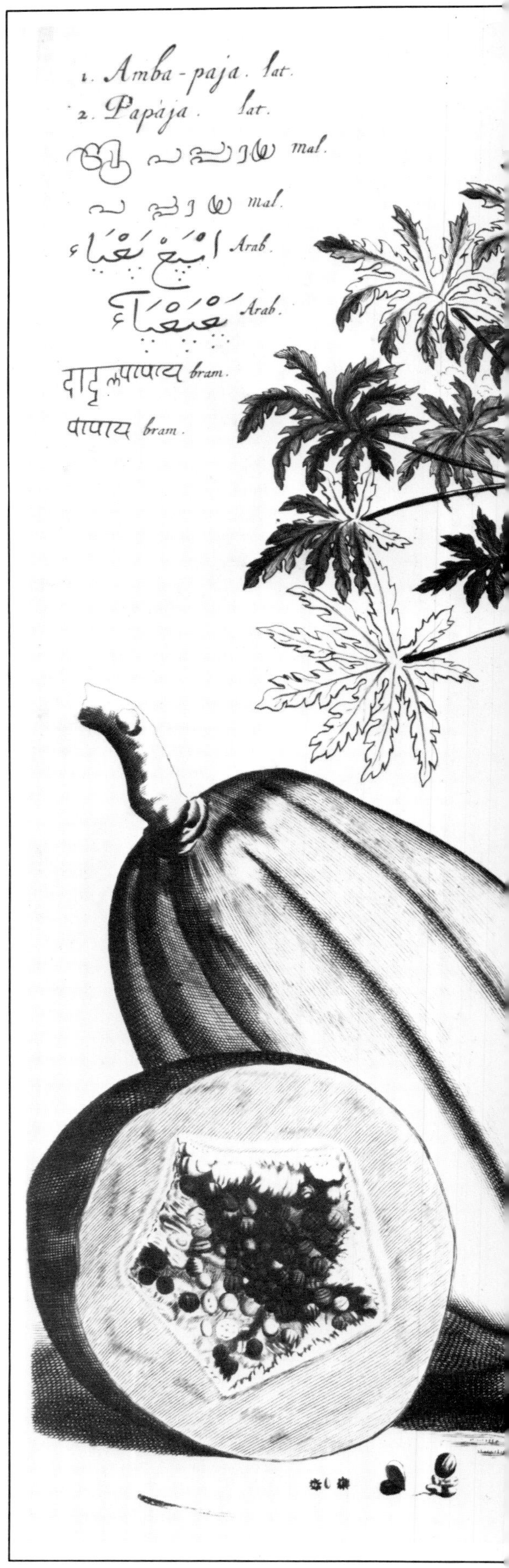

A succulent spurge, *Euphorbia antiquorum*, which is a small shrub that is native to India. Engraving from *Horti Medici Amstelodamensis*... (Part I, 1697) by Jan Commelin.

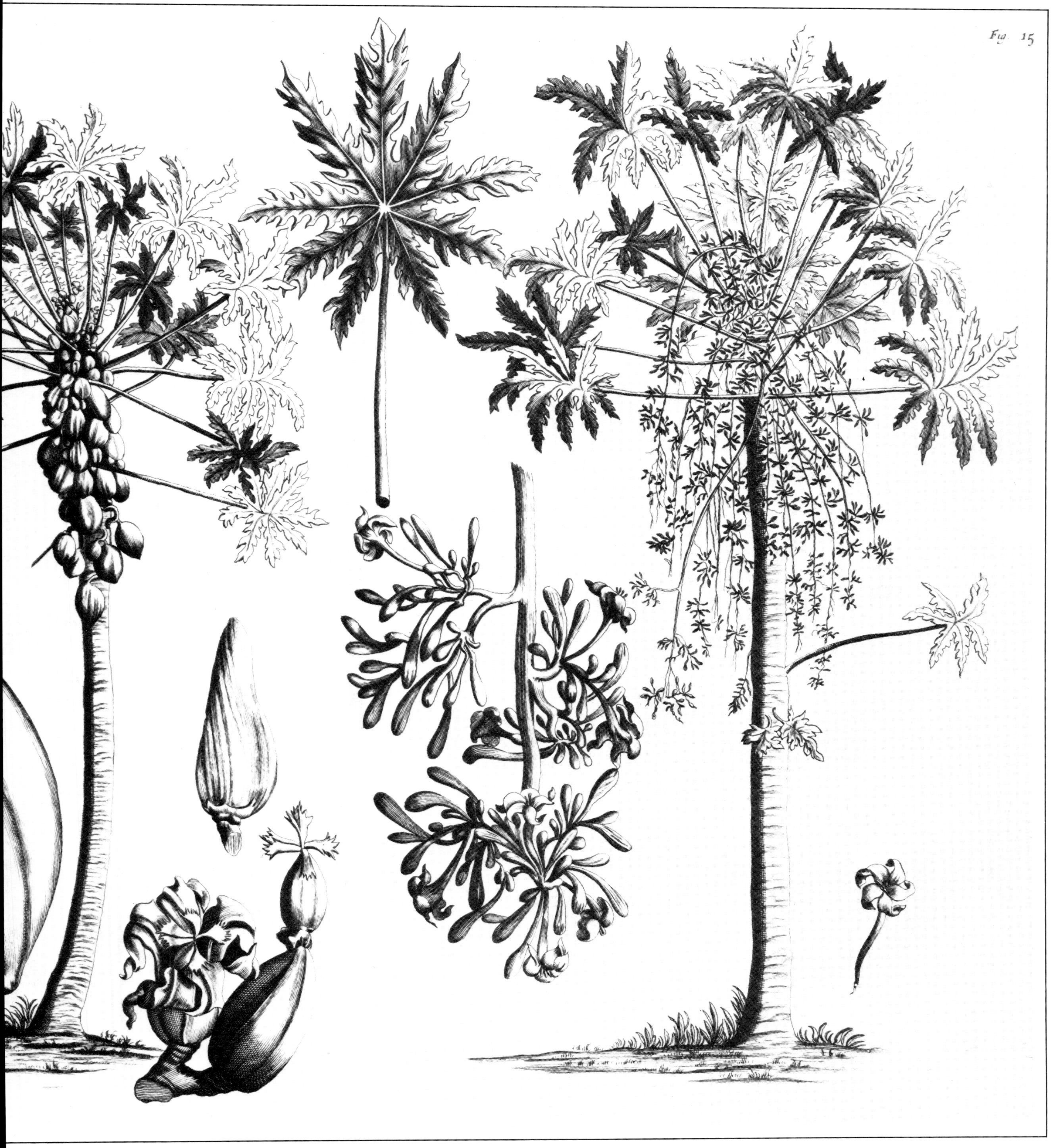

Paw paw (*Carica papaya*), showing male and female trees, and details of the flowers and fruit. Engraving from *Hortus Indicus Malabaricus* (1678–1703) by Henrik Adriaan van Rheede tot Draakestein.

product of the region, the cardamom, which forms low, leech-infested thickets in the tropical forests which cover the hills near the coast.

Another naturalist who worked for the Dutch East India Company was George Eberhard Rumpf (1628–1702), known as Rumphius, who spent most of his life in Indonesia on the island of Amboyna, arriving there in 1653 and staying until his death. He went blind in 1670, his wife and one of his daughters were killed in an earthquake in 1674, but he still persevered with his botanical work. Around 1686, he sent a voluminous manuscript to Holland to be printed, but the ship that carried it was attacked by the French. With the help of his remaining daughters, he started to rewrite it, but a fire in 1687 destroyed all his drawings. Finally, in 1690, the first three

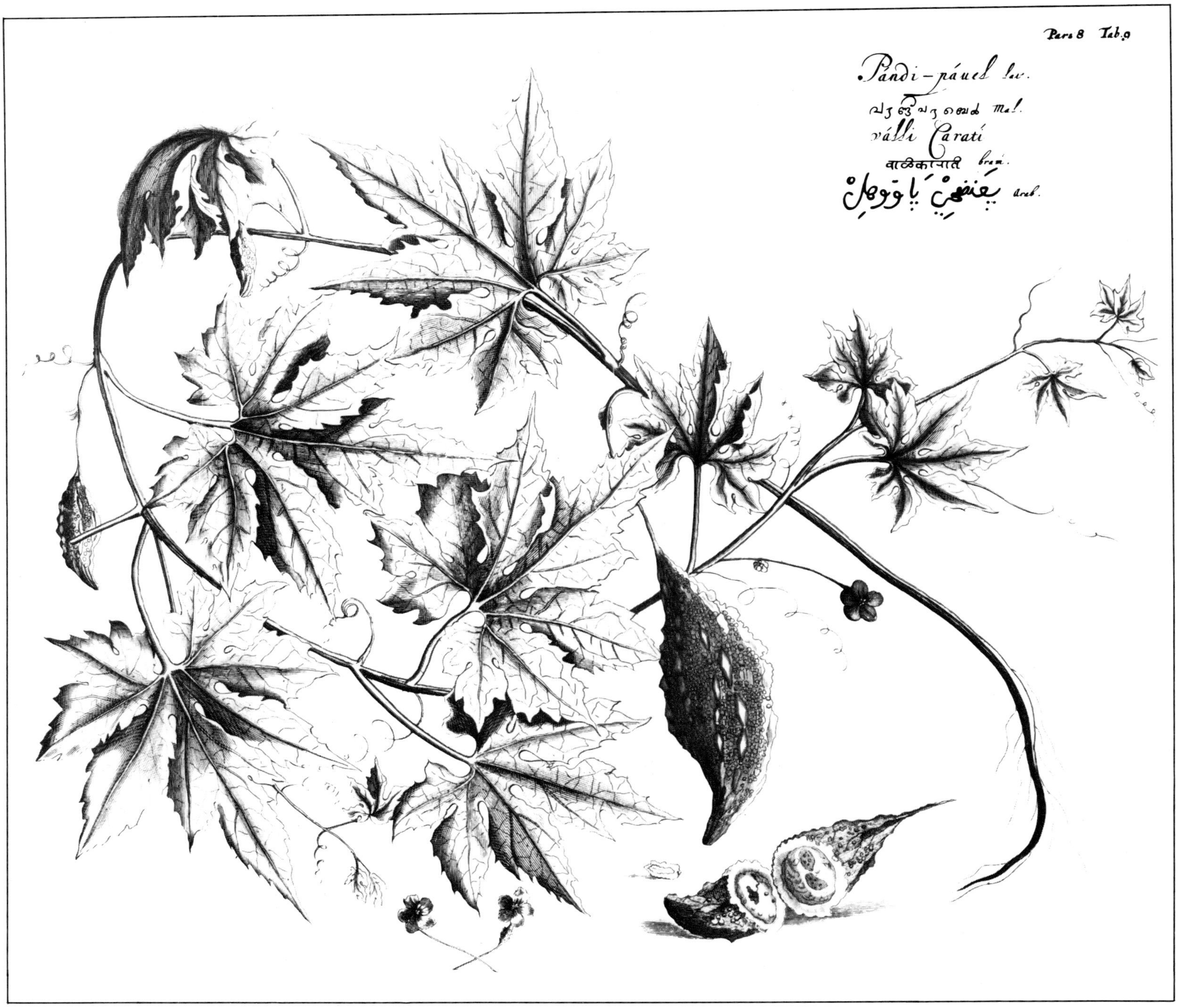

Bitter cucumber (*Momordica charantia*), which is frequently cultivated in the tropics and is particularly popular in India and Ceylon, where it is added to curries. Engraving from *Hortus Indicus Malabaricus* (1678–1703) by Henrik Adriaan van Rheede tot Draakestein.

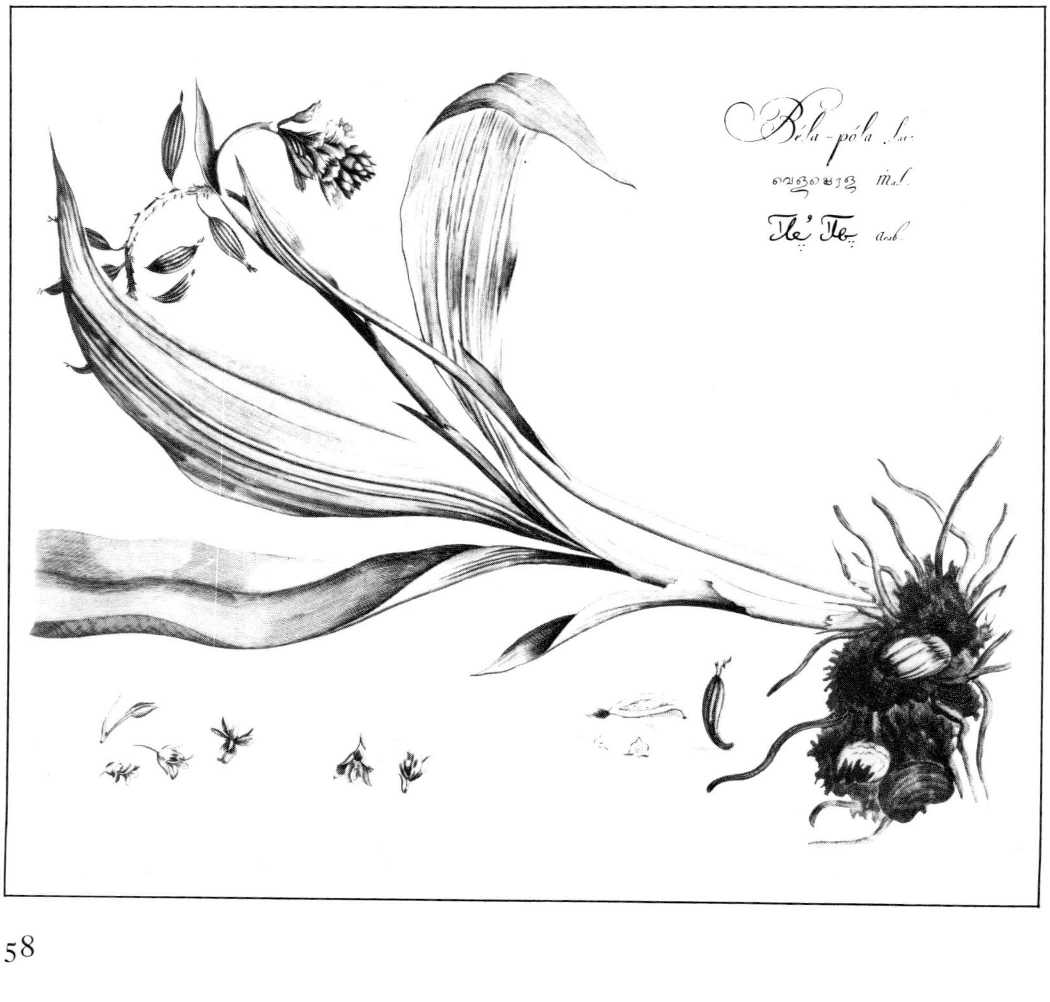

An orchid. Engraving from *Hortus Indicus Malabaricus.*

volumes of his botanical works were ready for publication, and nine more to follow, but Rumpf did not live to see them published. His zoological work. *Amboinsche Rariteitkamer*, was published in 1705, but it was not until 1741 that the twelve volumes of his *Herbarium Amboinense* appeared under the editorship of Johannes Burmann.

In France, the seventeenth century saw the establishment of a tradition of royal patronage for gardening and flower painting that extended unbroken for two hundred years to the most famous flower painter of all, Pierre-Joseph Redouté. The French court in the early 17th century was dominated by Marie de Médicis, who was married to Henri IV in 1600 and became Regent of France after his assassination in 1610. She brought with her from Florence the custom of providing patronage for natural history and science which was a feature of 16th-century Italy.

The first botanical painter to the court was Pierre Vallet, who produced the first important florilegium to be published, *Le Jardin du très Chrestien Henry IV, Roi de France et de Navarre, dédié à la Reyne* (1608). The garden was that of the Louvre which had been set up by the famous Parisian gardener, Jean Robin; it contained 'exotic plants which had been brought back from Guinea [western Africa] and Spain by Jean Robin the younger in 1603.' There is no text. The plates are very informal but botanically accurate in spite of its purpose, as a pattern book for embroidery. In addition to *Lilium canadense* and *Yucca gloriosa* from North America and the Jacobean lily from Mexico, there are such new arrivals as the belladonna lily from South Africa and *Haemanthus* from western Africa, and *Ramonda myconi*, a mauve-flowered alpine from the Pyrenees which may have been one of the introductions of Robin the Younger.

Vallet's successor as court painter was Daniel Rabel (1579–1637), who is equally well known for his designs for ballet; in his time, he was both

Two engravings from *Herbarium Amboinense* (1741–50) by George Eberhard Rumpf (Rumphius). *Below: Curcuma viridiflorum*, a ginger-like plant from southern India; the aromatic roots are used in curries. *Right: Cycas circinnalis*, a primitive palm-like plant found commonly in parts of southern India and Indonesia.

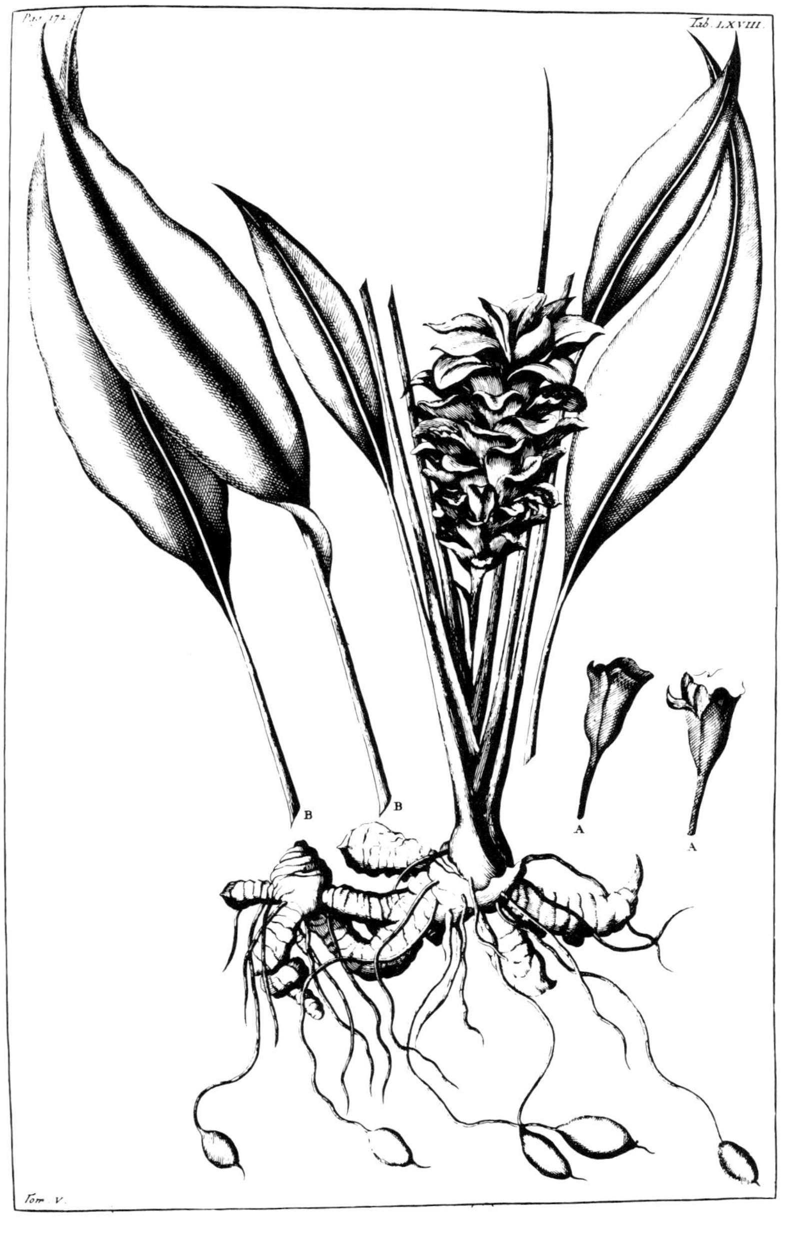

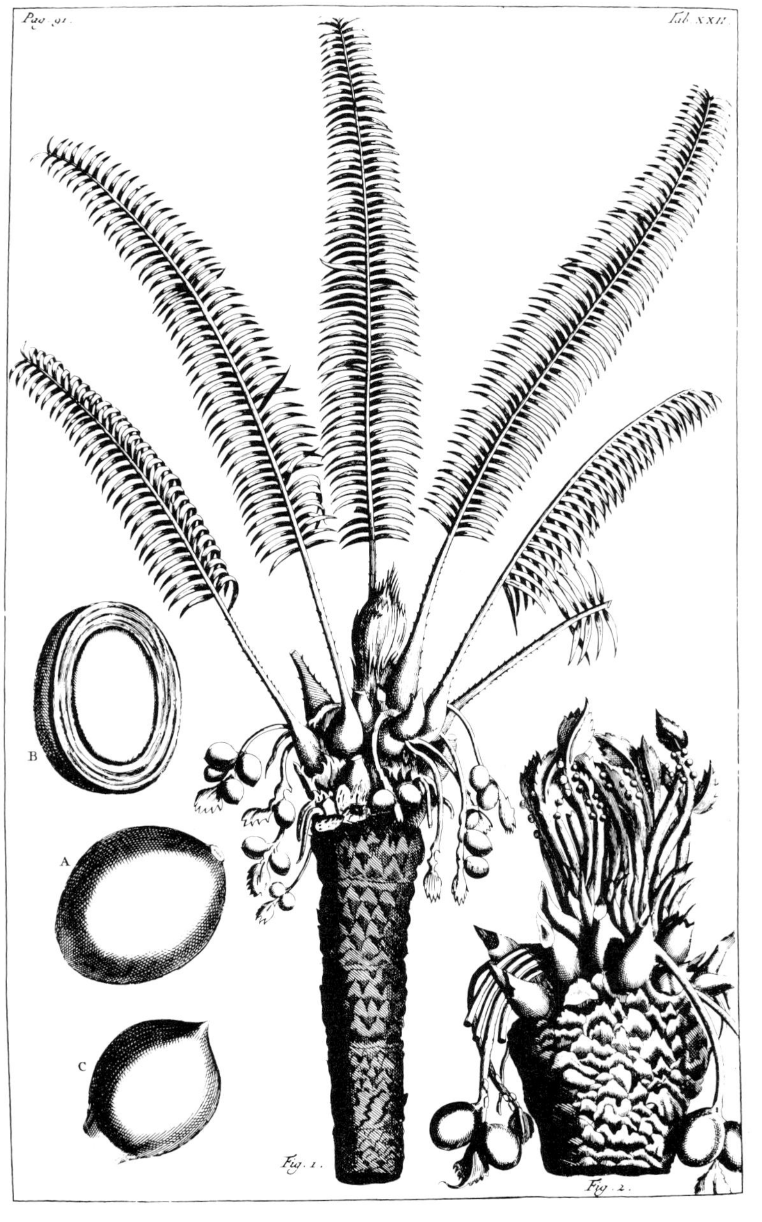

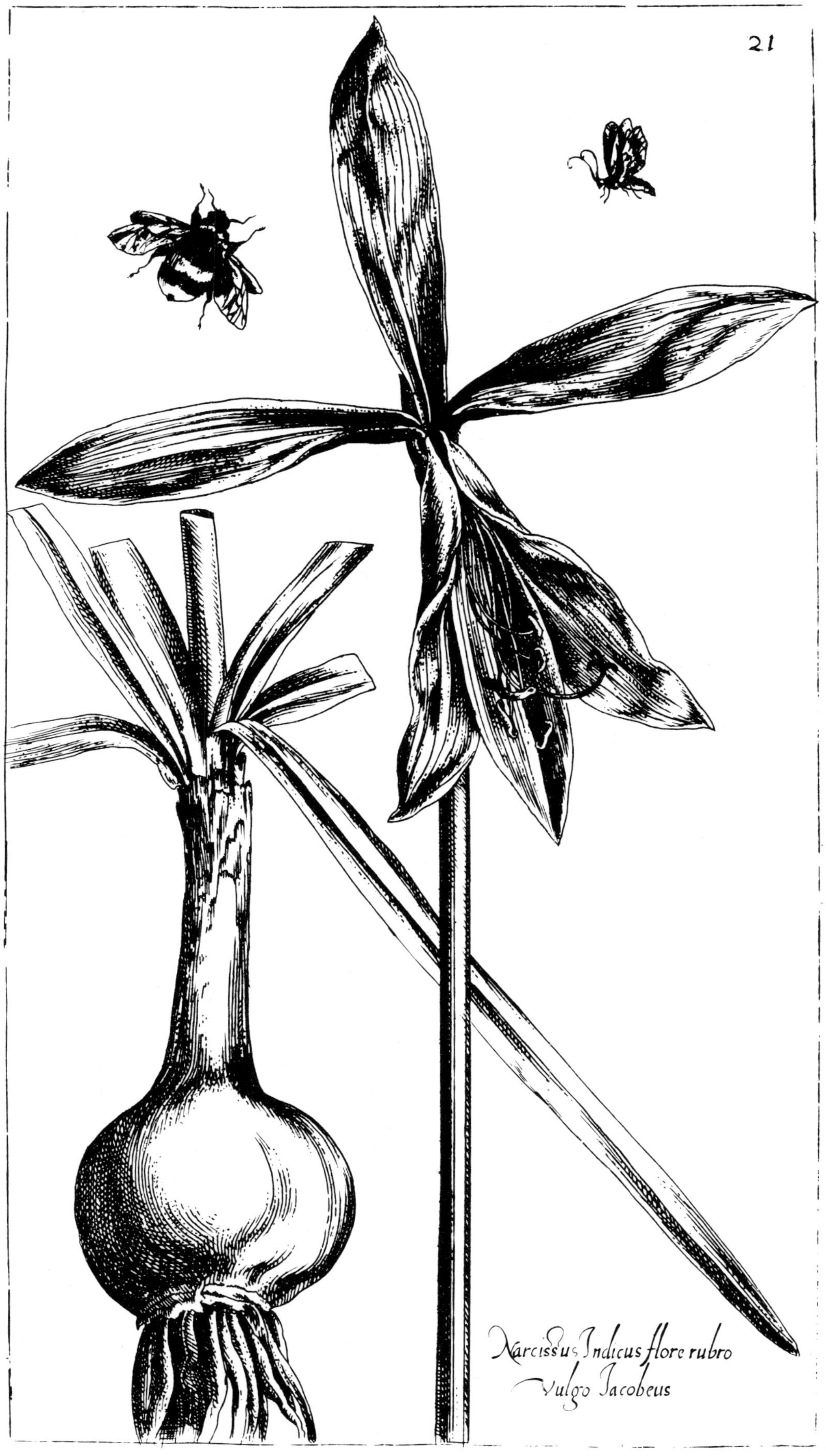

Sprekelia formosissima. This strange Mexican bulb was commonly called *'Narcissus indicus jacobaeus'*, a name given to it by Clusius because its flower reminded his friend, Dr Simon Tovar, of the crimson sword worn as a badge by the Knights of the Spanish Order of St James. Engraving by Pierre Vallet from his florilegium, *Le Jardin du très Chrestien Henry IV* (1608).

landscape painter and portraitist, in which rôle he was sent by Marie de Médicis to the Spanish court to produce a portrait of Anne of Austria, the fiancée of Louis XIII. His excellent botanical work survives in an album of paintings that is now in the Bibliothèque Nationale in Paris and is described by Wilfrid Blunt as 'one of the miracles of early flower painting'. Rabel's pictures were used in his *Theatrum Florae* of 1622, but although the plates are very finely engraved, the book is surprisingly similar to that of Vallet, with the same range of species and very few additions.

The career of Nicholas Robert (1614–85) had a romantic beginning when he illustrated the *Guirlande de Julie*, a manuscript book of flower pictures and poems which the Baron de St Maur commissioned to give to his fiancée, Julie d'Angennes, the beautiful daughter of Mme de Rambouillet, whose

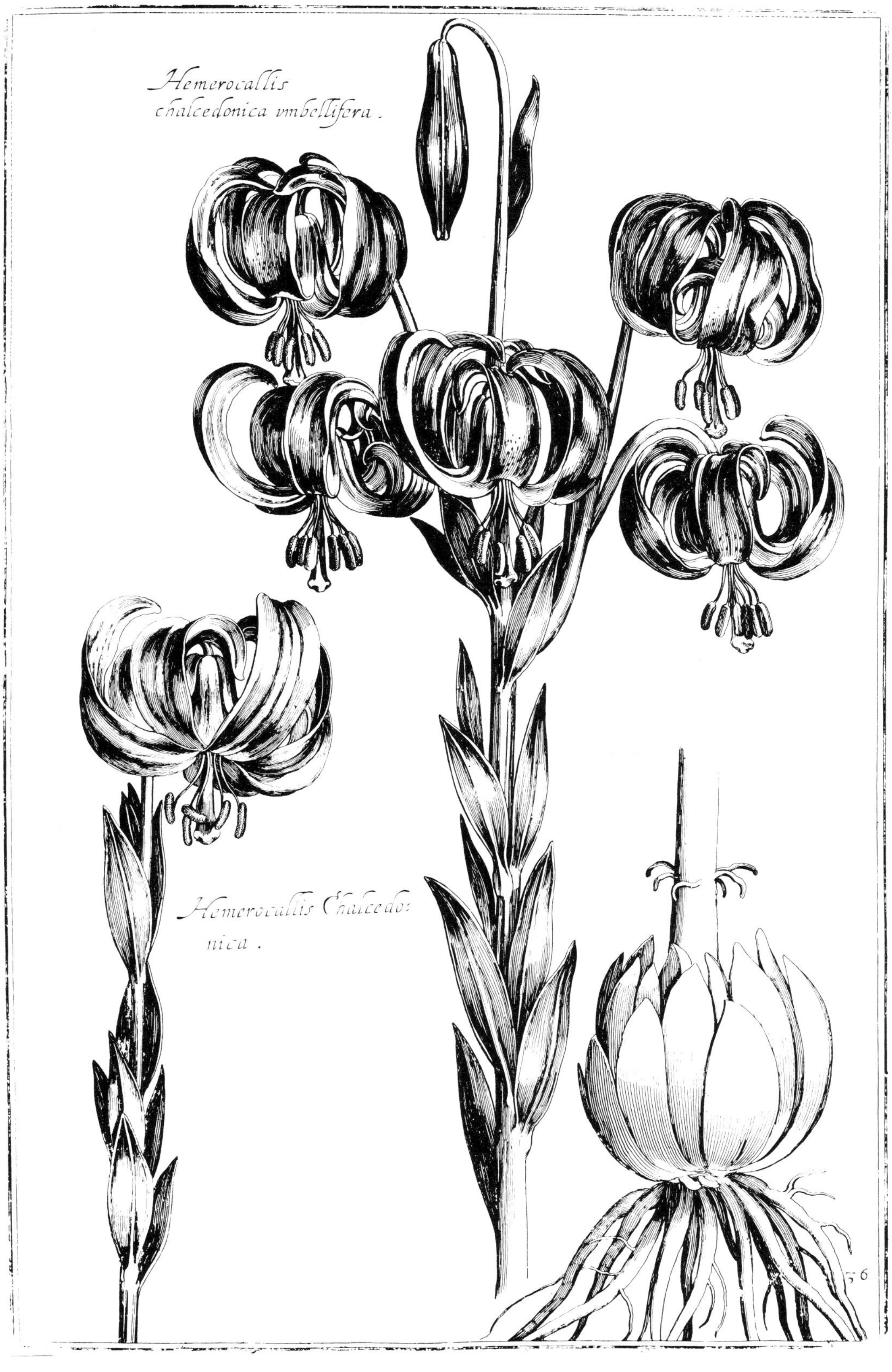

Lilium chalcedonicum and a single flowered species, *Lilium heldreichii*. These natives of Greece were introduced to the rest of Europe from Constantinople. Engraving from *Theatrum Florae* (1624) by Daniel Rabel.

salon was a centre of Parisian society. Royal patronage soon followed: Louis XIII's younger brother, Gaston d'Orléans, chose Robert to record the flowers in his collection at Blois. These paintings, together with those of Rabel, formed the starting point of the magnificent royal collection of paintings on vellum, the *vélins*, to which successive court painters contributed both animal and plant subjects. On the death of Gaston d'Orléans in 1660, Robert became court painter to Louis XIV and so continued to add to the *vélins*.

Robert had already published a small florilegium, *Fiori Diversi*, in Rome in 1640, and this was followed by his *Diverses Fleurs* in 1660. His masterpiece was to be the *Receuil des Plantes* with its huge plates of unparalleled quality – most were both drawn and engraved by Robert himself, but there are some by Abraham Bosse and by Louis de Châtillon. The text was to be by various members of the Académie Royale des Sciences, which was to be the publisher. By the time that the bulk of the plates were ready for the printer in 1692, the wars in which France had been involved since 1689 were resulting in financial strains, which delayed publication. The plates alone were issued in 1701, but the complete work did not appear until 1788, over a century after Robert's death.

Robert's successor, Jean Joubert, has not been accorded wide acclaim, and, although he continued to add to the *vélins*, none of his work seems to have been published. However, he did discover the young Claude Aubriet, who was to become the most famous botanical painter of the early eighteenth century. Claude Aubriet was born at Châlons-sur-Marne in 1665. After he arrived in Paris, he assisted Joubert in painting the *vélins* and worked for the great botanist Joseph Pitton de Tournefort (1656–1708), producing engravings for the influential *Eléments de Botanique* of 1694.

March lily (*Haemanthus coccineus*), a South African bulb of the family Amaryllidaceae. Engraving after a drawing by Nicolas Robert from *Mémoires pour Servir à l'Histoire des Plantes* by Dionys Dodart in the version that was printed in 1788, over a century after Robert's death.

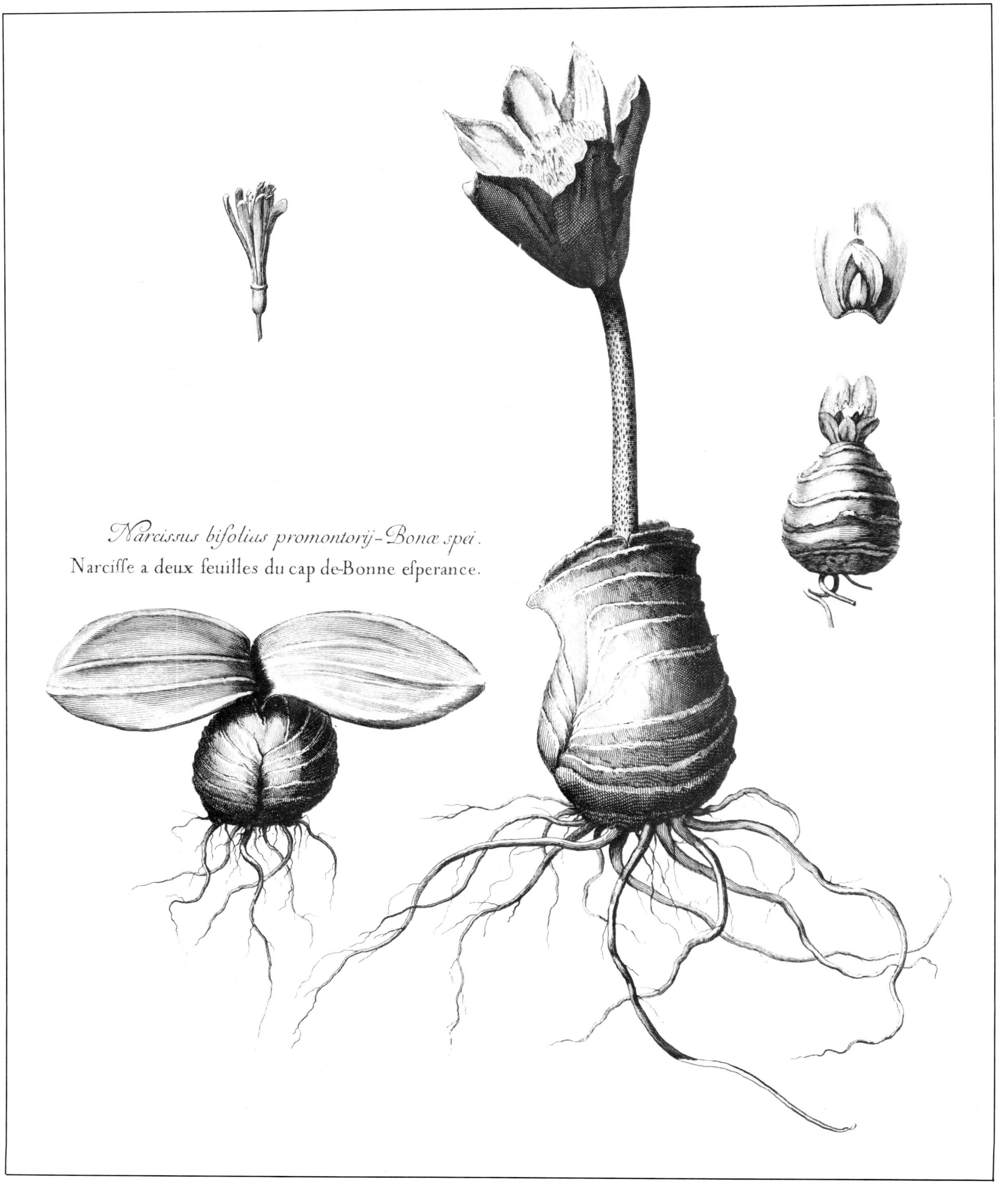

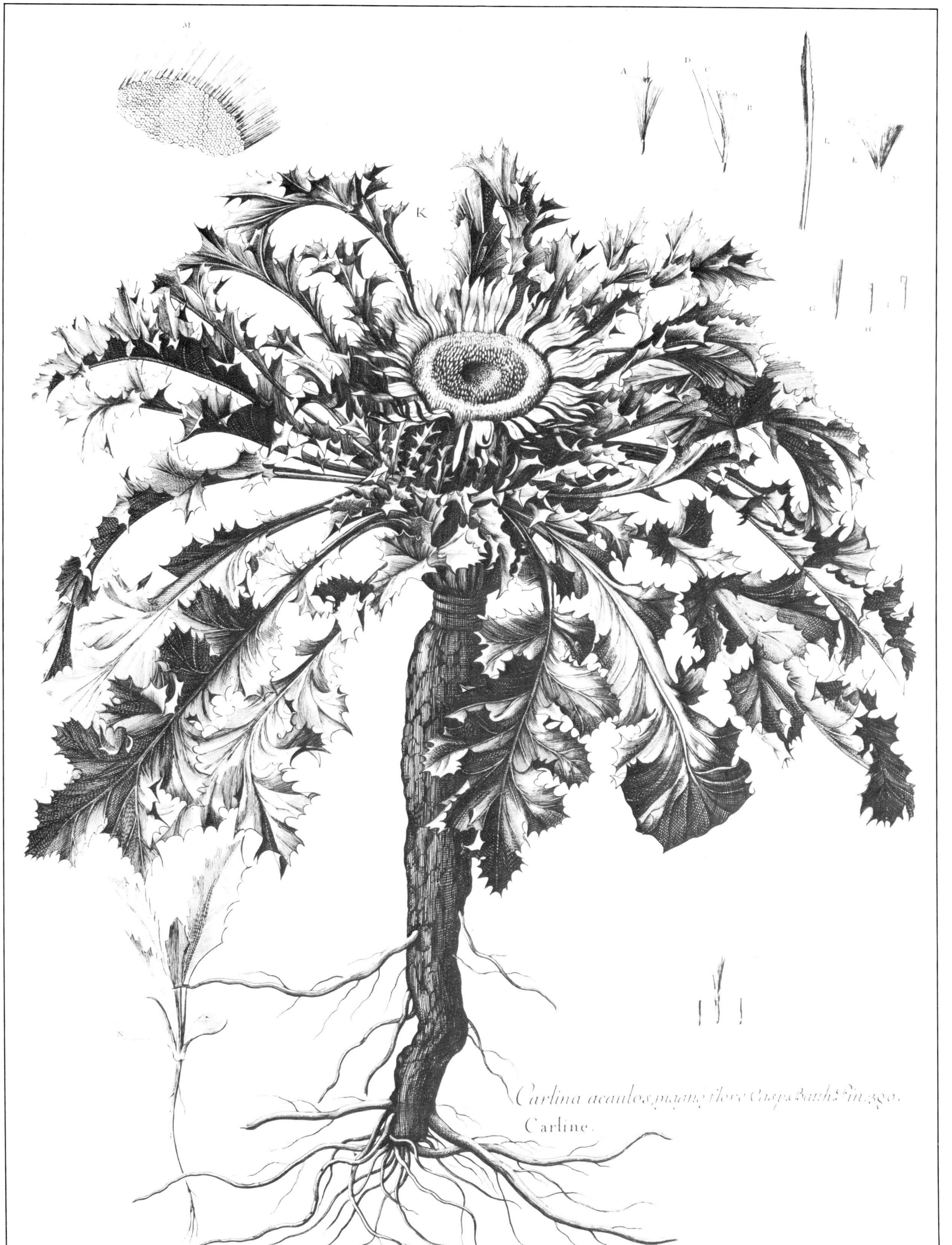

A carline thistle, *Carlina acaulis*. This striking thistle-like plant is found wild in the Alps and southern Europe. Engraving after a drawing by Nicolas Robert from *Mémoires pour Servir à l'Histoire des Plantes* (1788 edition).

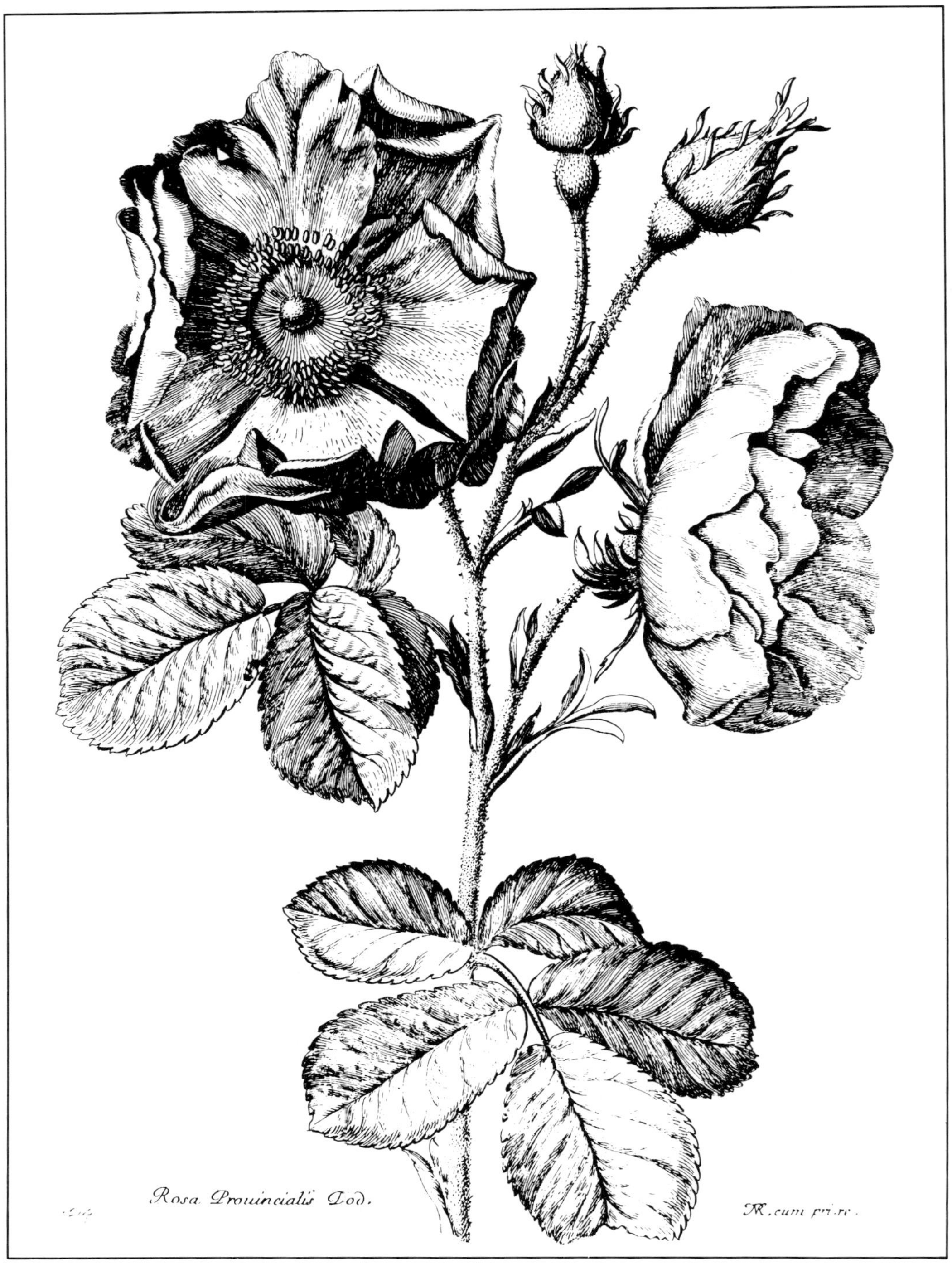

Rose of Provins (*Rosa gallica officinalis*), one of the oldest cultivated roses. Engraving by Nicolas Robert after his own drawing in the undated *Variae ac multiformes florum species appressae ad vivum et aeneis tabulis incisae*, which was printed around 1660.

When Tournefort, then Professor of Botany at the Jardin du Roi, was sent on a major expedition to the Levant, he took Aubriet with him as artist. Tournefort, accompanied by his friend Dr Andreas Gundelscheimer and Aubriet, sailed from Marseilles in 1700 and landed first at Canea in northwestern Crete. After three months spent exploring the island, they went on to visit as many of the Aegean islands as possible before wintering on Mykonos. In the spring, they took a Turkish ship to Lesbos and thence to Constantinople, where they prepared to visit the interior. They bought Armenian dress, as it was not wise to travel in European clothes, and were fortunate in finding a high Turkish official, a Pasha, travelling to Erzurum.

Although he already had a physician from Burgundy and a Provençal apothecary with him, the Pasha welcomed two more doctors to look after his family on the arduous journey. The party started by boat along the Black Sea coast and reached Trebizond (now Trabzon) on 23rd May. The three botanists visited the famous monastery of Sumela, which was perched on the cliffs and reached only by a most primitive ladder. Tournefort was so impressed by the rich forests as to write that he could gladly have spent the rest of his days there. He records having found the purple *Rhododendron ponticum* and the yellow azalea (*Rhododendron luteum*) in flower. Tournefort picked a large bunch of the yellow azalea for the Pasha, but discovered that its smell caused headaches. They left Trebizond at the beginning of June with the Pasha and various merchants, in all about six hundred, with many donkeys, horses and camels. The caravan climbed slowly up into the snows

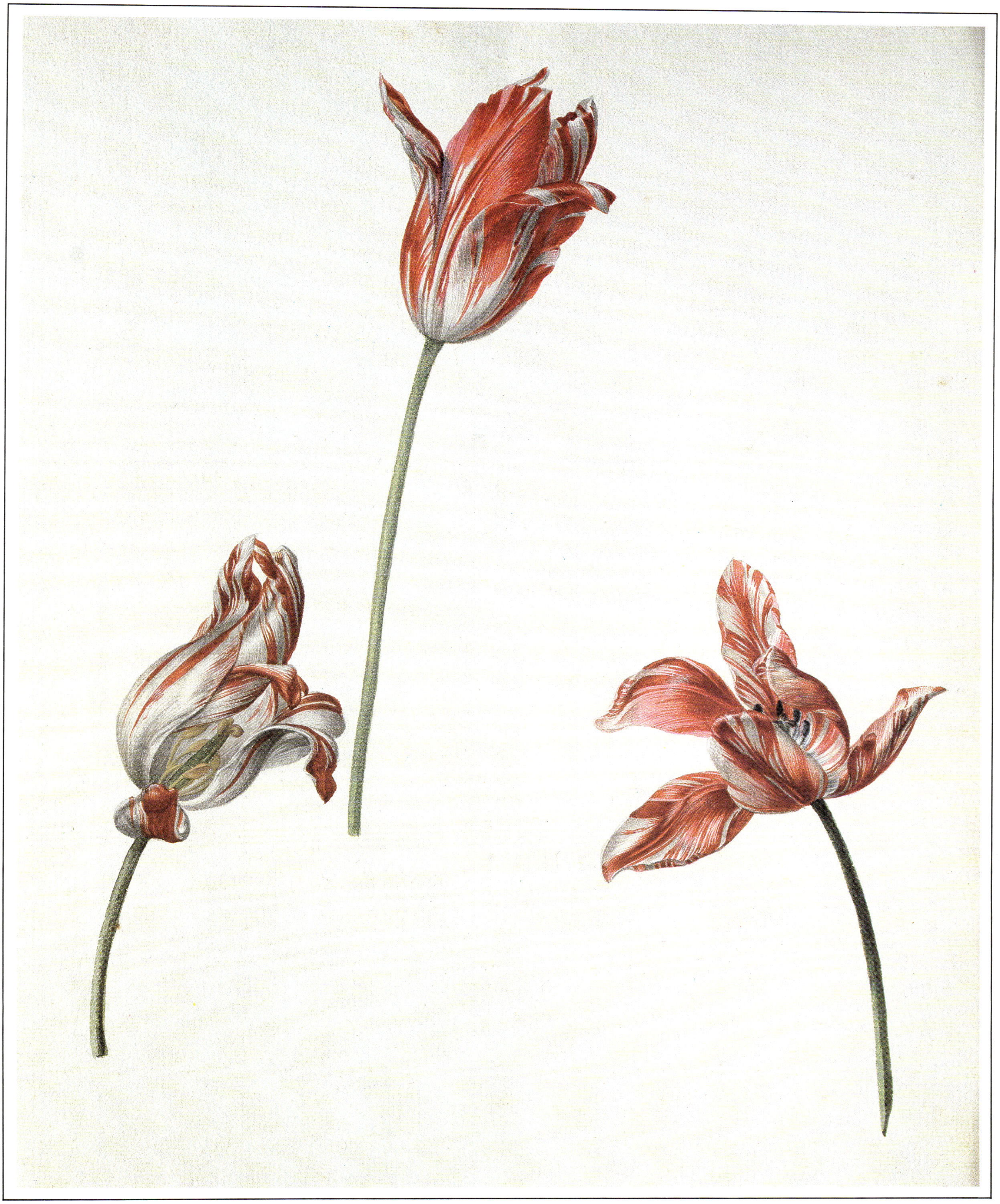

Broken tulips (*Tulipa sp.*), opaque watercolour on vellum, by Peter van Koowenhoorn, around 1630, when tulipomania in Holland was approaching its height. Lindley Library, Royal Horticultural Society, London.

of the Pontic mountains and after two days at high altitude they descended into Anatolia. 'From Trebizond hither,' wrote Tournefort in his *Voyage into the Levant* (published in 1718 in French in Amsterdam and in English in London), 'the country looked like the Alps and Pyrenees, but now the face of the earth seemed of a sudden altered, as if a curtain had been drawn, and a new prospect opened to our view. We descended into little valleys covered with verdure, intermixed with charming streams, and full of so many fine plants, so different from what we had been used to, that we knew not which to fall on first.'

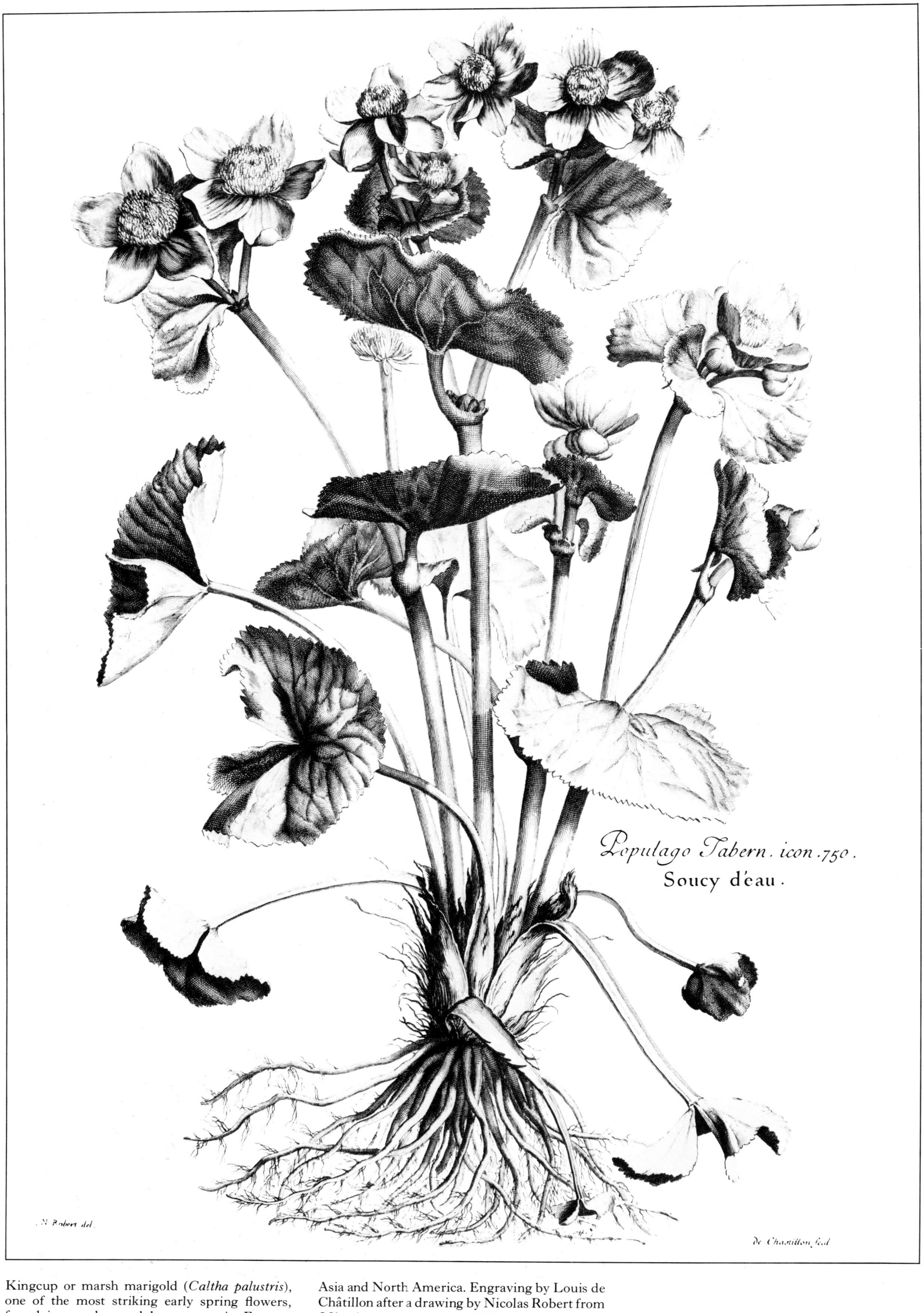

Kingcup or marsh marigold (*Caltha palustris*), one of the most striking early spring flowers, found in marshes and by streams in Europe, Asia and North America. Engraving by Louis de Châtillon after a drawing by Nicolas Robert from *Mémoires pour servir à l'Histoire des Plantes*.

Trebizond (today Trabzon in northeastern Turkey). It was here that the overland caravan route to Persia started, and the town is still an important port. Engraving from *Voyage au Levant* (1718) by Joseph Pitton de Tournefort.

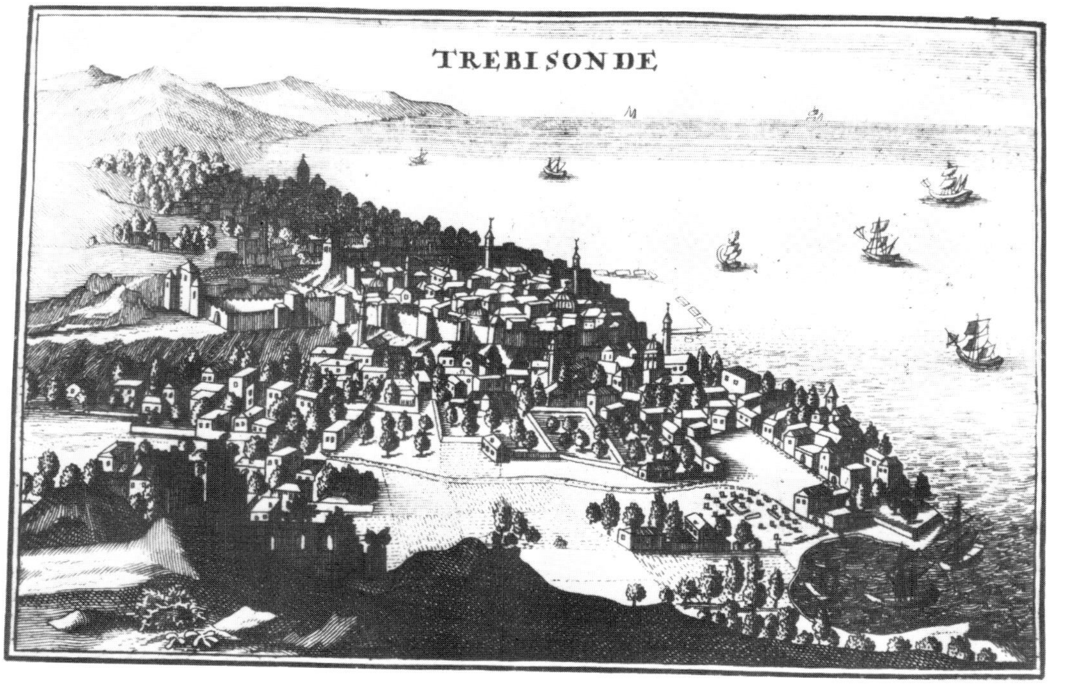

Their route took them past Bayburt, where they remarked on cushions of a vetch, *Onybrychis orientalis*, two or three feet in circumference, and discovered what Tournefort enthusiastically called 'one of the finest plants in the whole Levant'. It was named *Gundelia* in honour of Dr Gundesscheimer, who was the first to see it. It is most like a very green sea holly with reddish-purple flower heads and succulent young shoots that are still much eaten by shepherd boys.

The caravan mainly moved by day, allowing the botanists plenty of time to collect, but, when the moon was full, it 'shone so bright, that it invited the Turks, who had done nothing but snore the live-long day, to prosecute their journey: but how could we simple [i.e. collect simples, or herbs] by moonlight? We, however, omitted not to fill our bags, our merchants laughing all the while to see us three groping about . . .'

They reached Erzurum on 15th June and were surprised to find it so cold, with patches of snow still on the surrounding hills. Erzurum was then primarily an Armenian town, a resting place on the road from Persia and Arabia to Trebizond, deriving most of its wealth from taxing the merchandise that passed through. Silks from the East, rhubarb from Uzbekistan, and much caviar from the Caspian, which Tournefort considered 'a most odious dish [which] burns the mouth with its high seasoning, and poisons the nose with its nasty smell.'

The Kurds lived in the mountains around the city, ready to prey on caravans and unprotected travellers. Tournefort and his companions

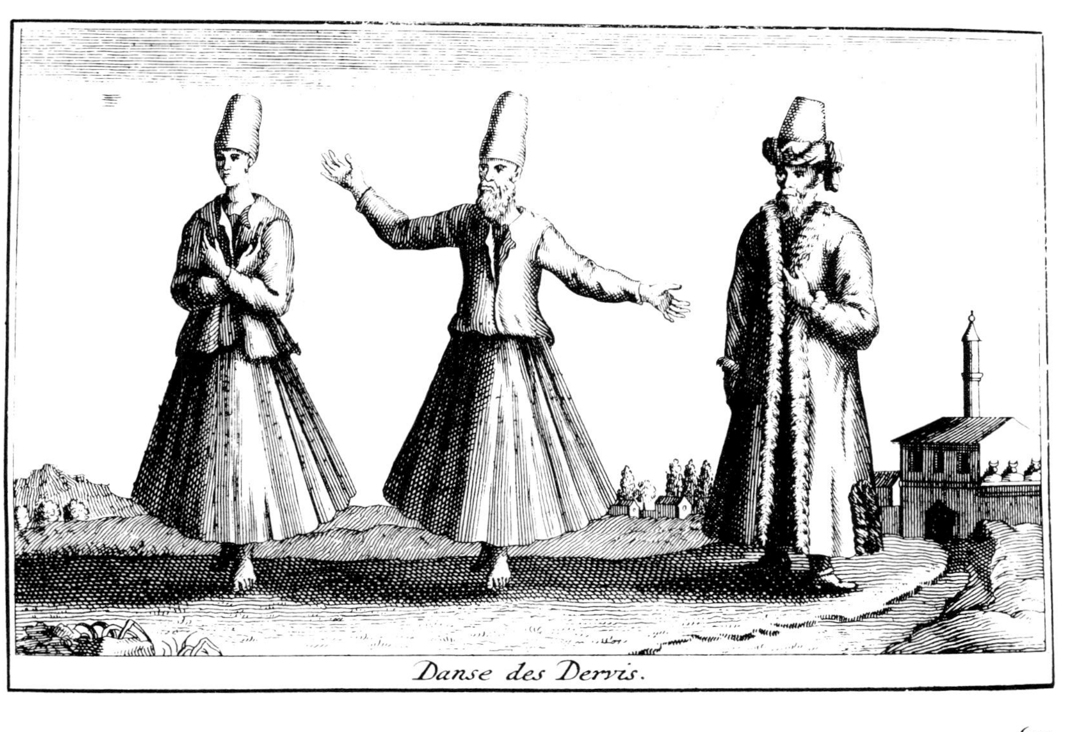

Whirling dervishes, an Islamic monastic order, whose headquarters was, and still is, Konya, the ancient city of Iconium, in Turkey. Engraving from *Voyage au Levant* (1718) by Joseph Pitton de Tournefort.

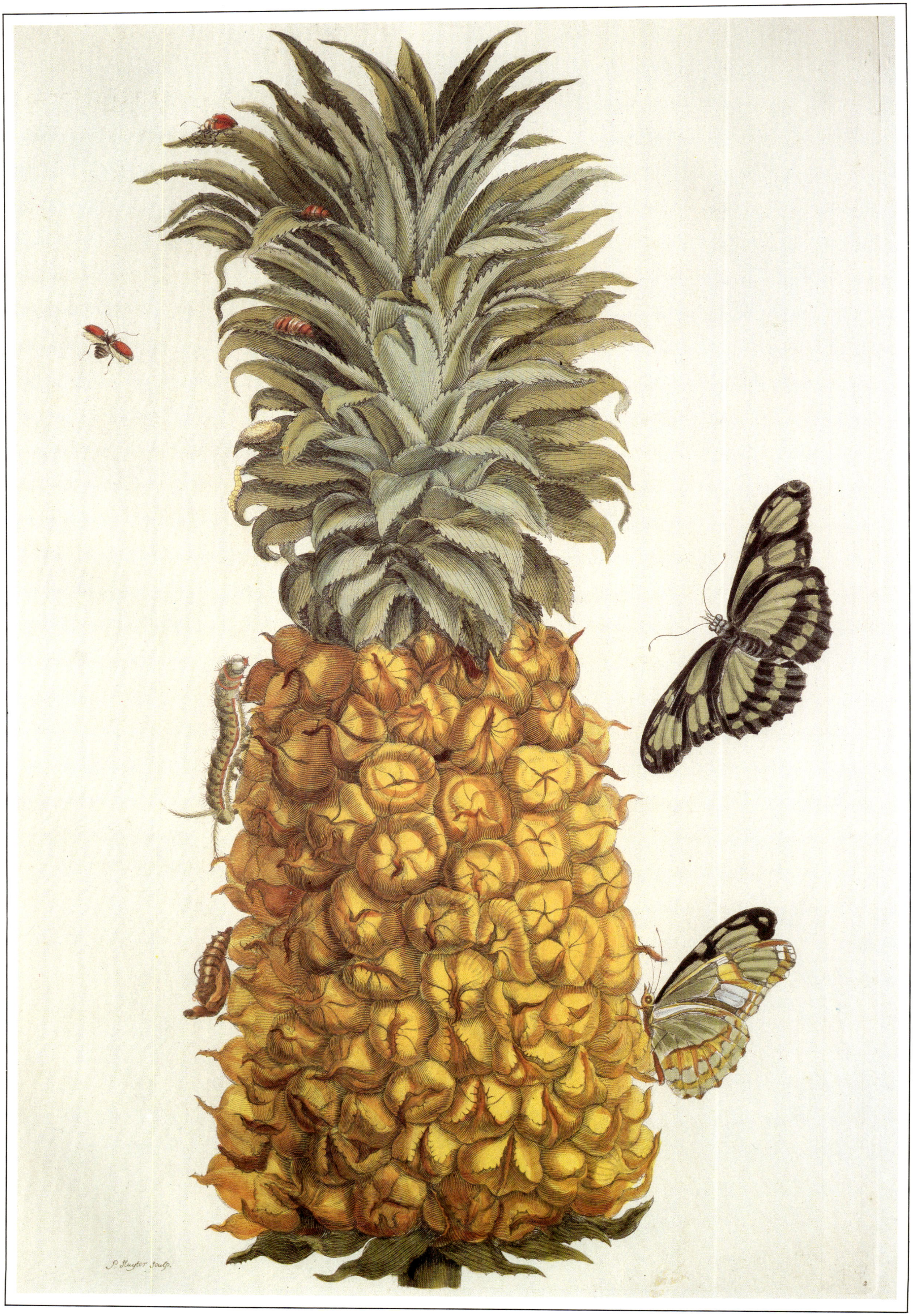

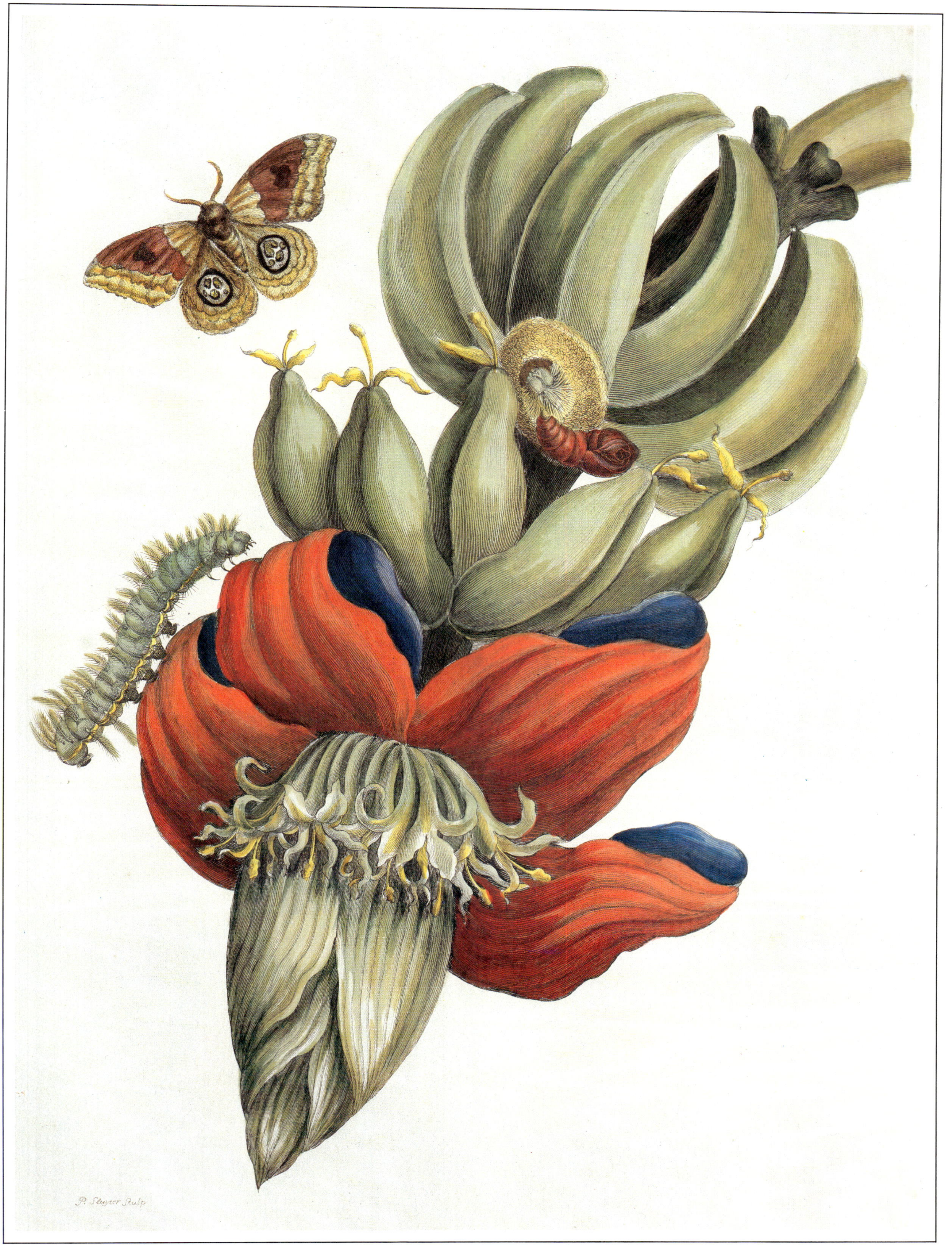

Pineapple (*Ananas comosus*). Hand-coloured engraving by P. Sluyter from *De Metamorphosibus Insectorum Surinamensium* (1714 edition) by Maria Sibylle Merian.

Inflorescence of banana (*Musa × sapientum*) with male flowers surrounded by conspicuous red bracts and female flowers which develop into bananas without fertilisation. Hand-coloured engraving by P. Sluyter from *De Metamorphosibus Insectorum Surinamensium* (1714 edition) by Maria Sibylle Merian.

visited an Armenian monastery in the hills to the east of Erzurum and they managed to persuade the bishop there to take them to see the sources of the Euphrates, as he was friendly with the local Kurds, an expedition for which they put on their very worst clothes, and gave their money to the bishop for safe keeping. They dined on freshly caught trout from the streams, and drank much wine to warm them and 'wash away the terror, which the dreadful name of Kurds had, notwithstanding all our care, struck upon our spirits,' expecting at any moment to be attacked and stripped. ''Twas a great way from thence to the monastery to go in one's shirt, and who knows but these People, who are used to making of Eunuchs, might have taken it into their heads to have metamorphosed us in the same manner, that we might be sold to more advantage.' But the bishop protected them and presented the Kurds with a bottle of brandy. Here they collected seed of the oriental poppy and of *Morina orientalis*, which they named in honour of M. Morin, who raised the seed which they sent back. Near Kars, they found the bellflower that Tournefort called *'Campanula orientalis'* but is now known as *Campanula crispa*.

From Erzurum, they joined another caravan bound for Tiflis (Tbilisi) in Georgia and on the way visited Yerevan and Three Churches (today Echmiadzin), where they kissed the Armenian patriarch's ring, hoping that he would help them ascend Mount Ararat. They found Ararat very dusty, and were disappointed to find so many familiar plants: they found two new ones, however, a large campion and a white-flowered 'Geum' (*Saxifraga sibirica*). The mountain was inhabited only by a few shepherds with scrawny flocks, and by tigers. When they saw two tigers, 'we laid ourselves along upon the sand, and let them pass by very respectfully.'

The travellers then went back to Erzurum. After passing through Tokat and Ankara, they reached Smyrna (Izmir) on 18th December 1701. They visited Ephesus and Samos before they finally returned home via Italy. Landing at Leghorn, they had to spend four days in the quarantine station

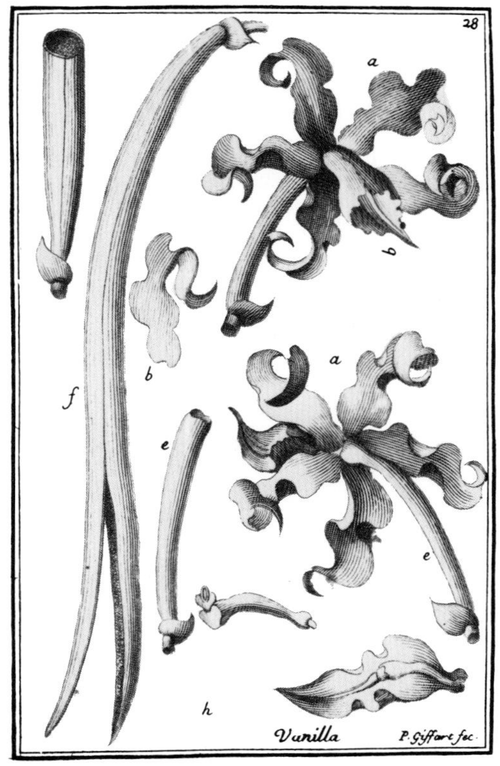

Flowers and fruit of vanilla (*Vanilla planifolia*), an orchid. Engraving by P. Giffort after a drawing by Charles Plumier in Plumier's *Nova Plantarum Americanarum Genera* (1703).

Below:
Two engravings after drawings by Claude Aubriet from *Voyage au Levant* (1718) by Joseph Pitton de Tournefort: *Aristolochia hirta*, found by Tournefort on the island of Chios, also occurs in Turkey and Cyprus; *'Campanula orientalis'*.

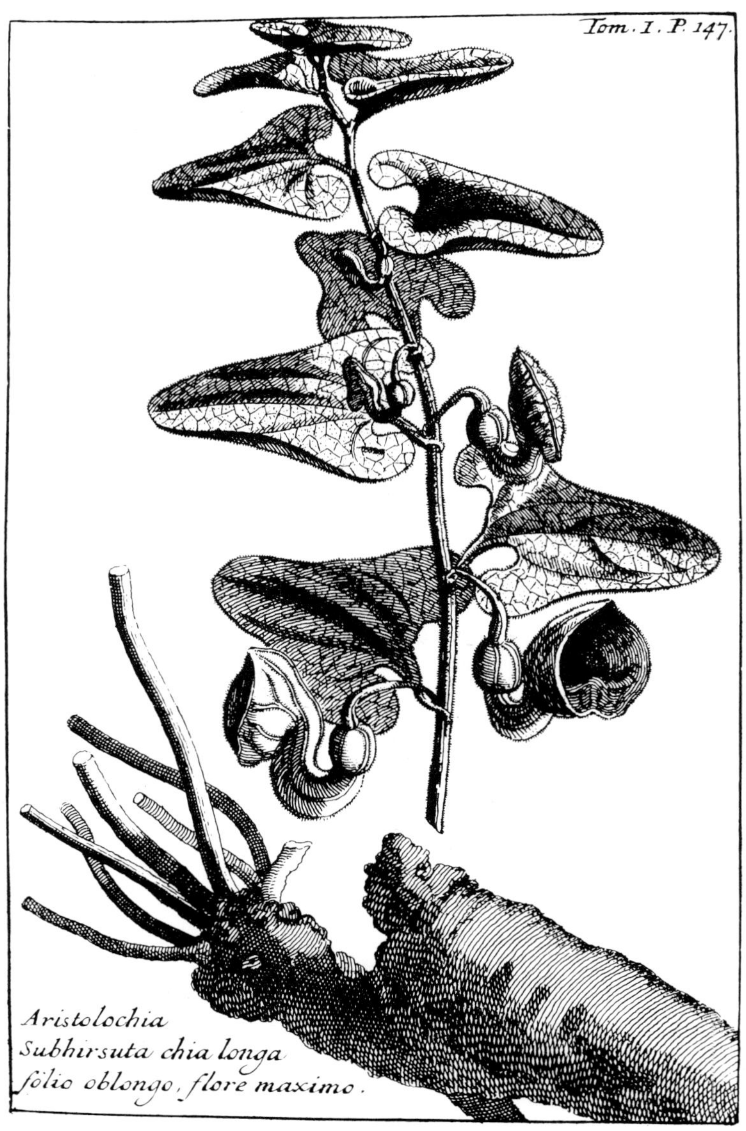

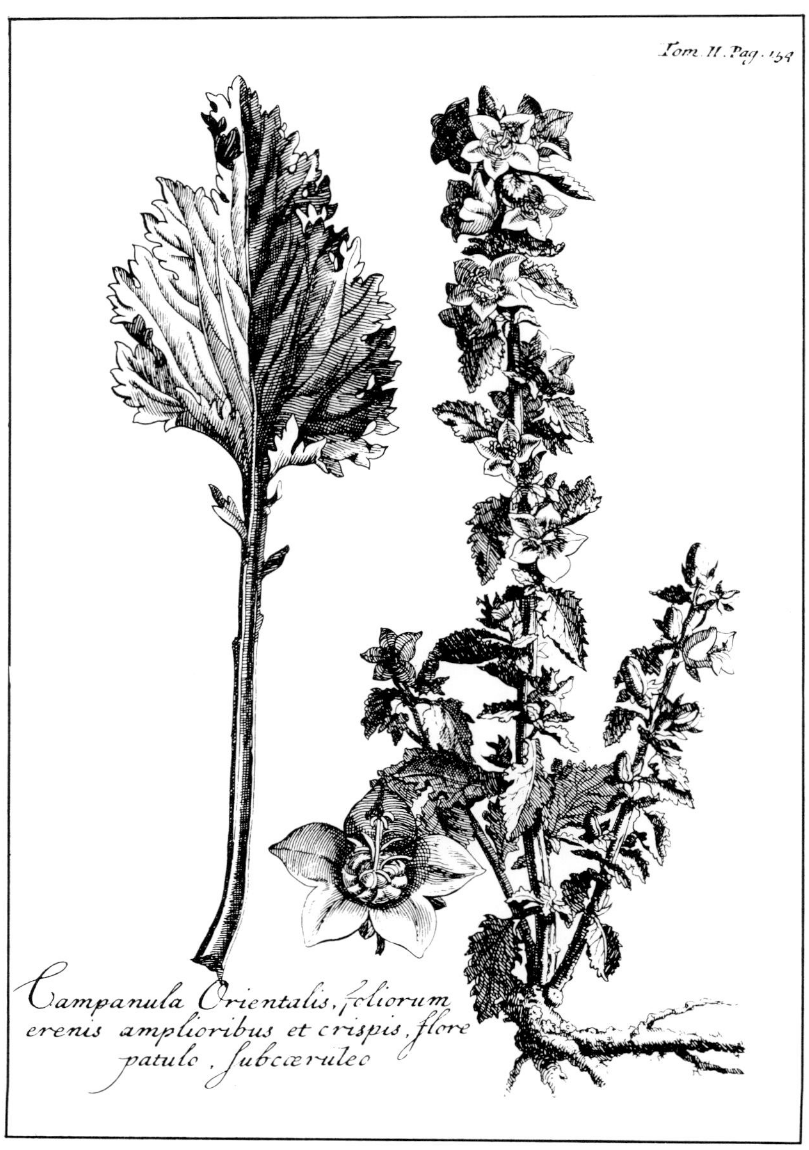

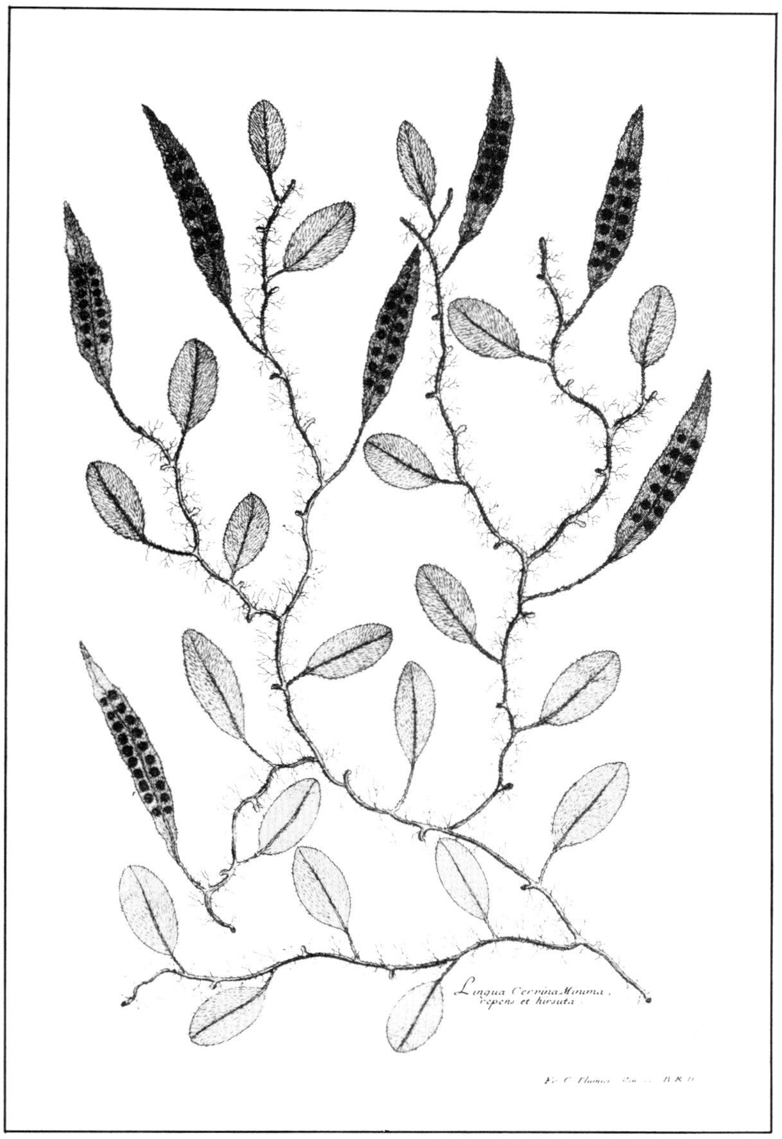
Lingua Cervina Minima,
repens et hirsuta

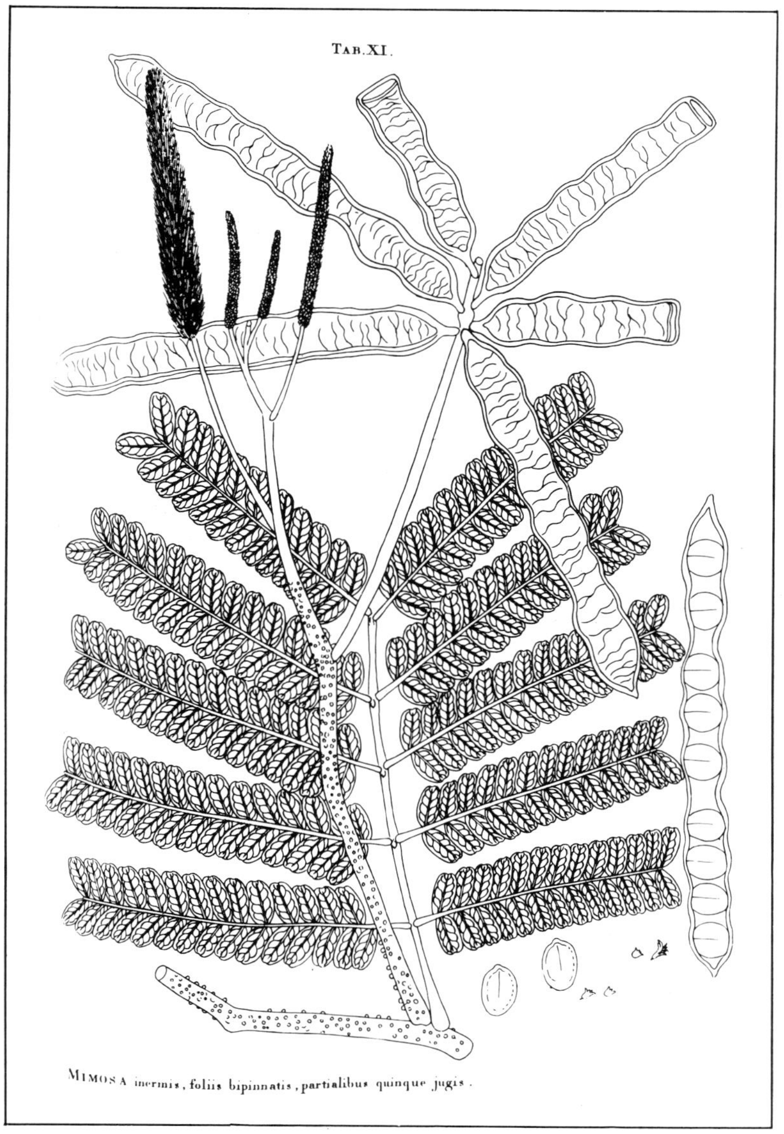
TAB. XI.
MIMOSA inermis, foliis bipinnatis, partialibus quinque jugis.

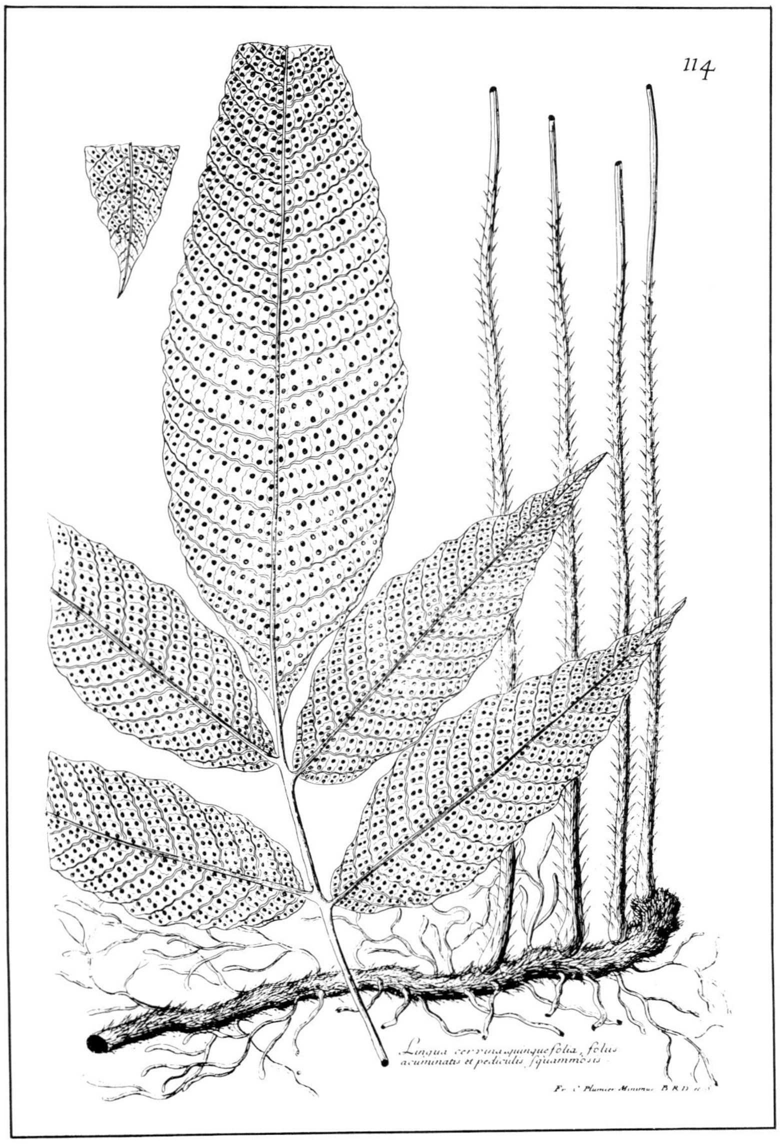
114
Lingua cervina quinquefolia, foliis
acuminatis et pediculis squammosis
Fr. C. Plumier Minimus B.R.D. et S.

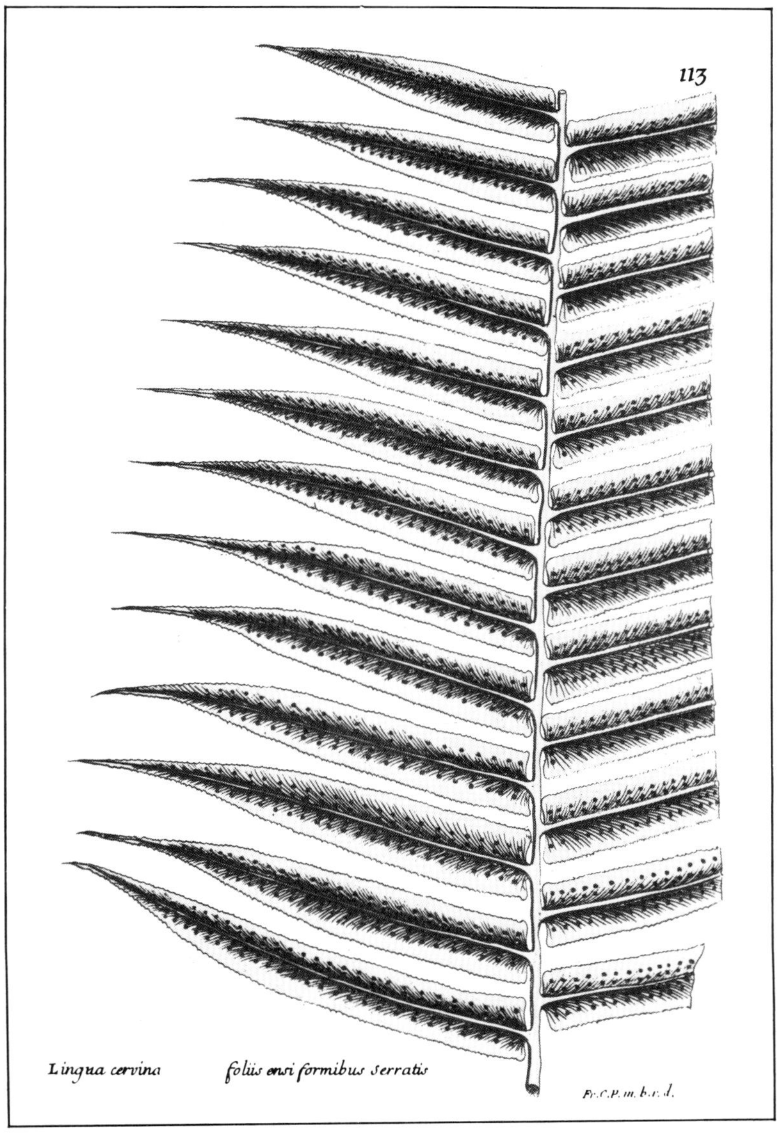
113
Lingua cervina foliis ensi formibus serratis
Fr. C.P. m. b.r.d.

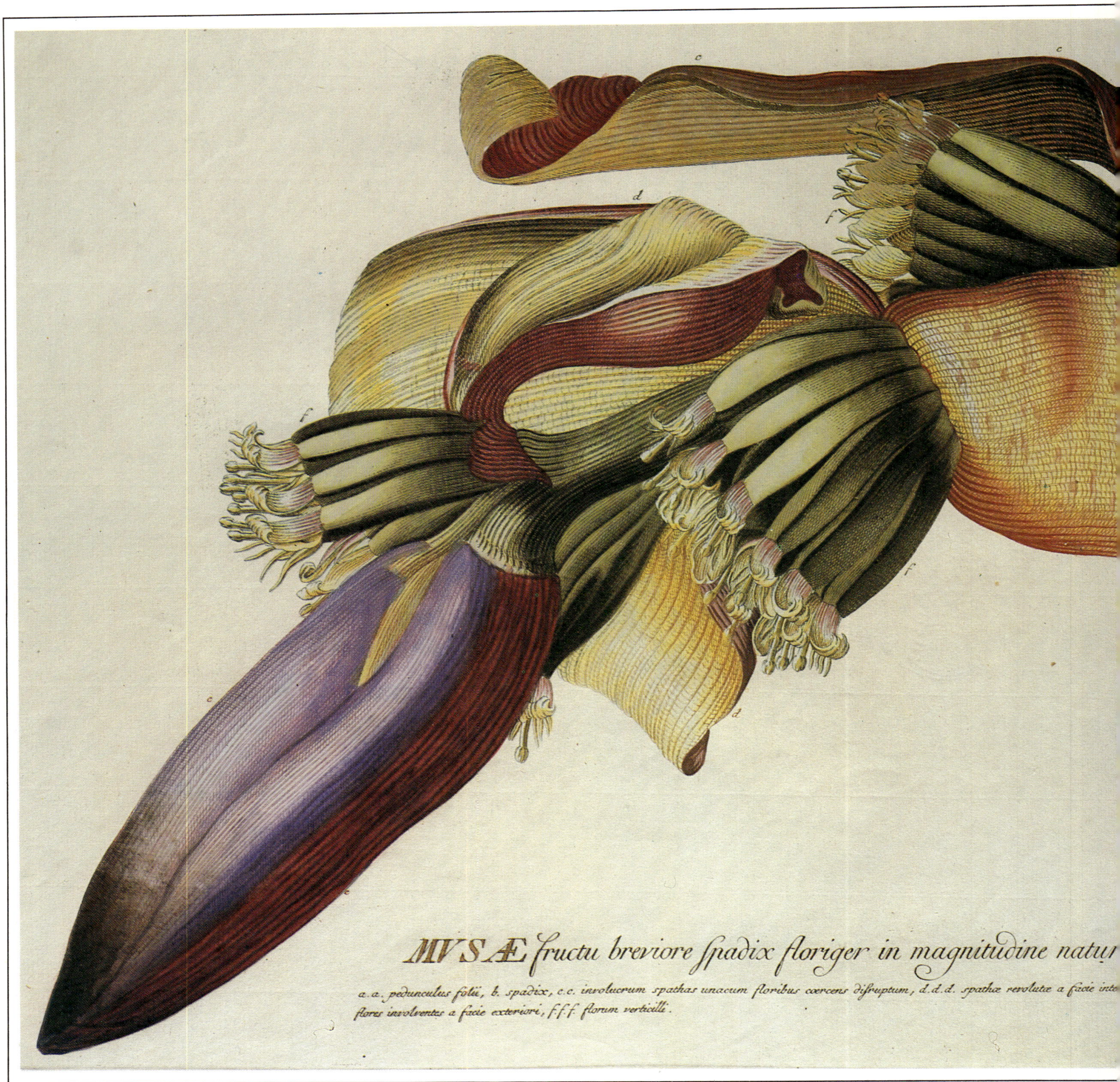

at San Lazzaro, one of the precautions against bubonic plague, which was endemic in Turkey and still a scourge in the cities of Italy. A felucca carried them on the last stage of their journey back to Marseilles.

The leading botanist in England in the seventeenth century was John Ray. Although he did not travel as far as Tournefort, he knew many of his European contemporaries and translated some of their works into English. Ray was born in 1627, the son of a blacksmith in the small village of Black Notley in Essex. He went up to Cambridge University in 1644, first to Catherine Hall and later to Trinity where he became a fellow in 1653.

Most of Ray's early botanical work was around Cambridge, where he was one of a small group who became interested in 'natural philosophy' and started to practise experimental science. He published his first book in 1660, a catalogue of plants growing around Cambridge; many of the rarities he mentioned can be found in the same places today. In 1622, Ray left Cambridge and travelled widely, first in the British Isles and later in Europe, where he visited many of the foremost botanists of the time. From 1664 to 1666, he was mostly in France and Italy, going as far south as Sicily and

Previous page:
Top right: Acacia nudiflora, a tree from Antigua, Haiti and St Thomas. Engraving after a drawing by Charles Plumier from the posthumous *Plantarum Americanarum . . .* (1755–60), edited by Johannes Burman. The other three engravings are by Plumier himself from his *Traité des Fougères de l'Amérique* (1705) and include *Polypodium piloselloides* (*top left*) from the West Indies.

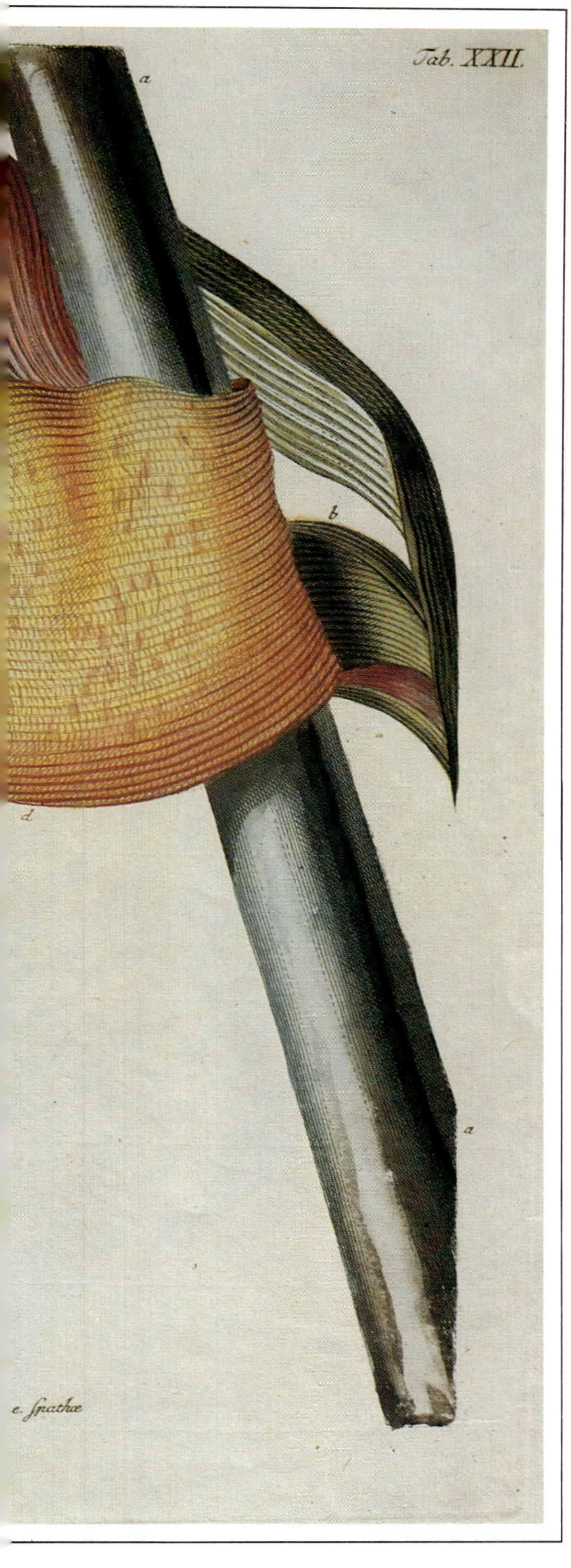

Inflorescence of banana (*Musa sp.*). Hand-coloured engraving after a painting by G. D. Ehret, from C. J. Trew's *Plantae Selectae* (1750–73).

Overleaf:
Aspidium triangulum, a fern of mountain rocks in Jamaica. Engraving by Charles Plumier after his own drawing from *Traité des Fougères de l'Amérique* (1705).

Page 75:
A storksbill, *'Erodium incarnatum'*. Engraving by Baron after a drawing by James Sowerby from *Geraniologia* (1791) by Charles-Louis l'Héritier de Brutelle.

Malta. His catalogue of the English flora was published in 1670, and later works include books on mammals, reptiles, birds, fishes and insects which were no less influential in the development of zoology. Ray's particular contribution foreshadowed that of Linnaeus a hundred years later: he brought order to the confusion of synonyms and vague descriptions that filled earlier botanical books. Ray died in 1704, and it was not until after the great stimulus of Linnaeus's *Species Plantarum* in 1753 that there were further significant developments made in British botany and finely illustrated botanical books began to be published regularly in England.

No less famous than Tournefort and Ray, but a specialist in the Americas was the priest Charles Plumier (1666–1706). He visited the West Indies three times, in 1689, 1693, and 1695 and published four very important works, of great significance in the development of botany. *Descriptions des Plantes de l'Amérique* was published in 1693, and *Nova Plantarum Americanarum Genera* in 1703; from this, Linnaeus took many generic names that are still in use today, among them *Fuchsia*, (after Fuchs), *Caesalpinia* and *Magnolia*. Plumier provided a concise generic description, with detailed drawings and dissections of each flower which contributed greatly to the scientific value of his work. Plumier's *Traité des Fougères de l'Amérique* (1705) is a large work and one of the first to be devoted exclusively to ferns. It has 170 excellent plates drawn and engraved by Plumier himself; each plant is described with ecological notes and details of locality. Finally, after his early death when *en route* for Peru, the *Plantarum Americanarum . . .* appeared between 1755 and 1760 under the editorship of Johannes Burmann and contained, as the full title says, 'plants which Plumier himself discovered, and himself painted in the Antilles.' It was Plumier's example which fired the enthusiasm of Maria Merian, and encouraged her to set out for Surinam in 1698.

Maria Sibylle Merian (1647–1717) had a remarkable career. Her father, Matthäus Merian the Elder, a Swiss engraver, was the son-in-law and successor in business of Johann Theodor de Bry, whose *Florilegium Novum* of 1611 he re-issued in an enlarged edition in 1641. Matthäus died around 1650, and Maria's mother married a Dutch flower painter, Jacob Marrell. Maria worked for various of her stepfather's pupils and married one of them, Johann Graff of Nuremberg. From her first published work (on European insects) in 1679, she was helped by her younger daughter, Dorothea. In 1680, she produced *Neues Blumen Buch*, with hand painted engravings of garden flowers taken from Robert's *Diverses Fleurs* and designed to be used as models for embroidery. In 1685, Maria left her husband after joining the Labadists, a protestant sect that did not recognise marriage with outsiders as binding, and she took her two daughters to live in the Labadists' community house in Friesland. There she found a collection of insects from Surinam, which together with the example of Charles Plumier, inspired her to visit Surinam herself.

Setting sail in 1698, Maria and Dorothea spent two years collecting and drawing plants and insects. The result is her most famous book, *De Metamorphosibus Insectorum Surinamensium*, published in Amsterdam in 1705; an enlarged edition appeared in 1714 with extra plates based on material provided by Maria's elder daughter, Johanna, who settled in Surinam. The book has a style very much of its own with strange and sometimes mythical insects and spiders on and around the plants on which they are usually found – plants which can themselves have certain mythical qualities, like the cashew which has the nut and the fruit attached to the stalk in the wrong order. The engraving uses shading made up of a large number of very finely drawn parallel lines, which gives the plates a rather heavy appearance when they are coloured. However, the boldness of the design and the confidence with which it is executed ensure the success of the illustrations, though more as works of art than works of science. In this respect, Maria Merian's book looks back to the seventeenth century rather than forward to the eighteenth, when accuracy was very much more at a premium in botanical art.

72

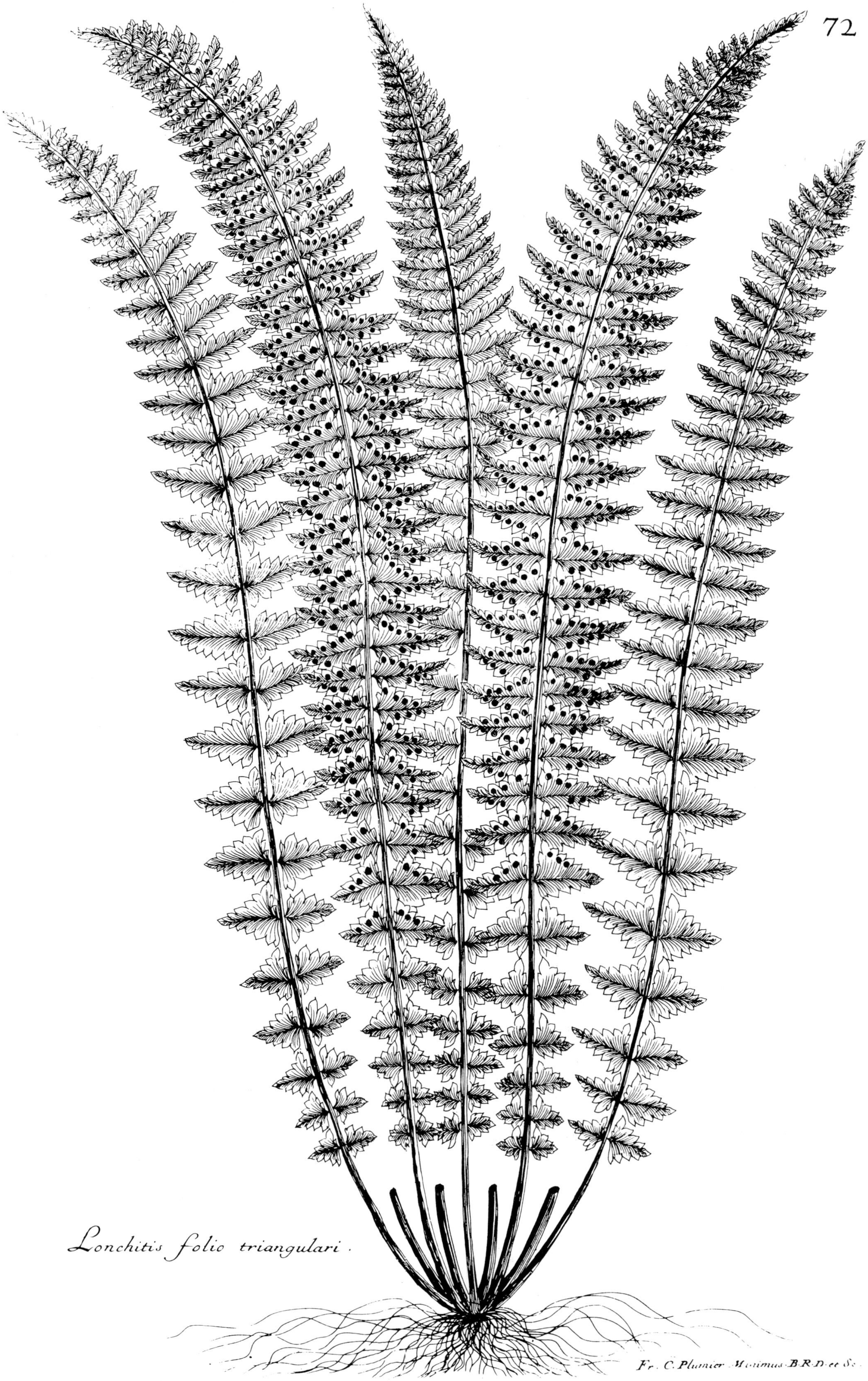

Lonchitis folio triangulari.

Fr. C. Plumier Minimus B.R.D. et Sc.

PART TWO

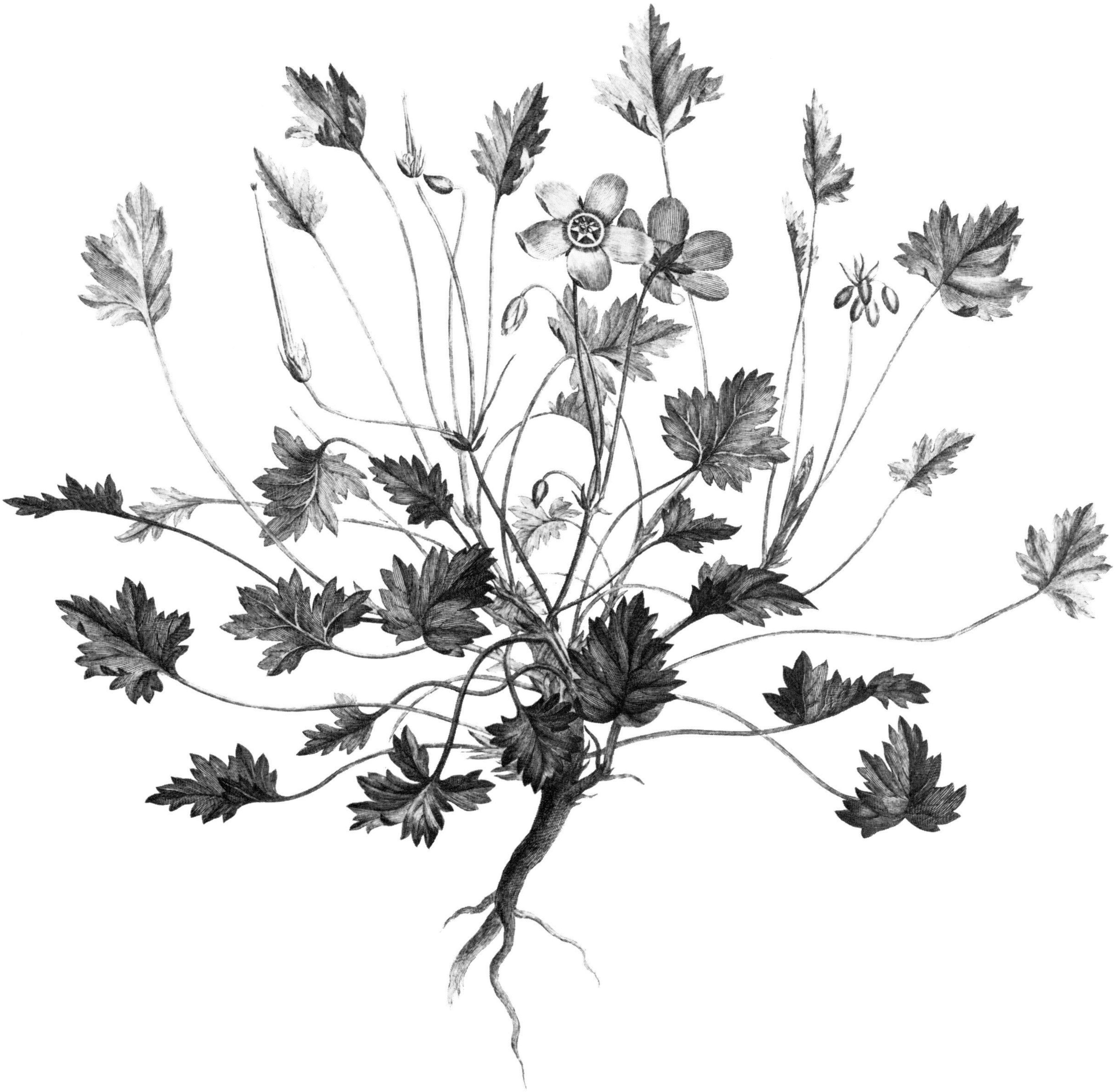

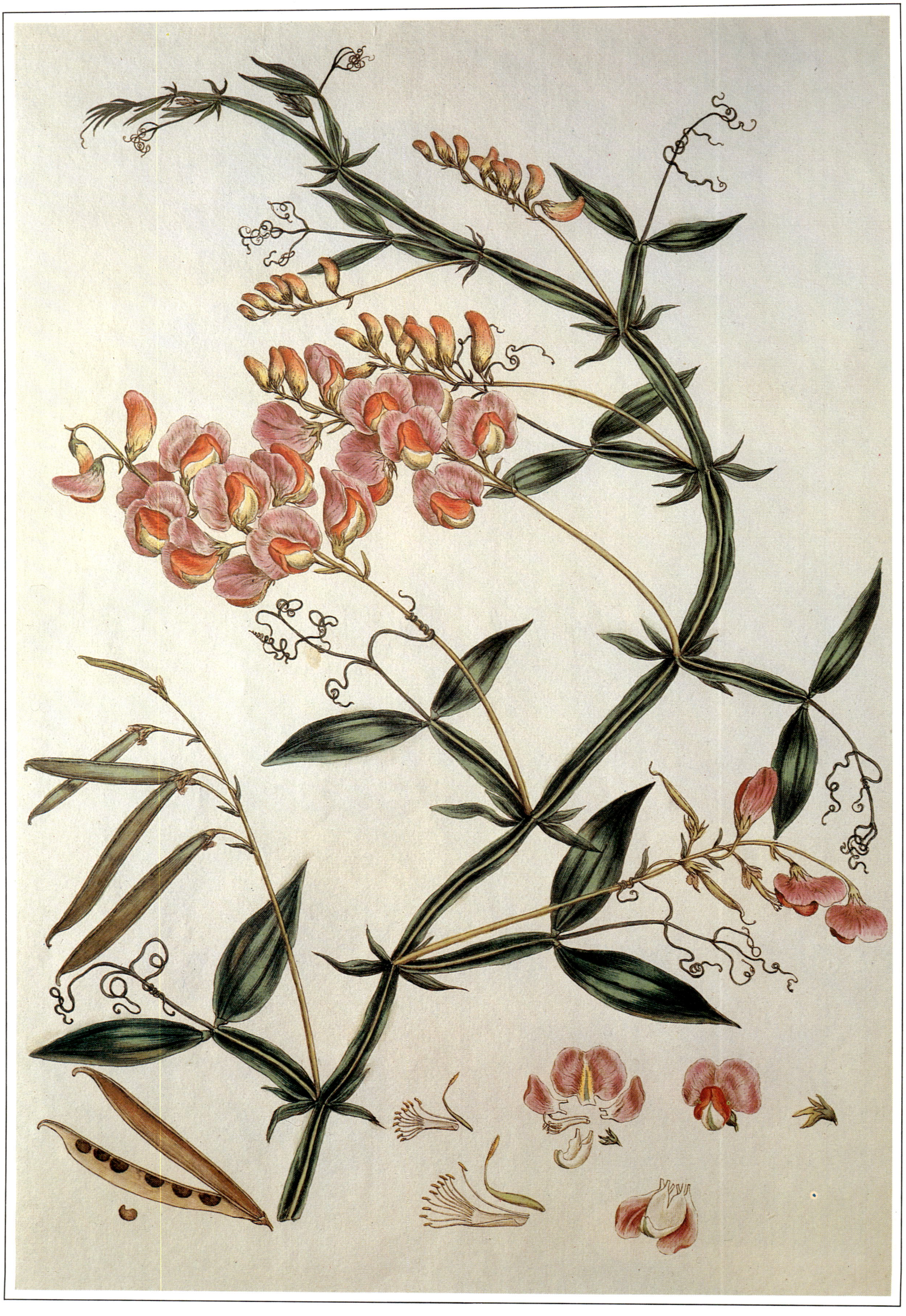

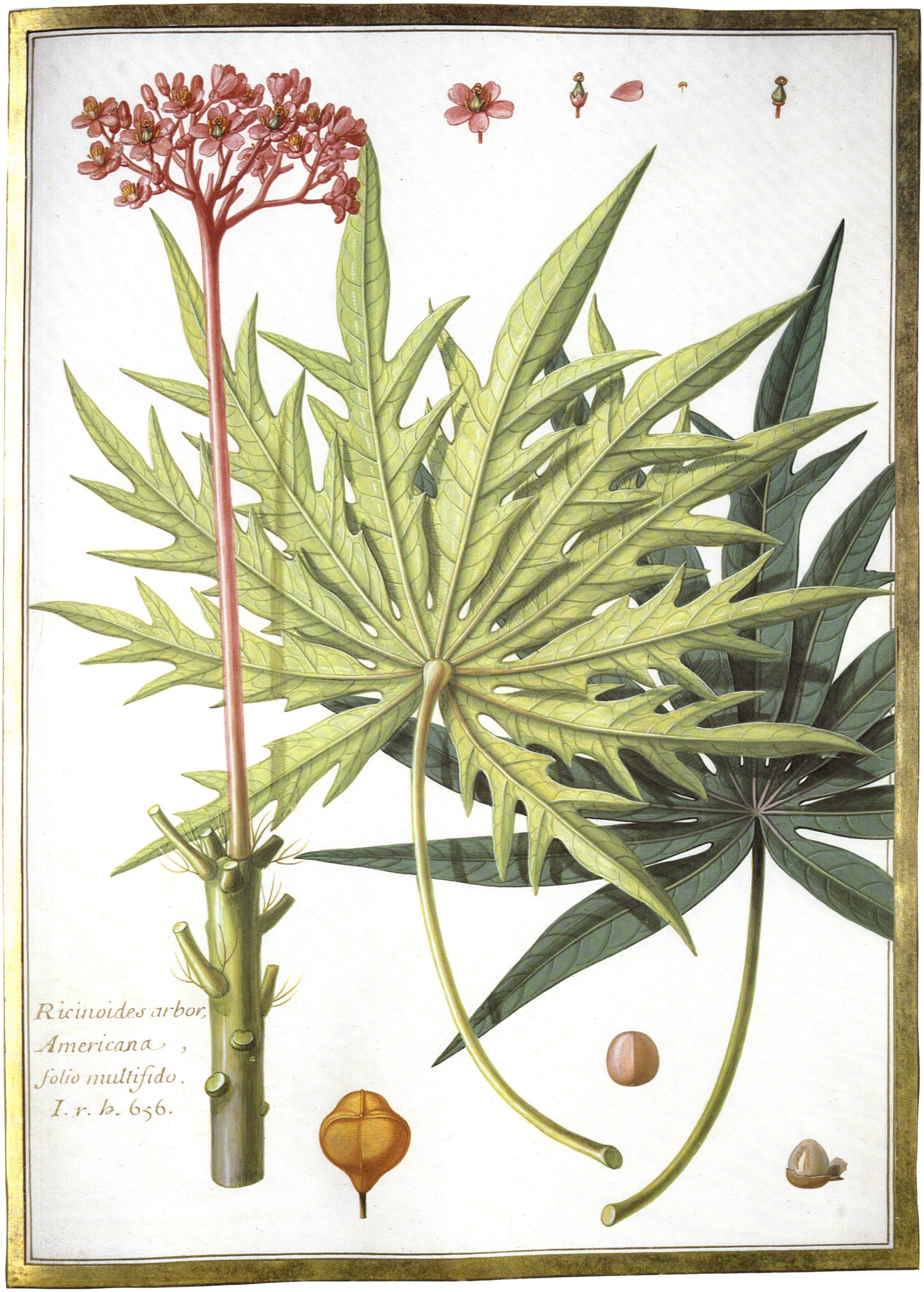
Ricinoides arbor,
Americana ,
folio multifido .
I. r. h. 656.

CHAPTER V

Linnaeus and Ehret

The eighteenth century and the first three decades of the nineteenth were the golden age of plant illustration. This was the era of Ehret and the Bauers in England and of Aubriet, the Redoutés and Turpin in France. They overshadow all the plant painters who came before or after them with their combination of artistic vision, botanical accuracy and sheer quantity of output.

Botany itself was dominated by the great Swedish taxonomist Carl Linnaeus (1707–78). The son of a poor clergyman, Linnaeus studied medicine at the universities of Lund and Uppsala before a treatise on sex in plants gained him the post of deputy to the Professor of Botany at Uppsala. In 1732, he was sent to collect plants in Lapland, a journey that took him some 4,600 miles and included a crossing of the Scandinavian Peninsula on foot to the Arctic Ocean, a remarkable feat of endurance. On the way, he found a hundred new plant species. Although his predecessors, notably Ray and Tournefort, had made valuable efforts at plant classification, it was Linnaeus who successfully devised a system that encompassed all living

Page 76:
Everlasting pea (*Lathyrus latifolius*). Hand-coloured engraving by John Miller from his *Illustratio Systematis Sexualis Linnaei* (1777). Miller, originally Johann Sebastian Müller, came to England from Nuremberg and worked for Philip Miller of the Chelsea Physic Garden as well as starting on large-scale publications of his own. His *Illustration of the Sexual System of Linnaeus* is a precursor of the even more grandiose enterprise on the same theme by Robert Thornton.
Page 77:
Jatropha multifida, a member of the spurge family, Euphorbiaceae, here described as *'Ricinoides arbor, Americana, folio multifido'*. Opaque watercolour on vellum with a border of gold leaf by Claude Aubriet, around 1700. The tree is a native of America, from southern Texas to Brazil; the seeds, which were known as *nuces purgantes*, were imported into Europe from Jamaica. Lindley Library, Royal Horticultural Society, London.

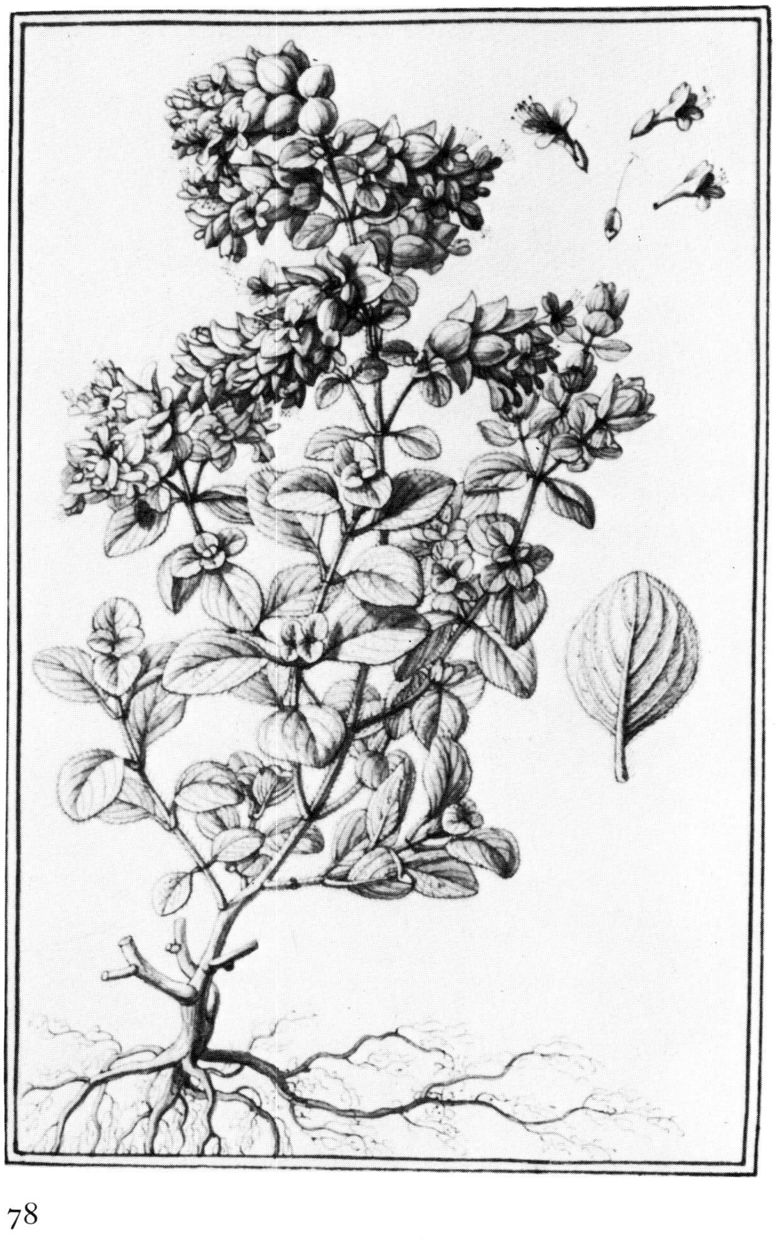

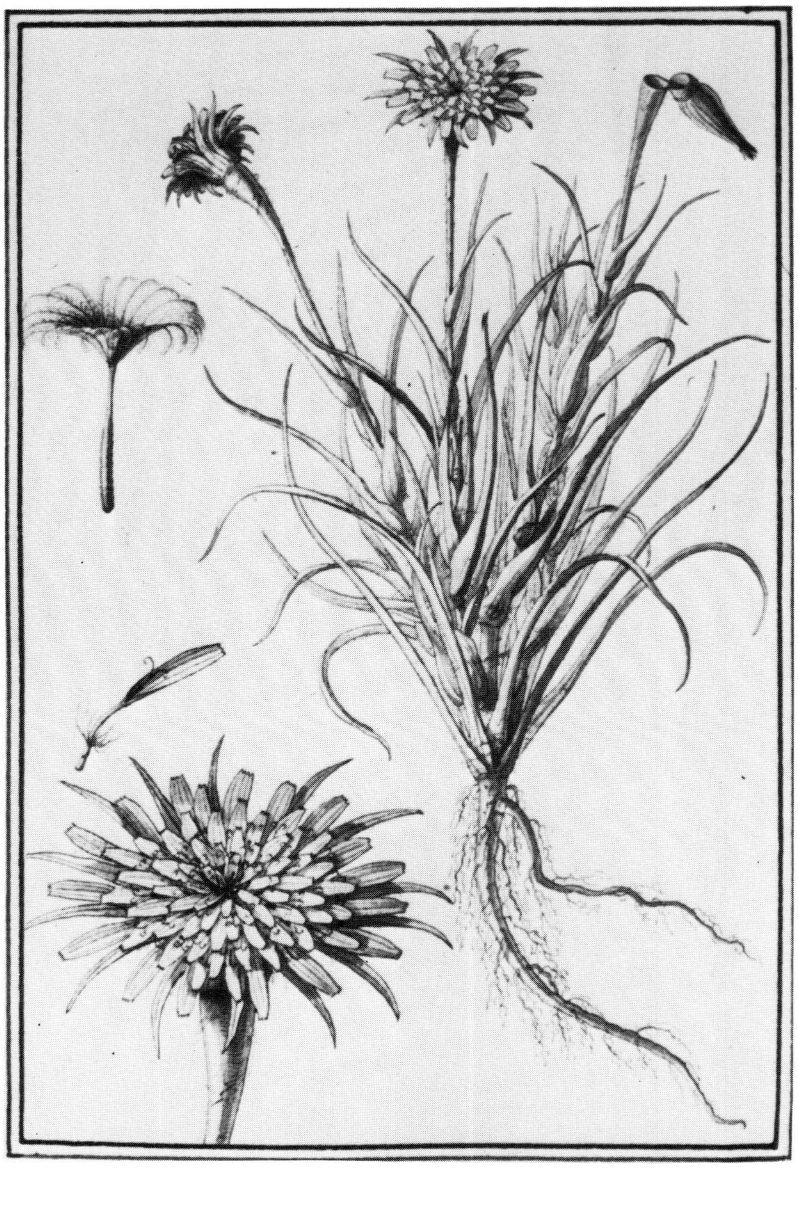

Overleaf:
Pasque flower (*Anemone pulsatilla*). Watercolour on paper by G. D. Ehret. Collection of Lord Derby.

things. He first enunciated it in his *Systema Naturae*, which was published in Holland in 1735. Linnaeus's artificial system of classification based on the sexual parts of the flower allowed the many plants that were being discovered and coming into cultivation to be put into some sort of order. The binomial system of nomenclature, by which every plant or animal has two names – a generic and then a specific name – was first used extensively in the *Species Plantarum* of 1753. In it, Linnaeus named every plant that was known to him, some 7,300 species. The book soon became the standard reference work, and the binomial system established itself as the accepted method of naming plants and animals.

Systematic in all things, Linnaeus also brought method to the scientific description of living organisms by introducing a prescribed sequence of presentation and a set of rules to be used by the describer. Such a method may have been the enemy of literary endeavour, but the benefits were enormous: in formalised description, the scientific virtues of tidiness, thoroughness and precision replaced whim, inconsistency and verbosity. Having created the verbal tools of the botanist's trade, Linnaeus, who had become Professor of Natural History at Uppsala, fired his students with the urge to travel far and wide in search of new animals and plants to describe and classify. The realities of exploration in the eighteenth century, it should be remembered, were not at all glamorous; the traveller had as a matter of course to accept great risks of hardship and deprivation, disease and attack – it has been estimated that a third of the Linnaean disciples died during the course of their expeditions.

The increasing popularity of gardening and the colonising efforts of the British, Dutch and Russians both served as further stimuli to botanical exploration. Most of the voyages of discovery now took scientists with them, and often a plant illustrator as well. Some of them had been students of Linnaeus, notably Daniel Solander, who was one of the scientists who

Four ink and wash drawings by Claude Aubriet, part of a large collection probably made around 1690 for Tournefort and mainly showing common northern European species or garden plants. *Left to right:* an oregano, *Oreganum tournefortii*; goat's-beard or Jack-go-to-bed-at-noon (*Tragopogon pratensis*); ginseng (*Panax quinquefolius*), a now well-known plant that is found wild in woods in North America and China; hemp (*Cannabis sativa*), the source of an important fibre and of narcotic resin. Lindley Library, Royal Horticultural Society, London.

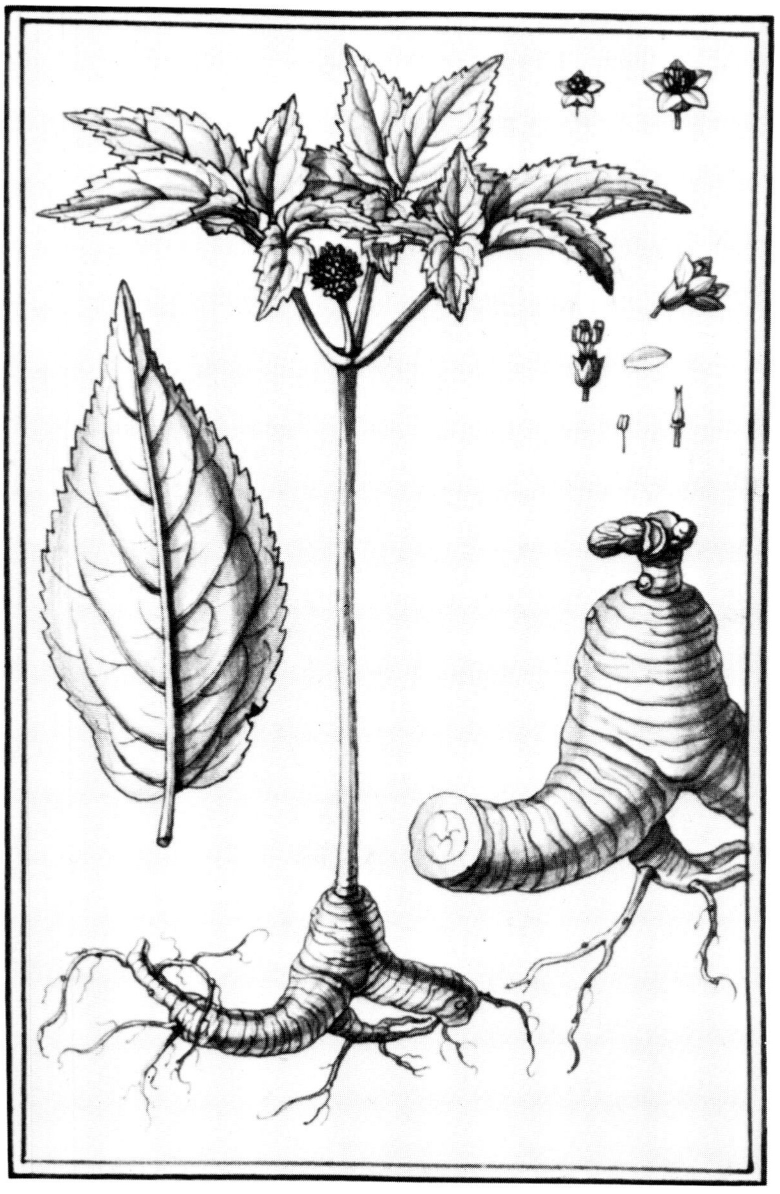

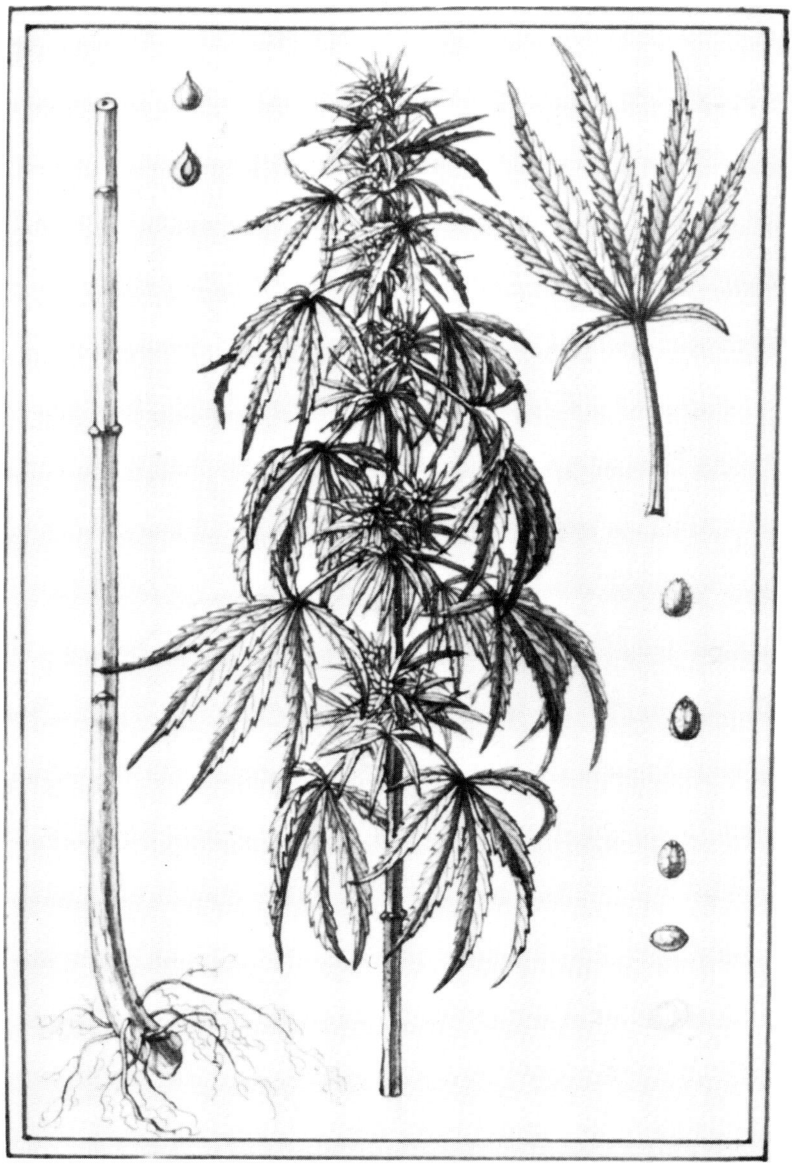

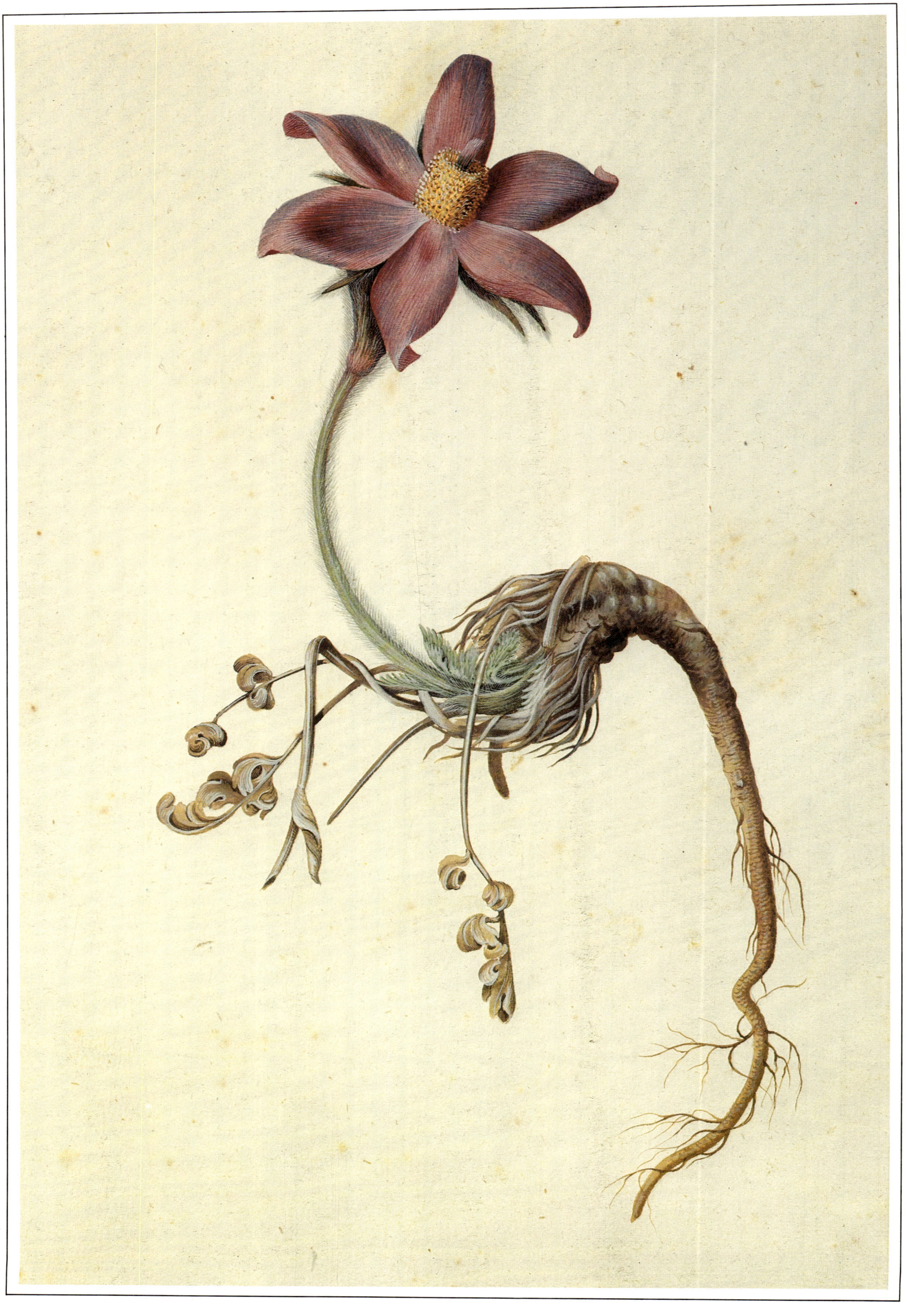

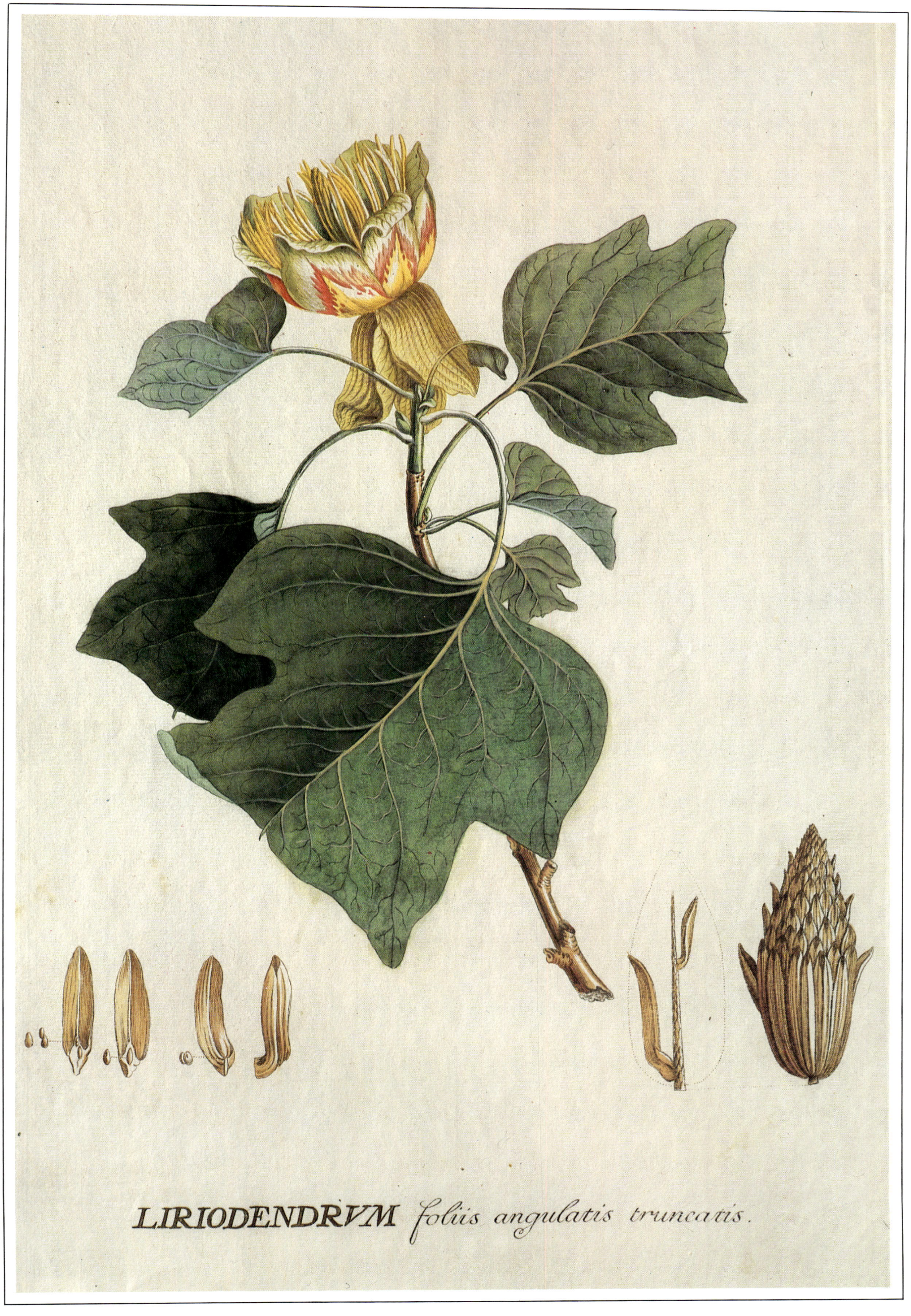
LIRIODENDRVM foliis angulatis truncatis.

accompanied Captain Cook on his famous voyage to Australia and New Zealand from 1768 to 1771.

Not all the artists who recorded the new species travelled far afield. Some of the greatest stayed at home and worked from specimens sent back by explorers and travellers. Thus, after his expedition with Tournefort, Claude Aubriet settled down in France, succeeding Jean Joubert as painter of the *vélins*. Although he drew the illustrations for Sébastien Vaillant's *Botanicon Parisiense* (1727), and some of his drawings were used in later books such as René Louiche Desfontaines's *Choix des Plantes du Corollaire des Instituts de Tournefort* (1808), his strengths are best judged by the mass of original paintings and drawings that he produced. The small ink and monochrome wash sketches that were intended for use in a book by Antoine de Jussieu show a delightfully spontaneous feeling for the growing plants, while Aubriet's paintings on vellum can be boldly and dramatically stylised without loss of botanical accuracy, as in his painting of *Jatropha multifida*.

Whereas Aubriet had undertaken one major expedition as a young man, another of the great eighteenth-century illustrators, Georg Dionysius Ehret (1708–1770) never ventured outside Europe and the British Isles. Ehret was born in Heidelberg, the son of a gardener who, before his early death, taught his young son to draw. As a youth, Ehret worked as a gardener to the

Previous page:
Tulip tree (*Liriodendron tulipifera*). Hand-coloured engraving after a painting by Georg Dionysius Ehret, from C. J. Trew's *Plantae Selectae* (1750–73).

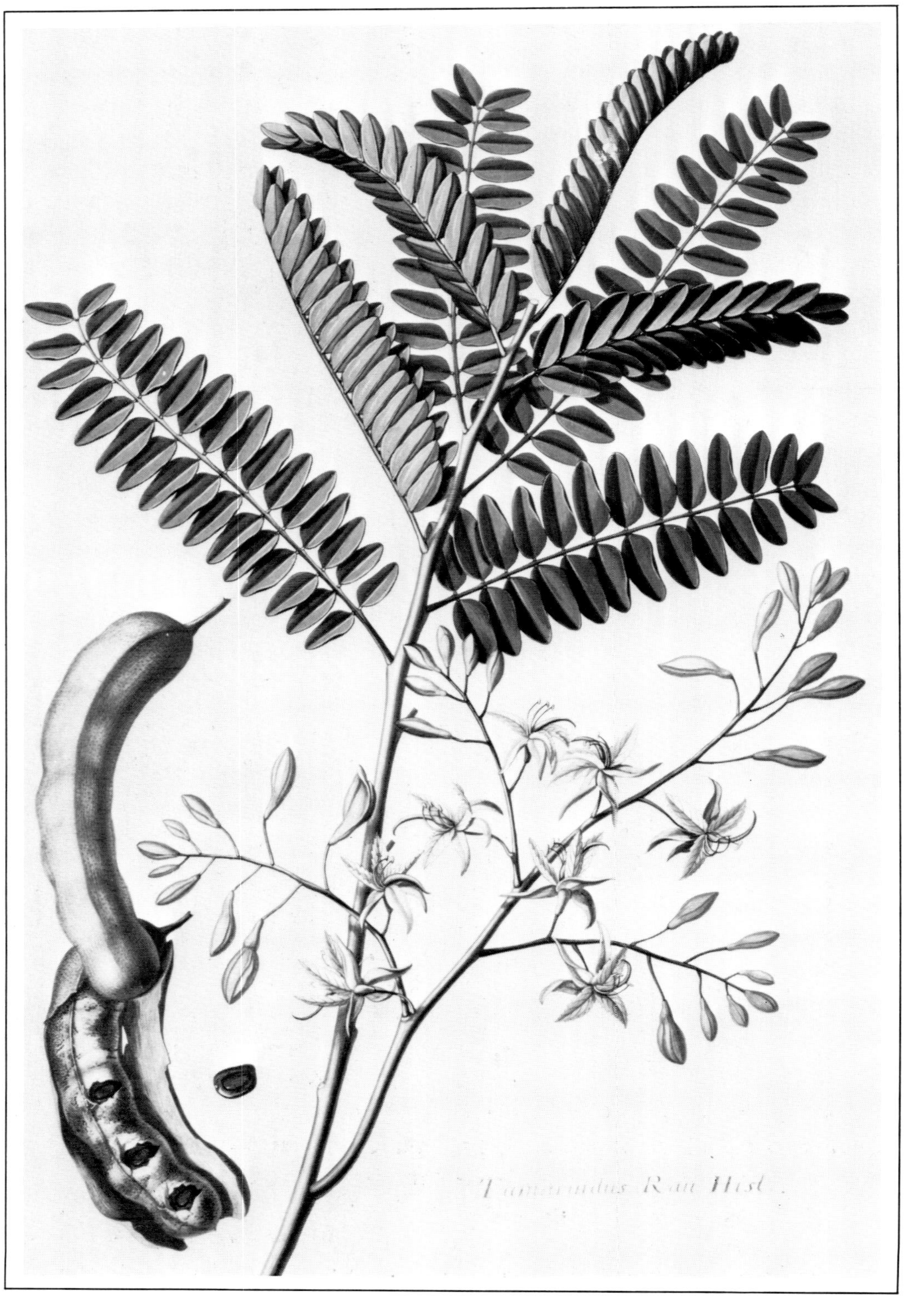

Tamarind (*Tamarindus indica*), a leguminous tree; the pulp around the seeds is an important flavouring in Indian cooking – its taste is sour and fruity. Opaque watercolour on vellum by Claude Aubriet, around 1700. Lindley Library, Royal Horticultural Society, London.

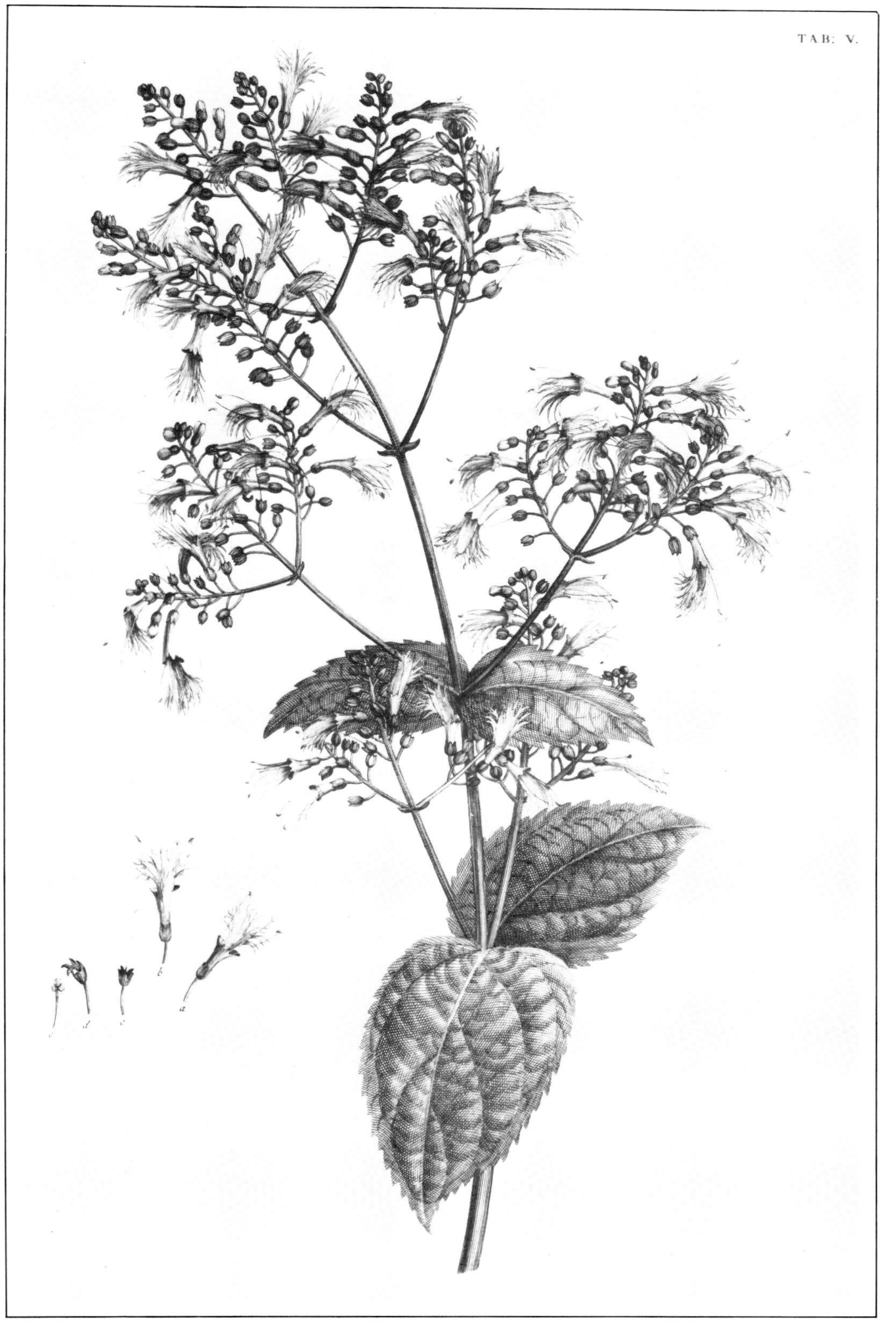

Collinsonia canadensis, a native of moist woods in eastern North America, named in honour of Peter Collinson, an English botanist who received plants sent back from America by John Bartram. Engraving by Jan Wandelaar after a drawing by Georg Dionysius Ehret from *Hortus Cliffortianus* (1737) by Carl Linnaeus.

Margrave Karl III of Baden at Karlsruhe, which had been laid out in 1715 as a fine piece of baroque town planning. He soon took to painting flowers, winning the approval of the Margrave but arousing such jealousy in the other gardeners that he left in 1726 to go to Vienna. His first major commission came from a Regensburg apothecary, J. W. Weinmann, who asked him to produce a thousand drawings for a modest salary of 50 thaler per annum. By the end of a year, Ehret had completed some five hundred, but Weinmann was apparently not satisfied; he paid Ehret only 20 thaler, and the contract was broken off. Later, some of these pictures appeared in Weinmann's *Phytanthoza Iconographia*, an impressive work which contains over a thousand engravings of plants but is marred by inaccuracy and 'representations of fictitious plants'. Still in Regensburg in 1728 Ehret started working for a banker, using the descriptions in *Hortus Indicus Malabaricus* as a basis for colouring the plates. This proved so irksome that, after five years, Ehret had completed only the first three parts.

Meanwhile, however, Johann Ambrosius Beuer, a Regensburg apothecary's apprentice and a keen botanist, introduced Ehret to his uncle, Christophe Jacob Trew of Nuremberg, who was to become Ehret's first

Overleaf:
'*Erica sebana*', a Cape heath (*Erica coccinea*). Hand-coloured engraving by Mackenzie after a watercolour by Francis Bauer, from his *Delineations of Exotick Plants Cultivated in the Royal Garden at Kew* (1796).

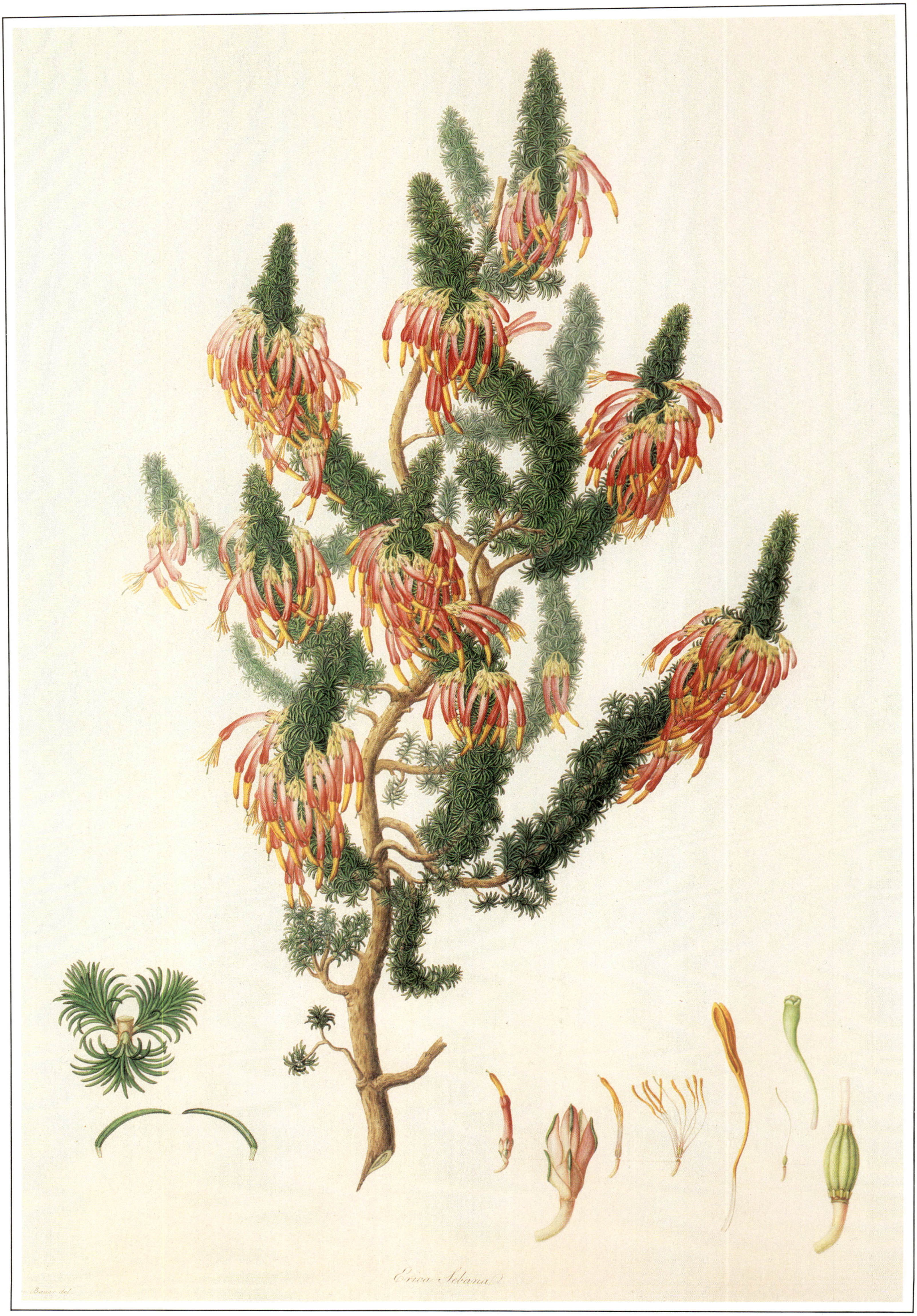
Bauer del.
Erica Sebana

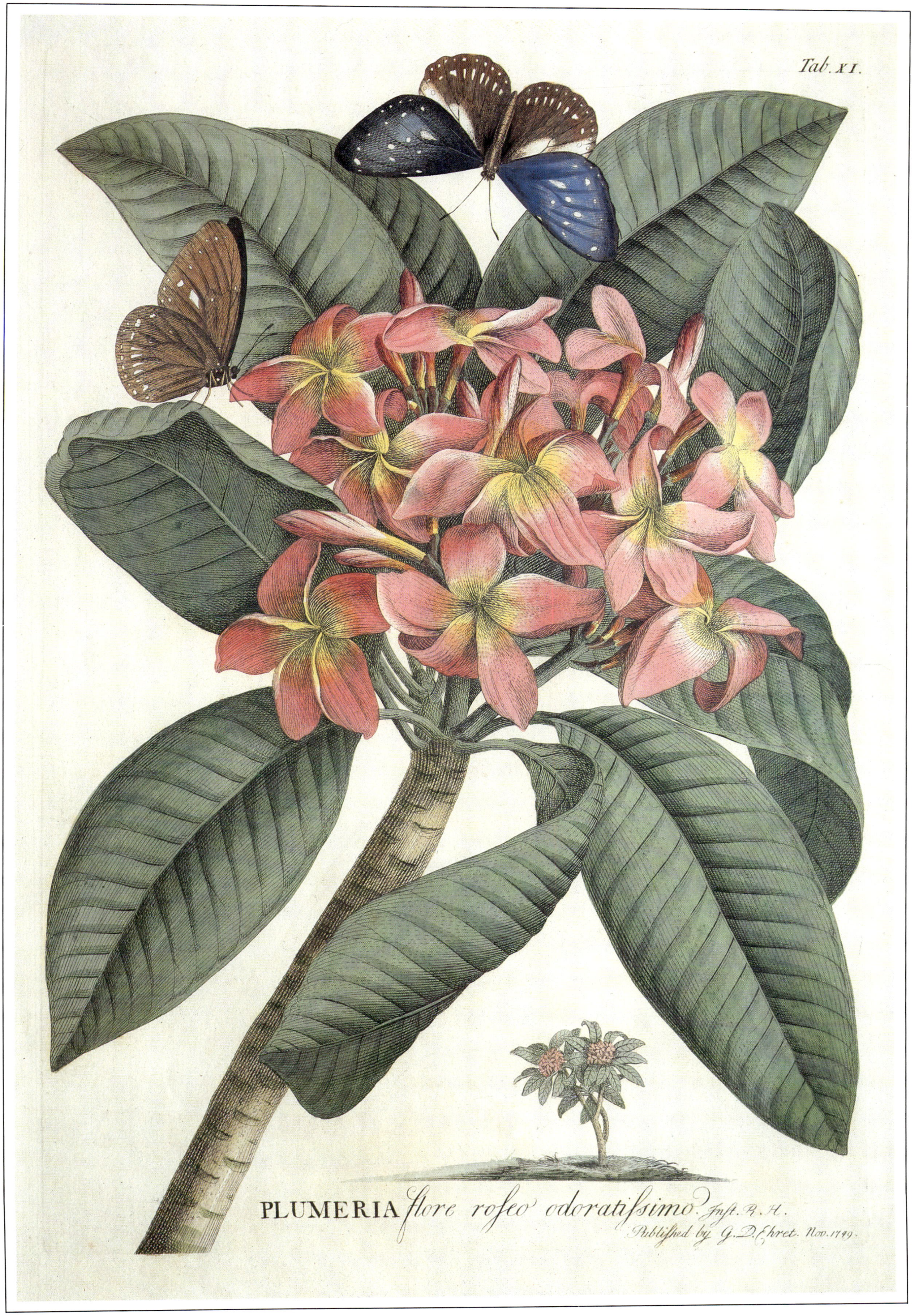
Tab. XI.
PLUMERIA flore roseo odoratissimo. Inst. R. H.
Published by G. D. Ehret. Nov. 1749.

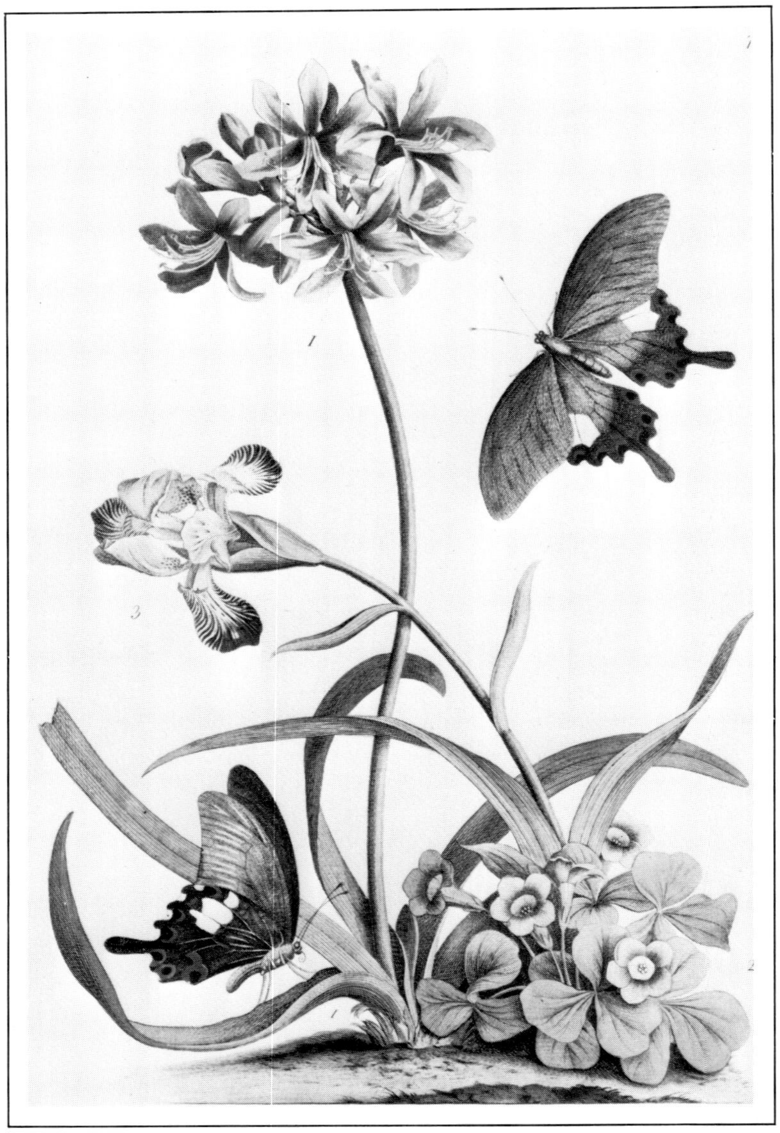

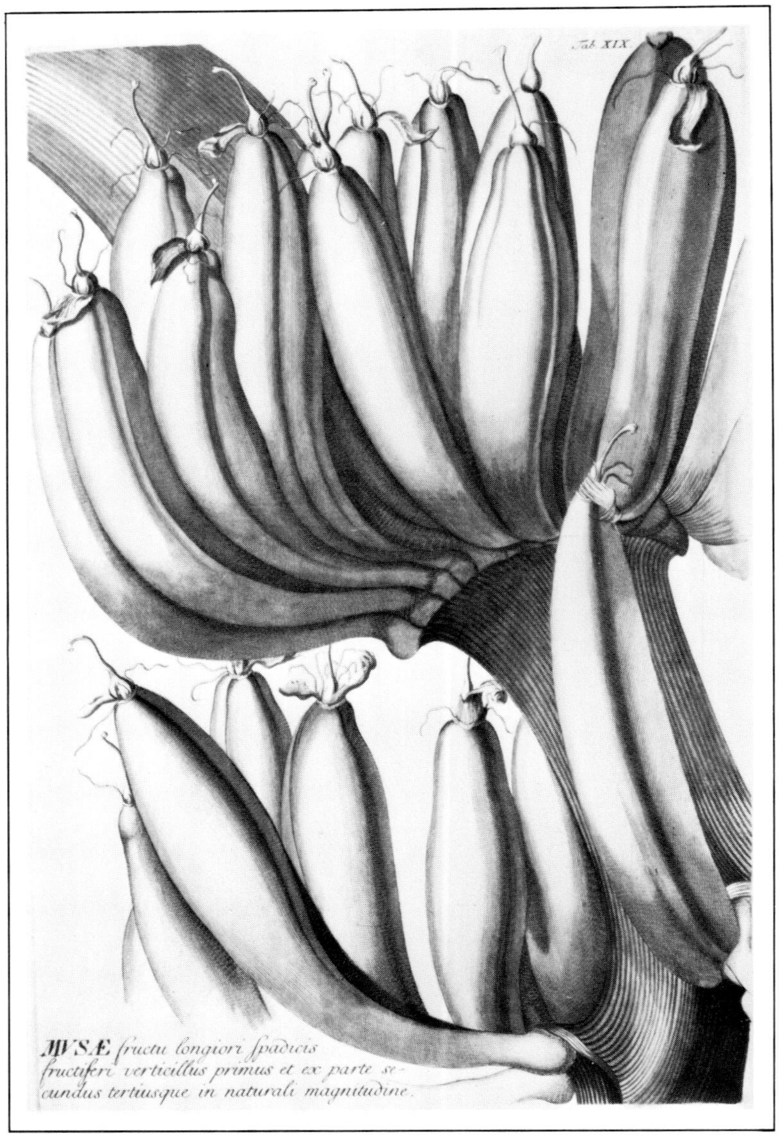

real patron, and his lifelong friend. Through Dr Trew, Ehret managed around 1732 to sell his first collection of paintings called *Herbarium Vivum Pictum* for 200 thaler. It consisted of six hundred paintings, mostly of wild flowers from the area around Regensburg. They are clearly and simply painted in watercolour, without the floral dissections which characterise much of Ehret's later work. An album of drawings of this period (possibly the *Herbarium Vivum Pictum* itself) is in the Earl of Derby's collection at Knowsley Hall.

During the next few years Ehret travelled through Europe, visiting botanic gardens and learning more about botany. His first visit to England lasted about a year, before he went on to Amsterdam. He had an introduction to an Amsterdam banker, George Clifford, and hence he met the young Linnaeus who had become Clifford's personal physician and garden manager. When Linnaeus compiled a descriptive catalogue of the rare plants in Clifford's garden, the *Hortus Cliffortianus* of 1737, it contained twenty plates by Ehret. In keeping with the Linnean system of classification that was used in the book, Ehret included for the first time in his drawings details of flowers and other plant parts that were to be a feature of much of his later botanical work. The plates are simple and well engraved by Jan Wandelaar.

Ehret returned to England late in 1736 and received commissions to illustrate all manner of botanical and zoological books, but continued to send flower paintings to Dr Trew. He often worked at the Chelsea Physic Garden, where the great gardener Philip Miller was curator, contributing illustrations to Miller's *Figures of the most Beautiful, Useful and Uncommon Plants* (1755–60); he also married Miller's sister-in-law. He travelled around England visiting interesting gardens and painting new plants as they came into flower; thus he saw *Magnolia grandiflora* at Fulham in 1737 and *Arbutus andrachne*, which had been brought from Aleppo by Alexander Russell and flowered at Upton in Essex in 1766. Ehret's practice of signing

Left:
Agapanthus africanus, *Iris variegata* and a species of *Oxalis*. Hand-coloured engraving by Georg Dionysius Ehret after his own drawing from his *Plantae et Papiliones Rariores* (1748–59).

Above:
Young bananas. Hand-coloured engraving after a painting by Georg Dionysius Ehret from C. J. Trew's *Plantae Selectae* (1750–73).

Previous page:
Frangipani (*Plumeria rubra*). Hand-coloured engraving dated November 1749 by Georg Dionysius Ehret for his *Plantae et Papiliones Rariores* (1748–59). The frangipani, a member of the same family, the Apocynaceae, as oleander and periwinkles, was originally a native of tropical America, from Mexico to Ecuador, but is now common throughout the tropics and is often planted near temples because of its very sweet-scented flowers. The generic name commemorates Charles Plumier, a French priest who made pioneering studies of the flora of the Caribbean in the late seventeenth century.

An Arabian mayweed, *Cladanthus proliferus*, here called *'Buphthalmum'*. Engraving by Jan Wandelaar after a drawing by Georg Dionysius Ehret from Carl Linnaeus's *Hortus Cliffortianus* (1737).

TAB: XXIV.

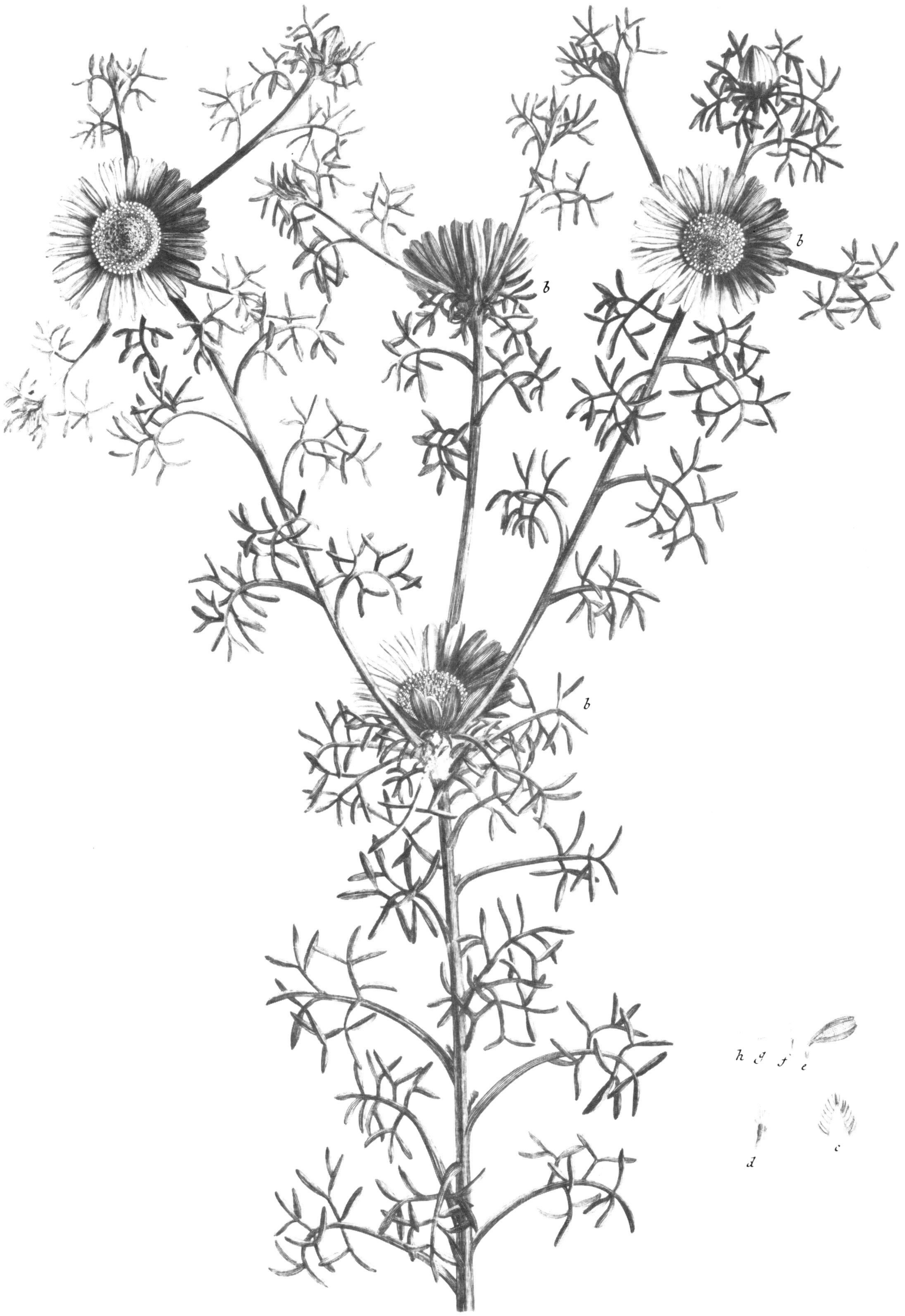

BUPHTHALMUM caule decompoſita, calycibus ramiferis. *Hort. Cliff.* 413. *ſp.* 1

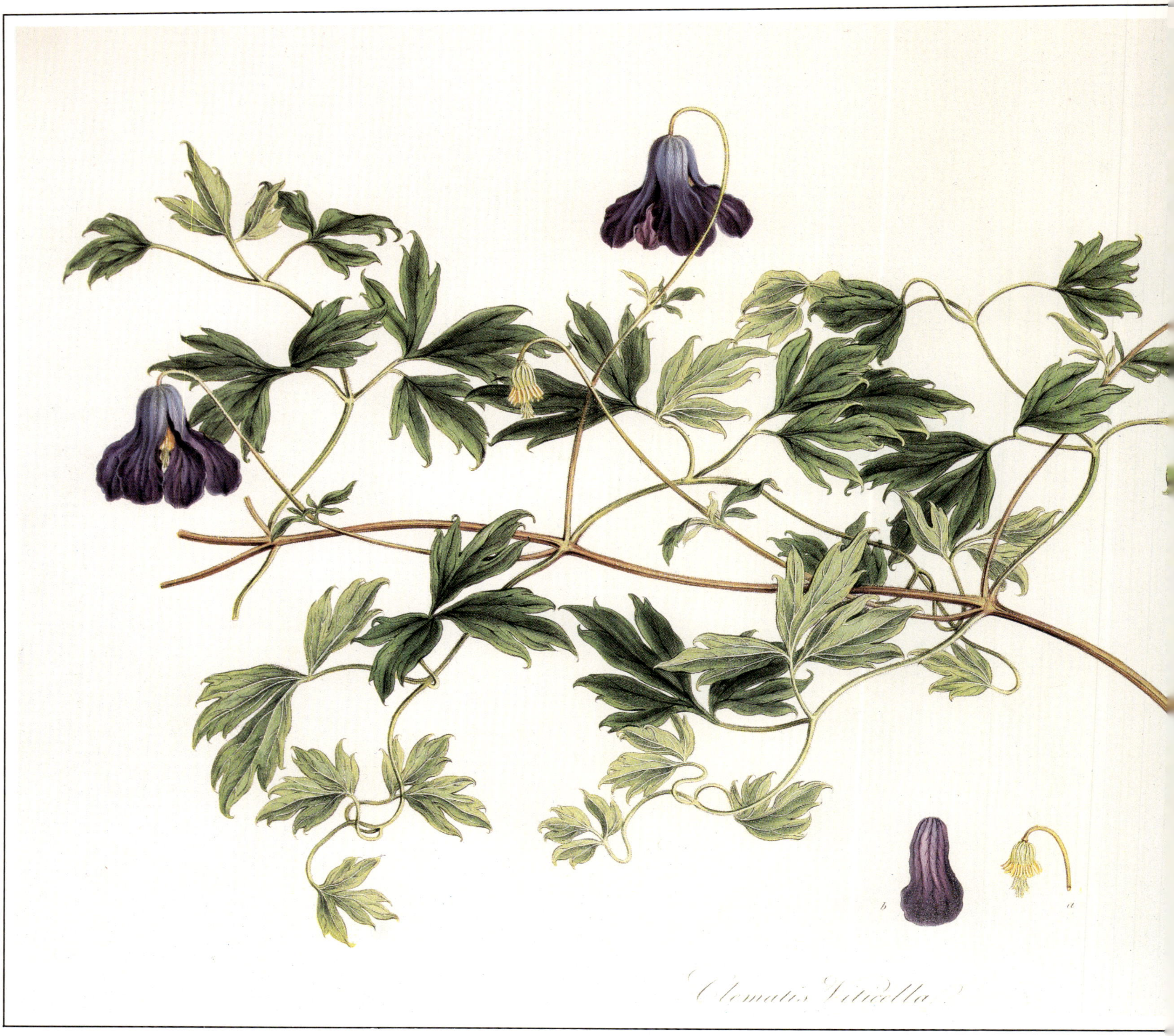

and dating all his work and often also detailing where the plants were grown adds immeasurably to the historical value and interest of the pictures.

In England, Ehret began to engrave from his own drawings and made a good living by selling the plates. One of his first engraving projects was *Plantae et Papiliones Rariores* (1748–59). The butterflies mentioned in the title play a distinctly minor part in the enterprise and are not even named. The engraving is quite bold, and the plates are hand-coloured. Also included are tables of botanical details showing such features as dissections of an Iris flower and fruit. Dr Trew used many of the flower drawings Ehret had sent him in his *Plantae Selectae*, printed in ten parts in Nuremberg between 1750 and 1773, as well as in *Hortus Nitidissimus*, which concentrated more on garden plants, between 1750 and 1792. *Plantae Selectae* contains a large number of drawings of North American plants that had recently been introduced into Europe, for example *Lilium superbum*, *Dodecatheon meadia* and many species of cactus. There is also an emphasis on tropical fruit, with bananas, pineapples, pawpaws and guavas all illustrated in detail.

Ehret also published what were known as Tabella, including an interesting one of the snowdrop tree, *Halesia carolina*, dedicated to 'the Revd. Steven Hales F.R.S, the author of the incomparable Vegetable Statiks: this rare shrub lately sent over by Alexander Garden M.D. of Charlestown from

Clematis viticella. Hand-coloured engraving after a watercolour painting by Ferdinand Bauer from *Flora Graeca* (1806–40) by John Sibthorp and J. E. Smith. *Clematis viticella* is a climber that is common in gardens and grows wild in Turkey (where Sibthorp and Bauer collected it) and the eastern Mediterranean region.

Saluda in the north west part of Carolina, is humbly dedicated by his much obliged friend and servant, John Ellis.'

One or two examples of Ehret's work appeared in other publications: a very fine plate of *Pinus pinaster* in Aylmer Bourke Lambert's monograph, *Description of the Genus Pinus* (1803–24), was taken from a drawing that belonged to Sir Joseph Banks. Some of Ehret's drawings, along with others by Francis Bauer and J. Sowerby, were among the small number of illustrations in William Aiton's *Hortus Kewensis* (1789), a list in three small volumes of plants grown in Kew and other gardens in England. This is especially interesting to students of horticultural history because it gives the dates of introduction and the introducers of most of the plants mentioned.

Ehret never produced any major botanical works like those of the great flower painters of the next generation. He had a short and unsuccessful spell as curator at the Oxford Botanic Garden, under Professor Humphrey Sibthorp, father of the author of *Flora Graeca*. He resigned after just a year; Ehret wished to continue his private commitments, while Sibthorp wished him 'to be subject and obedient to his orders and at his directions' and would not even allow him to correspond with other botanists without his permission, and so Ehret resigned. After this, he continued to teach drawing to the daughters of the nobility and, judging by the number he taught, must have been very successful. One of his most famous patronesses was the Duchess of Portland, a keen gardener, botanist and collector of natural objects. Ehret taught her daughters, and the Duchess added his paintings to her collection. He visited her house at Bulstrode many times and in 1768 was painting a collection of wild flowers for her, although his sight was failing; he died in 1770 at the age of 62. After her death in 1785, her collection was broken up and sold, many of Ehret's paintings fetching what were then large sums.

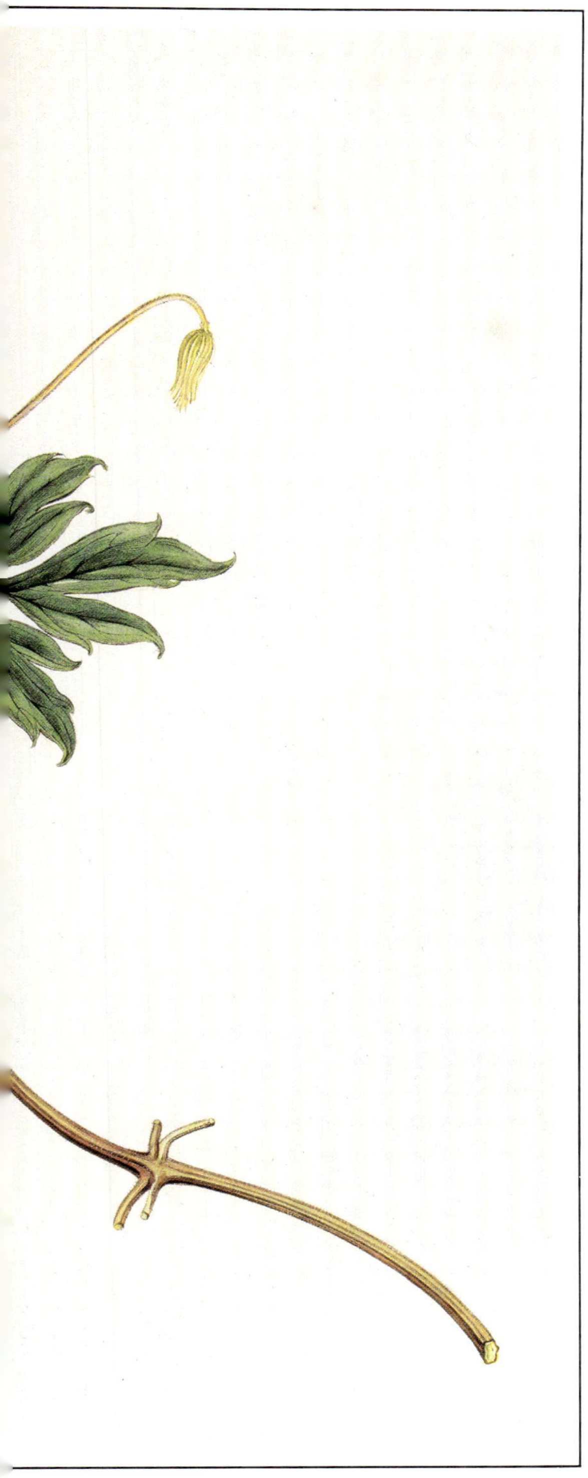

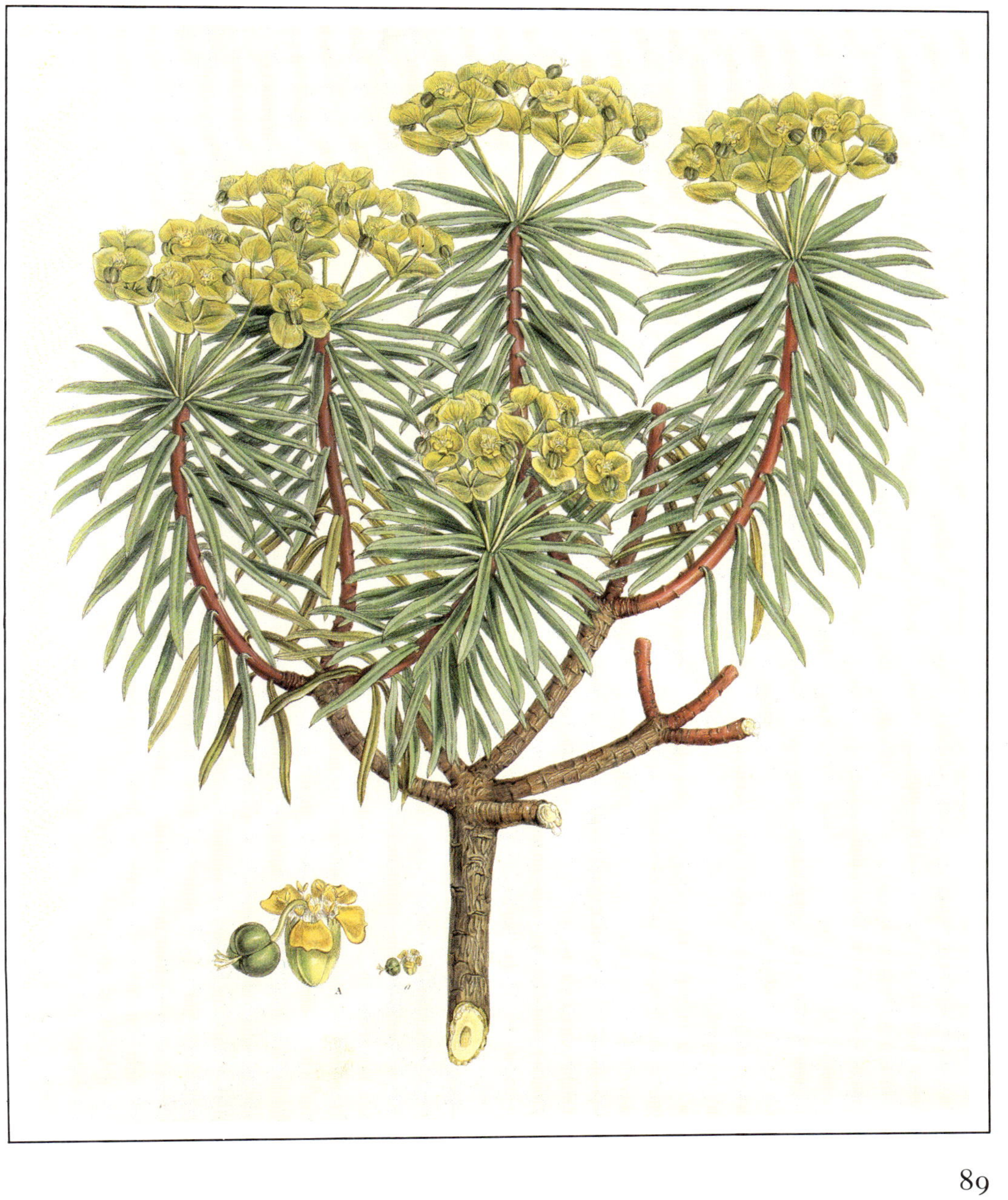

A spurge, *Euphorbia dendroides*, which forms a small shrub up to six feet high and usually grows on rocky slopes and cliffs near the sea in Yugoslavia and Greece. Hand-coloured engraving after a watercolour by Ferdinand Bauer from *Flora Graeca* (1806–40) by John Sibthorp and J. E. Smith.

CHAPTER VI
The Eastern Mediterranean

The journeys of Rauwolff and Tournefort to the Eastern Mediterranean had served to reveal the richness of its flora and to stimulate further botanical exploration. Rauwolff had been to the Lebanon in the 1570s, Tournefort had travelled through Greece and Northern Turkey in 1700–02, and the English apothecary and naturalist James Petiver had visited Egypt in 1717, but the botany of Syria and the Holy Land was still almost unknown.

In the early 1740s, Linnaeus was lecturing in the University of Uppsala and enumerating to his students the geographical gaps in his knowledge of natural history. Among them was Palestine, and one of Linnaeus's students was the young Frederick Hasselquist (1722–1752), who determined to visit the area and record its flora after he had completed his studies. He duly set out in 1749, landing first at Smyrna in Western Turkey where he made a base with the Swedish consul. From here, he travelled in Western Anatolia where he found pale and deep yellow crocuses, hyacinths and *Ornithogalum* near Magnesia. He also visited Egypt, Palestine, Cyprus and Rhodes, returning to Smyrna with a great collection of plants, animals and curiosities. But, as he awaited a passage home, he became ill: 'his strength

A henbane (*Hyoscyamus*) with an unidentified creeping herb, and a succulent Asclepiad. Two engravings by P. Haas after drawings by G. W. Baurenfeind from *Icones Rerum Naturalium* (1776) by Peter Forskal, a Danish naturalist who, with André Michaux, joined a Danish expedition to the Middle East. Arriving in Alexandria in 1761, the expedition spent a year in Egypt before moving to the Red Sea to explore southern Arabia, but Forskal died of malaria on the journey. His work was edited by Carsten Niebuhr and published in 1775 and 1776.

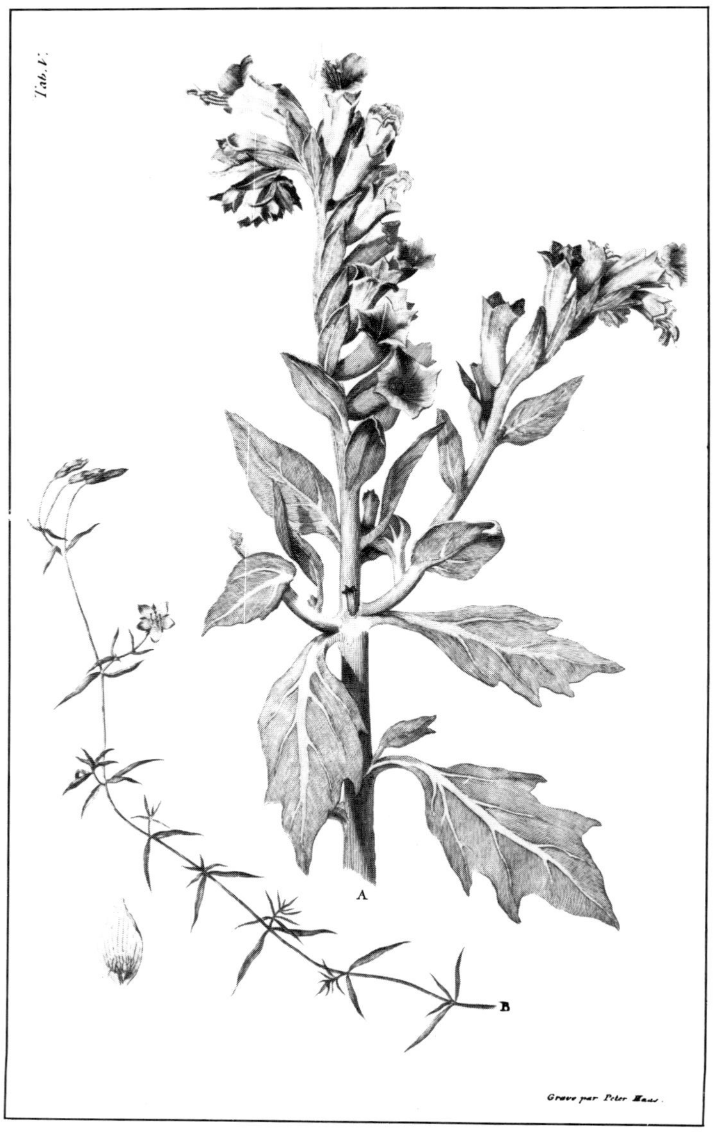

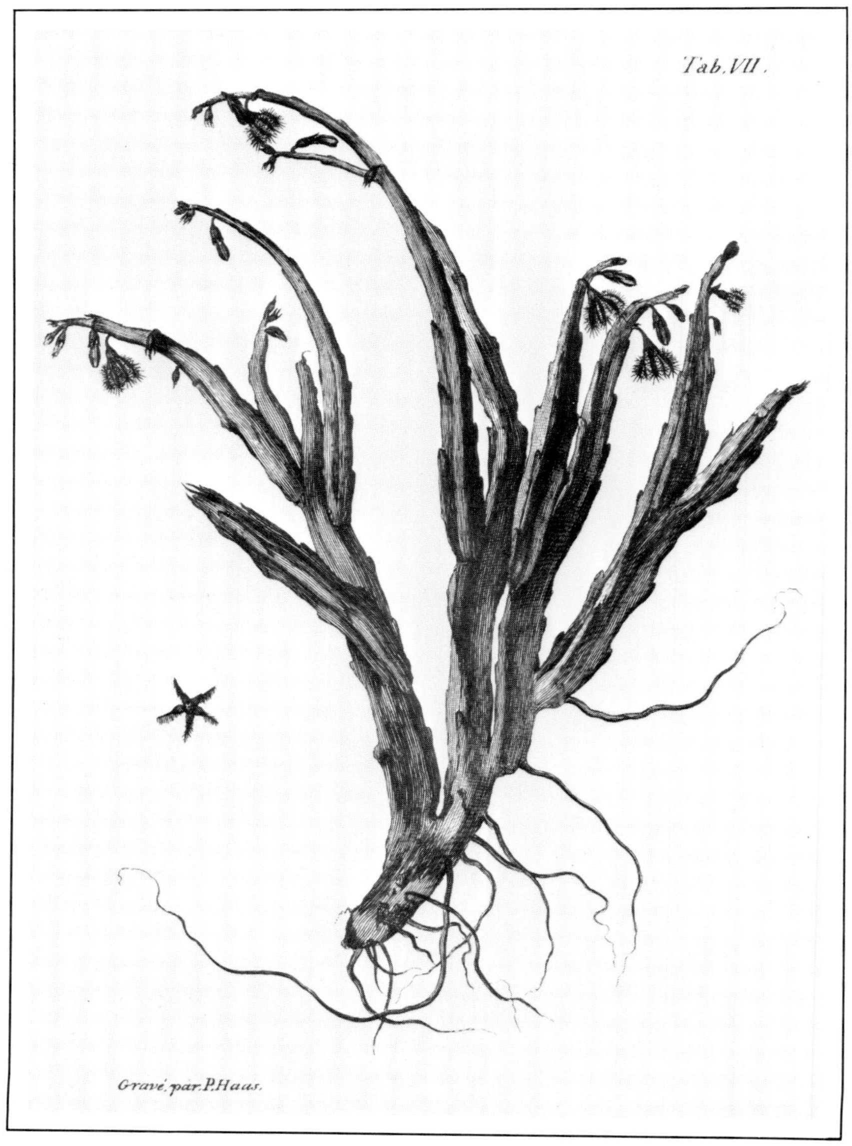

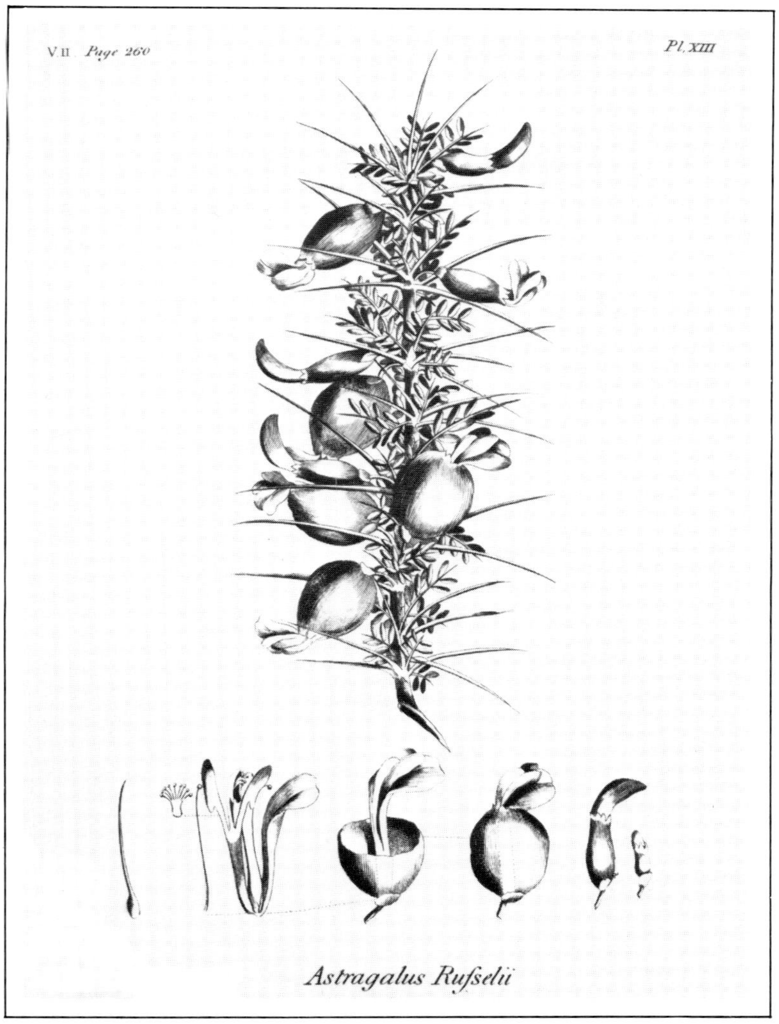

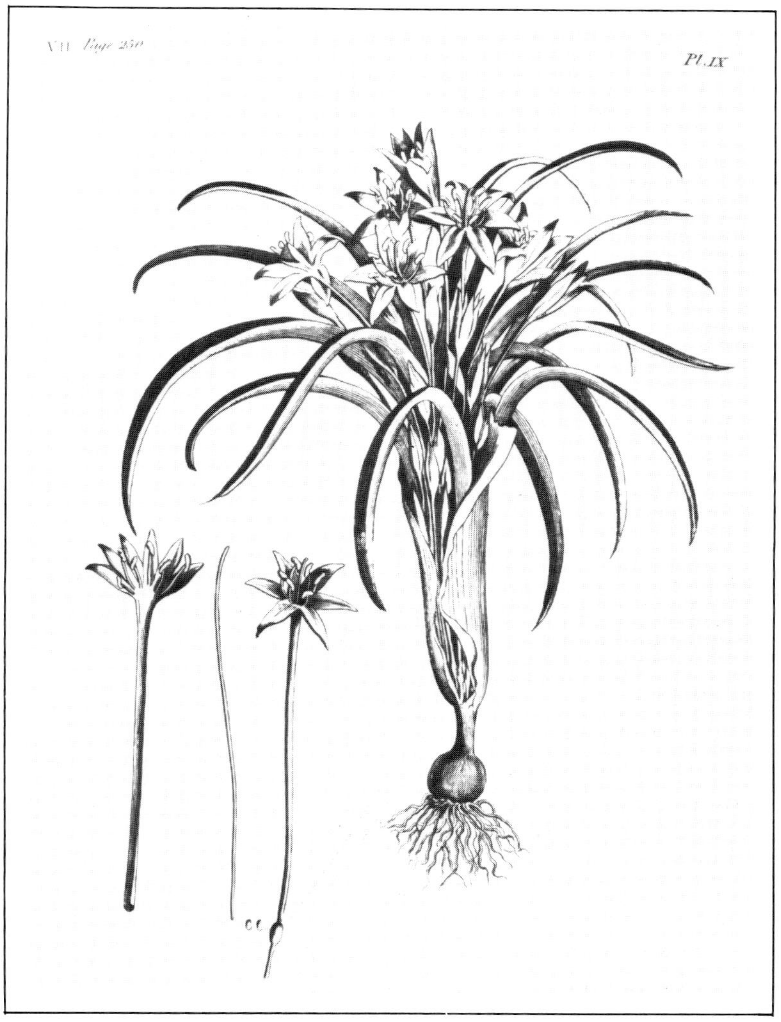

Two engravings by I. Mynde after drawings by G. D. Ehret from Alexander Russell's *Natural History of Aleppo and Parts Adjacent* (1756). *Above: Astragalus russelii*, a dwarf, much-branched leguminous shrub found only around Aleppo and in southern Turkey. *Right: 'Hypoxis fascicularis'* – the plant illustrated, as noted by Fr. Mouterde, has the basal ovary and single style of *Monocaryum fasciculare*, but the many leaves and flowers of *Colchicum fasciculare*, both of which occur around Aleppo.

was spent by the difficulties he had undergone in his travels, as well as by the violent heat of the climate of Palestine, and whilst he intended to recover it by tarrying in Smyrna, the serpent which he had long harboured in his bosom awoke, and a consuming fever confined him to his bed. He desired to be removed out of the city into the country to enjoy the free air, and to use milk as his only resource which was accordingly done; but notwithstanding all this, our beloved Dr Hasselquist wasted away daily like a lamp whose oil is spent, and departed this life, the 9th February 1752, to the inexpressible grief of all who knew him.'

His creditors in Smyrna seized all his belongings, but they were rescued and returned to Sweden by his friends, notably Linnaeus, who persuaded the Queen, Louisa Ulrica to 'pay his debts out of her own purse, and redeem his collection.' This duly arrived in the palace and Linnaeus was summoned to examine it. 'I was astonished at the sight of so many unheard of curiosities . . . The collections of dried plants from Anatolia, etc . . . the many rare fishes from the Nile, and poisonous serpents of Egypt, the rare insects, the extensive collections of oriental drugs, Arabian manuscripts, Egyptian mummies, etc., could not but excite the admiration of the beholder.' Linnaeus also edited and published Hasselquist's journal, and used his plant collection as the basis of *Flora Palestina* (1756).

No further illustrated botanical books on the Eastern Mediterranean were published until the two journeys of Sibthorp in 1786 and 1794 which resulted in the magnificent *Flora Graeca*. Ehret produced one or two illustrations for books of travels. *Descriptions of the East* by Richard Pococke, who visited the Lebanon in 1743–45, has eleven engravings of plants by Ehret. A later visitor, Alexander Russell, published his *Natural History of Aleppo and Parts Ajacent* in 1756. Russell was a medical doctor who lived for three years in Aleppo during a particularly bad epidemic of bubonic plague, and his intention was to give a scientific account of the outbreak. But the book ended as a general natural history, with details of the gardens and the plants grown in them, with a long list of wild plants in order of flowering. It also contains accounts of the sheep, goats, fish and birds of the area and of the customs of the various inhabitants. The plates

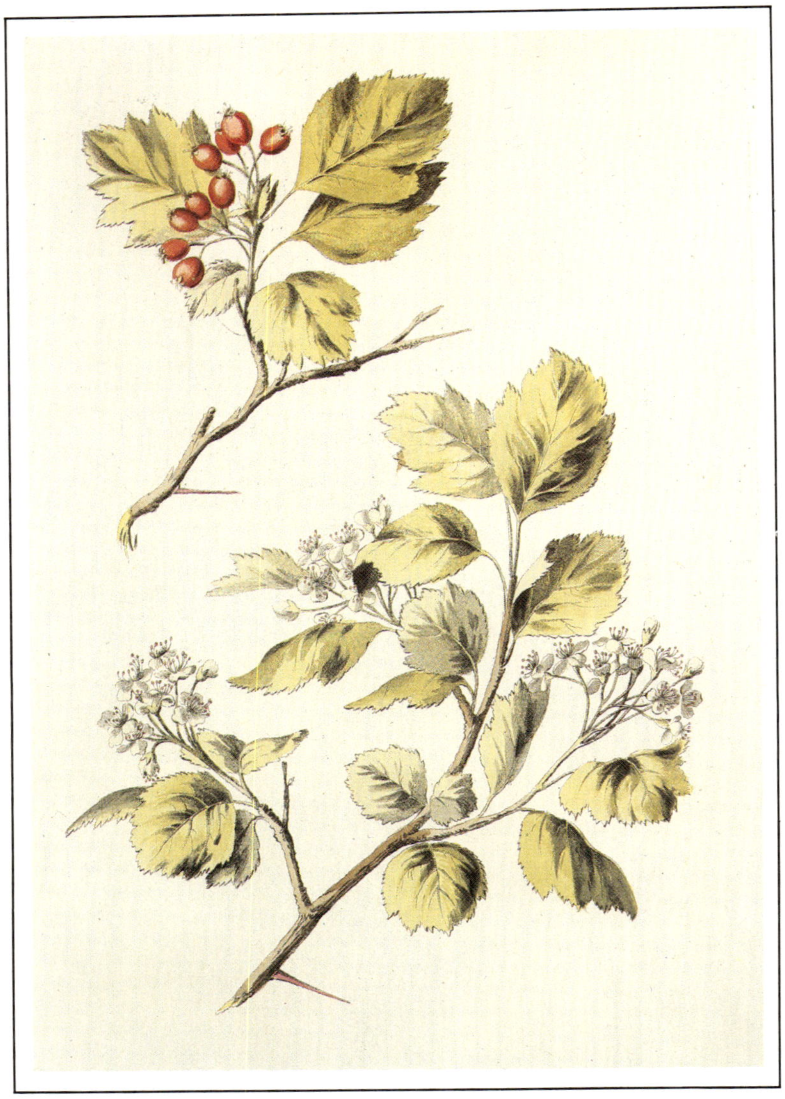

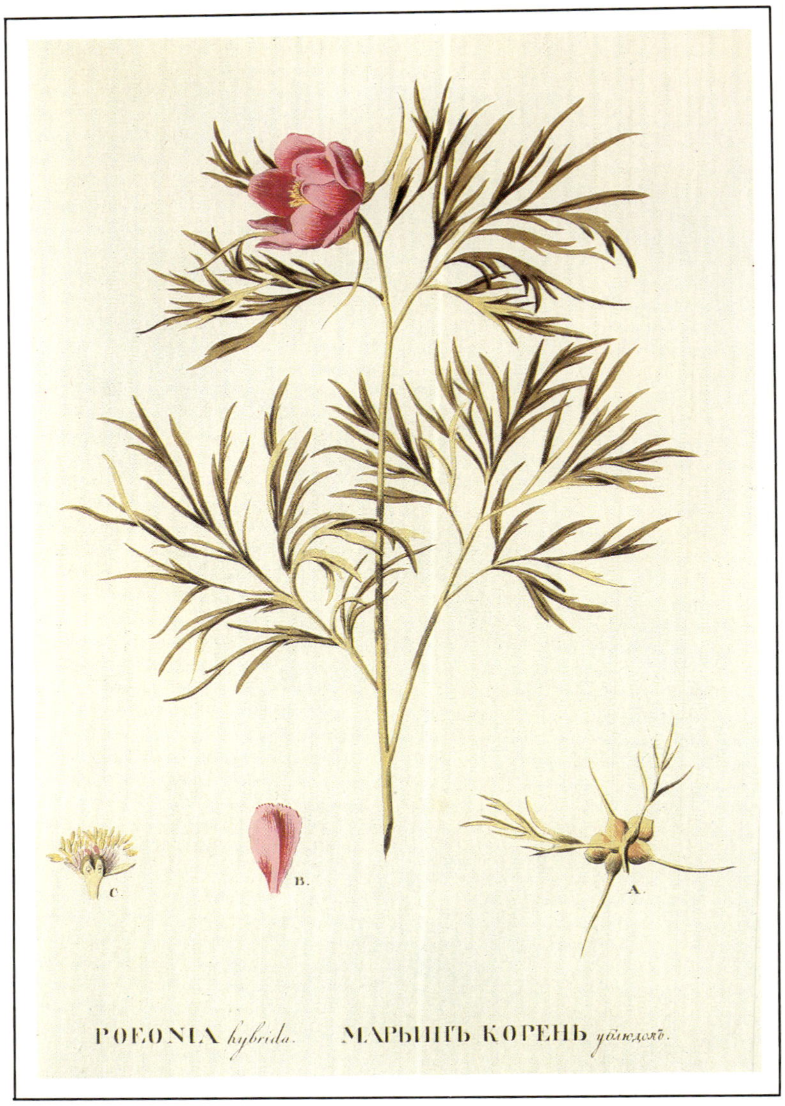

Two hand-coloured engravings by Karl Friedrich Knappe from *Flora Rossica* (1784–88). The peony, *Paeonia hybrida*, is a native of south-western Siberia and central Asia, growing on steppes and rocky hills. The hawthorn, *Crataegus sanguinea*, is a small tree that is native to European Russia and Siberia.

were drawn by Ehret and engraved by T. S. Miller. Ehret also engraved a few plates for other travel books, but work of this type formed a small part of his output in comparison to recording garden plants and new introductions from the Americas.

Dr John Sibthorp (1758–1796) was the youngest son of Humphrey Sibthorp by his second wife, Elisabeth Gibbs. Humphrey was also Professor of Botany at Oxford (and briefly Ehret's employer), but resigned in favour of John on inheriting Canwick Hall in Lincolnshire from his brother in 1783. During his thirty years in the Sheradian chair of botany, Humphrey is said to have delivered only one lecture, and that an unsuccessful one, but he corresponded with Linnaeus, and it is after him that the genus *Sibthorpia* is named. John Sibthorp studied medicine at Edinburgh and went on to Montpellier, then a centre of botanical studies and to Göttingen, where he received a doctorate. In 1785, he visited Vienna to study the ancient *Codex Vindobonensis* of Dioscorides and it was there that the young Ferdinand Bauer was introduced to him by the great botanist Nikolaus von Jacquin.

Ferdinand and his almost equally famous brother Francis were sons of Lucas Bauer, court painter to the Prince of Liechtenstein. Their early botanical talent was fostered by the Prince's physician Father Boccius, the Abbot of Feldsburg, near Vienna, who was a keen botanist and was himself assembling a collection of flower paintings. In Vienna, the brothers came under the influence of Jacquin, who was Professor of Botany at the university; some of their earliest botanical work was for his publications on the rare plants in the gardens of the Schönbrunn Palace, but both spent most of their lives in England, Ferdinand working on *Flora Graeca* and later on his Australian collections, and Francis at Kew, where he had been invited by Sir Joseph Banks, who personally provided a generous salary.

In spite of the travels of Belon, Tournefort and others in the Levant, many of the plants mentioned by Dioscorides remained unknown to contemporary botanists. Their rediscovery or identification, Sibthorp's

Title page to Volume 6 of *Flora Graeca* (1826) by John Sibthorp and J. E. Smith. Hand-coloured engraving after a watercolour landscape painting by Ferdinand Bauer. The flowers shown are (left to right) *Consolida regalis* and *Nigella sativa* (top), *Helianthemum nummularium*, *Capparis spinosa*, *Cistus incanus* ssp. *creticus*, *Cistus salviifolius*, *Geum coccineum* and *Peganum harmala*. All are found in Greece.

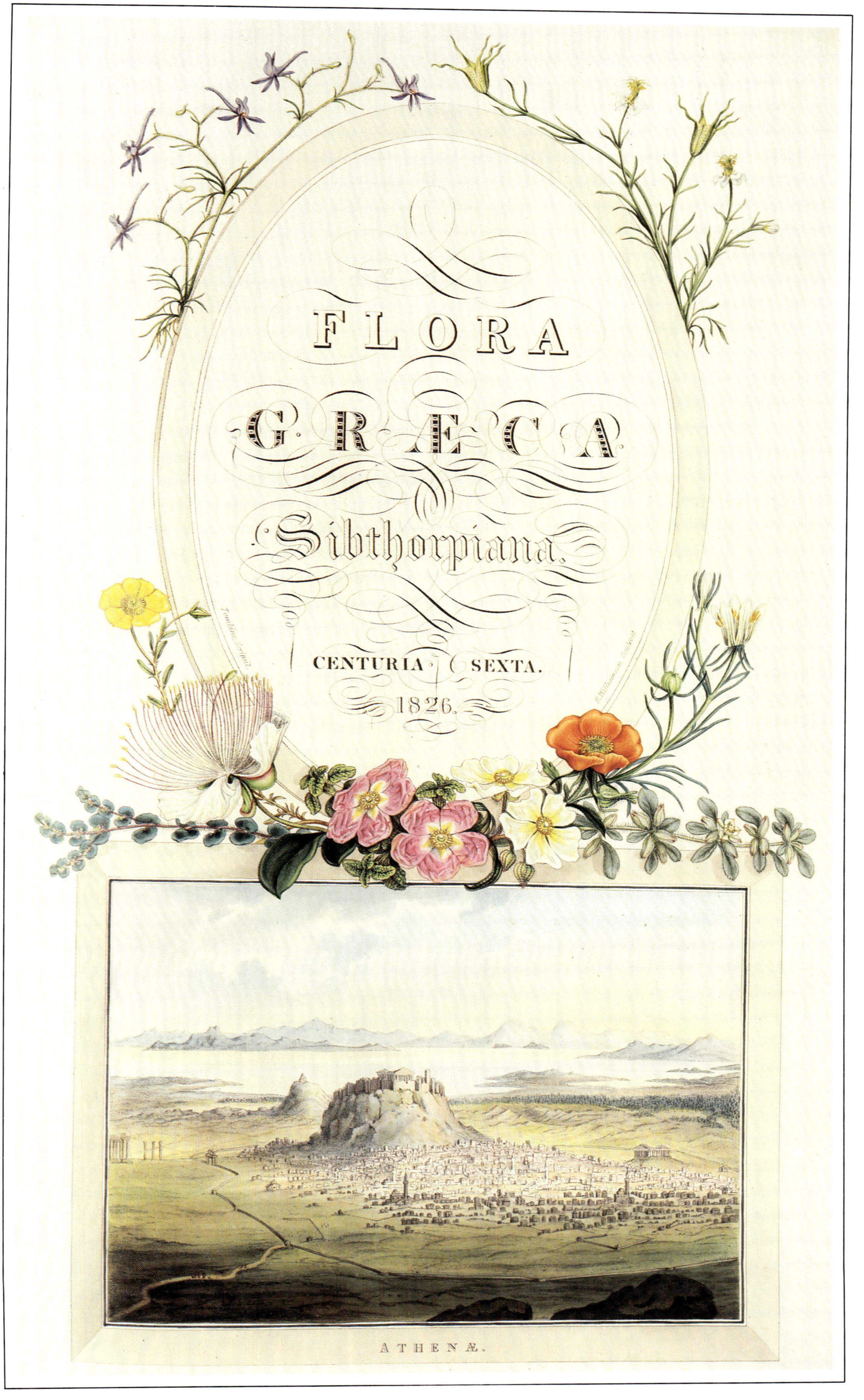
FLORA
GRÆCA
Sibthorpiana.
CENTURIA SEXTA.
1826.
ATHENÆ.

primary aim in undertaking his journey via Vienna to the East, would have seemed conspicuously less eccentric in an era when classical antiquity was still one of the dominant forces in European culture than it does today. Nevertheless, the aims of the expedition are dwarfed by the achievement of the *Flora Graeca*.

John Sibthorp and Ferdinand Bauer set out together in 1786, travelling overland to Naples where they set sail for Crete, landing first at Messina and Milos. The disturbed state of Greece, the possibility of a Russian war, the rebellion of the Turkish pashas and an outbreak of plague at Larissa made a land journey through Greece impossible. They reached Crete in June and visited other Aegean islands as well as Athens and Smyrna. From there, they travelled overland northwest to Bursa and climbed the Bithynian Mount Olympus (Uludag), on which they found the purple flowered *Clematis viticella*. They spent the winter in Istanbul, making some short journeys before setting out again on 14th March 1787. With them was John Hawkins, a friend of Sibthorp and a keen amateur botanist and classical scholar; he had studied at Cambridge and at the universities of Freiburg in Saxony and of Berlin. His ability as a linguist was to prove useful on the journey and in recording the vernacular Greek names of the plants that were found.

Hawkins helped Robert Walpole to compile his *Memoirs Relating to European and Asiatic Turkey* (1817) and his *Travels in Various Countries of the East* (1820), both of which contain many extracts from Sibthorp's journals. As Sibthorp's executor, he was to play a most important part in the production of the great Flora. The three men visited Lesbos, Skiros and Cos and on 28th March, while trying to reach Rhodes, were forced by contrary winds into the harbour of Porto Cavalieri, near present day Marmaris in southwestern Turkey. They visited the ruins of an ancient town, probably Cressa, and found 'a diversity of plants whose beauty was not less striking than their variety.' It was here they collected *Fritillaria sibthorpiana*, such a great rarity that it was not seen again for nearly two hundred years. The day after this discovery the wind changed, and they reached Rhodes to find the town, which had once been one of the main strongholds of the Knights of St John, half ruined and almost deserted. From Rhodes, they sailed to Cyprus where the town of Famagusta was in a similar state. Then they sailed north again to Athens with the intention of climbing Mount Parnassos. Although it was by now 30th June, there was still much snow lying near the summit. Sibthorp was interested to find that the plants were still given many of the same names by the shepherds as were used by Dioscorides and that 'their virtues were faithfully handed down in the oral traditions.'

At Delphi, there was little for them to see, as the site had not yet been excavated, but they found the beautiful dwarf shrub *Daphne jasminea* growing on the rocks there. Here, too, Sibthorp recorded that pigs were eating the tubers of the wild arum which the shepherds called *drakontion*. Truffles, which were considered a great delicacy, were not rooted out by pigs but found with the aid of a sort of divining rod.

On 24th September, when the party finally sailed for home from Patras, they had amassed drawings and specimens of over 2000 plants.

Sibthorp and Hawkins made a second journey to Greece in 1794, while Bauer stayed in Oxford to work on his drawings. On this journey they were accompanied by a young Italian, Francesco Borone, who had been trained as a botanical assistant and had already accompanied the botanist Adam Afzelius on a trip to Africa. In October, when they were in Athens, Borone fell from an upstairs window while walking in his sleep. His death at the early age of 25 cast a shadow over the expedition, and Sibthorp, after spending the early spring of 1795 exploring the Peloponnese, sailed for England at the end of April. After encountering contrary winds, and being detained at Otranto by quarantine restrictions because of bubonic plague, he travelled overland to reach England in the autumn, complaining of 'a nasty low fever, with a cough that alarms me, from some affection of the lungs.'

On his travels Sibthorp was not just interested in botany; he looked at and recorded birds and fish, and local agriculture, medicine and customs. Much of his journal is concerned with these. He records that the spurges *Euphorbia dendroides* and *Euphorbia characias* were used to make fish poison as a method of fishing, a purpose to which he notes that hemlock and cyclamen tubers were put in other places. We learn, too, that the Turks planted '*Amaryllis lutea*' (now called *Sternbergia lutea*) on graves, and that they used the roots of various species of *Orchis* and *Ophrys*, as they still do, for a drink, salep, which was esteemed for its aphrodisiac properties. Cannabis was used as a medicine, being drunk in an infusion as an anaesthetic for surgery, or boiled in oil for use against rheumatism. Sibthorp did not think highly of the local retsina, wine flavoured with resin from the pine, *Pinus pinaster*: $1\frac{1}{2}$ parts of resin were added to 20 parts of wine to act as a preservative and prevent it from becoming acidic.

Back in England, Sibthorp's fever and cough developed into tuberculosis and in spite of treatment in Brighton and baths in asses' milk he died in Bath in 1796 at the age of 38. In his will, he directed that 'all my freehold estate profits or rents of such estate [are] to be applied in the following manner, first for the publication of a work for which I have collected the materials to be entitled *Flora Graeca*, which is to consist of ten folio volumes, each volume to consist of 100 plates, and also a small octavo without plates entitled *Prodromus Florae Graecae*.' Hawkins was appointed chief executor, together with Thomas Platt and the Hon. Thomas Wenmar. Their first task was to find an editor for *Flora Graeca*. They managed to enlist J. E. Smith (1759–1828), one of the best and most prolific botanists of the period, who was given a salary of £150 per annum for eleven years. He had to start by correlating all Sibthorp's notes, specimens and journals, as well as Bauer's drawings. The enterprise was all the more difficult because Sibthorp had intended to rely to a considerable extent on his memory, and many of the specimens were without details of locality. Sibthorp's handwriting also caused Smith some difficulty. As Hawkins admitted to Smith, 'It is certainly a pity that Dr Sibthorp did not mark all his specimens or the drawings: but he trusted to his memory and dreamed not of dying.' The early part of the journal was written as a diary with each bird or plant mentioned in its place, but in Cyprus the plants and animals were brought together at the end; on the second tour, each area visited was written up as a separate entity. Smith persevered with the work and in 1805 wrote to Sir Joseph Banks, 'I am now hard at work on the *Prodromus* (having got the great work sufficiently forward) which would be easy enough if Sibthorp had referred from his manuscript habitats and catalogues to either specimens or drawings; but as he has not, it is a [matter] of laborious criticism and investigation.' But after Smith's death, his wife wrote, 'The *Flora Graeca* was the work in the compilation of which Sir James had particular pleasure. Those who have seen the thousand beautiful delineations of the flowers of that country by the hand of Mr F. Bauer may conceive that it was no dull employment, it was a work of relaxation.' Smith himself wrote all the descriptions of the plants. Sibthorp had compiled two lists of Greek plants to which he added citations of Linnaeus and Tournefort; Smith amended these lists and took them as the basis of the *Prodromus*, which was published in 1806, the same year as the first part of the large *Flora Graeca*.

Smith died in March 1828, after six volumes had been published and he had prepared the first half of the seventh. Robert Brown finished Volume 7 and the task of completing the work was entrusted to John Lindley, ironically a vigorous critic of Smith, but a friend of Hawkins. He produced the remaining three volumes over the next ten years. To the last volume he appended a list of species arranged by the natural classification of Augustin de Candolle which by 1840 was superseding the artificial sexual system of Linnaeus that had been used in the rest of the *Flora*. Lindley also added a list of the Greek names of Dioscorides and of the vernacular names of the plants in use at the time. Volume 10 was published in late 1840, and Hawkins died in 1841, his duty as an executor well and truly finished. The original

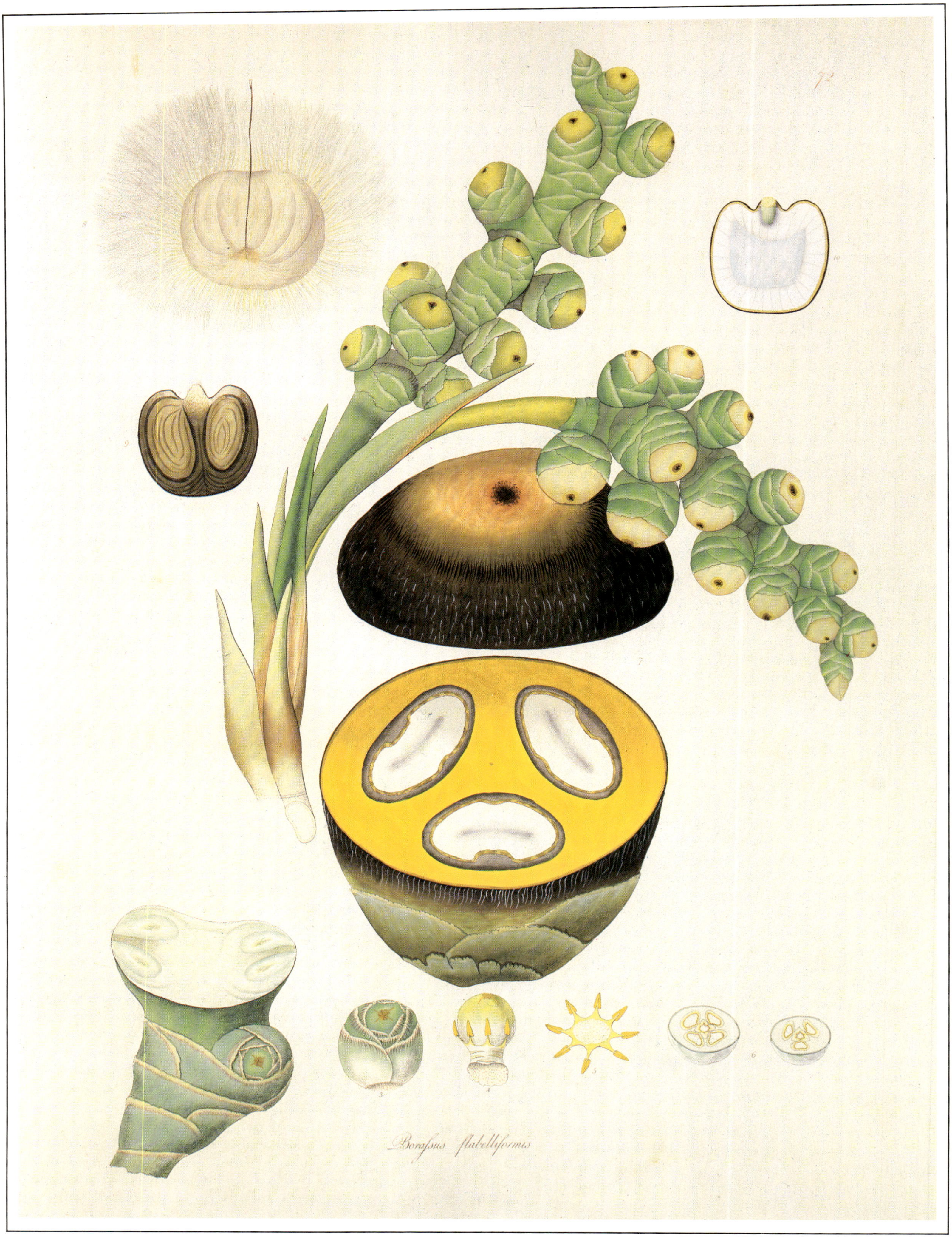

intention had been to produce fifty copies for subscribers at £254 a set; by 1840, only twenty-five of the original subscribers were left. The production had cost £15,576 6s. or £620 per set. Demand for the work continued, and between 1845 and 1856 Henry G. Bohm produced a reissue of about forty copies at only £63 per set.

Flora Graeca itself is one of the most magnificent floras ever produced. In the ten volumes, one thousand of Ferdinand Bauer's paintings are reproduced in folio size, the majority of them engraved by James Sowerby and his family; in addition, there are charming title pages with small vignettes of some of the places visited and garlands of appropriate flowers. The standards of botanical accuracy and aesthetic appeal are maintained throughout – no mean feat where a thousand pictures are involved.

While Sibthorp and Hawkins were visiting Greece and western Turkey, a Frenchman, Jacques Julien Houton de La Billardière decided to visit the area of southern Turkey around the city then called Alexandria Minor (later Alexandretta, today Iskenderun). He sailed from Marseilles in 1787, on a boat bound for Syria, hoping to go from there overland to southern Turkey, but when he landed on Cyprus, he heard that plague had been raging in Antioch (Antakya) and the area around it for several months and that some villages between Aleppo and Alexandria Minor had been completely wiped out. He changed his plans and instead went to the Holy Land, Mount Lebanon and Damascus, finally going as far north as Latakia and visiting Mount Cassius, on the present Turkish-Syrian border. After two years in the East, he returned to Paris. Descriptions of the new plants he found were published between 1791 and 1812 in *Icones Plantarum Syriae Rariorum*, a quarto containing extremely elegant engravings after drawings by Pierre-Antoine Poiteau, H-J. and P-J. Redouté, and Pierre-Jean-

A juniper, *Juniperus drupacea*, a tree native to the mountains of Greece, Turkey and Syria. Engraving by Milsan after a drawing by Pierre-Joseph Redouté from *Icones Plantarum Syriae Rariorum* (1791–1812) by Jacques-Julien Houton de La Billardière.

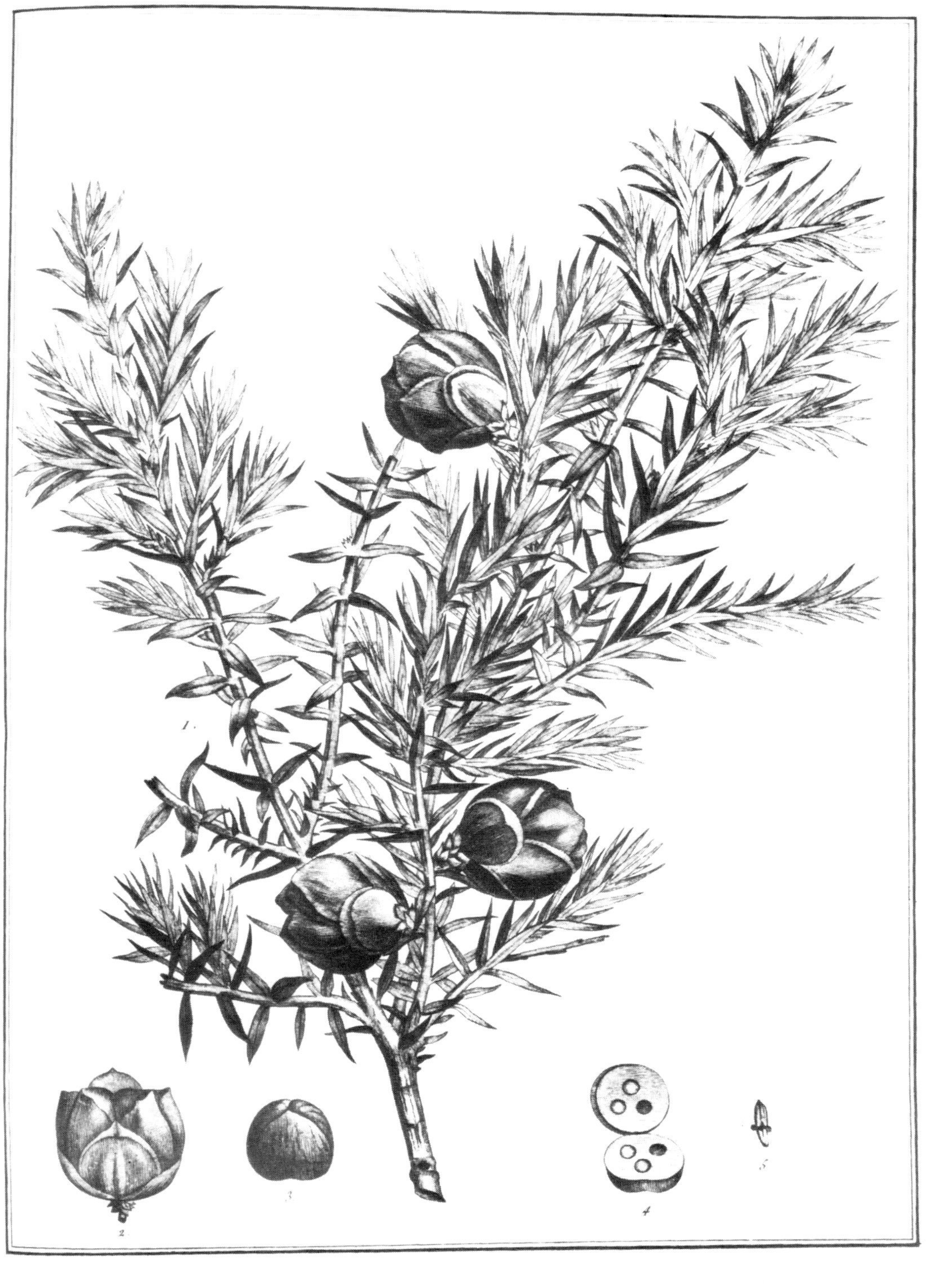

Left:
Palmyra palm or toddy palm (*Borassus flabelliformis*). Hand-coloured engraving by Mackenzie after a painting by an unknown Indian artist from *Plants of the Coast of Coromandel* (1795–1818) by William Roxburgh. The paintings used in this book had been accumulated by the East India Company. The palm, which is similar in habit to a coconut palm, but with a thicker trunk and more drooping leaves, is found in southern India, Burma and Malaysia. The text notes that the juice collected after the inflorescence has been cut off is fermented to make a toddy, which is also distilled into a potent spirit.

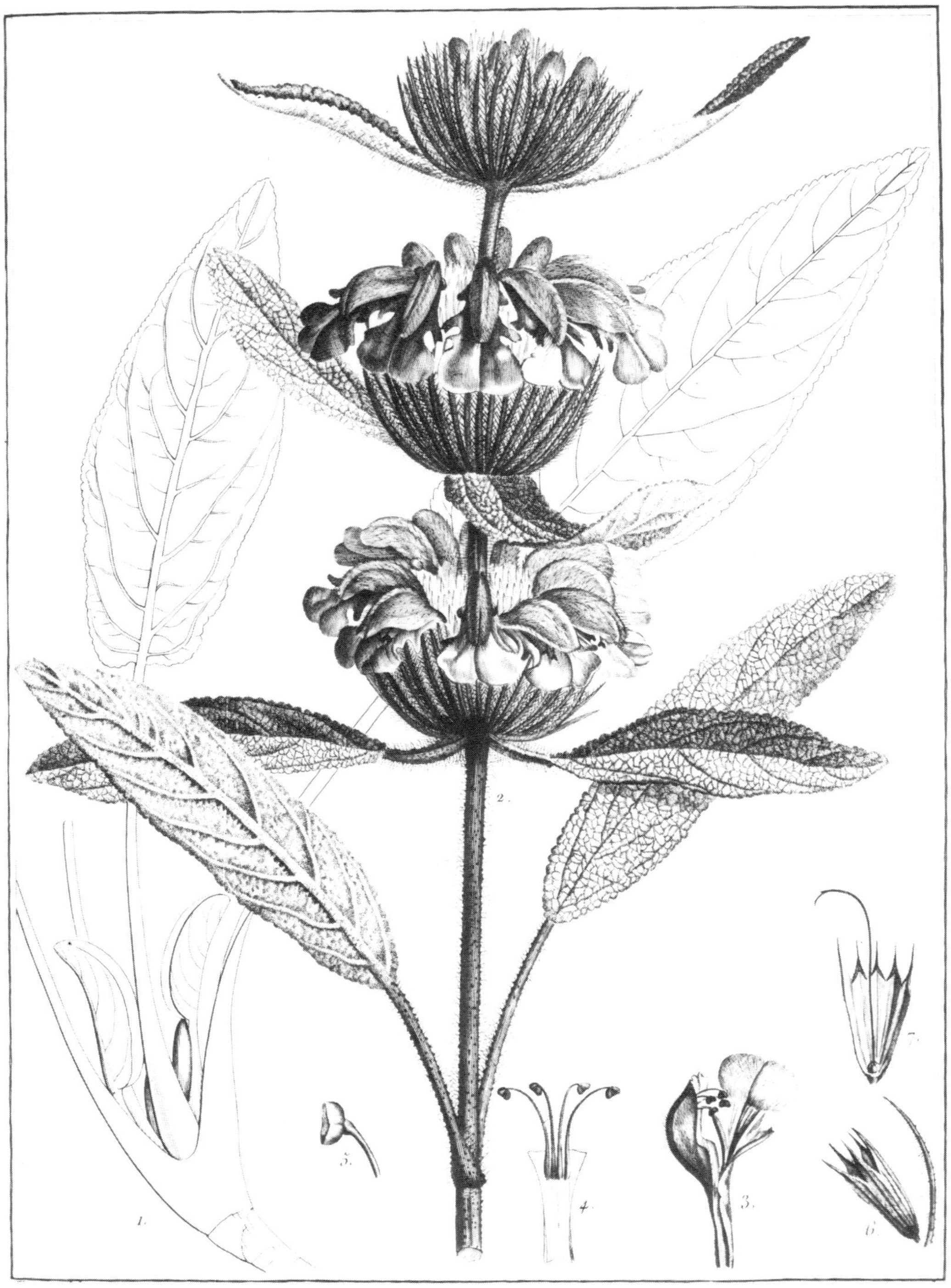

Four illustrations from *Icones Plantarum Syriae Rariorum* (1791–1812) by Jacques-Julien Houton de La Billardière. *Left:* a small herbaceous labiate, *Phlomis rigida*, found on mountains from Turkey to Lebanon; this engraving by Dien after a drawing by Pierre-Jean-Francois Turpin is a fine example of the highly sophisticated work that was being produced in France in the late eighteenth and early nineteenth centuries. *Right:* a dwarf knapweed, *Centaurea pygmaea*, engraving by Dien after a drawing by Turpin. *Below:* *Asperula capitata*, a member of the bedstraw family, Rubiaceae, showing the long twiggy stems that enable it to survive on mobile screes, on which it occurs from Turkey and Iron to Lebanon; engraving by Voisar after a drawing by Pierre-Joseph Redouté. *Far right:* *Sorbus trilobata*, a native of mountains in Lebanon; engraving by Dien after a drawing by Pierre-Antoine Poiteau.

François Turpin. These four plant illustrators were working together during this period, and their achievements can be seen in several scientific botanical works produced in France at the time. These works also used the talents of a number of engravers, notably Sellier and Dien, who worked together and employed a combination of stipple and line engraving with particular skill. The plate of the hairy *Phlomis rigida* for La Billardière is one of the very finest examples of plant engraving. Among new species figured in this work are *Prunus prostrata* from the Lebanon, *Juniperus drupacea* from Mount Cassius, and *Fontanesia phyllerioides*, which was named in honour of the botanist René-Louiche Desfontaines.

A similarly illustrated work by Desfontaines himself, the *Flora Atlantica*, was published in 1798. Desfontaines had a childhood friend who was French Consul in Algiers, and for two years he lived there collecting plants, 'not without molestings and difficulties'. He covered the area between Tripoli and Algiers, studying the vegetation, crops, geology and rivers. The plates in his flora are very delicately drawn by the Redouté brothers, Maréchal and Marin, and engraved by Sellier.

Desfontaines was also instrumental in publishing in 1808 the *Choix de Plantes du Corollaire des Instituts de Tournefort*, a memorial volume to mark the centenary of Tournefort's death. It was illustrated with stipple engravings by Lambert from paintings by Aubriet in the royal collection of *vélins*, which at the Revolution had been transferred to the Musée d'Histoire Naturelle. The seventy plates include the lenten rose (*Helleborus orientalis*),

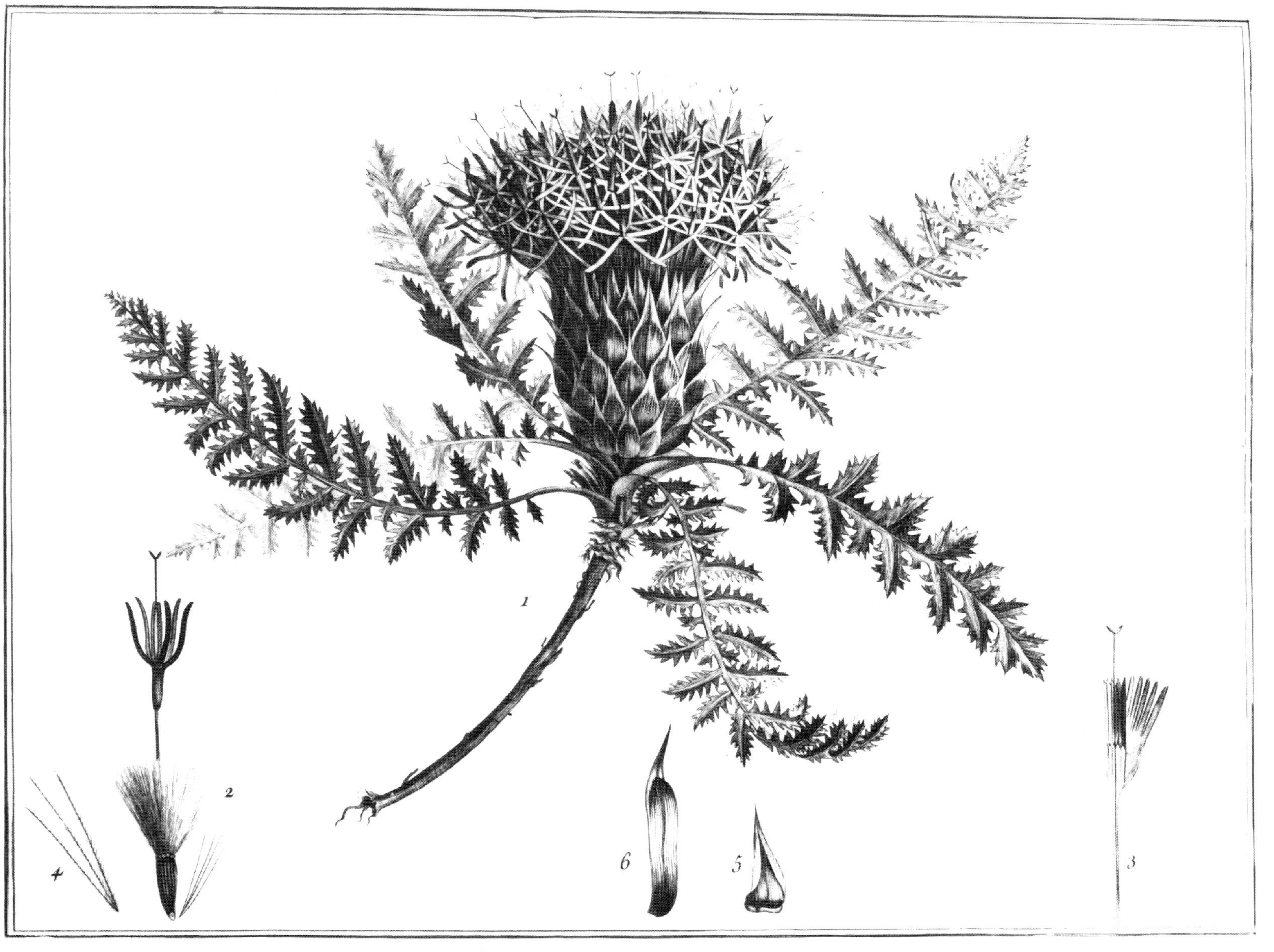

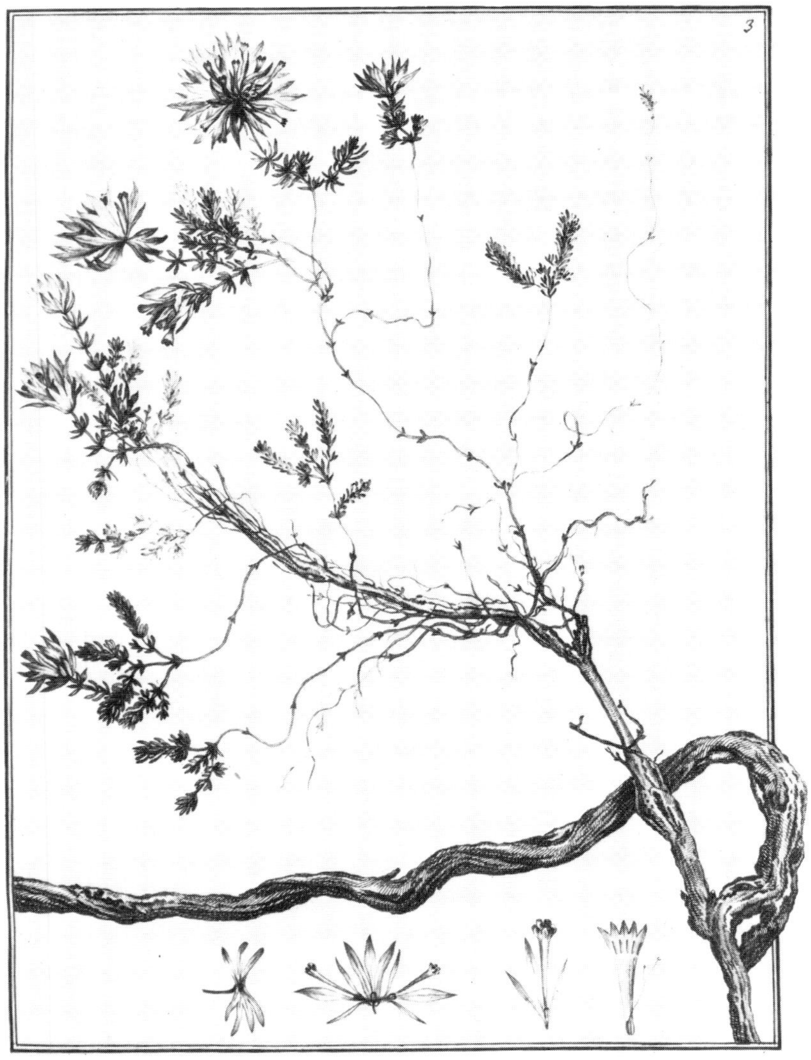

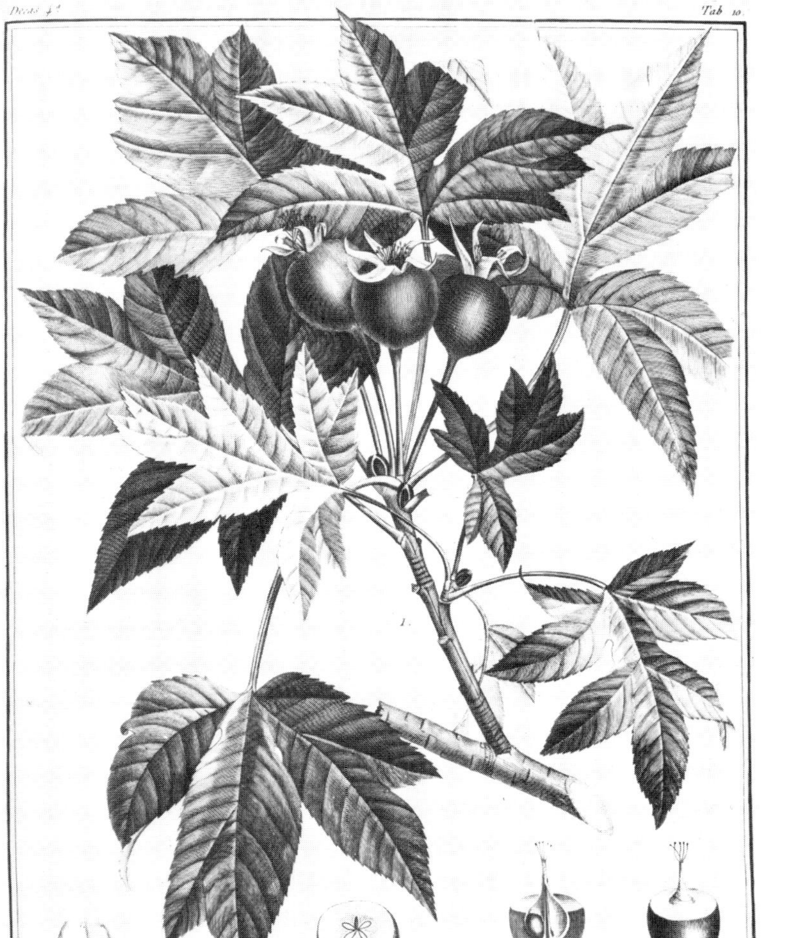

Poiteau del. Dien sculp.

Crataegus trilobata.

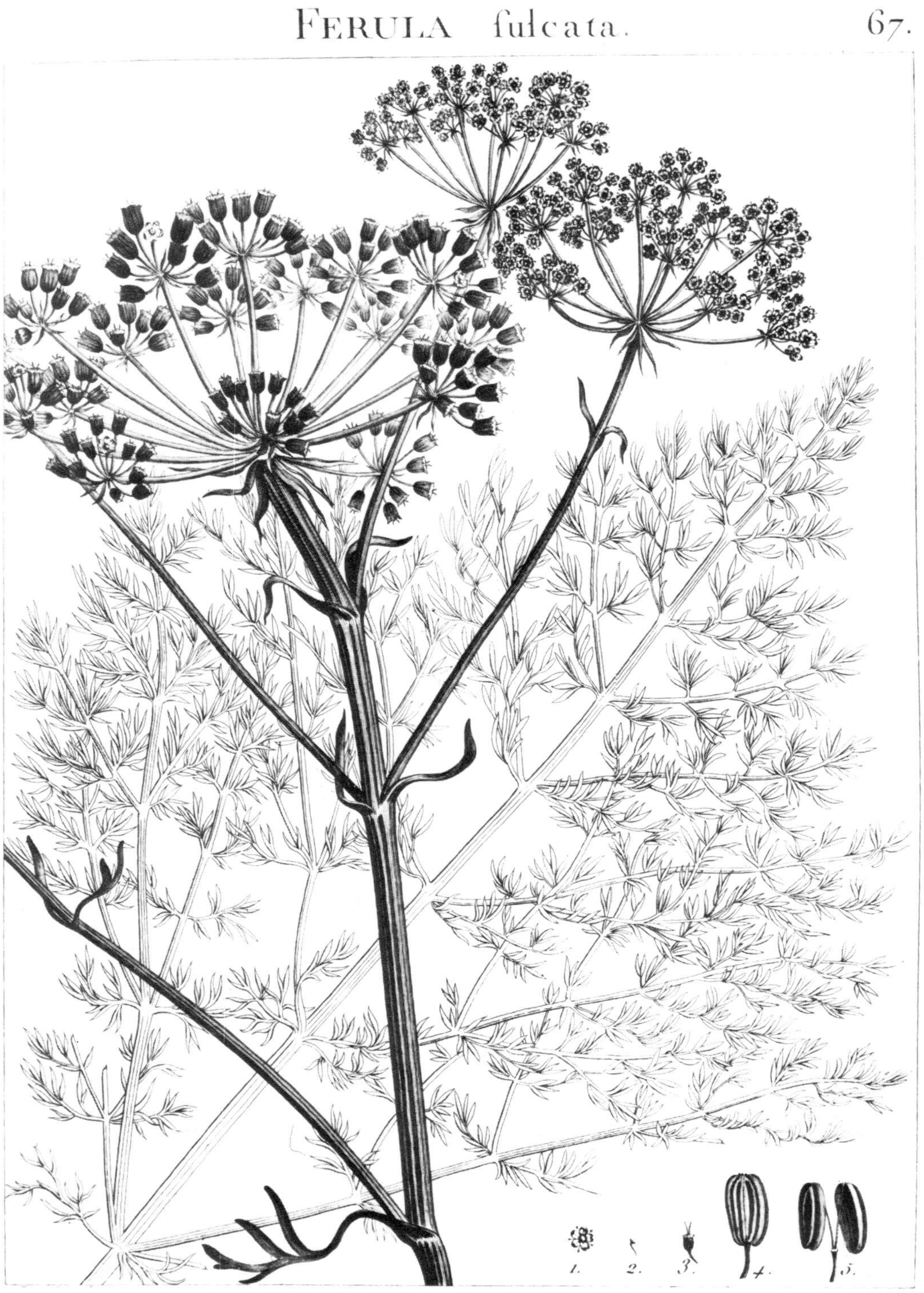

A fennel, *Ferula lutea*, which is found in North Africa, Spain and Portugal. Engraving by Sellier after a drawing by Marechal from *Flora Atlantica* (1798) by René Louiche Desfontaines.

and the strange *Phelipaea tournefortii*, named in honour of the Phelipeaux family, '*protecteurs de Tournefort et des sciences naturelles*'. It is a parasite on the composite *Achillea* and is still a common sight in the hills of Armenia with its scarlet velvety flowers like stiff pansies among the wiry grass.

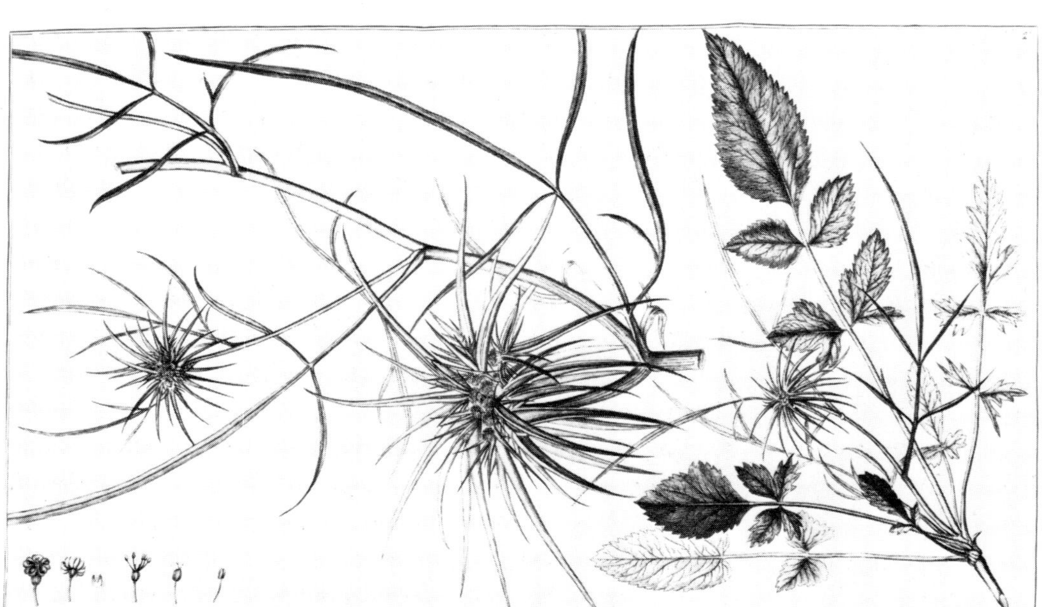

Exoacantha heterophilla, a Syrian umbellifer. Engraving by Aubry after a drawing by Henri-Joseph Redouté (the younger brother of Pierre-Joseph) from La Billardière's *Icones Plantarum Syriae Rariorum*.

CHAPTER VII
Asia and Africa

Russia

The botanical exploration of Russia was initiated by Peter the Great when he sent a German, Dr Daniel Messerschmidt, on a journey across Siberia in 1720. Messerschmidt made a large collection of plants but nothing of it has survived, and the results of the expedition remained unpublished. The Imperial Academy of Sciences in St Petersburg was founded by Peter's widow, Catherine, in 1725, and the first botanical expedition which had any lasting results was that led in 1733–43 by another German, Johann Georg Gmelin, who was born in Tübingen in 1709. Its purpose was to explore and collect across Siberia from the Urals to Kamchatka, describing the main rivers, deserts and lakes, the climate and the plants. The party included two other academicians, an ethnologist G. F. Muller, and an astronomer, L. de L'Isle de la Croyère. At the same time, a naval expedition led by Captain Vitus Bering was to survey the eastern coastline of Russia and Kamchatka and keep the scientists supplied.

Gmelin's party left St Petersburg on 8th August 1733 and did not return for ten years. By 1735 they had reached Lake Baykal; in 1736, they arrived on the shores of the Sea of Okhotsk where they found Bering and the naval contingent already established. Here Gmelin lost most of his belongings and money and some of his collections in a fire, and transport could not be found to take supplies on the next stage of the journey to Okhotsk. Instead, the expedition returned to Irkutsk in March 1738 and sent to St Petersburg for permission to return. Meanwhile, however, another botanist, Georg W. Steller (1709–1746) had been sent from St Petersburg to join them. He arrived in Siberia in 1739. Steller was a tougher man than Gmelin and prepared to undergo much greater hardship in the cause of botany. He went on eastward to Kamchatka, where he forced himself upon Bering's expedition to America. Because of bad weather and scurvy among the crew, only one landing was made on Alaska, but Steller was there and collected 140 species of plants. The results of the expedition were described in Gmelin's *Flora Sibirica*. A long preface in Latin recounts some of the vicissitudes the travellers encountered; Gmelin is unstinting in his praise for Steller. The first volume was published in 1747, the fourth and last in 1769. The drawings, engraved on copper, were by Johann Christian Berkhan and Johann Wilhelm Lursen, who had accompanied the expedition throughout. The style of drawing is free and rather untidy, but a large number of species are described and illustrated, including the ghost orchid (*Epipogium aphyllum*) and the arctic orchid (*Calypso bulbosa*). Gmelin and Muller finally returned to St Petersburg in 1743, and Steller reached the Urals in 1746. Arrested and unfairly charged, Steller was sent back to Siberia and, although finally exonerated and freed, he died of fever on his way back to St Petersburg in November 1746.

The second great Siberian journey for plant exploration was that of Peter Simon Pallas (1741–1811) and J. G. Gmelin's nephew, Samuel Gottfried Gmelin (1744–1774). Pallas, too, was a German, the son of a Berlin surgeon whose reputation at the age of 23 was sufficient for him to be offered the post of Professor of Natural Science at St Petersburg by Catherine the Great. The Russian Academy of Sciences was planning an expedition to observe the

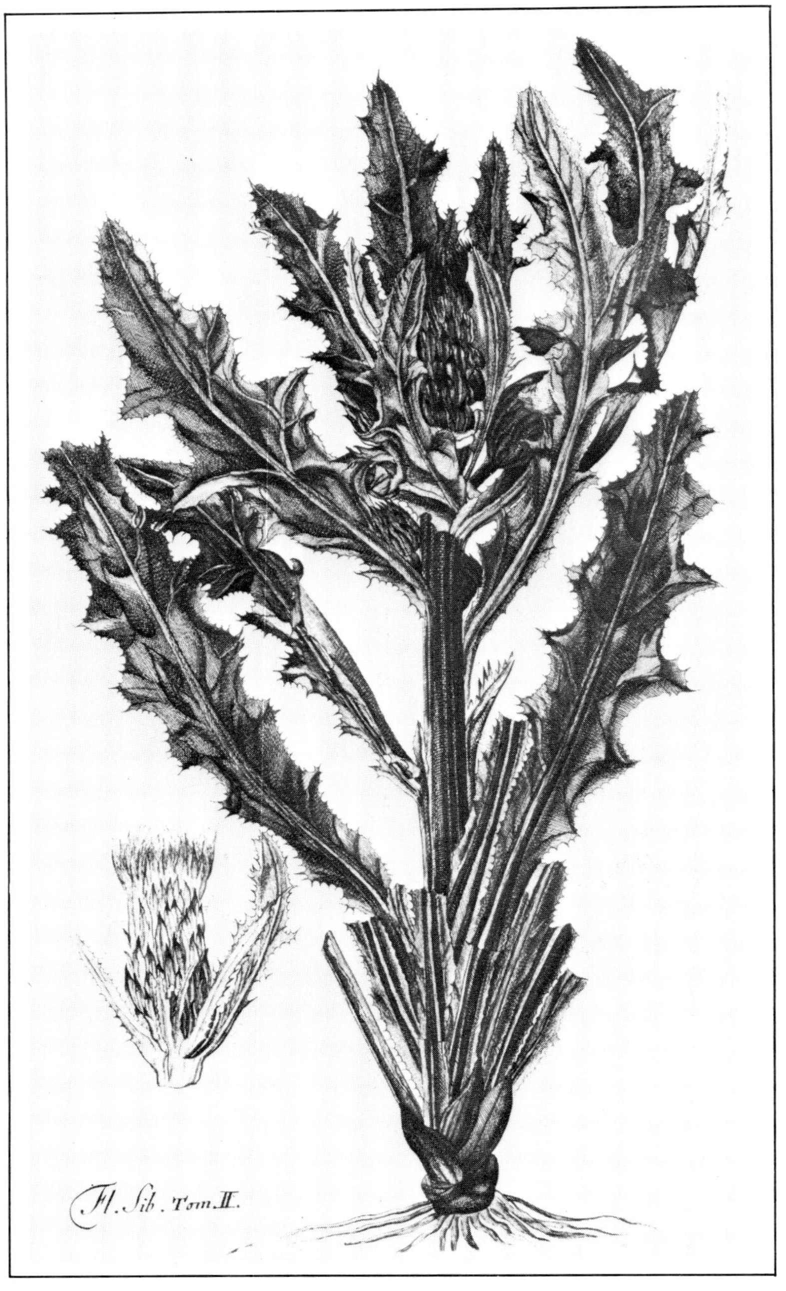

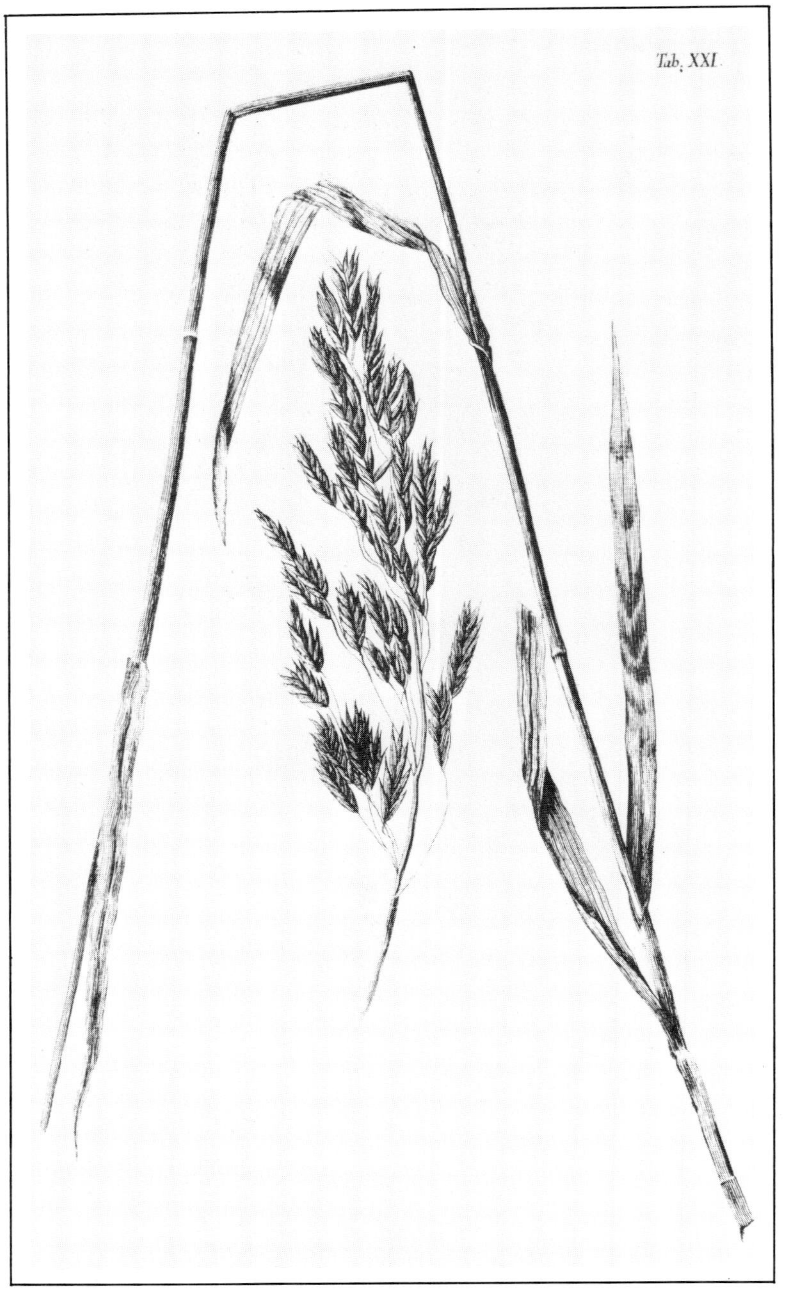

A thistle (*Cirsium*) and a grass. Engravings from *Flora Sibirica* (1747–69) by Johann Georg Gmelin.

transit of Venus (which was also one of the aims of Captain Cook's great voyage), and Pallas and Gmelin decided to join them. The journey lasted six years and was described by Pallas in detail in the stout, dull volumes of *Travels*, which covered every branch of natural history. After his return, Pallas continued to live in St Petersburg and published many works of natural history. However, his main botanical work, the beautiful *Flora Rossica*, was never finished although the first two volumes were published in 1784 and 1788. He made a second journey to Astrakhan and the Crimea in 1793–94 and, because of his courtly reports to Catherine of the beauty of the country, was granted an estate in the Crimea on which to end his days. This proved to be no more than an exile, as the place was unhealthy and much of the countryside had been ravaged by Russian soldiers. According to the Cambridge mineralogist E. D. Clarke who visited Pallas in the Crimea for two months in 1800, the drawings for the last two volumes were finished, but under the tyrannical rule of the mad Tsar Paul they could not be published in Russia. The Russian authorities confiscated the proof sheets sent to Pallas from Leipzig; only twenty-five extra plates were ever published, and they are very rare.

The drawings for *Flora Rossica* were by Karl Friedrich Knappe and are mostly of central Asian and Siberian plants, as well as a few arctic species.

Pallas corresponded with Sir Joseph Banks at this time and sent him specimens of seeds of several plants, including a pear, *Pyrus salicifolia*, and *Paeonia lactiflora*, which is the ancestor of the cultivated herbaceous peonies. He also managed to complete a folio monograph on Russian species of *Astragalus*, with coloured illustrations by the German C. G. H. Geissler (1770–1844). This is an attractive book and shows much of the variation in

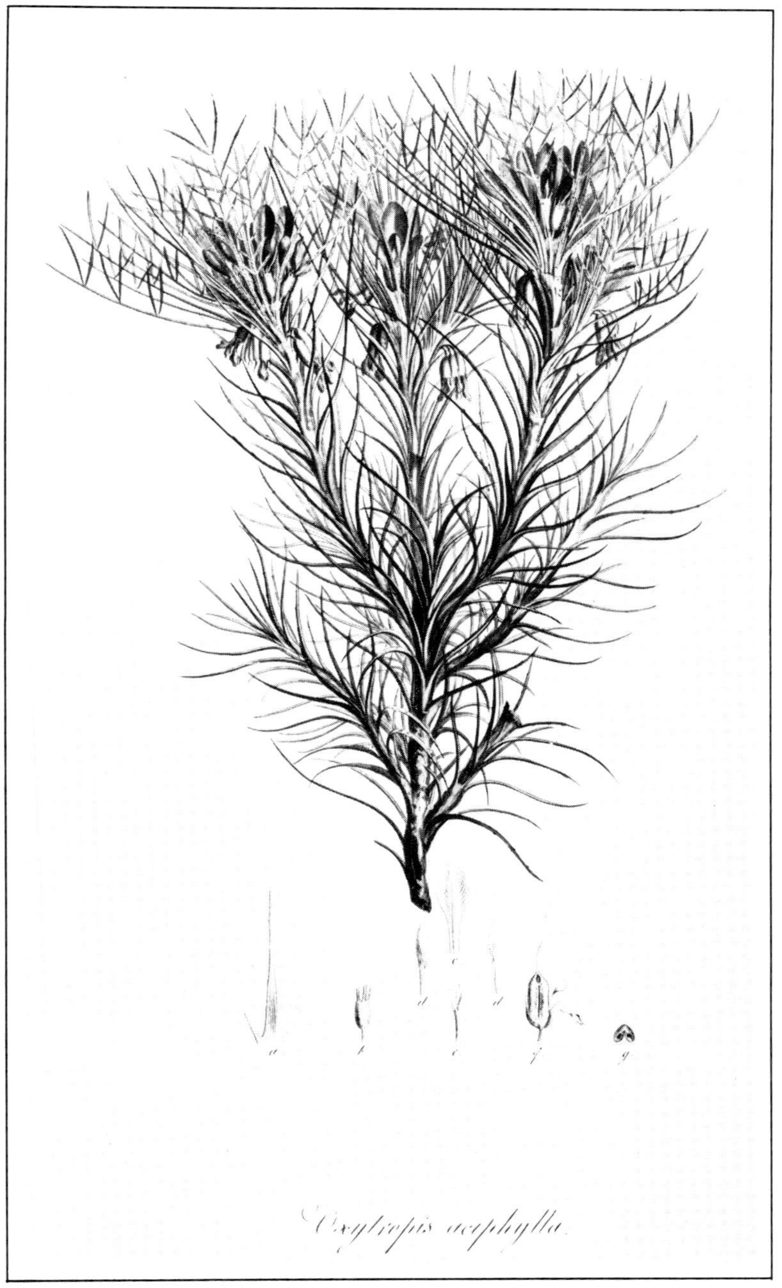

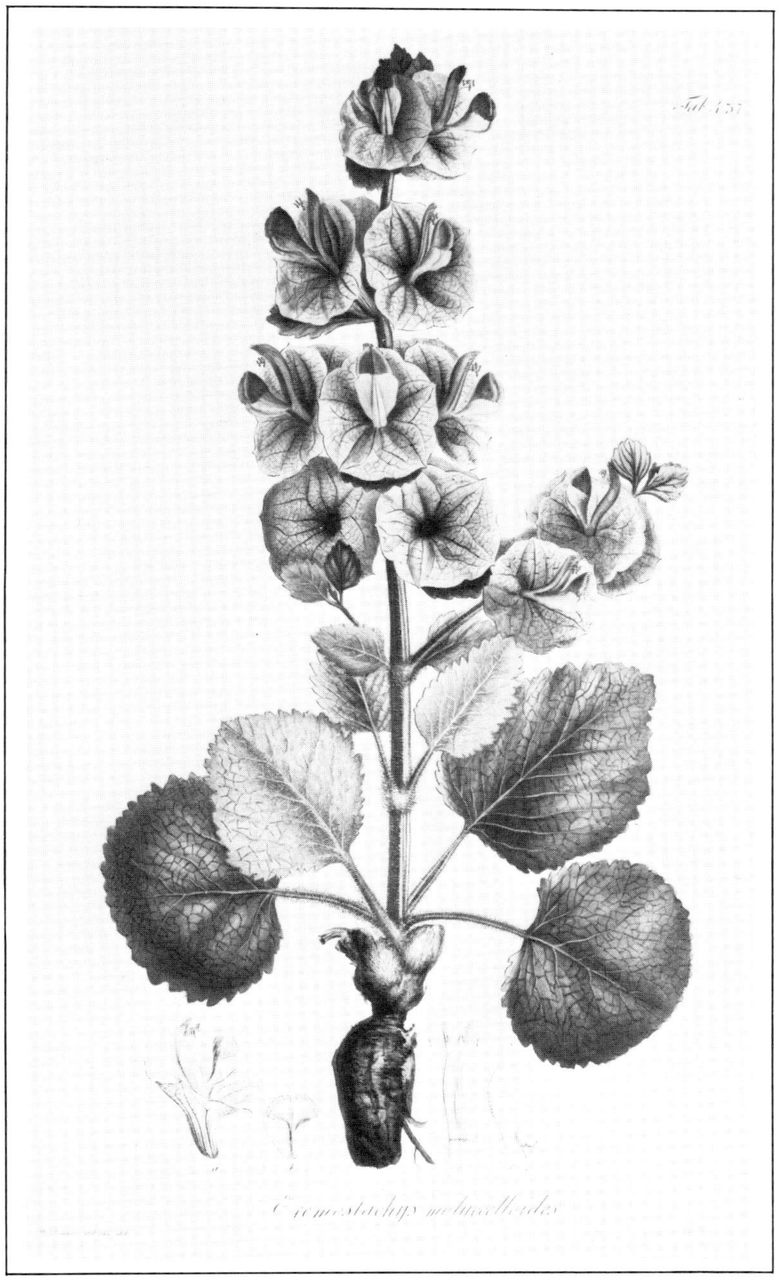

Two illustrations from *Icones Plantarum, Floram Rossicam, Imprimis Altaicam, Illustrantes* (1829–34) by Karl Frederick von Ledeboer, published in Riga, Latvia, engraved by W. Siegrist after drawings made from dried specimens by T. Schefner and W. Krüger respectively. *Above: Oxytropis aciphylla*, a spiny vetch from Siberia and Outer Mongolia. *Right: Eremostachys moluccelloides*, an unusual member of a primarily Central Asian genus, with large calyces similar to those of *Moluccella*.

this huge genus, including many of the tall, hairy-fruited species from the central Asiatic steppes.

A botanist whose collecting was more local, but whose influence on botany and horticulture was as great as that of Pallas, was Baron Friedrich August Marschall von Bieberstein (1768–1826). He was a German who joined the Russian army and was stationed in the Caucasus between 1792 and 1795. At this time the Russians were extending their power in the area, fighting both the Turks and the Persians for possession of Georgia and the Caucasus.

In 1795, Bieberstein joined an expedition to Persia under Count Valerian Zubov, the 24-year-old brother of Catherine the Great's latest lover. The expedition conquered the Persians and took over the territory of Azerbaijan. Later, Bieberstein made extensive explorations throughout the western end of the Caucasus and Georgia, and his main botanical work was the *Flora Taurico Caucasica* published between 1808 and 1819. His very rare illustrated work on the region was published at the same time under the title of *Centauria Plantarum Rariorum Rossiae Meridionalis.* Part 1, of which only seventy copies were produced, was published in 1810 in Charkov. Parts 2 and 3 were published in St Petersburg in 1842–43. The delicate hand coloured engravings include such familiar garden plants as *Iris reticulata*, *Campanula lactiflora* and *Lilium monadelphum*.

The third large illustrated botanical book to appear in Russia after the *Flora Sibirica* and the *Flora Rossica* was by Karl Frederick von Ledebour, Professor of Botany at the University of Dorpat in Livonia (a Baltic province of Russia now divided between Latvia and Estonia). In 1826, he made an expedition to central Asia, taking with him two pupils who were later to become famous as botanists, Alexander von Bunge and C. A. Mayer. The results of this journey across the steppes and into the Altai Mountains of Mongolia were published between 1829 and 1834 as a travel book, a *Flora Altaica* and a magnificent folio in five volumes, *Icones Plantarum . . . Floram Rossicam Imprimis Altaicam Illustrantes.* The illustrations for this are mostly by E. Bommer, some from living plants and some from dried specimens. The plates were lithographed by W. Siegrist in Germany. Only a few of the plants have become familiar to gardeners, such as *Fritillaria verticillata* and *Polemonium pulchellum*; many plates show species of two genera that are very well represented in central Asia, *Allium* (onions) and *Astragalus.*

Japan

Throughout the seventeenth and eighteenth centuries, Japan staunchly protected itself from outside influence; it was almost impossible for foreigners to see much of the country, a state of affairs that was mainly the fault of Spanish and Portuguese missionaries, who had at first been well received, but had finally been banished in 1639 because of their intolerance and interference in politics. The Dutch were allowed a single trading station at Nagasaki, and their yearly embassy via Kyoto to the Court at Tokyo provided the only glimpses that Europeans had of the interior of the country. The first Western botanists to make any observations at all in Japan were both physicians attached to the Dutch East India Company, a German, Engelbert Kaempfer (1651–1715), and a Swede, Carl Peter Thunberg (1775–1828).

Kaempfer was 39 when he landed at Nagasaki in 1690. He stayed for two years at the company's post on Deshima Island. He was born in the north German town of Lemgo and studied medicine at Krakow and Königsberg (now Kaliningrad); in 1683, he was in Uppsala, where he got himself appointed secretary to the Ambassador leading a Swedish mission to Russia; this gave him the chance to go to Moscow, Astrakhan and later Isfahan. When the Ambassador returned to Sweden, Kaempfer remained in Persia, travelling and making notes of anything that interested him, while practising as a physician to pay his way. In 1688, he joined the Dutch fleet, which was then in the Persian Gulf and obtained the post of chief surgeon. The fleet was bound for Java and called at Ceylon and Bengal on the way. In 1689, it reached Batavia, where Kaempfer obtained his appointment to the Company's Governor at Nagasaki.

Deshima was an artificial island measuring only 236 paces by 82 paces, surrounded by a high fence and linked to the mainland by a closely guarded bridge. The Dutch were forbidden to go ashore, although they were allowed to be attended by Japanese servants and prostitutes. By trading his knowledge of mathematics and astronomy, and by the use of his medical skill and liberal supplies of liquor, Kaempfer won the confidence of his interpreters and persuaded them to bring him specimens of plants growing around Nagasaki and to tell him about the country. For the yearly embassy, the journey to Tokyo in the early spring took two months; the route went across the island of Kyushu, by sea to near Osaka, then to Kyoto and finally along the coast, past Mount Fuji, to Tokyo. There was little opportunity for botanising, as the party was closely guarded, but Kaempfer was able to collect plants that were growing along the roadside and in the gardens of inns. Kaempfer gave up his post and returned to Holland in 1693. He took a degree at the University of Leiden and acquired a small estate near his birthplace, becoming doctor to the ruling Prince.

During his lifetime he published one large volume of travels in obscure Latin, the *Amoenitates Exoticae.* As a contemporary wrote in 1726, 'scarcely can one find anywhere more curious things concealed under dark language. Reading him is like travelling over those craggy rocks and rough mountains that he went over to gather his observations and compose his treatise.' The book was in five parts, of which the fifth was a skeleton flora of Japan with the earliest descriptions of many now familiar plants such as camellias, hydrangeas and the maidenhair tree (*Ginkgo biloba*).

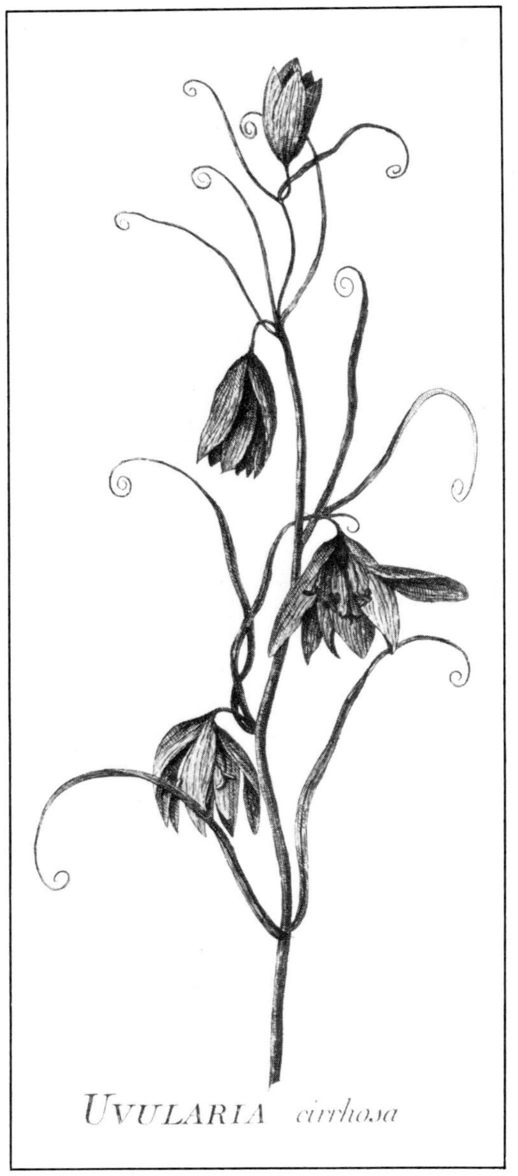

'*Uvularia cirrhosa*', a Chinese fritillary (*Fritillaria thunbergii*), which is commonly grown in Japan and is used in cough medicines. Engraving from *Icones Plantarum Japonicarum* (1794–1805) by Carl Peter Thunberg.

The maidenhair tree (*Ginkgo biloba*), showing fruits and seeds. Engraving from *Amoenitates Exoticae* (1712) by Engelbert Kaempfer. Originally a native of China, *Ginkgo* was grown in Japan long before it was introduced from there to Europe around 1730. It arrived in England in 1754, and the old tree in the Royal Botanic Gardens at Kew was planted in 1762. Male and female flowers are formed on separate trees, the producing cherry-like yellow fruits with edible seeds.

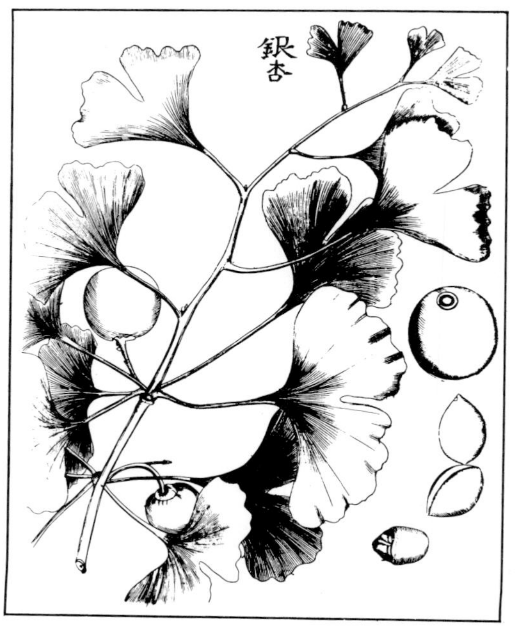

Tea (*Camellia sinensis*), showing flowers and fruits (A) and a young shoot suitable for picking to make tea-leaves. Engraving from *The History of* Japan (1728) by Engelbert Kaempfer, translated into English by J. G. Scheuchzer.

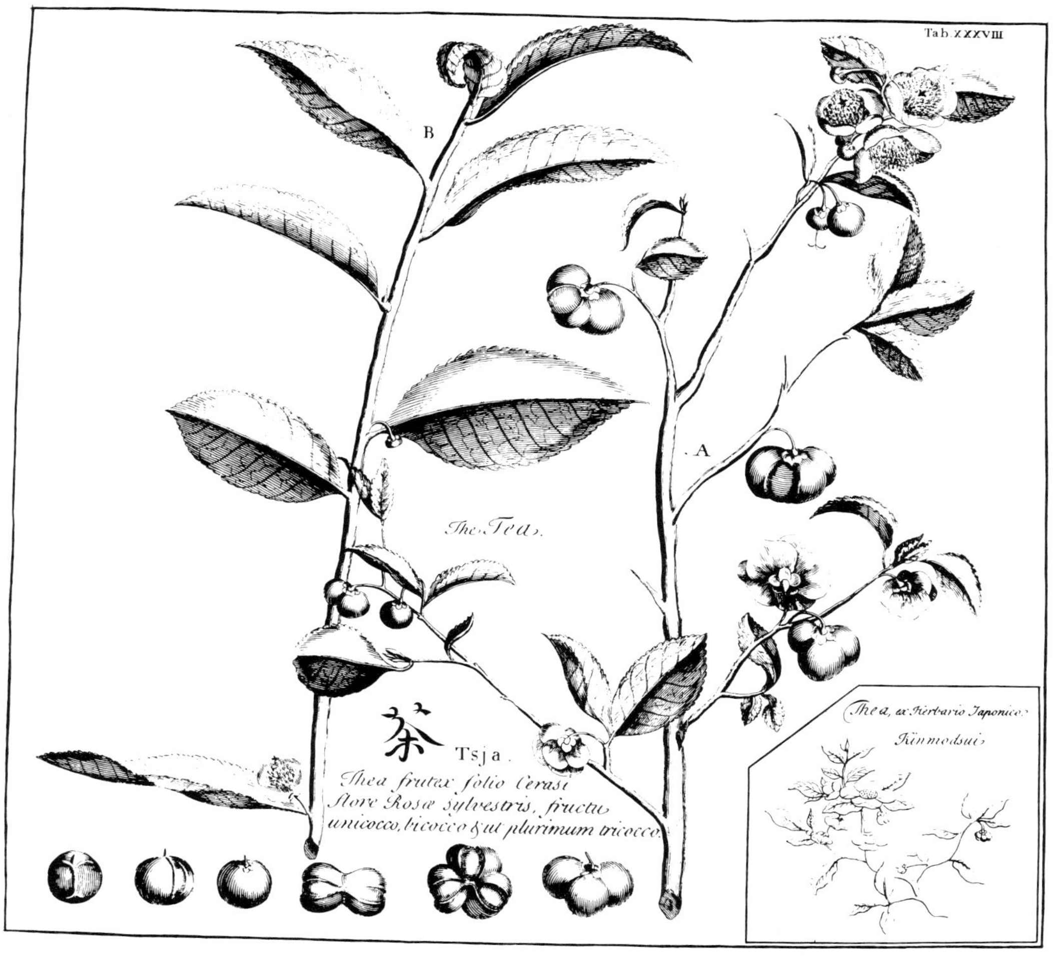

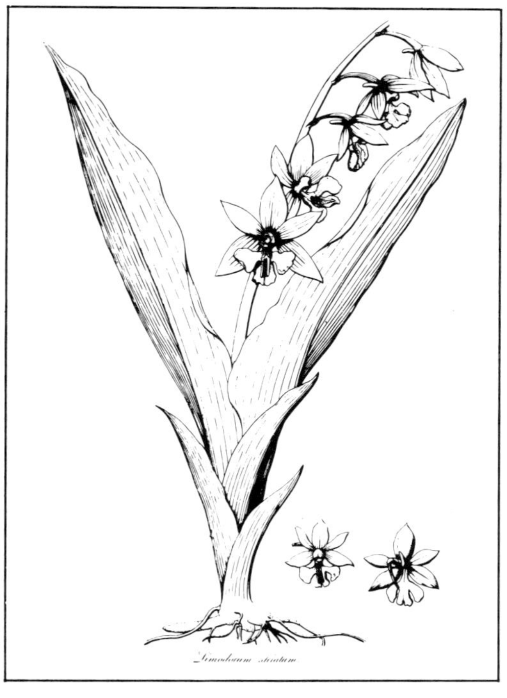

'*Limodorum striatum*', an orchid. Engraving from *Icones Selectae Plantarum quas in Japonia collegit et delineavit Engelbertus Kaempfer* (1791) edited by Sir Joseph Banks.

After Kaempfer's death, Sir Hans Sloane bought all his papers from his nephew; these are now in the British Museum. In 1791, Sir Joseph Banks published engravings from a selection of Kaempfer's drawings, *Icones Selectae Plantarum quas in Japonia Collegit et Delineavit Engelbertus Kaempfer*; many of the species illustrated are again now familiar in gardens, including *Lilium speciosum*, *Magnolia kobus*, *denudata* and *liliiflora*, *Euonymus japonicus* and *Hosta sieboldiana*.

The drawings are elegant, simple and accurate; they show Kaempfer to have been a skilled and observant draughtsman. They are later and more successful than the much smaller engravings published in Kaempfer's own work.

The conditions in which the Europeans at Nagasaki had to exist had changed little 85 years later, when Carl Thunberg arrived in Japan in 1775. Thunberg was 27 when he set out from Uppsala, going first to Amsterdam where he saw Professor Johannes Burmann, the botanist who had edited posthumous works by both Rumpf and Plumier. Burmann persuaded several wealthy amateur plant-lovers to send Thunberg to Japan to collect plants. These patrons gave him the means and the letters of recommendation necessary for the voyage, but do not seem to have expected a quick return. Thunberg entered the service of the Dutch East India Company, and, after spending two years at the Cape in South Africa, he finally reached Japan in August 1775. He was unable to set foot on the mainland until Spring 1776 and relied for plant specimens on what he could persuade the interpreters to bring him and on what he could find in the fodder collected to feed the cattle that the Dutch were allowed to keep for meat. Astonishingly, he managed to collect over nine hundred different species. Together with the plants he managed to grab from the wayside as he went with the Ambassador on the annual embassy to Kyoto, these formed the basis of his *Icones Plantarum Japonicarum* which was published between 1794 and 1805. Among the novelties illustrated in the rather crude engravings are some plants that are now widely grown in gardens, notably shrubs and trees such as *Acer japonicum* and *palmatum*, *Magnolia stellata*, *Mahonia japonica* and *Aucuba japonica*.

Thunberg left Japan in 1778 and returned to report to his sponsors in Amsterdam; he rewarded them by naming new genera in their honour. He

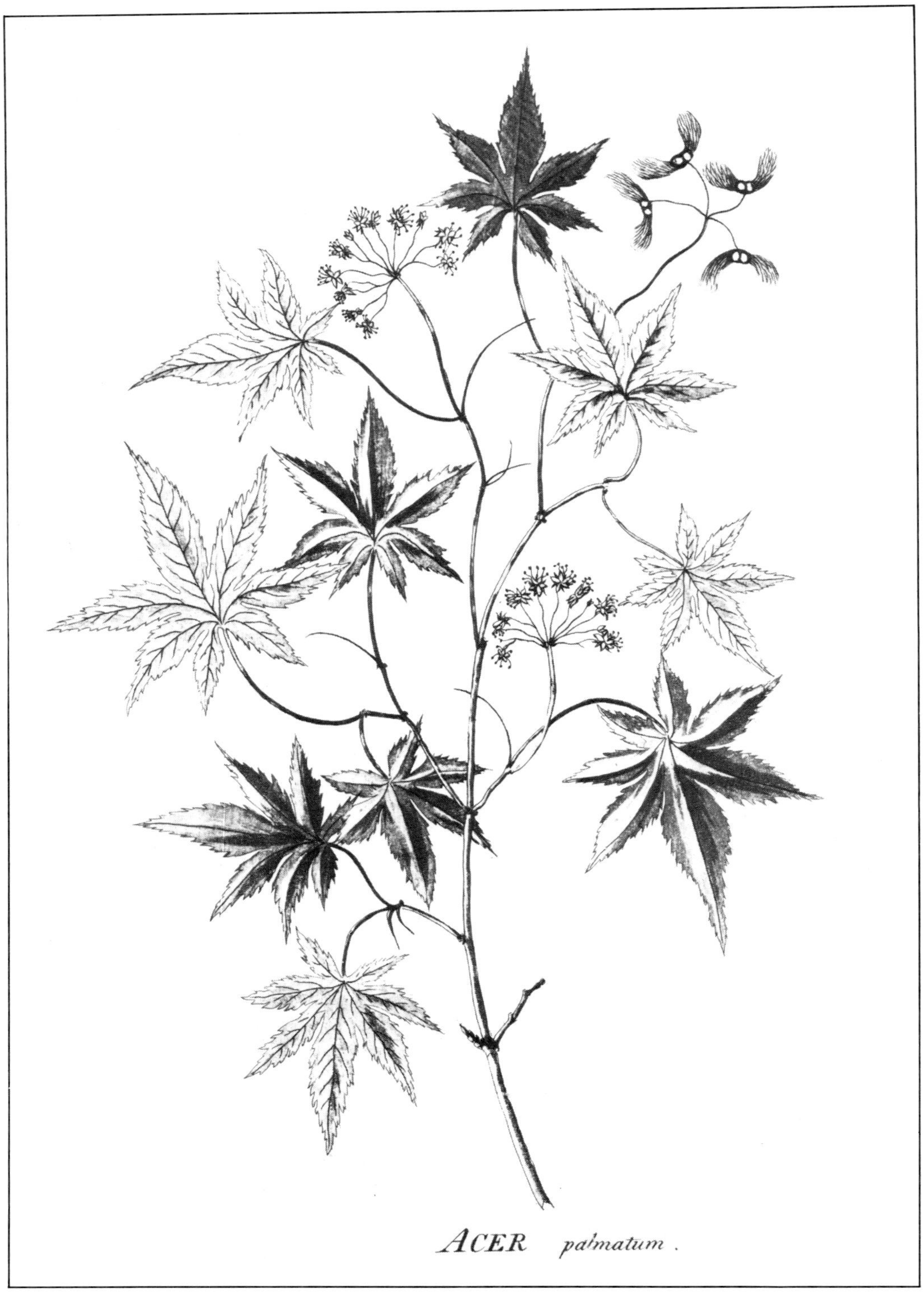

A Japanese maple (*Acer palmatum*), a species that is greatly venerated in Japan, where it is widely cultivated and has given rise to hundreds of varieties. It was introduced to cultivation in England in 1820. Engraving from *Icones Plantarum Japonicarum* (1794–1805) by Carl Peter Thunberg.

also had the pleasure of seeing some of the plants that he had sent back growing happily in Holland.

China

The early exploration of China by Europeans was also beset by difficulties. The first plants came back to Europe in ones and twos, mostly from the tropical coast of China and from the Treaty ports such as Macao and Canton. It was not until 1803 that Sir Joseph Banks sent out William Kerr (after whom *Kerria japonica* is named), at a salary of £100 a year, to collect plants and send them back to England. Kerr remained in China for eight years and was then appointed Director of the Botanic Garden in Ceylon, where he died shortly after, in 1814, without having produced any writings. Four years later, however, there appeared *Icones Pictae Indo-Asiaticae Plantarum* containing twenty-four lithographs of Chinese plants by Charles Ker, based on Chinese paintings from the library of Andreas Everard van Braem-Honckgest and published without a text, 'rather to illustrate Chinese pictorial art, than further botanical knowledge.' The style is flat and decorative, rather after the manner of Japanese woodcuts. Most of the plants shown are cultivated, like the water chestnut (*Trapa natans*), *Rosa laevigata* and *Narcissus tazetta*, a Mediterranean species long grown in Chinese and Japanese gardens.

In 1816 John Reeves, a tea inspector in the British East India Company, began sending back both plants and flower paintings by Chinese artists to the Horticultural Society in London. Many of the paintings were prepared under his personal supervision. In 1859, when the Royal Horticultural Society's library was sold, the five volumes fetched £70, the highest price reached by any lot. They were returned to the library in 1936 as part of a large bequest by Reginald Cory, and another three large folio volumes came on the market in 1953 and were bought by the Society. These volumes form the Reeves collection in the Society's library and mainly depict camellias, peonies, chrysanthemums, *Wisteria* and other garden plants which were then unknown in the West.

India

In spite of the wonderful richness of the Indian flora, the botanical exploration of the subcontinent began slowly. After Rheede tot Draakstein's *Hortus Indicus Malabaricus* of 1678–1703, there was little of significance for a hundred years, until the publication between 1795 and 1818 of William Roxburgh's *Plants of the Coast of Coromandel, selected from drawings and descriptions presented to the Hon. Court of Directors of the East India Company, published by their order, under the direction of Sir Jos. Banks.*

Roxburgh was a Scot, born in Ayrshire in 1751, who was director of the East India Company's garden in Calcutta from 1793 to 1814. Before that, he had been stationed in Madras, where he had arrived in 1766, meeting John Gerard Koenig, a Dane who was employed by the Danish court, partly as physician to the Danish settlement, partly as naturalist to the Nabob of Arcot. Koenig was an excellent field botanist and geologist, as well as a good companion, and acted as guide and teacher to Roxburgh. His contribution is fully acknowledged in the preface to *Plants of the Coast of Coromandel*, the text of which was compiled from the two men's notes after Roxburgh's death in 1815. Three hundred of the drawings that Roxburgh assembled are

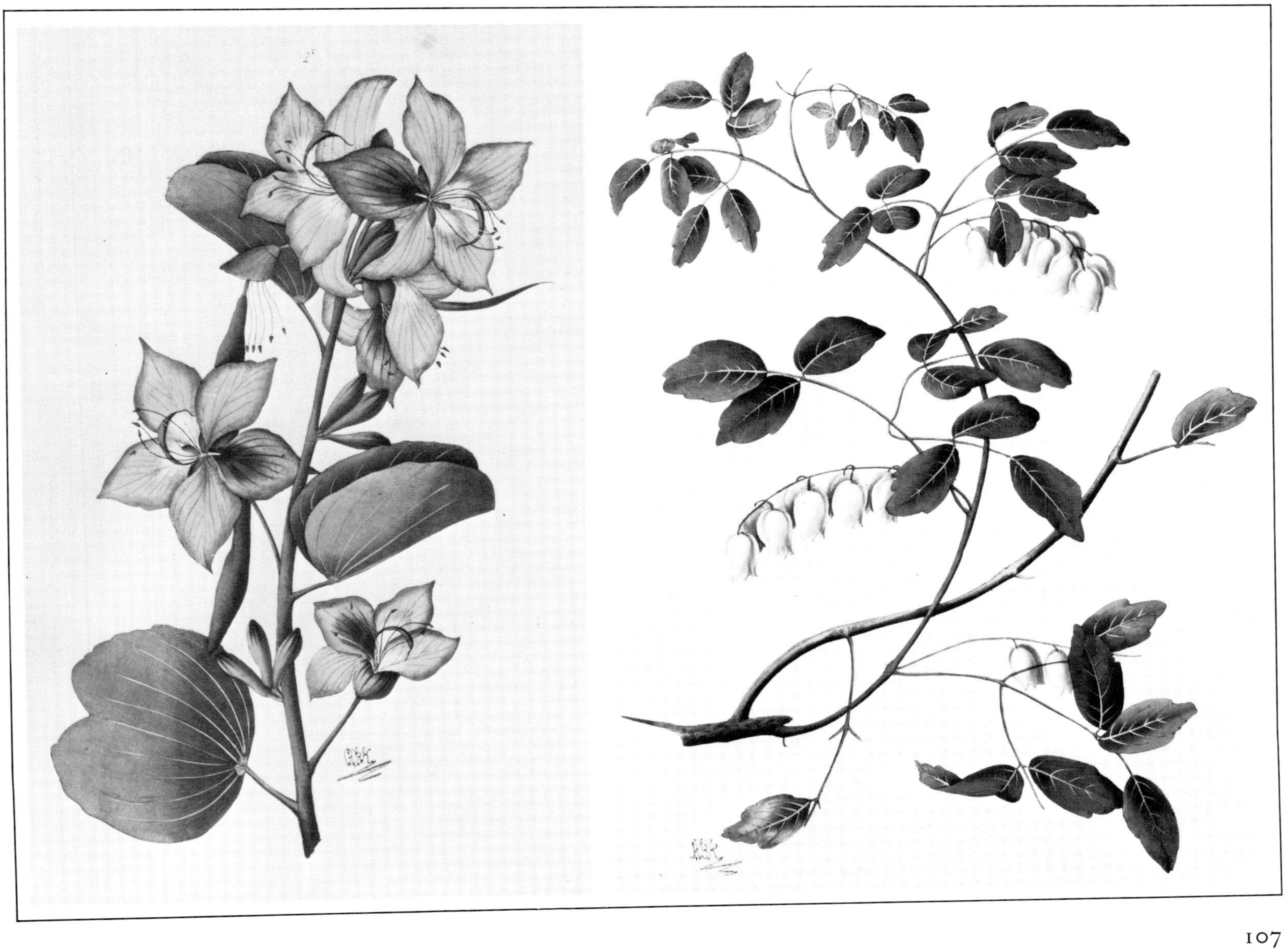

A tropical tree, *Bauhinia*, and another Chinese plant. Two colour plates by Charles Henry Bellenden Ker after Chinese originals from *Icones Plantarum sponte China nascentium è bibliothêca braamiana Excerptae* (1821).

reproduced in folio size. They are hand coloured line and stipple engravings of often strikingly composed paintings by various Indian artists. Generally the flowers are very simply drawn, and there is little attempt to show fine veins or modelling on the leaves.

Roxburgh's assistant and eventual successor was Nathaniel Wallich (1786–1854). In his long career, he was to lay the foundations of Indian botany in the Victorian era and to initiate the exploration of the Himalayas with their unsurpassed floral riches.

Africa

In the eighteenth century, Central Africa was still unexplored – David Livingstone did not make his famous journeys until the late nineteenth century. One of the first botanists to study the flora of tropical Africa was Michel Adanson (1726–1806), a pupil of the influential French botanist Bernard de Jussieu (1699–1777) and only 22 when he obtained a clerical post with the Compagnie des Indes in Senegal in West Africa. Life there was so dangerous that he was unable to venture far from the trading posts. Nevertheless, in five years, he amassed a large collection of new plants, which he found did not fit well into the European system of classification. The new system which he devised to accommodate them was published in his *Familles des Plantes* (1763). Although the system was not very successful, Adanson's contribution to botany is commemorated in the generic name of the Baobab tree, *Adansonia digitata.*

New species from South Africa had been finding their way back to Europe since the early eighteenth century with travellers returning from the East; as they flowered in Europe, the plants were illustrated in florilegia and later in monographs, such as those on *Oxalis* by Jacquin and on *Erica* and *Geranium* by Henry Andrews. Most of these plants were from the Cape region, where the Dutch East India Company had established a colony. An unusual journal of an early expedition further afield, northwest of the Cape into Namaqualand, between Cape Town and the Orange river, in 1685–86 was kept by a Dutch soldier, Simon van der Stale. One copy is now in the library of Trinity College, Dublin. Van der Stale made a collection of plants found on the expedition which were rather crudely drawn by Heinrick Claudius, an apothecary employed by the Dutch East India Company. They include some of the earliest pictures of *Lachenalia* and succulent spurges.

The two collectors who introduced the most new species from the Cape area were Carl Thunberg and Francis Masson. Thunberg spent nearly three years there *en route* for Japan. In 1772, he made a three month journey up country on horseback and by ox wagon, collecting plants, seeds and bulbs, which he sent back to the botanic gardens in Leiden and Amsterdam. In 1773, he set out again, this time with Francis Masson, a collector trained at Kew and sent out by Sir Joseph Banks to collect seed for the Botanic Garden. They made two journeys, lasting from September 1793 to January 1774. On the second journey they crossed part of the Karroo plateau, but were able to collect only a small sample of the wealth of succulents that they found there. Masson's introductions were extensive – Alice Coats records in her book *The Plant Hunters* that he is credited with introducing 50 species of *Pelargonium*, nearly 70 of *Mesembryanthemum*, 88 heaths, 48 *Oxalis*, 40 *Stapelia*, 11 *Lobelia* and 12 *Arctotis*.

Masson returned to England in 1775, but made a second visit to the Cape in 1786 and set up a garden at the foot of Table Mountain. He remained until 1795, sending back more plants and seeds to Kew. His main work, *Stapeliae Novae*, published in 1796–97, contains many of his own paintings of stapelias, 'drawn in their native climate, and although they have little to boast in point of art, they possibly exhibit the natural appearance of the plants better than figures made from subjects growing in exotic houses can do.' These are almost the only botanical illustrations of the time that were made from the living plants in South Africa. Nearly all the others which appear in the florilegia of the period were from plants grown in the hothouses of northern Europe.

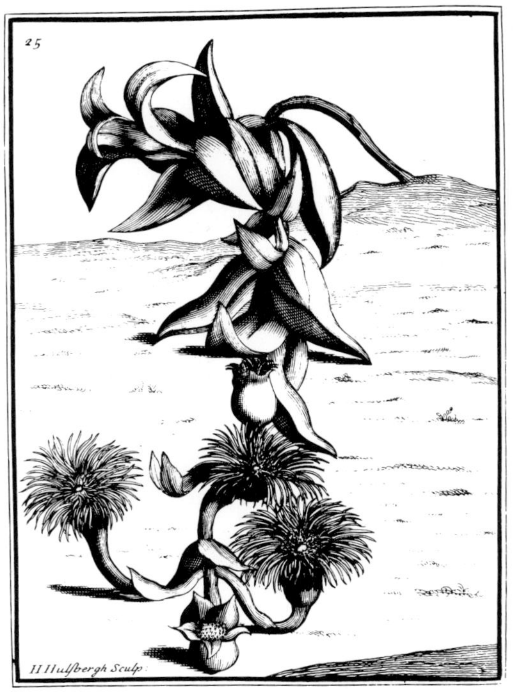

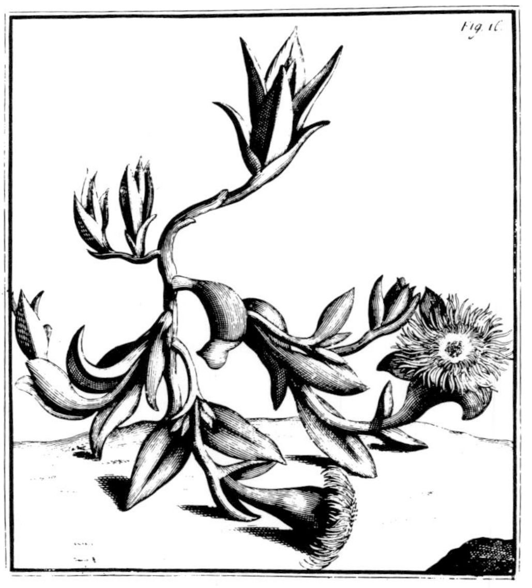

Two species of *Mesembryanthemum*. Engravings by H. Hulsbergh from *The History of Succulent Plants* (1716) by Richard Bradley.

The partridge-breasted aloe (*Aloe variegata*), a native of the Cape Province of South Africa. Engraving intended for an unpublished work by Johannes Burmann. It is clearly based on an unsigned watercolour made around 1685, possibly by Henrik Claudius, one of a group now in the Natural History Museum in London.

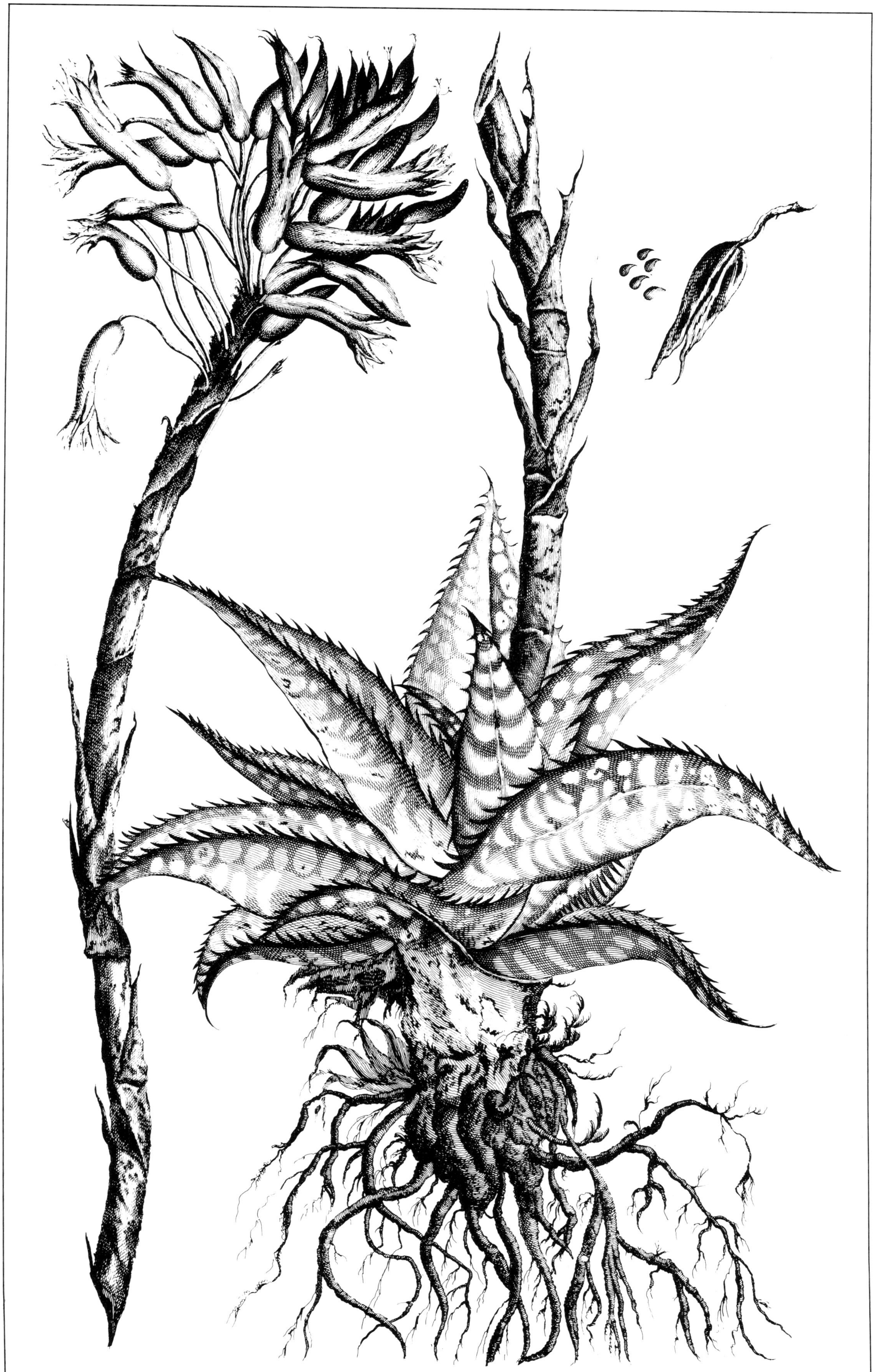

CHAPTER VIII
Australia and New Zealand

Of all voyages of exploration in the eighteenth century, those of Captain Cook to Tahiti, New Zealand and Australia are the most remembered today. The intention of his first voyage was both scientific and political, to observe the transit of Venus and so calculate more accurately the distance of the earth from the sun, but at the same time to discover the great southern continent believed to lie in the South Pacific, or to prove that it did not exist. The expedition was proposed by the Royal Society, which decided that Tahiti was a suitable place from which to observe the transit of Venus. The biological aspect of the expedition was secondary, and depended on the enthusiasm and fortune of the young Joseph Banks, then aged 25 and lately made a fellow of the Royal Society. Banks took with him a party of eight: the Swede Daniel Solander, who was a pupil of Linneaus, Alexander Buchan, a landscape painter, to record the scenery and natural features,

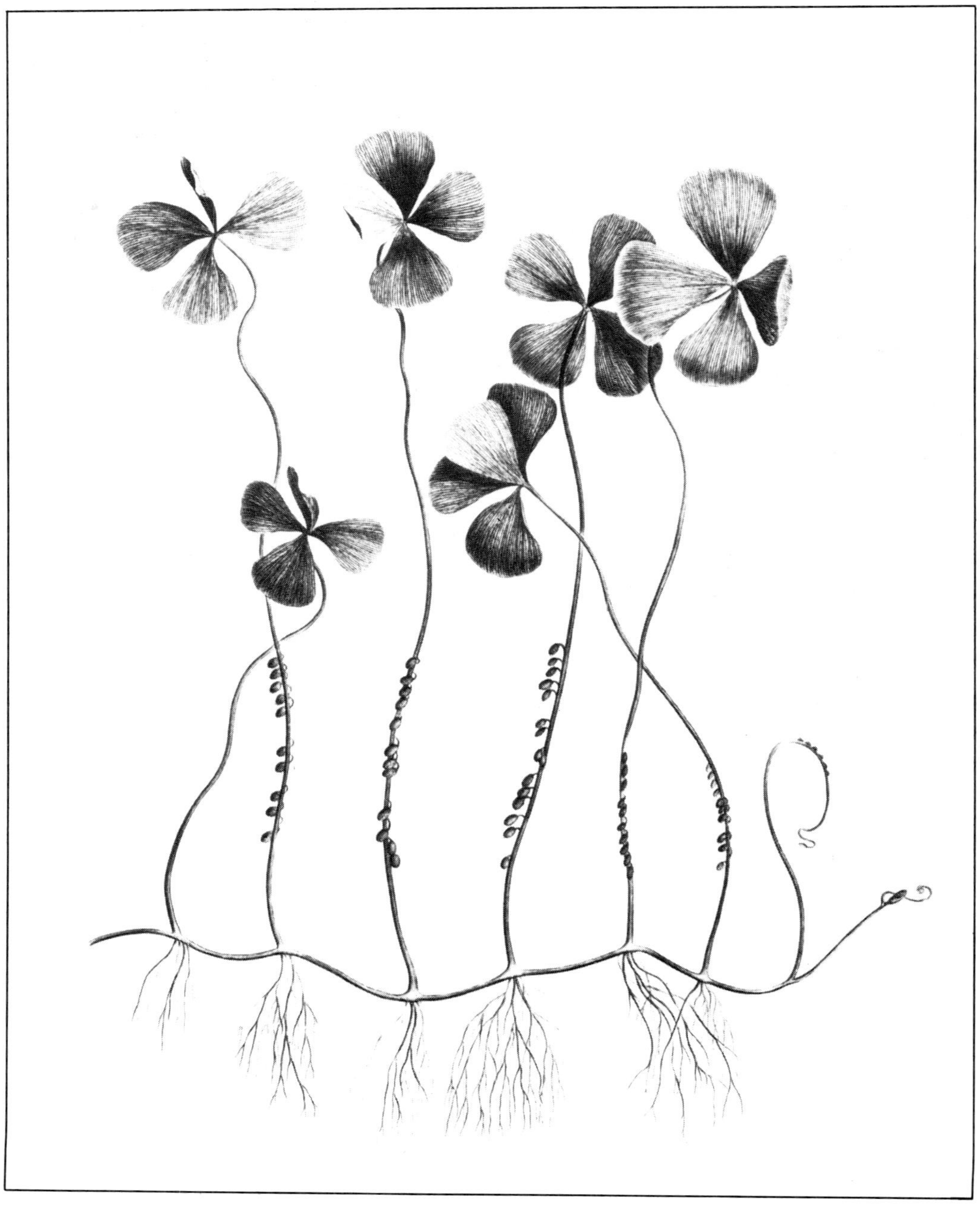

Proofs of two engravings made after the botanical sketches of Sydney Parkinson from Captain Cook's first voyage. The engravings remained unpublished until the twentieth century; the contemporary proof pulls reproduced here are from a volume preserved in the collection of the Linnean Society of London. *Left:* an aquatic fern, *Marsilea*, with spores in pods attached to the leaf stalks. *Right:* a species of *Banksia*.

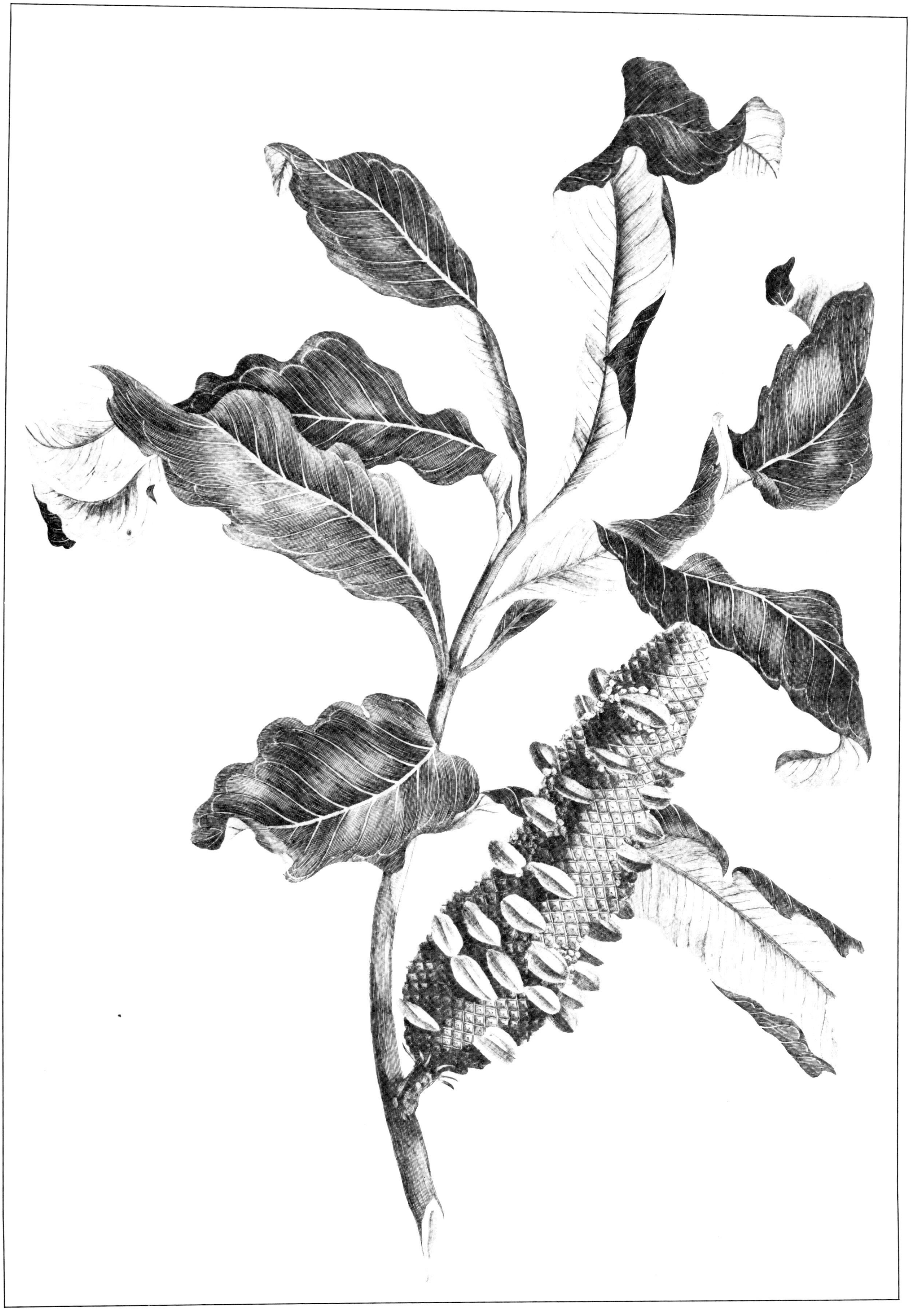

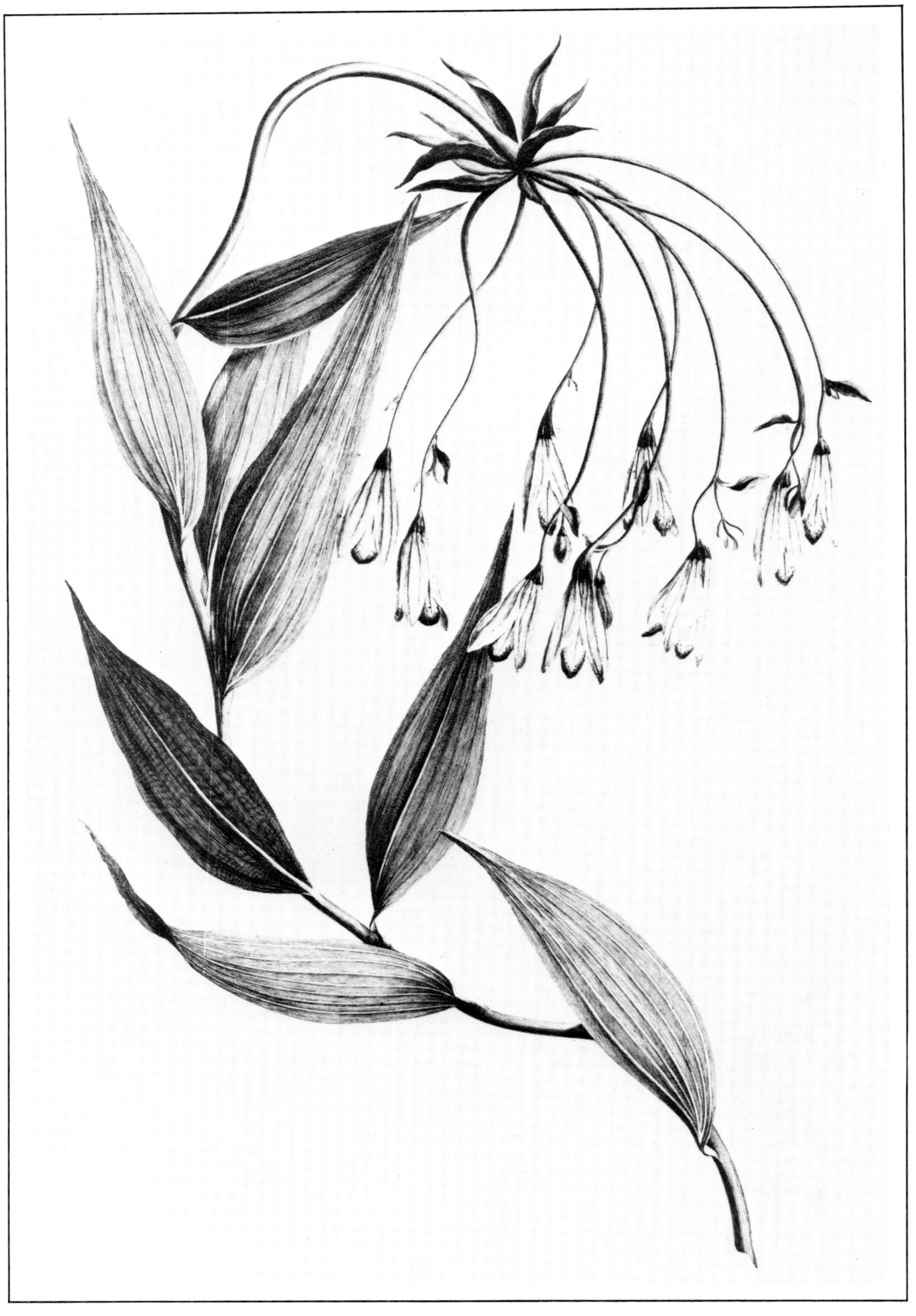

Bomarea, a climber related to *Alstroemeria*, from South America. Engraving after a drawing by Sydney Parkinson from Captain Cook's first voyage.

Sydney Parkinson, a botanical draughtsman, and Herman Sporing, a Finnish doctor and naturalist who was Banks's secretary; the other four were servants.

On 25th August 1768, they sailed from Plymouth on HMS *Endeavour*. After visiting Funchal in Madeira and Rio de Janiero, they anchored in the Bay of Good Success in January 1769 to make a trip into the interior of Tierra del Fuego. It is hard now to imagine the excitement they must have felt, first in Tierra del Fuego and even more in New Zealand and Australia, at seeing so many completely new and strange plants. Nearly all the species and many of the genera were new, and even several of the families were unknown to science. In Rio, the botanists found bromeliads and cacti, and in Tierra del Fuego, *Gunnera*, *Acaena* and the Antarctic beech (*Nothofagus antarctica*), as well as many new species in familiar genera. However, the expedition into Tierra del Fuego was a disaster, as the weather was much colder than expected, the party ran short of food, and one of the servants got drunk and died of exposure.

The *Endeavour* reached Tahiti in mid-April; here the natives lived mainly on breadfruit and sweet potatoes, which were both new to botany. Four days after arriving in Tahiti, Alexander Buchan died of an epileptic fit. From then on, Sydney Parkinson had to draw landscape studies as well as doing

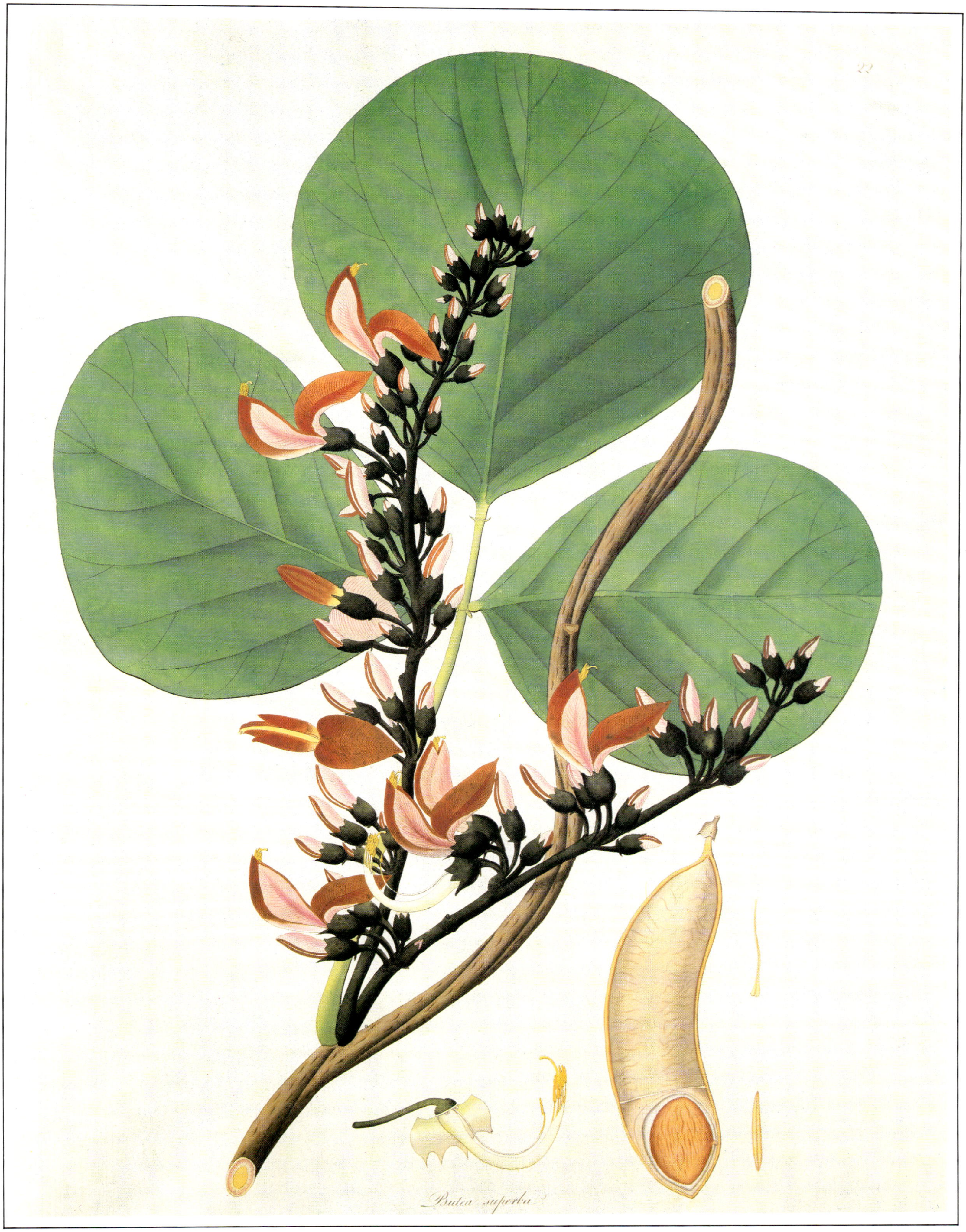

Butea superba. Hand-coloured engraving by Mackenzie after a painting by an unknown Indian artist from *Plants of the Coast of Coromandel* (1795–1818) by William Roxburgh. The plant, a woody climber belonging to the family Leguminoseae, is named in honour of the Earl of Bute (1713–92), an important patron of botanical publications, including the *Flora Londinensis*.

his own botanical work. The transit of Venus occurred on 3rd June and was duly observed by the astronomers. The *Endeavour* then sailed south to explore New Zealand, which had been sighted by Abel Tasman as long before as 1642. However, prevented by the Maoris from landing, Tasman had been unable to remain long enough to explore the country. Cook had brought with him a Tahitian chief called Tapara, who was

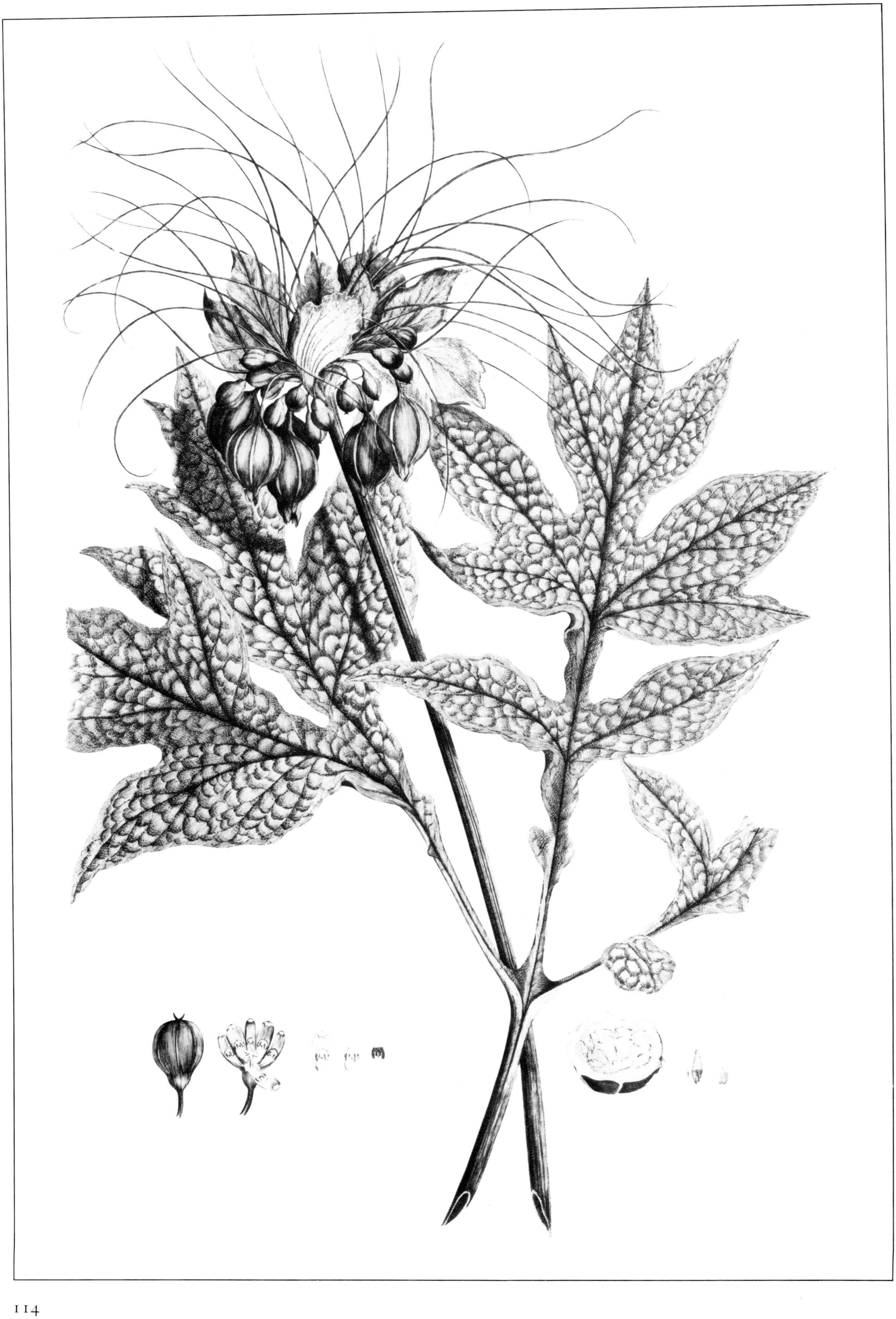

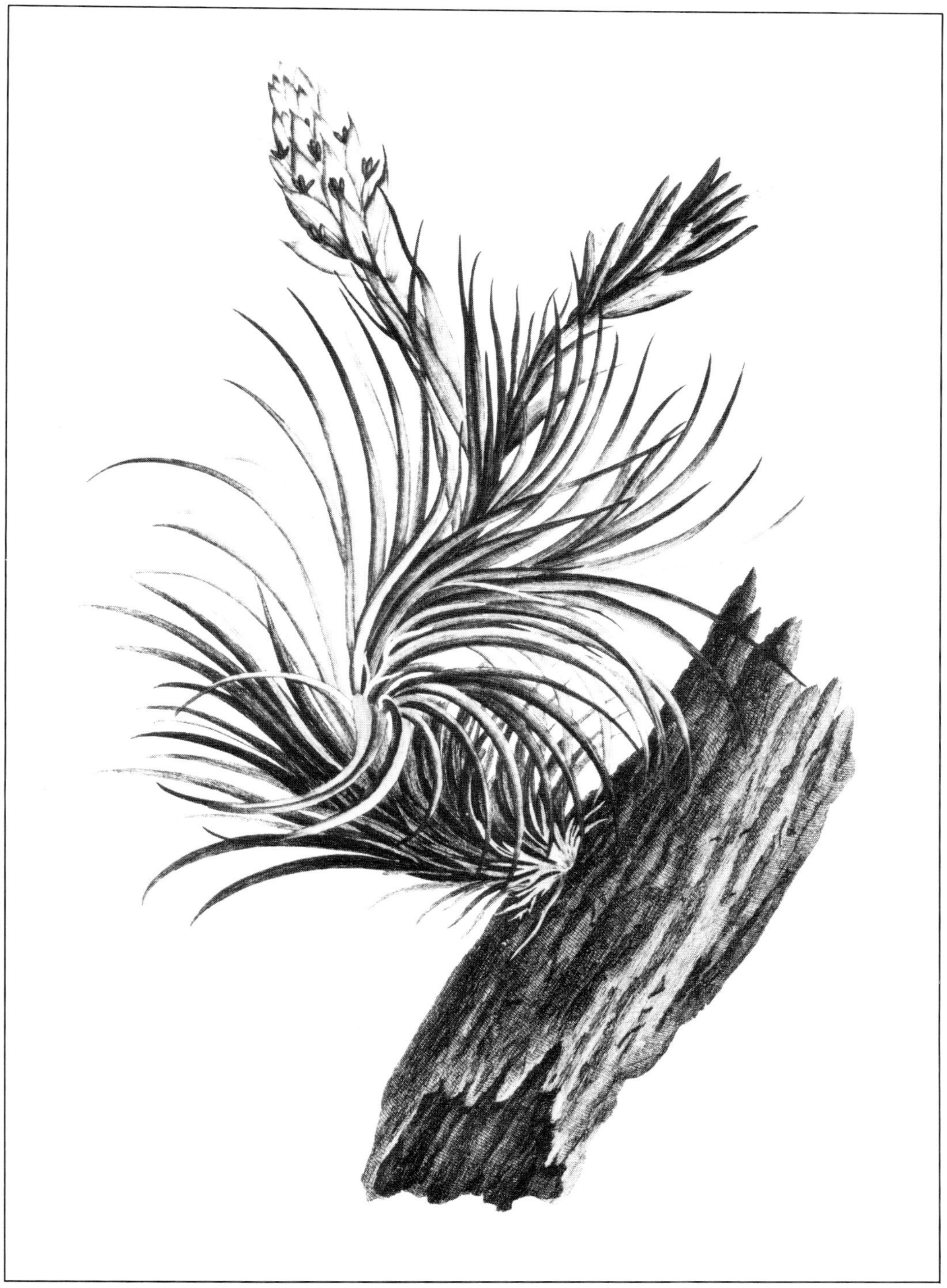

Two engravings after drawings by Sydney Parkinson from Captain Cook's first voyage. *Left*: a species of *Tacca*, a genus of tropical evergreen perennials which have characteristic hair-like 'tails' emerging from the inflorescences – some species are known as cat's whiskers. *Right*: an epiphytic bromeliad, *Tillandsia*.

able to understand something of the Maori language and acted as interpreter. Cook managed to prove that New Zealand consisted of two islands, while the botanists collected over four hundred new species including *Sophora tetraptera* and New Zealand flax (*Phormium tenax*) which provided the Maoris with fibre for clothing. The botanists were fascinated to find that some plants were similar to those they had seen on Tierra del Fuego. From New Zealand, the *Endeavour* sailed up the eastern coast of Australia, landing at a bay where the flora was the most fantastic they had yet seen; nearly all the plants were assigned to completely new genera. Cook called the place Botany Bay, a name that came to be associated with convicts rather than with flowers.

As the *Endeavour* sailed northward, trying to gain the open sea, she struck part of the Great Barrier Reef and had to be beached for extensive repairs to be made to her hull. This afforded the botanists a welcome interlude for further collecting before the voyage continued northward for Cook to establish that New Guinea and Australia were not joined. Until they landed in Java, Cook's crew had been healthy and free from disease, but there they met both malaria and dysentery. Parkinson and Sporing died, as did Tupara the Tahitian and seven of the crew. Almost twenty more sailors died between Batavia and Cape Town, where the sick were put ashore in mid-March 1771. The survivors finally reached England in July after a voyage of two and a half years.

Overleaf:
Grevillea banksii. Hand-coloured engraving after his own painting by Ferdinand Bauer from his *Illustrationes Florae Novae Hollandiae* (1813). Bauer himself engraved and coloured the plates for this work, of which only three parts, containing fifteen plates, were issued. *Grevillea banksii*, named after C. F. Greville and Sir Joseph Banks, is a native of Queensland that makes a very good greenhouse shrub, as it remains almost continuously in flower.

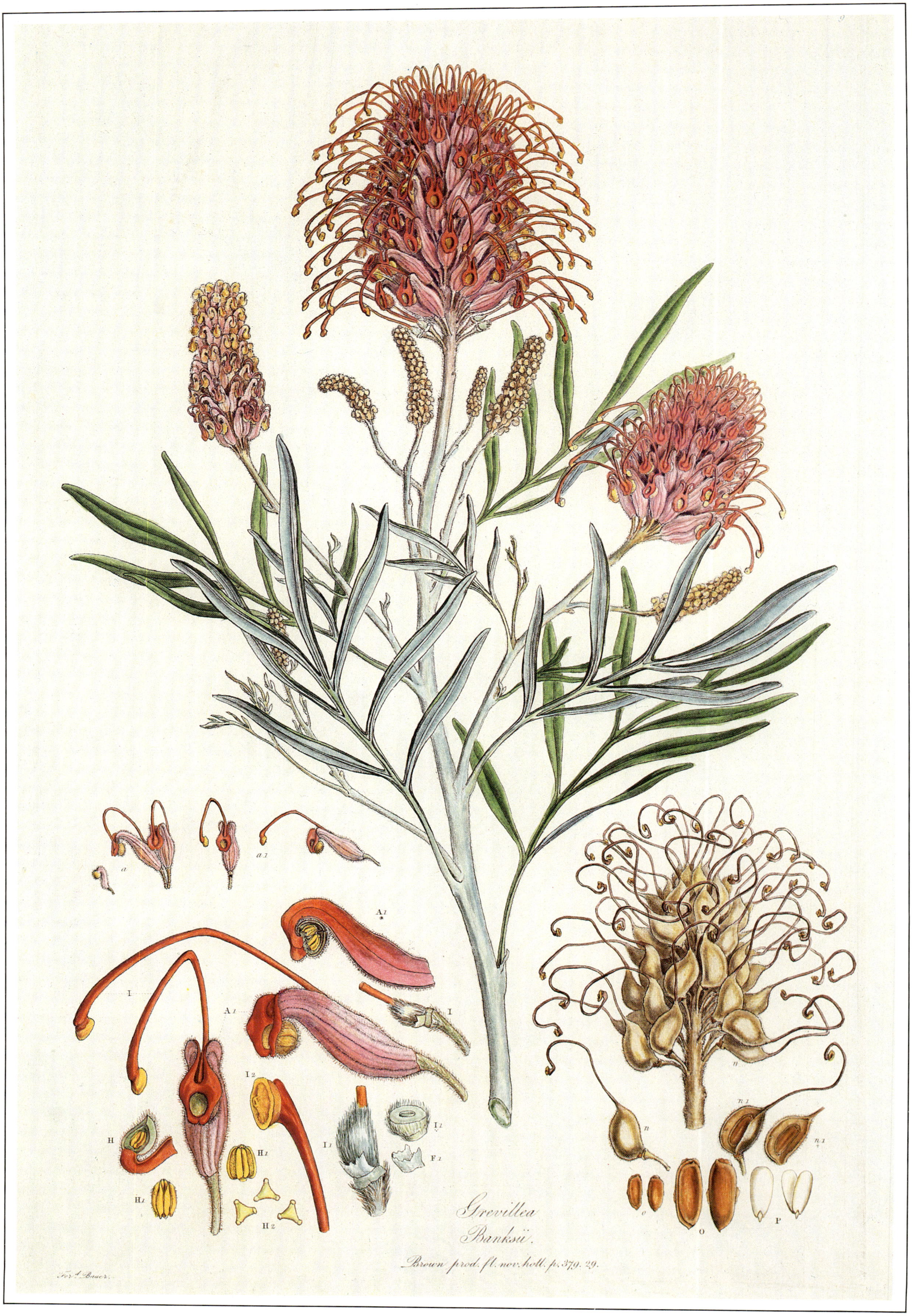
Grevillea
Banksii.
Brown prod. fl. nov. holl. p. 379. 29.
Ferd. Bauer.

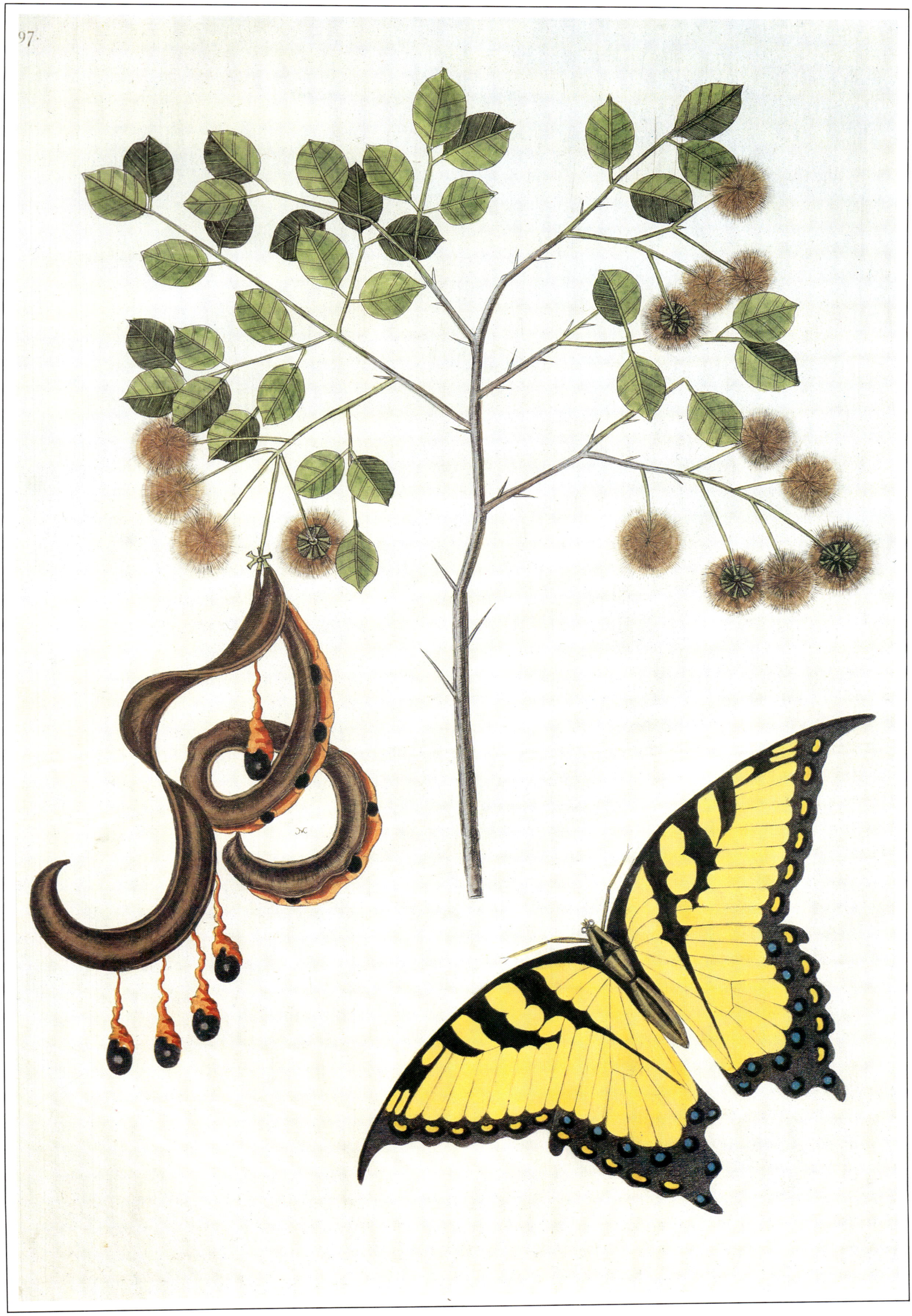

Before his death, Parkinson had made 955 drawings, 675 sketches and 280 finished botanical paintings. Banks succeeded in purchasing Parkinson's drawings and had many of the sketches finished and the drawings engraved by the brothers John Frederick and James Miller, and by Francis Nodder. By 1778, about 550 plates had been engraved and more were completed by 1782, when Solander died, leaving his text still unfinished. The plates remained unpublished until the twentieth century and, along with Banks's papers, came into the possession of the Natural History Museum. It was not until 1900 that the first were published by the Museum under the title *Illustrations of the Botany of Captain Cook's Voyage Round the World in H.M.S. Endeavour in 1768–71, by Sir Joseph Banks and Daniel Solander; with determinations by James Britten.* This included lithographic reproductions of 318 of Parkinson's plates. More recently, the Royal College of Art produced a volume of illustrations printed from the original engraved copper plates with an account of the voyage by Wilfrid Blunt and of the botanical explorations by W. T. Stearn. This huge tome, which took over ten years to produce, finally appeared in 1973. The plates are uncoloured and show the engraving to be rather heavy compared with the delicate

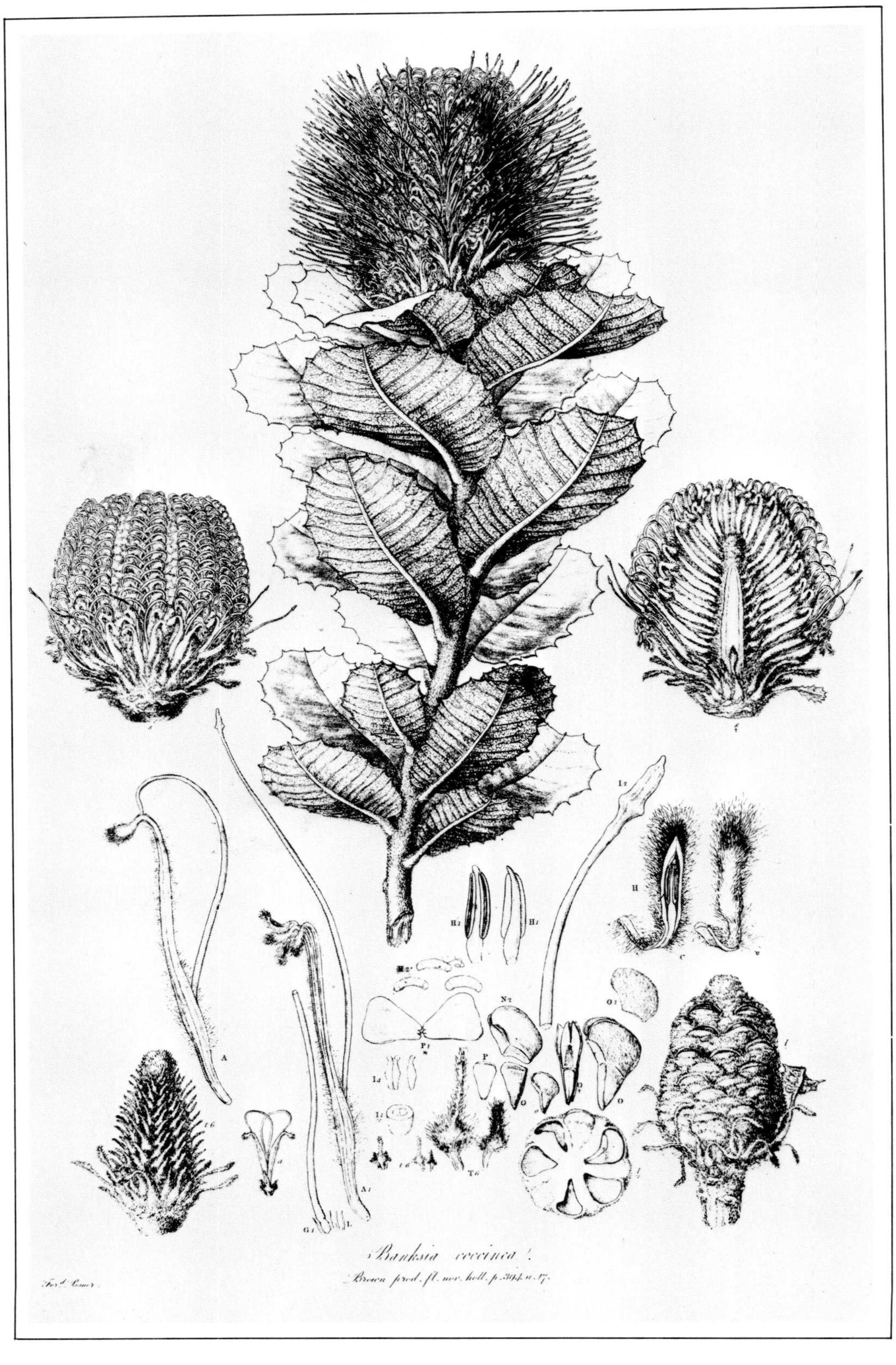

Banksia coccinea. Engraving by Ferdinand Bauer after his own drawing from an uncoloured copy of his *Illustrationes Florae Novae Hollandiae* (1813). The genus was named in honour of Sir Joseph Banks.

Previous page:
Florida cat's-claw (*Pithecolobium unguiscati*) and swallowtail butterfly. Hand-coloured engraving by Mark Catesby from *The Natural History of Carolina, Georgia, Florida and the Bahama Islands* (1730–47). This member of the mimosa family is a large shrub or small tree found in southern Florida, the islands of the Caribbean and tropical South America.

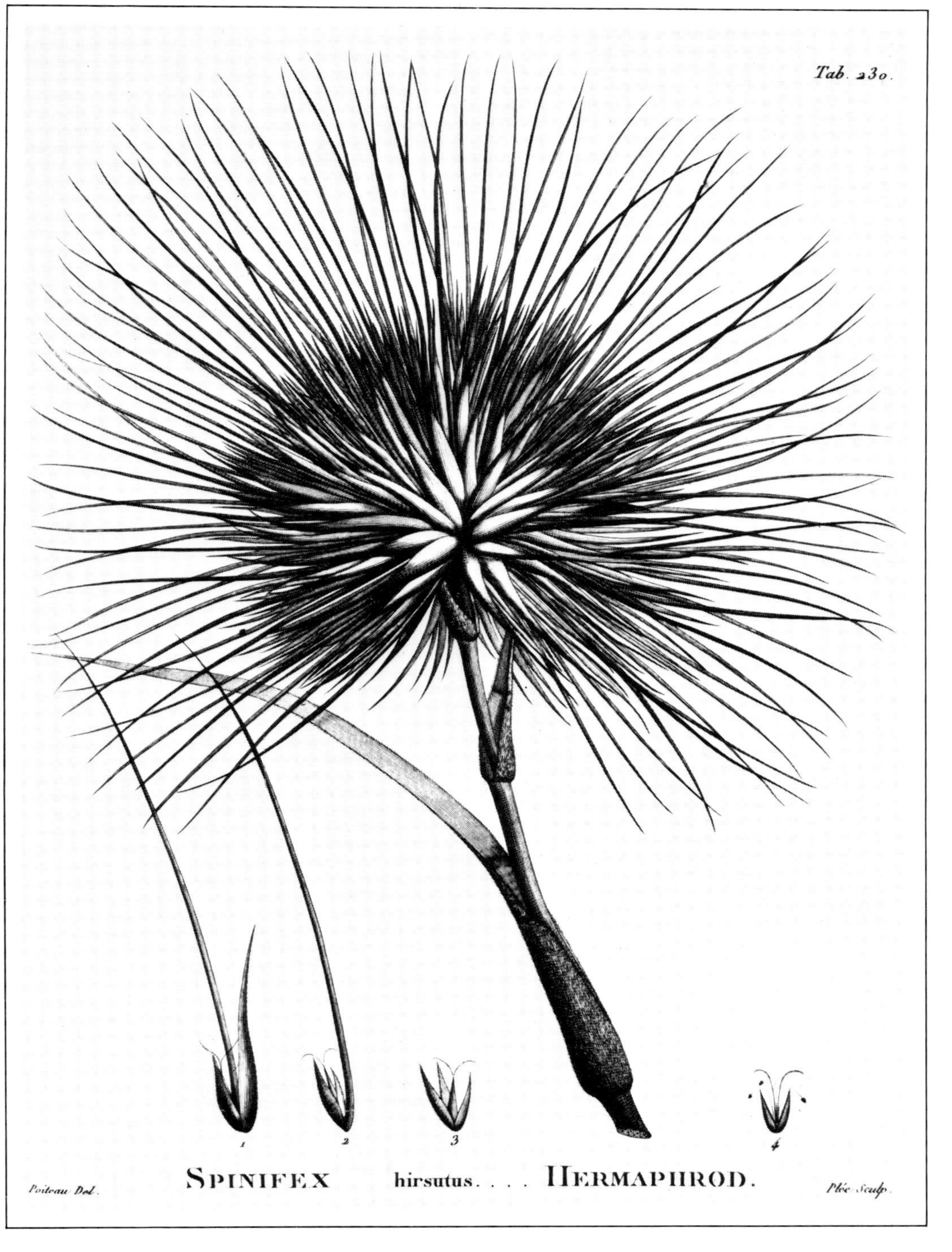

Spinifex hirsutus, a grass from Australia and the Pacific Islands. Engraving by Plée after a drawing by Pierre-Antoine Poiteau from *Novae Hollandiae Plantarum Specimen* (1804–06) by Jacques-Julien Houton de La Billardière.

French work of a few years later, or with the engravings done by Ferdinand Bauer of his own Australian paintings.

Bauer, travelling again after his twelve years' work on the *Flora Graeca*, was botanical draughtsman on the next great botanical voyage of exploration to Australia. HMS *Investigator* set out in 1801 with Matthew Flinders in command and Robert Brown as botanist. In the years since Cook's voyage, the penal colony had been established at Botany Bay and seeds of many plants had been sent back to Kew and to the nurserymen Lee and Kennedy by their collector, David Burton, who had introduced many new plants before his death in Australia in 1792 of gunshot wounds acquired while out duck shooting.

The sloop *Investigator* was an old and leaky vessel specially refitted for the voyage, but Flinders was an experienced navigator. He went to sea in his sixteenth year and was a veteran of many voyages to the Pacific, including Captain Bligh's second and successful expedition to Tahiti to collect the breadfruit tree and introduce it to the West Indies. Flinders had visited Australia in 1795, and in 1798 had made the first circumnavigation of Tasmania.

Robert Brown had studied medicine at Edinburgh and had become an army surgeon. On a visit to London, he made the acquaintance of Sir Joseph Banks, who allowed him to work in his library and study his collections. It was Banks who recommended Brown as expedition botanist and Brown took a duplicate set of Banks's specimens to study on the voyage out in order to be well prepared when the expedition arrived in Western Australia. As his assistant, Brown brought with him Peter Good, a foreman gardener from Kew.

The aim of the voyage was to circumnavigate Australia, to survey the south coast and look for any possible sea routes into the interior of the country. The expedition remained in Western Australia, near King George Sound, for nearly a month examining the particularly rich flora before sailing eastward. After overhauling and patching the ship at Sydney, they continued north along the east coast to explore the Gulf of Carpentaria, which they reached in November. Here they discovered a new *Grevillea*, which Brown named *Grevillea banksii* in honour of his patron. It is now grown in Europe and has made a particularly satisfactory greenhouse shrub with its silver leaves and almost continuous succession of red flowers.

During its time in the Gulf of Carpentaria, the *Investigator* was careened and found to be in such bad condition that Flinders dared not set sail for England. Instead, he decided to return to Sydney, after putting in at Timor for supplies. It was here that the previously healthy ship picked up disease, and when the *Investigator* finally reached Sydney in June 1803, some of the crew were dying of dysentery and many others were in bad shape. Flinders changed ships, leaving Bauer and Brown to continue their botanical work, and did not reach England until 1810. In the meantime, he had been shipwrecked on the Great Barrier Reef and had suffered six years' imprisonment by the French on Mauritius. He was still only 40 when he died in 1814, just before his book, *A Voyage to Terra Australis*, was published. This contains maps, topographical drawings and ten of Bauer's botanical illustrations.

Bauer and Brown stayed in Australia until 1805, by which time Bauer had made an expedition to Norfolk Island, and Brown had been to Tasmania. On their return to England, they set about publishing their botanical discoveries. The first volume of Brown's *Prodromus Florae Novae Hollandiae et Insulae Van Diemen* was published in 1810. Brown himself paid the cost of the 250 copies that were printed, but as he sold very few, publication ceased.

Bauer's production of the illustrations was scarcely more successful. His *Illustrationes Florae Novae Hollandiae* consisted of fifteen plates of new genera described by Brown. It was published in three parts, with parts I and II costing £1 11s. 6d. coloured or 7s uncoloured. More were planned, but Bauer found it necessary to do all the engraving and much of the colouring himself. Consequently, the work went very slowly. However, the plates that were produced are among the finest prints of the time. An uncoloured copy in the Lindley Library of the Royal Horticultural Society shows the very fine and delicate work, with much variation in tone and delicate shading. Bauer's colouring is of comparable excellence.

Because of Brown's influence, each plate contains many floral details at different magnifications, which did not appear in Bauer's *Flora Graeca* pictures but are characteristic of his Australian work. The plate of *Banksia coccinea*, one of the genus named in honour of Sir Joseph Banks, is a good example of Bauer's wonderfully detailed and delicate style, which is less obviously elegant than that of his French contemporaries. As Goethe wrote when he saw Bauer's work, 'it is a real joy to look at these plates, for nature is visible, art concealed.'

Over two hundred of Bauer's Australian paintings remained hidden in the library of the Natural History Museum which acquired them from the Admiralty in 1843. It was not until 1975 that 25 of them were published in a limited edition. These plates, which use the best modern colour printing techniques, show the delicacy and detail of Bauer's finished watercolour paintings, which are perhaps the finest botanical illustrations ever made.

At the same time as the *Investigator* expedition was in Australia, a French expedition with La Billardière as botanist was exploring the coast. La Billardière, whose previous travels in Syria had resulted in the fine *Icones Plantarum Syriae Rariorum*, published his Australian discoveries in the

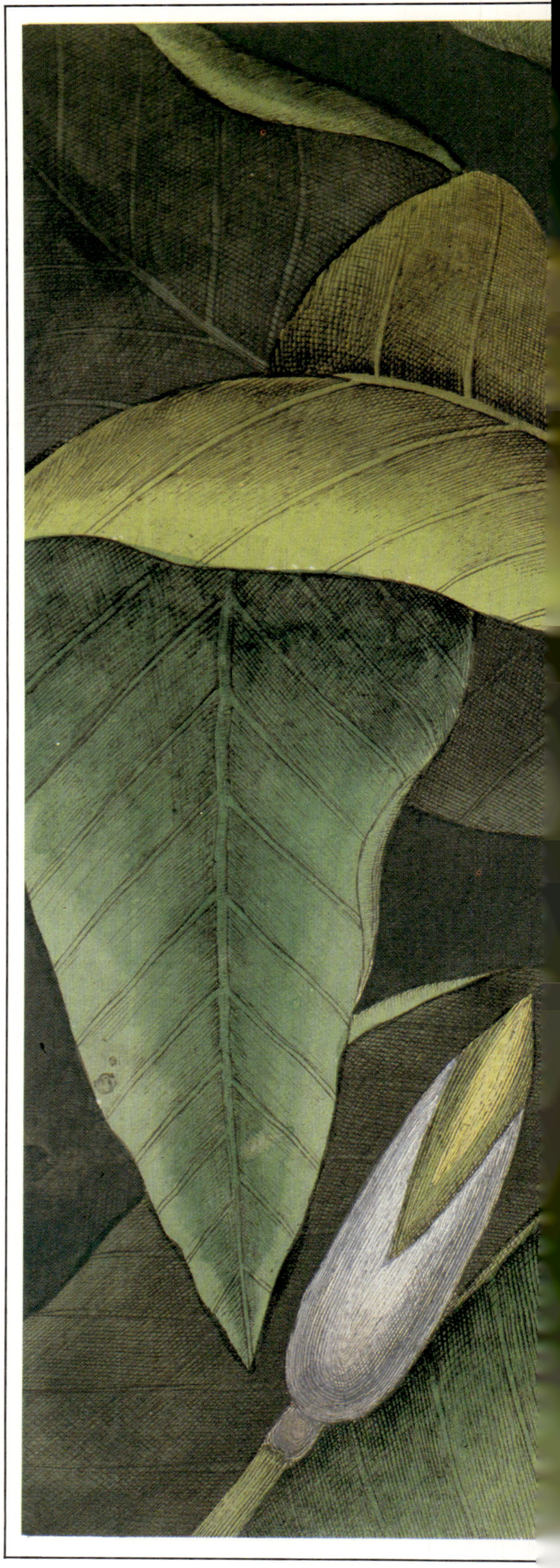

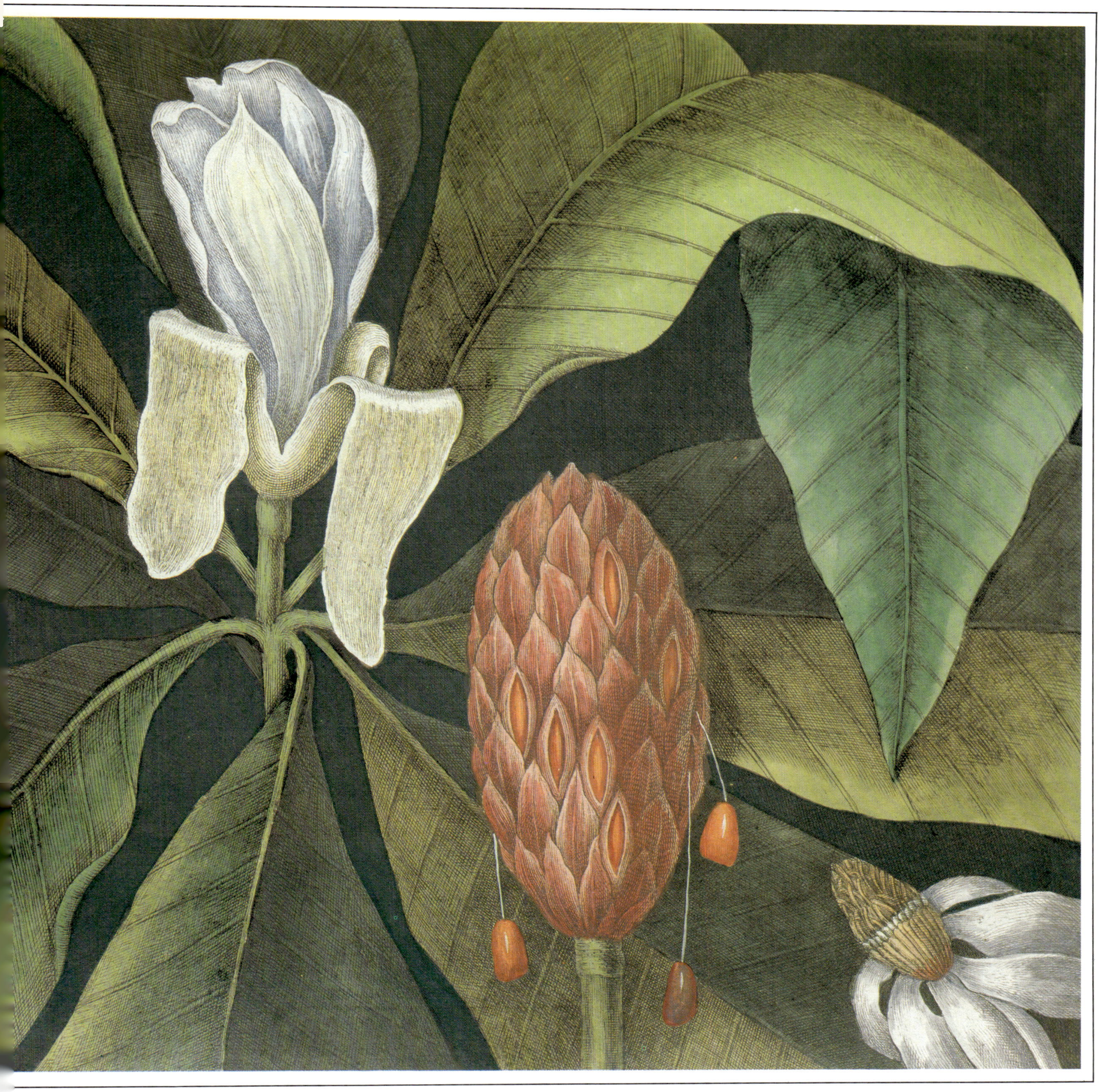

Umbrella tree (*Magnolia tripetala*), which grows wild beside streams in the southeastern United States but is planted as far north as New York. Hand-coloured engraving by Mark Catesby from *The Natural History of Carolina, Georgia, Florida and the Bahama Islands* (1730–47).

Novae Hollandiae Plantarum Specimen of 1804–06 with 265 engravings after Turpin, Poiteau, Sauvage and Piron. As these would have been made from dried specimens brought back by La Billardière, they lack the naturalism of Bauer's work or of Turpin's South American drawings. Later, in 1844–45, La Billardière's *Sertum Austro-Caledonicum* was published with eighty engravings after Turpin of plants from New Caledonia.

A small book on the Australian flora was published in 1793 by J. E. Smith, who went on to write the first six volumes of the *Flora Graeca*. He called it *A Specimen of the Botany of New Holland* and stated in his preface that it was 'to be considered only as what it pretends to be, a specimen of the riches of this mine of botanical novelty.' The plates were by James Sowerby and were taken from paintings made in Australia by John White Esq, Surgeon General to the colony, and from herbarium specimens sent back by White. Sixteen plates were issued, but the proposed continuation failed to materialise. The hand coloured engravings are clear and simple, in a style similar to that of Sowerby's other work.

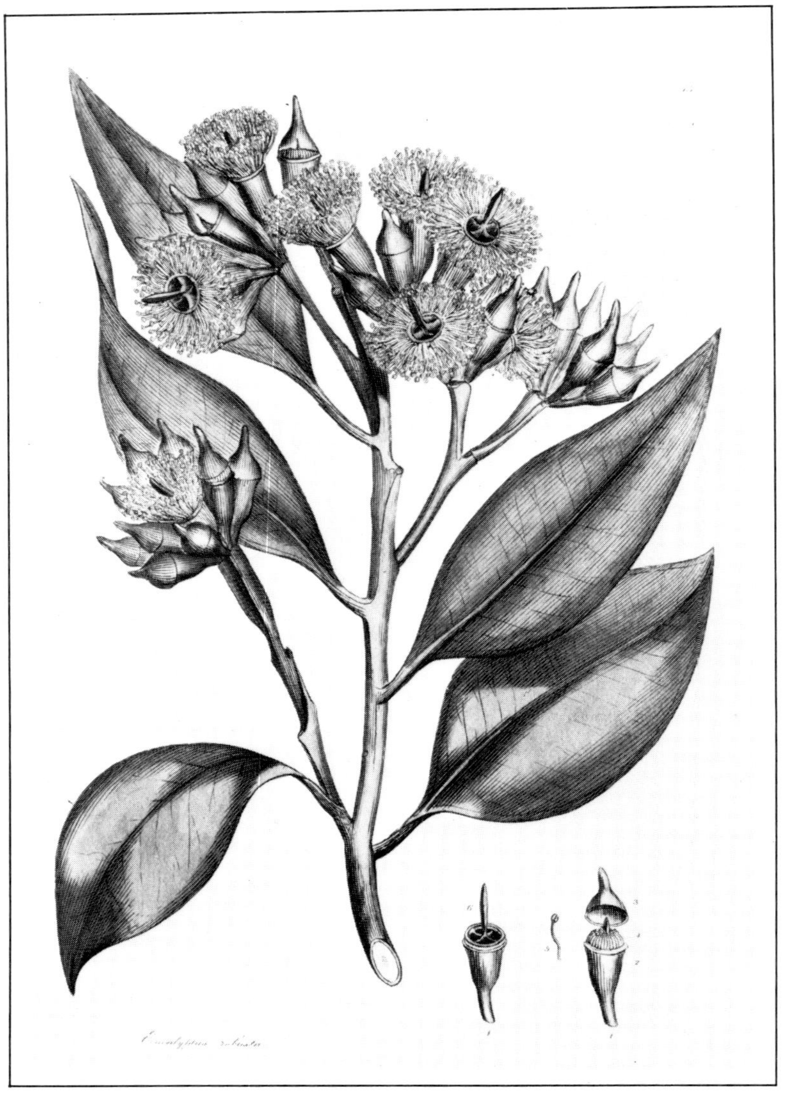

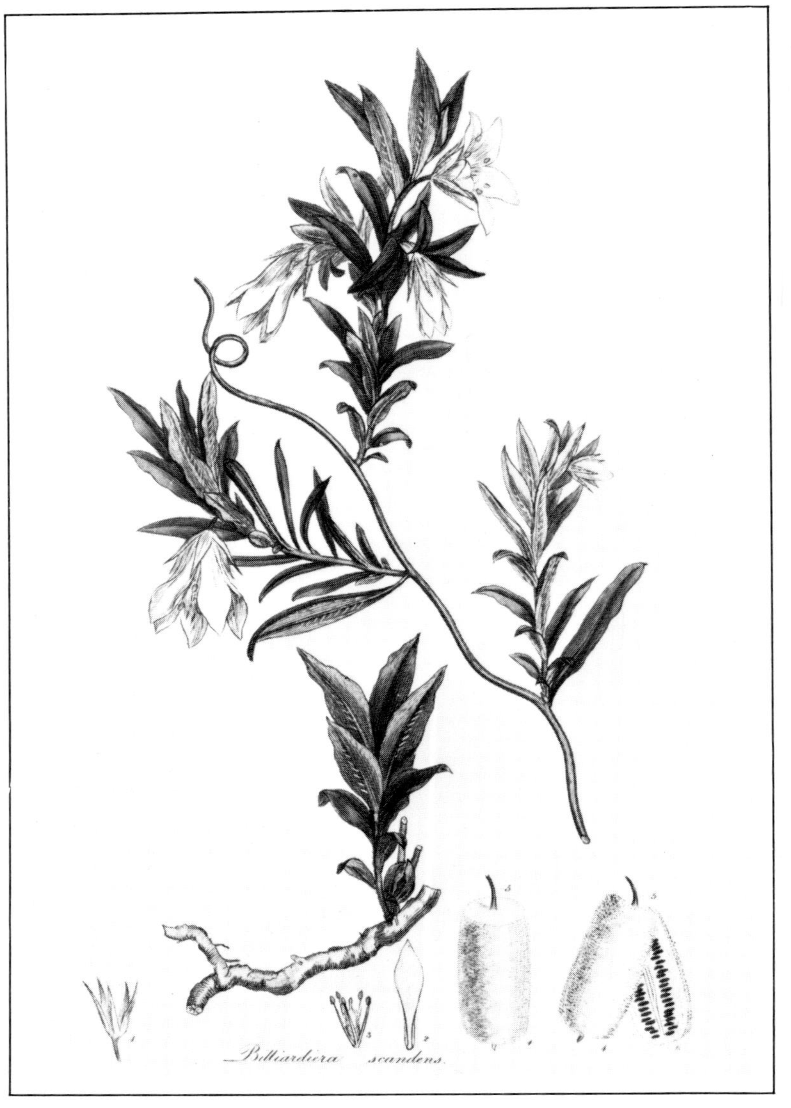
Billiardiera scandens

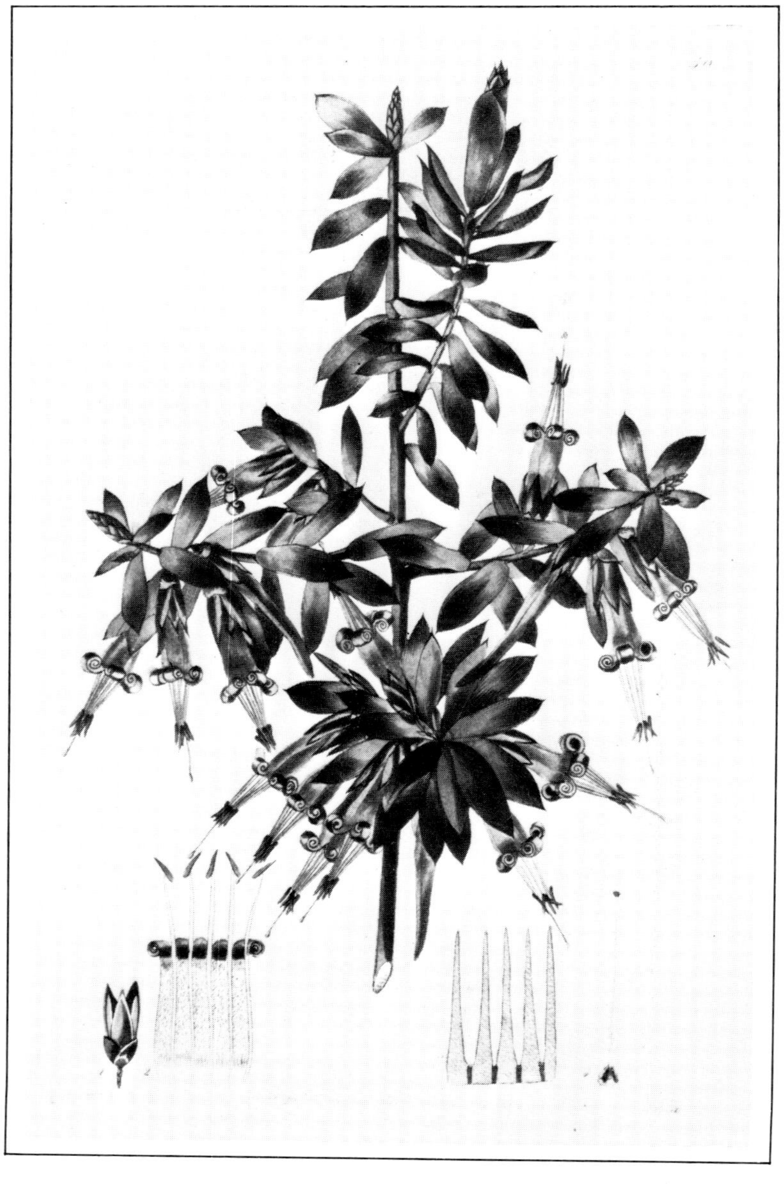

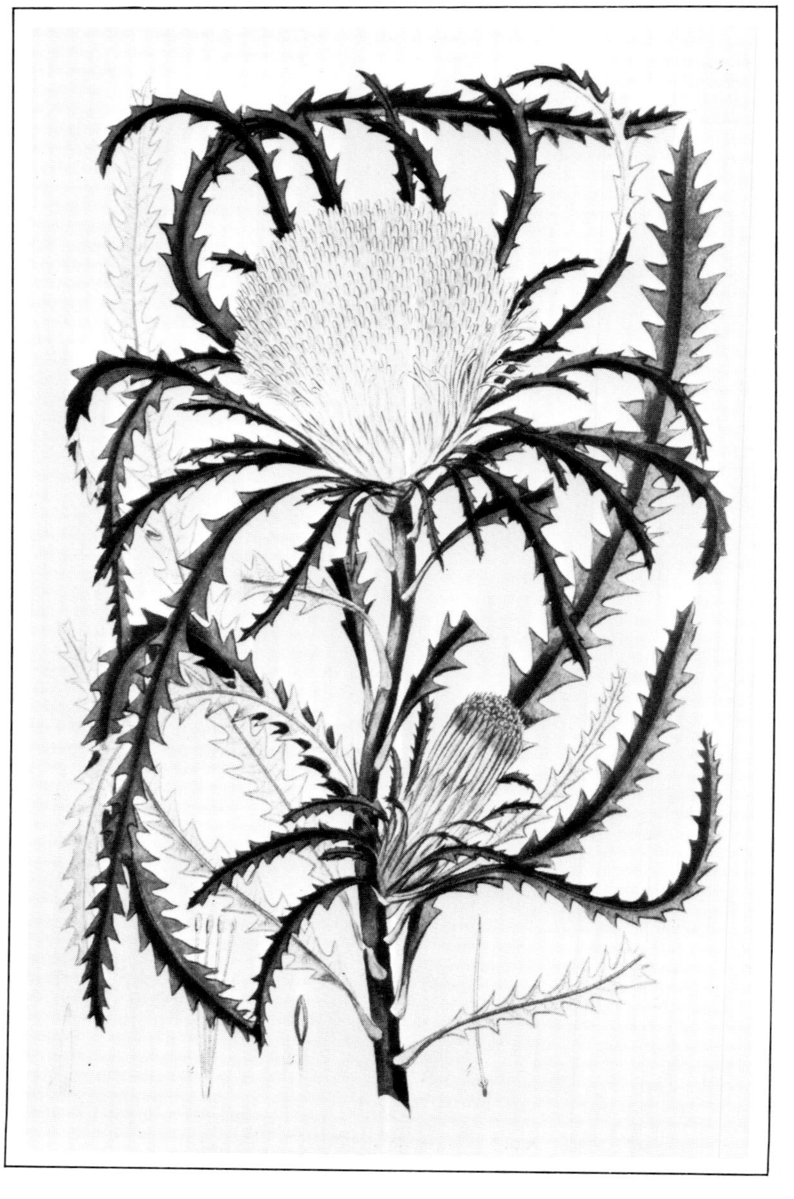

CHAPTER IX
America

As conditions in the North American colonies improved, botanists began to spend more time there and to penetrate slowly westward into Indian territory. This expansion continued gradually throughout the eighteenth century, with the Mississippi forming an effective western boundary to botanical exploration.

One of the first naturalists to live for any length of time in North America was John Banister (1652–1692). He arrived in Virginia in 1678 and collected many new plants, some of which he drew himself. He sent his lists to John Ray, who published them in Volume II of the *Historia Plantarum* (1688),

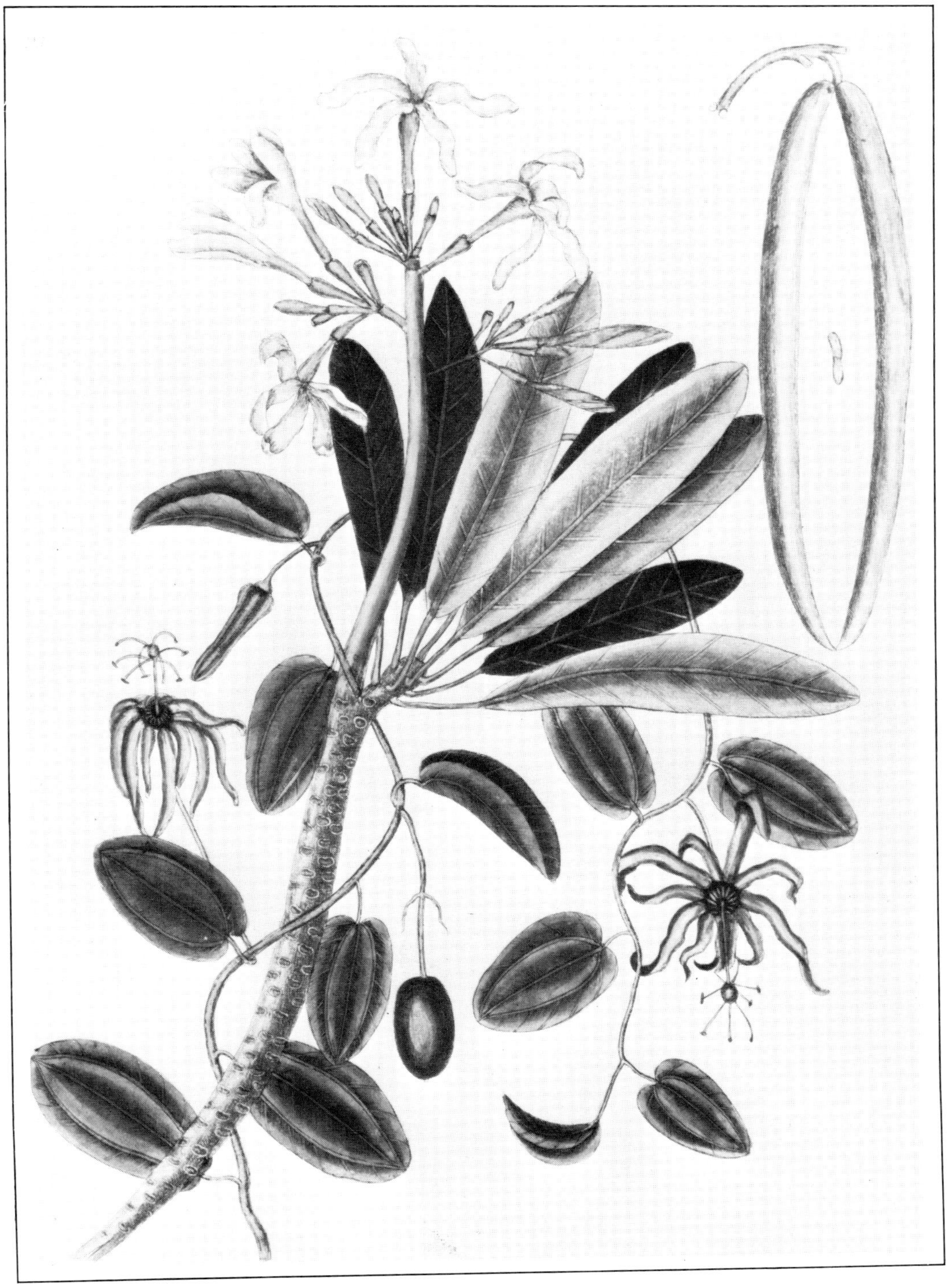

A *Plumeria* and a *Passiflora*. Hand-coloured engraving by Mark Catesby after his own painting from *The Natural History of Carolina, Georgia, Florida and the Bahama Islands* (1730–47).

Opposite page:
Top: two hand-coloured engravings from J. E. Smith's *A Specimen of the Botany of New Holland* (1793), published by James Sowerby; *left – Eucalyptus multiflora*, a large tree, up to a hundred feet tall, native to New South Wales; *right – Billardiera scandens*, a climber with white flowers and large purple fruits which can be grown out of doors in mild temperate areas of the northern hemisphere. *Bottom:* two hand-coloured engravings by J. Watts after paintings by E. D. Smith from *Flora Australasica* (1827) by Robert Sweet; *left – 'Styphelia viridiflora'*; *right – Banksia dryandroides.*

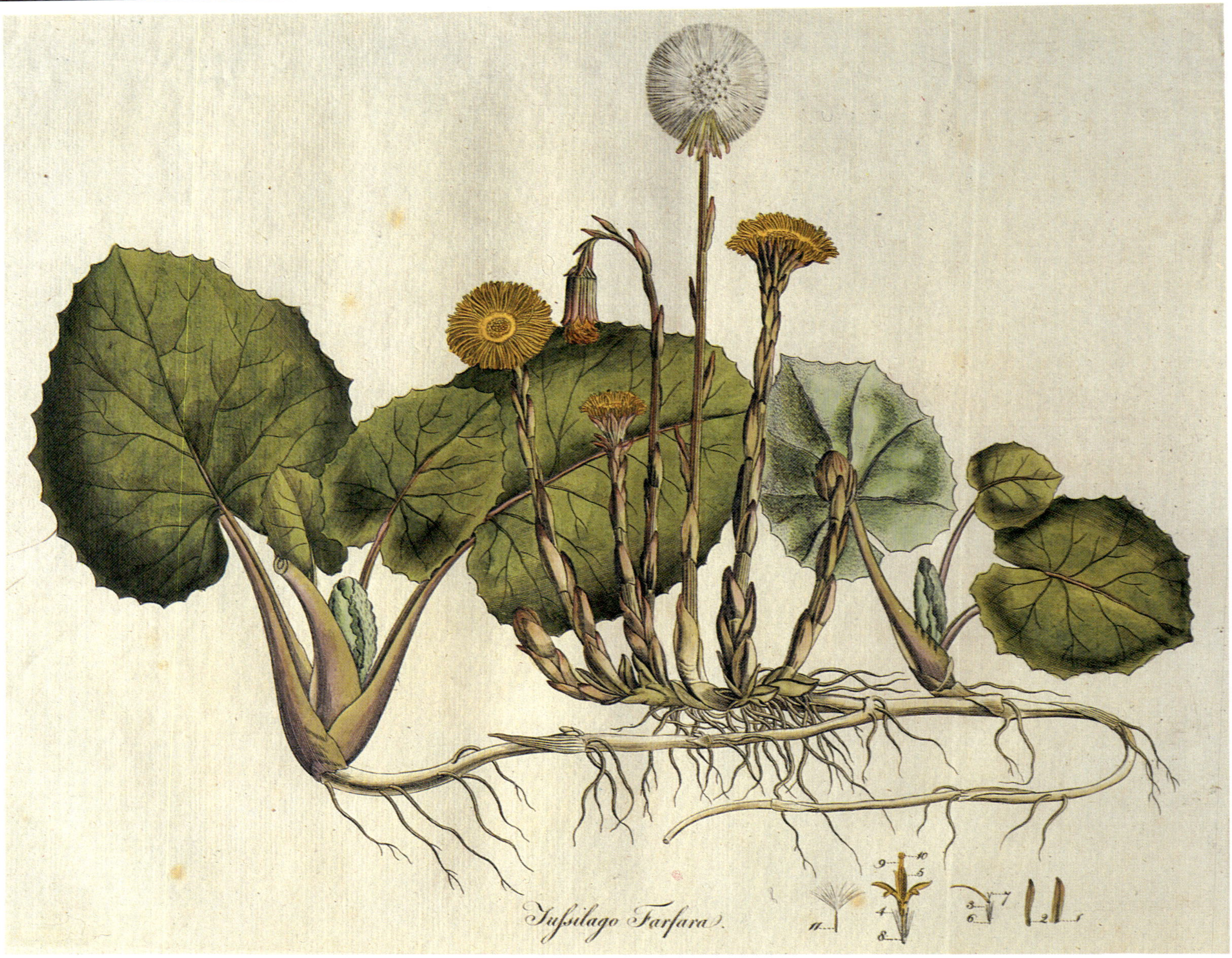

Coltsfoot (*Tussilago farfara*). Hand-coloured engraving from Volume II of *Flora Londinensis* (1775–87) by William Curtis.

but without illustrations. Banister died in uncertain circumstances while on an expedition into the interior in 1692.

The most famous of the early eighteenth century American botanists was Mark Catesby, who was born in Sudbury, Suffolk, around 1682. When Catesby was 23 his father died and left him a modest fortune. Inspired with a love of natural history by the elderly John Ray, Catesby decided to visit America. His sister, who had, against her parents' wishes, married a Dr William Cocke, was already living in Virginia. Catesby arrived in Williamsburg in 1712; in 1713, he began to send back small amounts of seed and dried plant specimens to England. On Catesby's return to England in 1719, William Sherard organised a group of subscribers to send him back to America, this time to Carolina, as a natural history collector at a salary of £20 a year. He embarked for Charleston in 1722, arriving on 3rd May. During his stay, which lasted until 1726, he made several journeys up country as far as Fort Moore (near present day Augusta, Georgia) and possibly a brief trip into the Appalachians. He also went to the Bahamas in January 1725. Catesby sent back boxes of seeds and plants to England throughout his stay and asked Sherard to send him bulbs for American gardens.

For his main work, *The Natural History of Carolina, Georgia, Florida and the Bahama Islands*, published between 1730 and 1747, Catesby taught himself engraving and produced all the plates himself. Many of the original paintings are in the Royal Library at Windsor Castle. The book contains 220 plates, and the industrious Catesby coloured many copies of the first two editions himself. Catesby's economical method of usually combining an animal and a plant in the same plate produced some charmingly incongruous results. One of the most striking plates shows the American bison and the

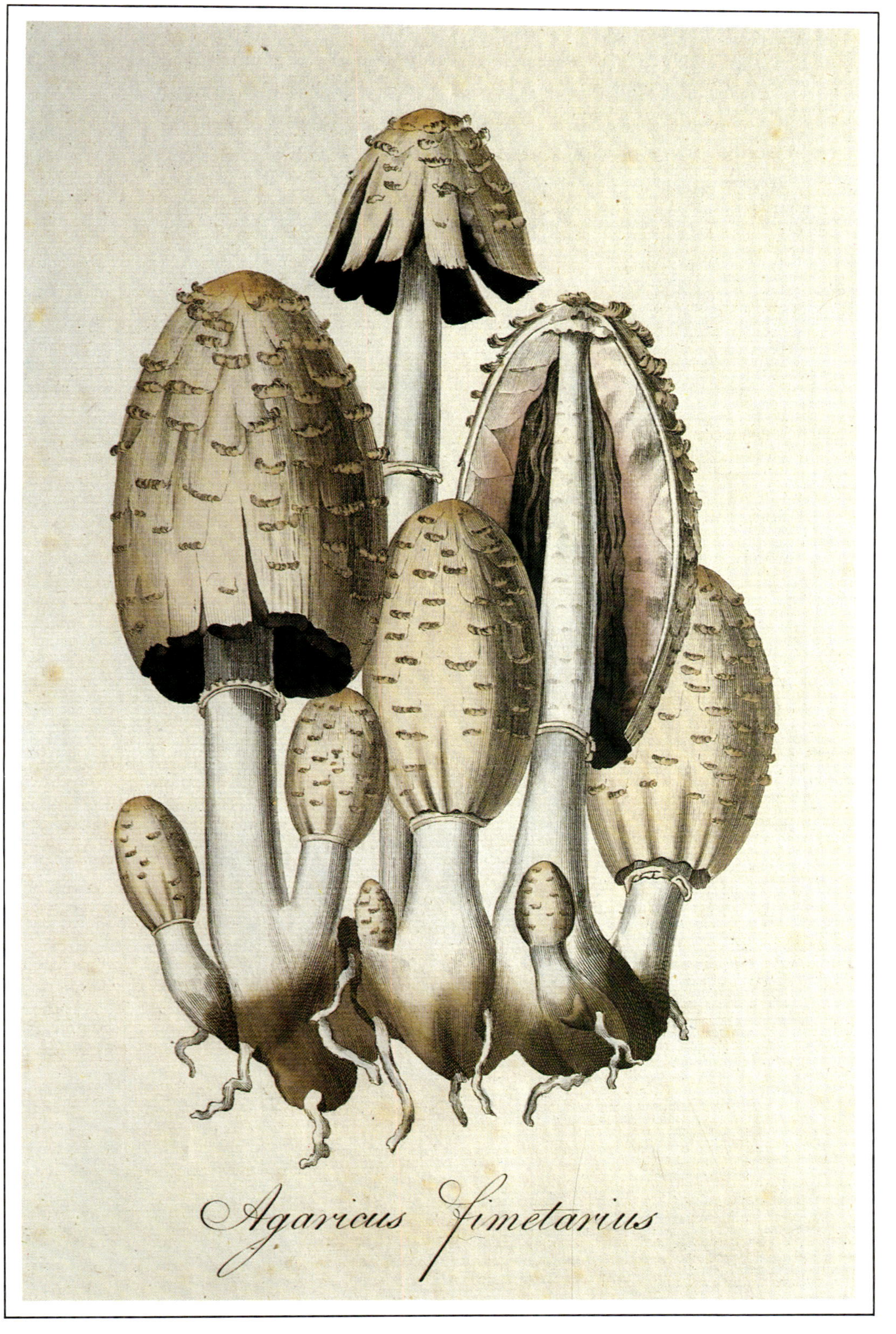

'*Agaricus fimetarius*', the shaggy cap or lawyer's wig (*Coprinus comatus*). Hand-coloured engraving from Volume II of *Flora Londinensis* (1775–87) by William Curtis.

pink acacia (*Robinia hispida*), which Catesby found flowering in 1714 on the James River in the foothills of the Appalachians. When he returned later in the same year to collect seed, he found that the Indians had burned the whole area; the tree was not introduced into cultivation until later. Many of the printed flower plates are rather stiff and awkward in effect, where, on the whole, the original paintings are not. Catesby's pictures of magnolias are particularly effective because their backgrounds are painted matt black to throw the white flowers into relief.

Catesby's second work, published in 1763, was specifically on American trees introduced into English gardens and was entitled *Hortus Britanno-Americanus : or, A Curious Collection of Trees and Shrubs. The produce of the British Colonies in North America ; adapted to the soil and climate of England, etc.* It includes careful instructions for the importation of *Magnolia* seed, to prevent it becoming dry and losing its viability, and for dealing with it when it has arrived. It contains 62 illustrations of trees, shrubs and climbers, including such familiar garden trees as *Magnolia grandiflora*, the tulip tree (*Liriodendron tulipifera*), *Cornus florida*, *Kalmia angustifolia* and *Hamamelis virginica*.

Other botanists and collectors were sent out from England, but the most prolific introducer of new plants to Europe was the Quaker John Bartram (1699–1777), who had been born in Pennsylvania, and lived for most of his life near Philadelphia. He was an indefatigable traveller and covered the coasts of Delaware, Maryland and Virginia, the Catskill mountains, New Jersey and the Appalachians. His favourite botanising area was the valley of the Shenandoah River in Virginia. His son William was a fine botanical painter – a good plate of one of Bartram's most interesting finds, *Gordonia altamaha*, is in the British Museum – but he did not contribute to any fine botanical books, and what little of his work was published was of poor quality.

The first important journey undertaken by André Michaux (1746–1802) had been to the Levant, from which he had sent back seeds of many new plants to Le Monnier in Paris and to Sir Joseph Banks in England. He arrived in New York with his fifteen-year-old son, François-André (1770–1855), in October 1785. He had instructions to buy a piece of land in New York and set up a nursery there, so that he could send plants to Europe in good condition. His sponsors in France were the Abbé Nolin of the Parc de Rambouillet, and Le Monnier and the Jardin des Plantes in Paris. Michaux soon realised that the climate of New York was too cold for the cultivation of many of the plants that were suitable for France. He therefore moved south to Carolina and, by May 1786, had bought a piece of ground near Charleston. Leaving the New York garden in charge of a gardener he had brought from France, he made his headquarters at Charleston. Here he met the Scotsman John Fraser (1750–1811), who had been sent out by a syndicate in England; the two made many journeys together. Michaux's first expedition, part of which was in company with Fraser, took him into Indian territory as far as the Tennessee River, and the following year he and his son visited Florida, exploring the St John's River to Lake George.

In the early spring of 1789, they visited the Bahamas, returning in May to Charleston, where they first heard of the French Revolution. That year, they visited the highlands of North Carolina. They remained in America in

Chestnut oak (*Quercus prinus*), a large tree, native to eastern North America. Engraving by Plée after a drawing by Pierre-Joseph Redouté from *Histoire des Chênes de l'Amerique* (1801) by André Michaux.

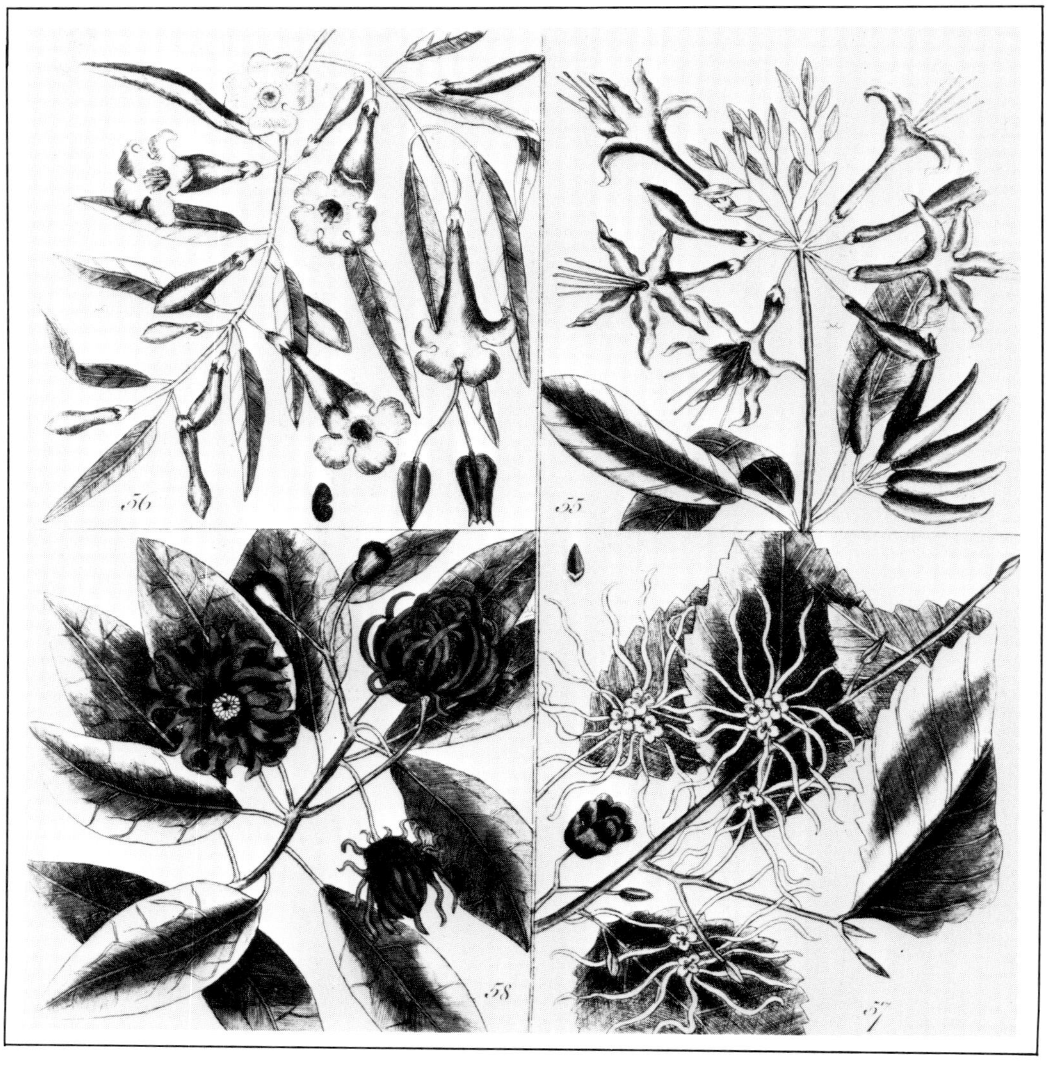

A page with four hand-coloured engravings by Mark Catesby from his *Hortus Britanno-Americanus* (1763). 55: *'Cistus virginiana*, upright honeysuckle' (an azalea, perhaps *Rhododendron calendulaceum*). 56: *'Jasminum luteum*, yellow jessamin'. 57: witch-hazel (*Hamamelis virginiana*). 58: Carolina allspice (*Calycanthus floridus*).

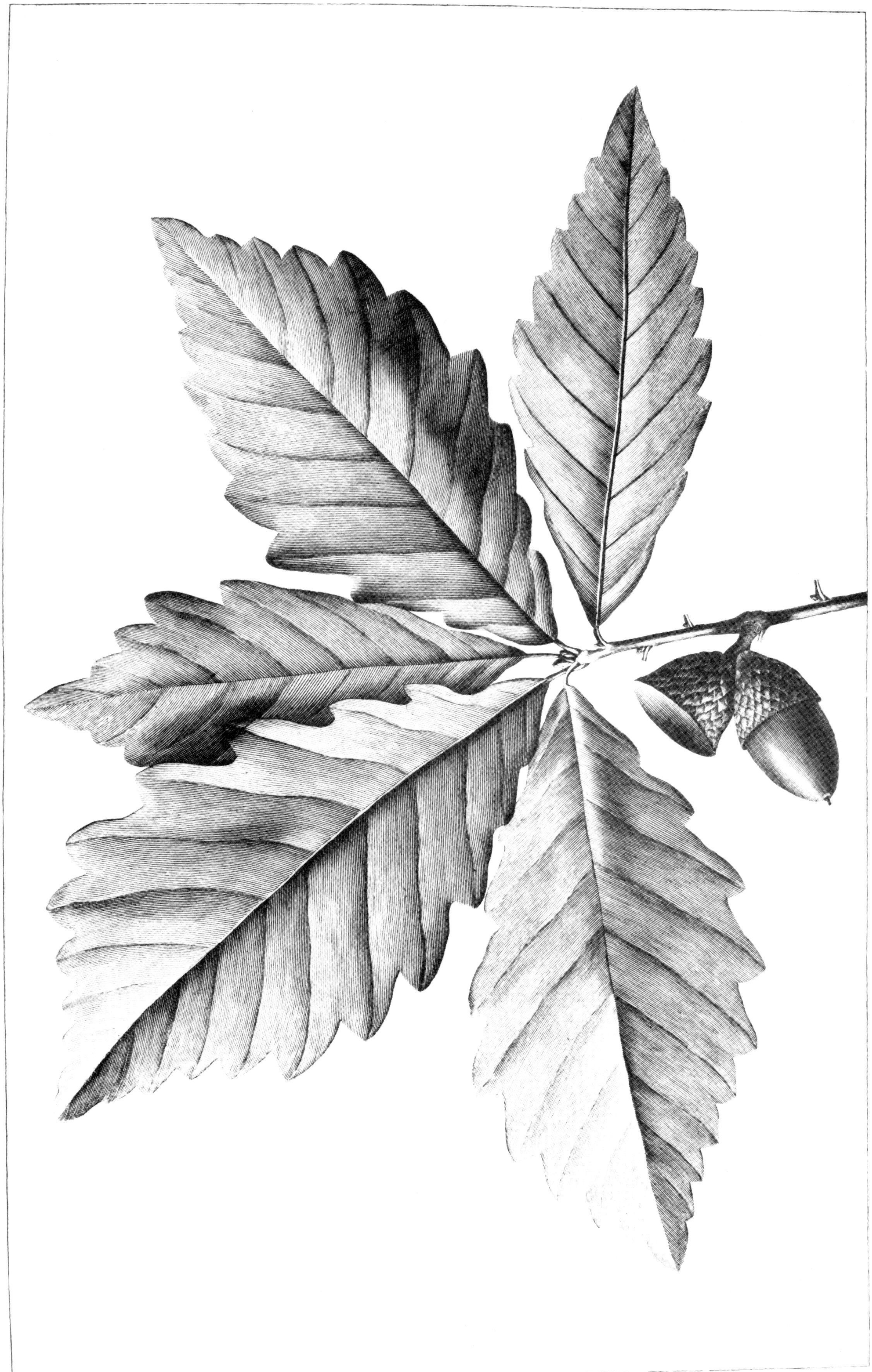

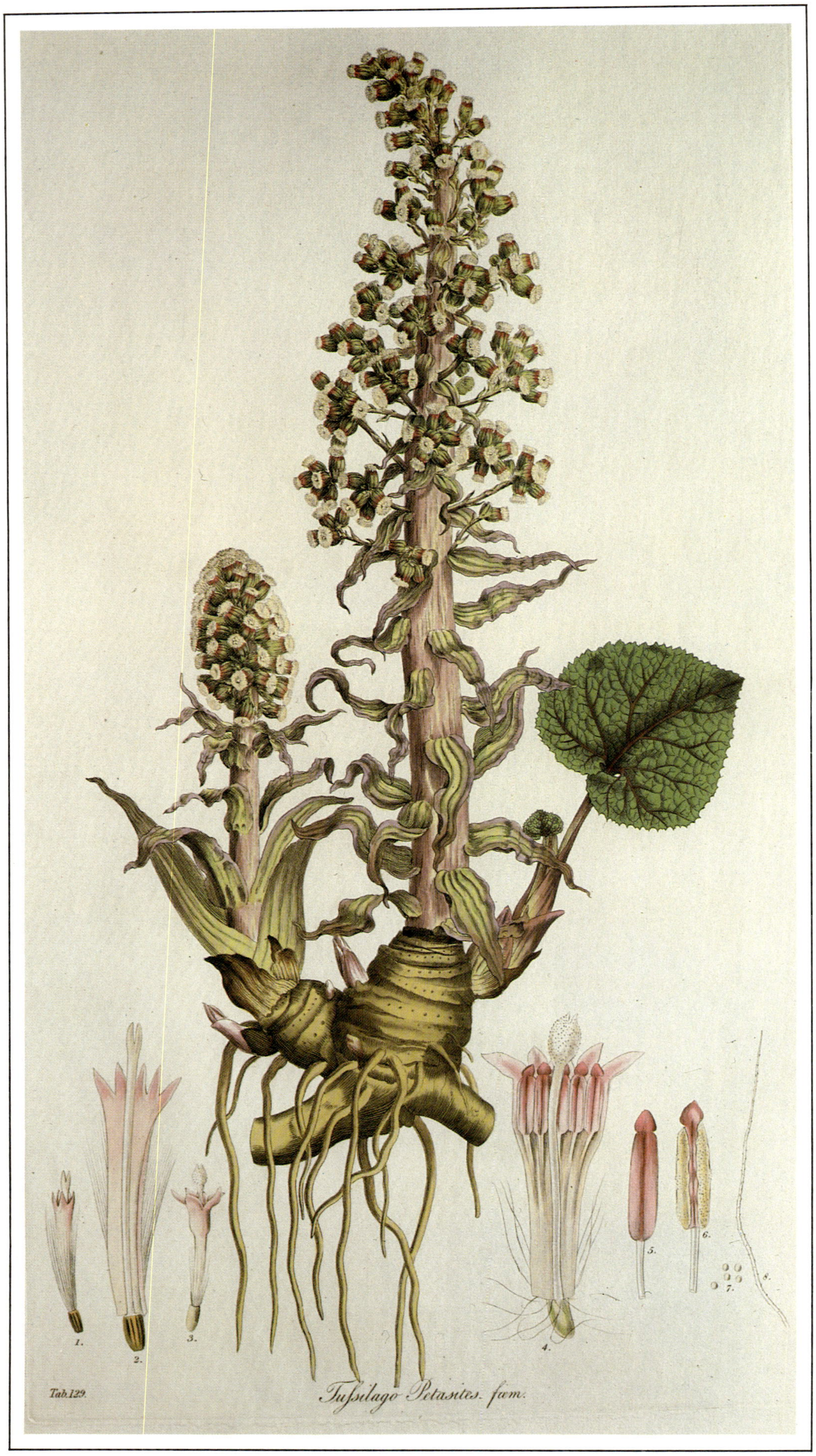

'Tussilago petasites', butterbur (*Petasites hybridus*). Hand-coloured engraving from Volume IV (1819) of the enlarged edition of *Flora Londinensis* (1817–28) with text by W. J. Hooker.

great financial straits for the next seven years. The salary of 12,000 livres, which Michaux was paid before the Revolution, had ceased, but he was not anxious to return to France. He made his longest journey in 1792, northwards to Quebec and towards Hudson's Bay, but was turned back by the cold of approaching winter. An expedition up the Missouri planned for the following year failed to take place. Michaux set off on his final American journey in April 1795. He started up the Catawba River and, crossing the Cumberland Mountains, reached Nashville and then Danville in Kentucky, which he had visited two years before on a diplomatic mission for the French minister. The next stage was a difficult journey through very dense

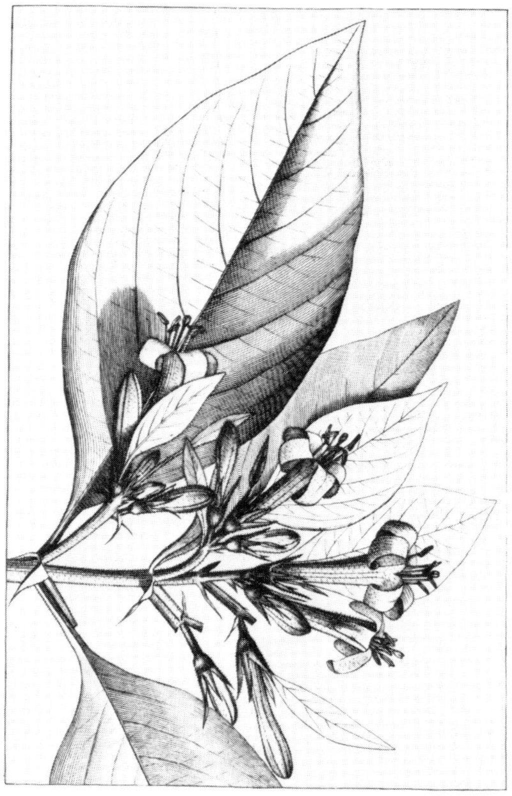

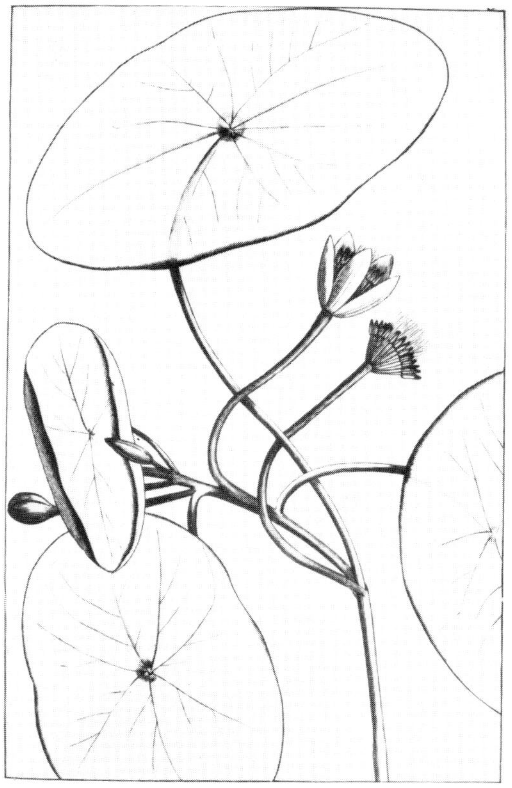

Two engravings by Plée after drawings by Pierre-Joseph Redouté from *Flora Boreali-Americana* (1803) by André Michaux. *Right:* bitter bark or fever tree (*Pinckneya pubens*), a small tree from the southeastern United States, named after General Charles C. Pinckney of South Carolina. *Far right: Brasenia schreberi*, a waterlily-like plant that is found in ponds and rivers throughout the tropics.

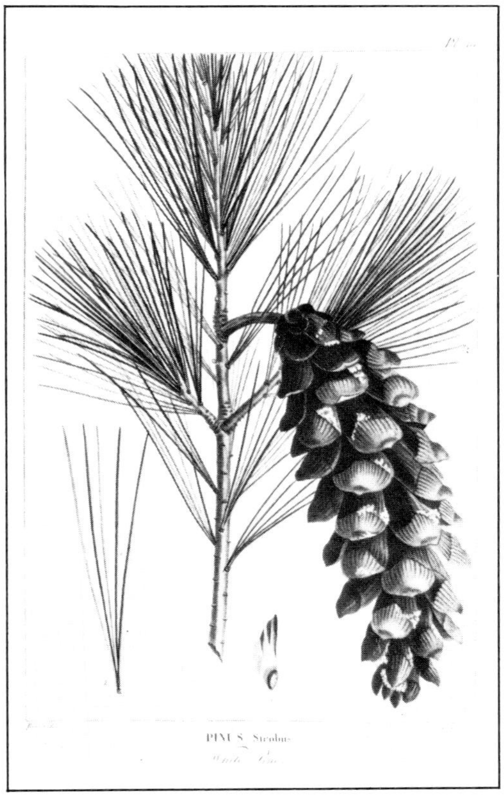

Weymouth pine or white pine (*Pinus strobus*), a native of eastern North America that is commonly planted in Europe. Stipple engraving by Gabriel, colour printed *à la poupée*, after a drawing by Pancrace Bessa from *Histoire des Arbres Forestiers de l'Amérique Septentrionale* (1810) by Francois-André Michaux.

forests to Louisville and then to Vincennes on the River Wabash. With an Indian couple as guides, he left Vincennes on 23rd August. In a week they reached Fort Kaskaskia on the Mississippi. He remained there for three and a half months, collecting along the Illinois River before setting out for home in mid-December. He eventually reached his nursery at Charleston in April.

With his salary still not paid and his inheritance now exhausted, Michaux had no choice but to return to France. After a shipwreck on the coast of Holland in October 1796 in which he lost all his personal belongings, although he managed to rescue most of his collections, he reached Paris. Refused his full arrears of salary by the government, he struggled in poverty to publish the result of his ten years' work in America. His *Histoire des Chênes de l'Amérique* was a folio, published in 1801, with line engravings after drawings by the Redouté brothers; a German edition was published between 1802 and 1804.

Michaux's main work was published after his death by his son. The *Flora Boreali Americana* is an octavo in two volumes with simple but very elegant engravings by P-J. Redouté. François-André himself made a second expedition to America between 1806 and 1808. On his return, he wrote the *Histoire des Arbres Forestiers de l'Amérique Septentrionale*, which was published between 1810 and 1813 as a quarto with coloured stipple engravings by Pancrace Bessa and P-J. Redouté, covering the majority of the trees native to eastern North America. An English translation was published in Philadelphia in 1818–19.

With the Lousiana purchase in 1803, President Jefferson bought from Napoleon territory that stretched from the Gulf of Mexico to Canada and indefinitely westwards; at a stroke, he more than doubled the size of the United States. The two leaders of the expedition in the following year to find a practicable route across the new territory to the Pacific were Captain Merriweather Lewis (1774–1809) and Captain William Clark (1770–1838). Both were soldiers and neither had any botanical knowledge, but Lewis was sent to Professor Barton in Philadelphia for some elementary training in natural history. The party left St Louis in May 1804. They worked slowly up the Missouri and camped for the winter in North Dakota. In April 1805, they sent home letters and the collections they had made before continuing. They finally crossed the Rockies by the Lolo pass, through the Bitterroot mountains and reached the upper waters of the Clearwater River. In canoes, they navigated the Clearwater and Snake rivers to the Columbia to reach the Pacific on 7th October.

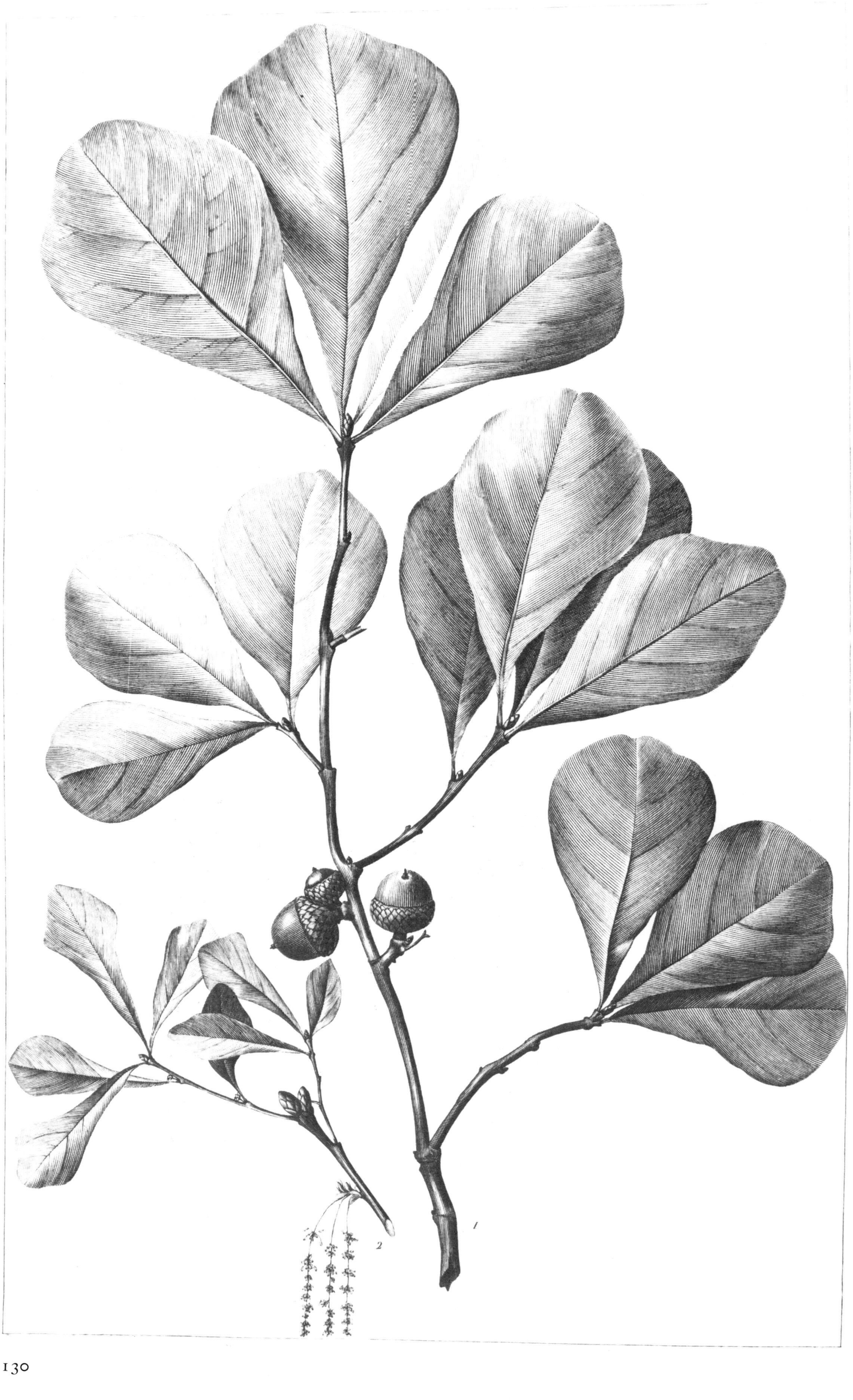
1
2

Water oak (*Quercus aquatica*), a large tree that is a native of the southeastern United States. Engraving by Plée after a drawing by Pierre-Joseph Redouté from *Histoire des Chênes de l'Amérique* (1801) by André Michaux.

They spent the winter near the mouth of the Columbia river before returning eastward again across the Lolo Pass to arrive at St Louis on 23rd September 1806. They had made a considerable collection on the outward journey, but all of it was lost, and only the small herbarium of about 150 specimens and some seeds that Lewis made on the return journey was saved. Even so, it was a fascinating collection; only about a dozen of the species were already known from North America, and at least six genera were new to science. Lewis gave his collection to the German botanist Frederick Pursh, and it was in Pursh's *Flora Americae Septentrionalis* that most of the new species and genera were published. These included *Lewisia*, the 'Bitter Root' after which Lewis had named the mountains, and *Clarkia*. A few of the new species were illustrated with drawings by William Hooker (1779–1832), who is best known for his paintings of fruit for the Royal Horticultural Society. They include *Clarkia pulchella* and *Mimulus lewisii* collected 'on the headsprings of the Missouri, at the foot of Portage Hill'.

Frederick Pursh himself was a native of Saxony who had left Dresden for America in 1799. He determined to get to know the plants of North America, and in doing so met the Bartram family and also botanised alone in the hills of Virginia, the Carolinas and New Hampshire. Then finding America at war an unsuitable place to write up his Flora, he went back across the Atlantic to London and prepared his work there with the co-operation of Sir Joseph Banks and A. W. Lambert, who had acquired Pallas's herbarium, which included plants from the Pacific coast of Asia. The first edition was published in 1814 and a second in 1816, both octavo with 24 engravings. Other new species that are now familiar but were illustrated for the first time in Pursh's Flora are *Mahonia aquifolium* (collected by Lewis) and *Gaultheria shallon*, which was first collected in 1792 by Archibald Menzies, surgeon on Captain Vancouver's ship exploring the northwestern coast. Pursh returned to America after the success of his book, determined to study the flora of Canada, but in 1820 died penniless

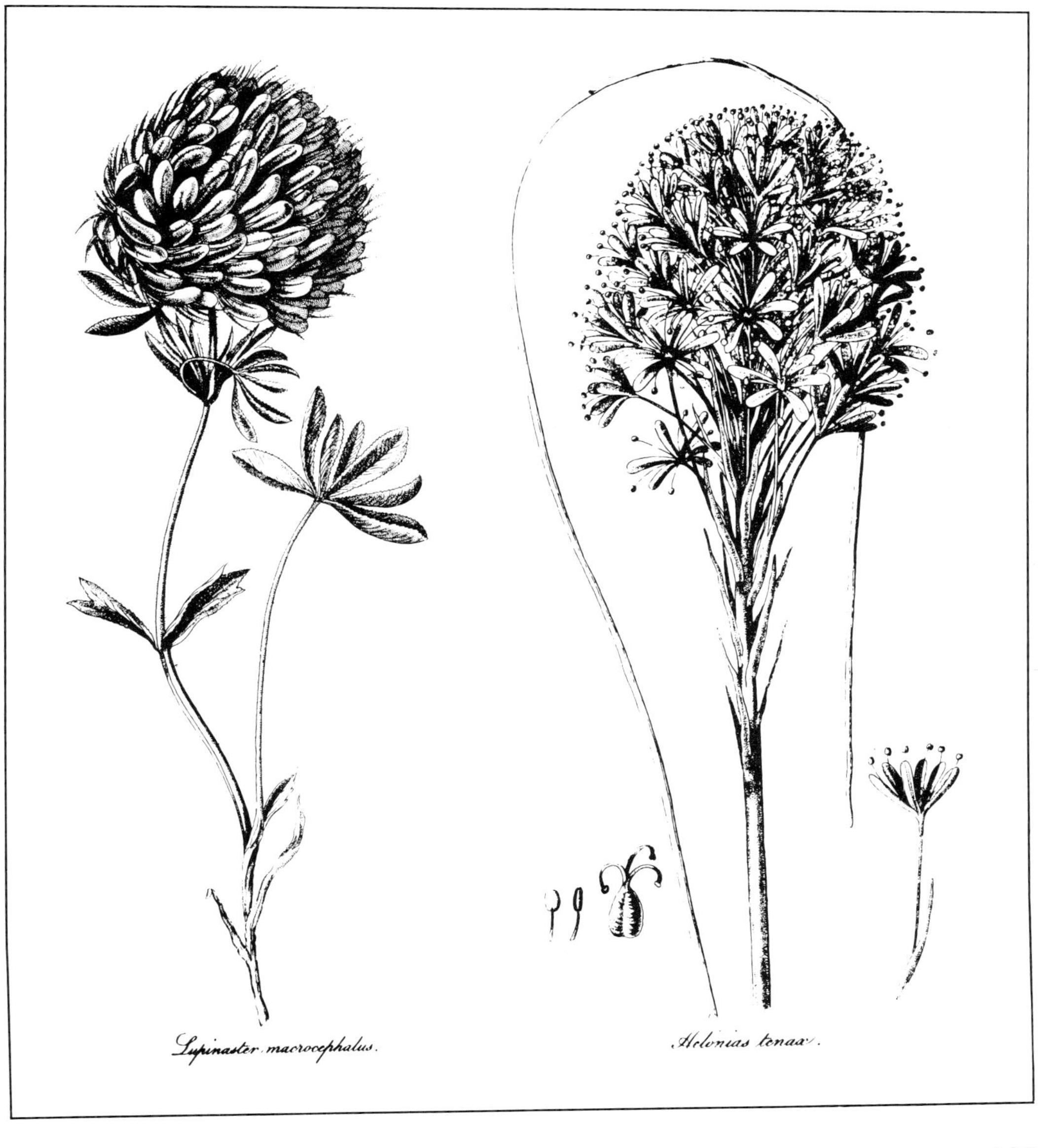

Two engravings from his own drawings by William Hooker (who was not related to the famous botanists, W. J. and J. D. Hooker) from Frederick Pursh's *Flora Americae Septentrionalis* (1814). *'Lupinaster macrocephalus'* is a clover, *Trifolium megacephalum*, from western North America. *'Helonius tenax'* is bear grass or Indian basket grass (*Xerophyllum tenax*), a striking member of the lily family from western North America; it grows up to five feet tall, often in great numbers.

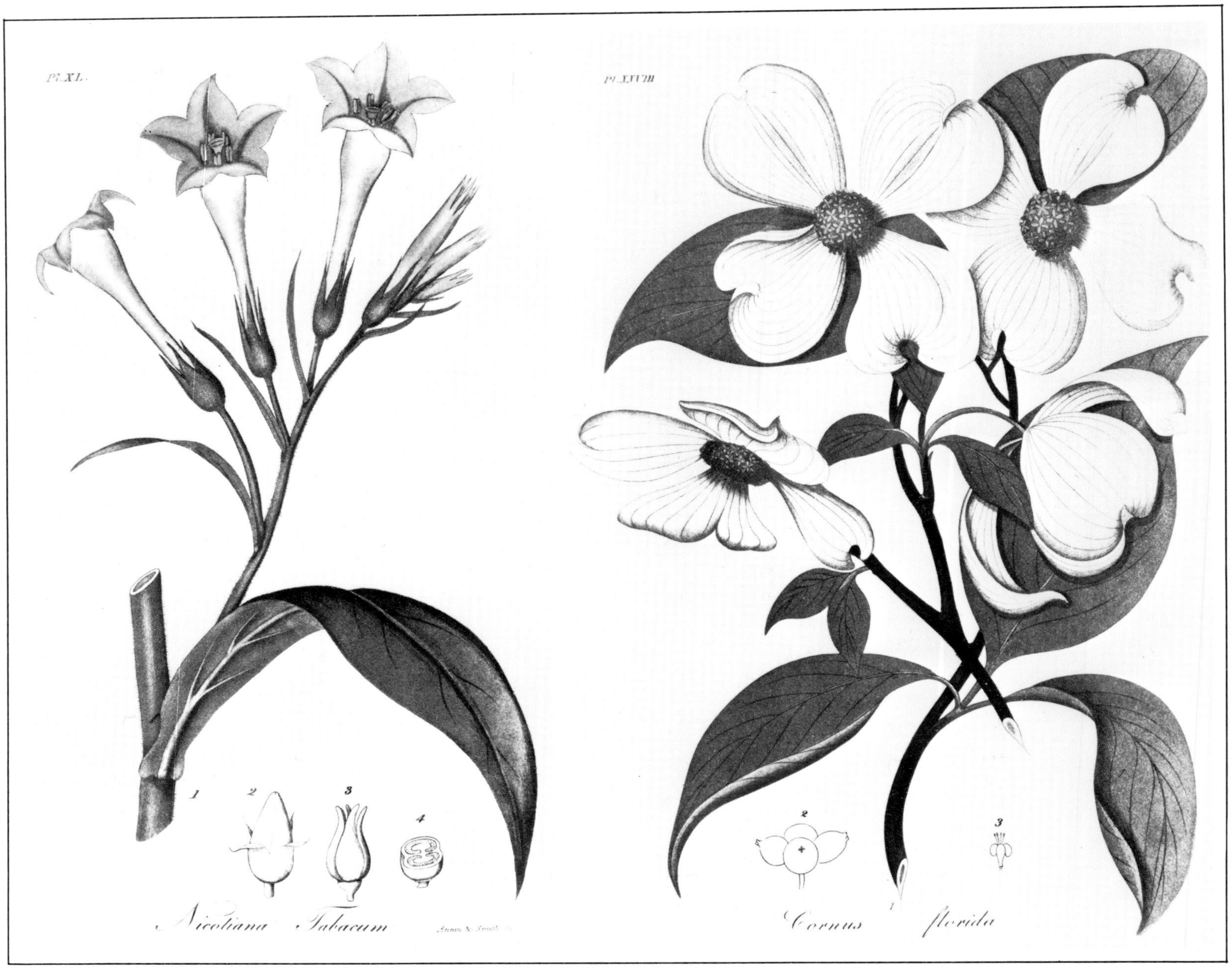

Two colour plates from Jacob Bigelow's *American Medical Botany* (1817–21): tobacco (*Nicotiana tabacum*), and *Cornus florida* from eastern North America.

in Montreal at the age of 46. In 1857, a fund was set up to raise a fitting memorial to him. Pursh has been criticised for his inaccuracy and for publishing the discoveries of others without their permission. Yet his work was the most comprehensive on American botany to be published for many years, and in his preface and often in the text, he acknowledges the work of other collectors and names the sources of his records.

In spite of the difficulties put in their way by the Spanish authorities, and later by repeated revolutions, botanists managed to investigate much of south and central America sooner than they did the northwestern part of the continent. Starting with Hernández in 1570, a steady trickle of brave or foolhardy naturalists succeeded in covering much of this vast area before the beginning of the nineteenth century.

In 1707, at a time when France and Spain were allies, a French priest, Louis Feuillée (1660–1732), managed to get permission to visit Chile and Peru. He reached Concepción in Chile in January 1709 and spent the next four years travelling, living in Lima, and finally sailing up the coast. His brief was to make all sorts of astronomical and geographical measurements and observations; his *Journal des Observations Physiques, Mathématiques et Botaniques etc.* was published in Paris in 1714. All the illustrations are by Feuillée himself, including the maps and drawings of cities. The plants which are arranged in a supplement, *Histoire des Plantes Médicinales du Pérou et du Chily*, which contains fifty simple and clear engravings of plants, including recognisable illustrations of *Salpiglossis*, *Alstroemeria* and *Bomearea*.

Botanists have often been suspected of being spies, but Amédée François Frezier was first a spy and only secondarily a botanist. In 1712, he studied the coast of Chile and brought back to France one of the ancestors of the modern strawberry, *Fragaria chiloensis*. His journal, published in 1717, includes pictures of the strawberry (possibly with the size of the fruit somewhat exaggerated) cotton and scenes of Indian life including a form of golf.

In the eighteenth century, the other main area of British interest apart from North America was the Caribbean. In 1750, Griffith Hughes produced a handsome folio, *The Natural History of Barbadoes*. Ehret contributed most of the botanical illustrations, together with an artist called George Bickham whose only recorded botanical work this is. Ehret's drawings include most of the common tropical fruits, such as bananas and pawpaws; each is very elegantly engraved, dedicated to a member of the aristocracy and emblazoned with the appropriate coat of arms – a reminder of just how much the publication of fine and expensive books depended on finding rich and noble patrons.

Nikolaus von Jacquin made his first long journey as a botanist to Central America in 1754, when he was 27, to collect plants for the Imperial Gardens at Schönbrunn, which had recently been laid out by Francis I. Jacquin and his companions, the Dutch head gardener, Richard van der Schott, and two Italian zoologists, first went to Grenada, Martinique and neighbouring islands, with one member returning in charge of a large consignment of plants every few months. Jacquin himself remained, but was captured by the British and kept prisoner for over a year. When he was released, he went on to Cuba and Jamaica to collect more plants before returning to Austria in

Two engravings from Louis Feuillée's *Histoire des Plantes Médicinales qui sont le plus en usage aux Royaumes de l'Amérique Méridionale, du Pérou & du Chily* (1714). *Herreria stellata* from Chile and the sweet potato (*Ipomoea batatas*), which is commonly grown in the tropics for its edible tubers.

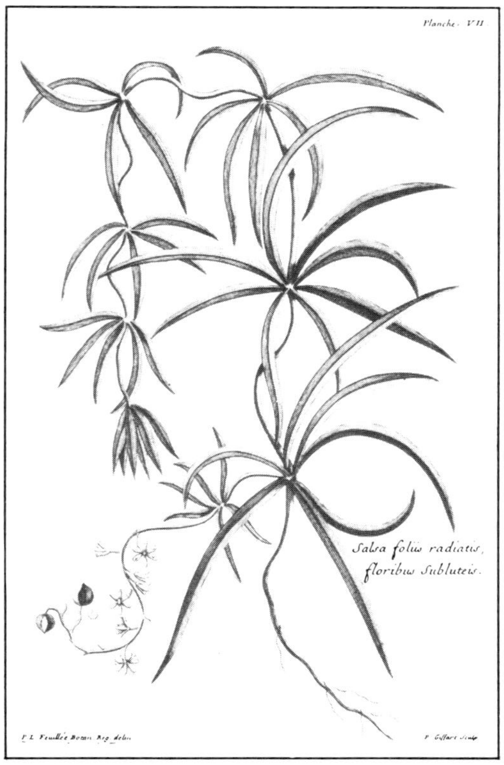

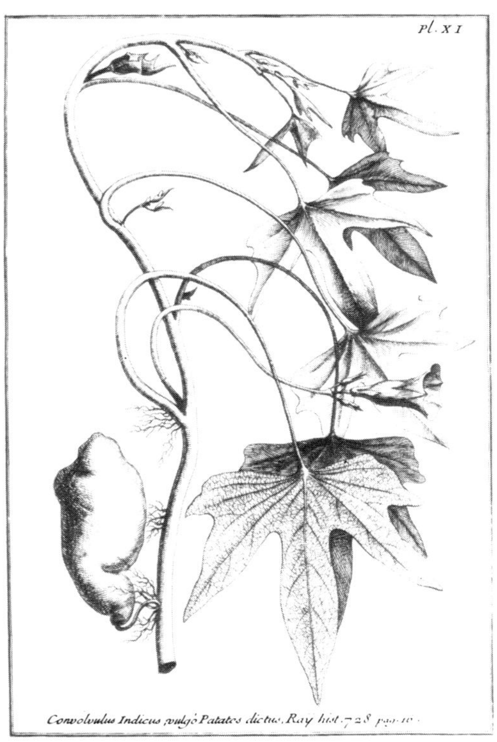

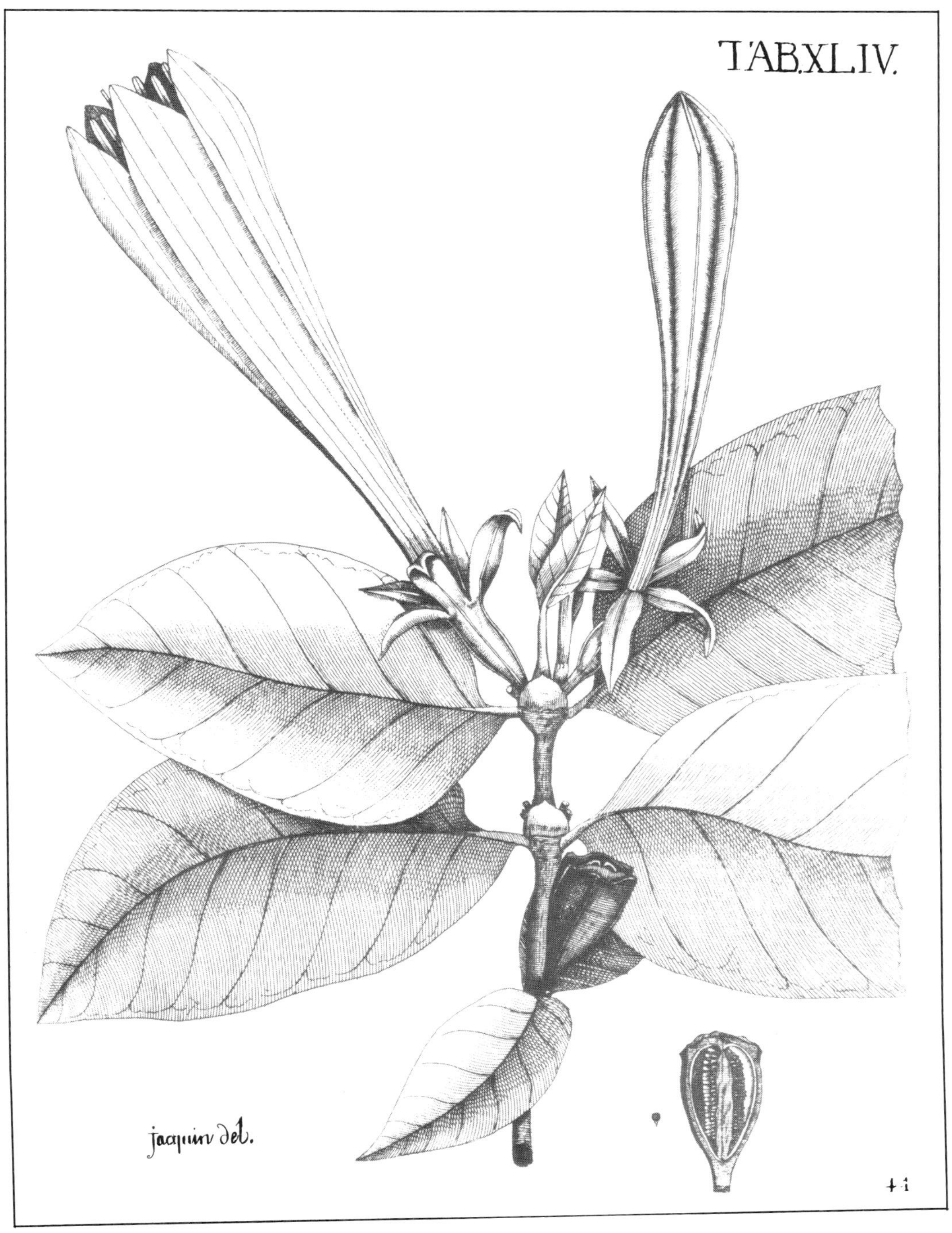

'*Portlandia grandiflora*'. Engraving after a drawing by Nikolaus von Jacquin from his *Selectarum Stirpium Americanorum Historia* (1763).

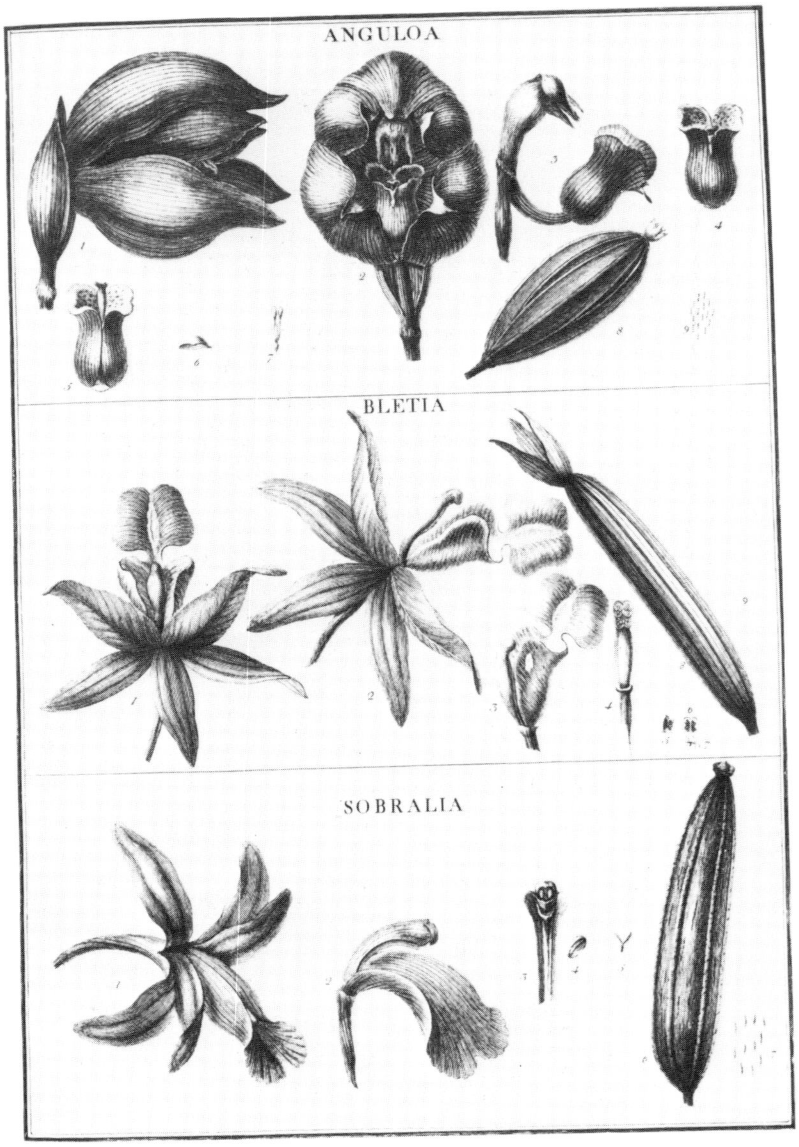

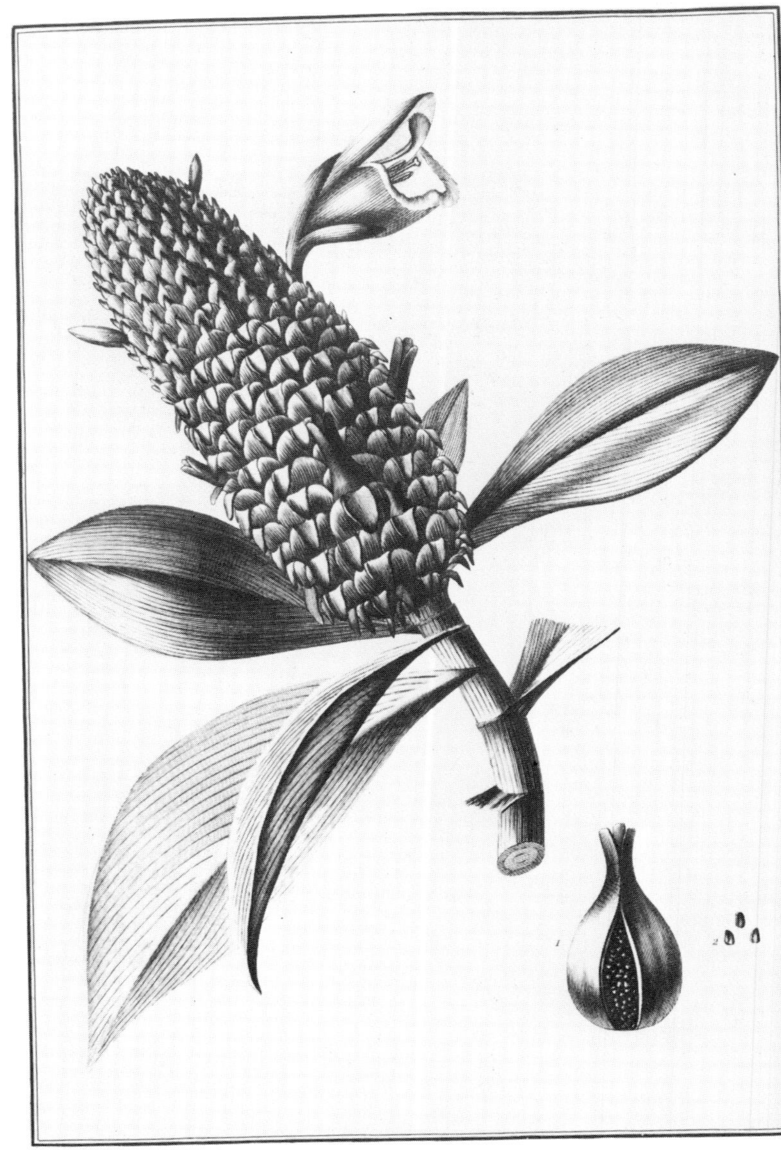

Left:
Flowers of *Anguloa uniflora*, *Bletia* and *Sobralia*, a plate accompanying the first published description of these three genera of orchids. Engraving from *Flora Peruviana et Chilensis Prodromus* (1794) by Hipolito Ruiz Lopez and José Pavon y Jiminez.

1759. He published an account of the botany of the trip, the *Selectarum Stirpium Americanarum Historia* in 1763. The plates are by Jacquin himself and are rather crude, not bearing comparison with work he commissioned later from the Bauers and other professional botanical painters or even with the sketches in his own letters, which are preserved in the British Museum.

The Spaniard José Celestino Mutis (1732–1808) came to Bogotá in 1760 as physician to the Viceroy, and remained in Colombia for forty years in various scientific pursuits which included botany. After many attempts, he obtained permission from the Spanish authorities to confine his attentions to botany. In 1783, he took charge of an '*Expedición Botanica*', a group of pupils who travelled widely throughout the country collecting plants for him. He is also said to have gathered together a team of draughtsmen, at one time as many as thirty, who completed six to seven thousand plant drawings for him. Mutis never published the results of his work; his collections and drawings were finally sent to Spain, where they were ignored – the first of his drawings to be published appeared 150 years later.

Slowness, incompetence and obstructiveness on the part of the Spanish government further complicated by wars between Britain and France form a recurring theme in the stories of botanists in South America during the eighteenth century. The expedition of Joseph Dombey (1742–1796), a Frenchman and pupil of Bernard de Jussieu, with Hipolito Ruiz Lopez (1754–1816) and José Pavon y Jiminez (1754–1840) to Peru and Chile was beset by all these difficulties. Alice Coats has described their lives and problems in some detail. Dombey, the most experienced of the three, was 'gay, charming and extravagant'; he augmented his salary by gambling and was very popular with the ladies of Lima, attending all their card parties. He is also said to have rescued the town of Huánuco from a rebellion by the force of his character. During a serious outbreak of plague in Concepción, he 'temporarily abandoned botany' and dedicated his whole time to ministering to the poor, paying for nurses, food and medicine himself. After this he was

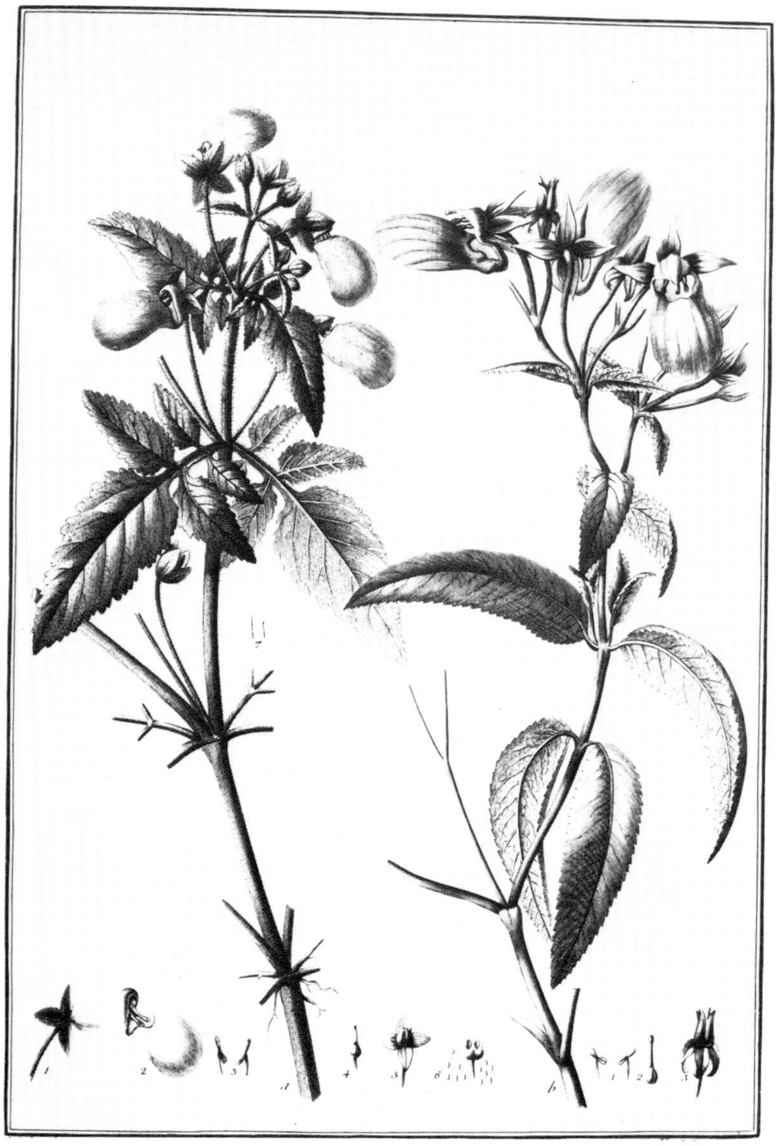

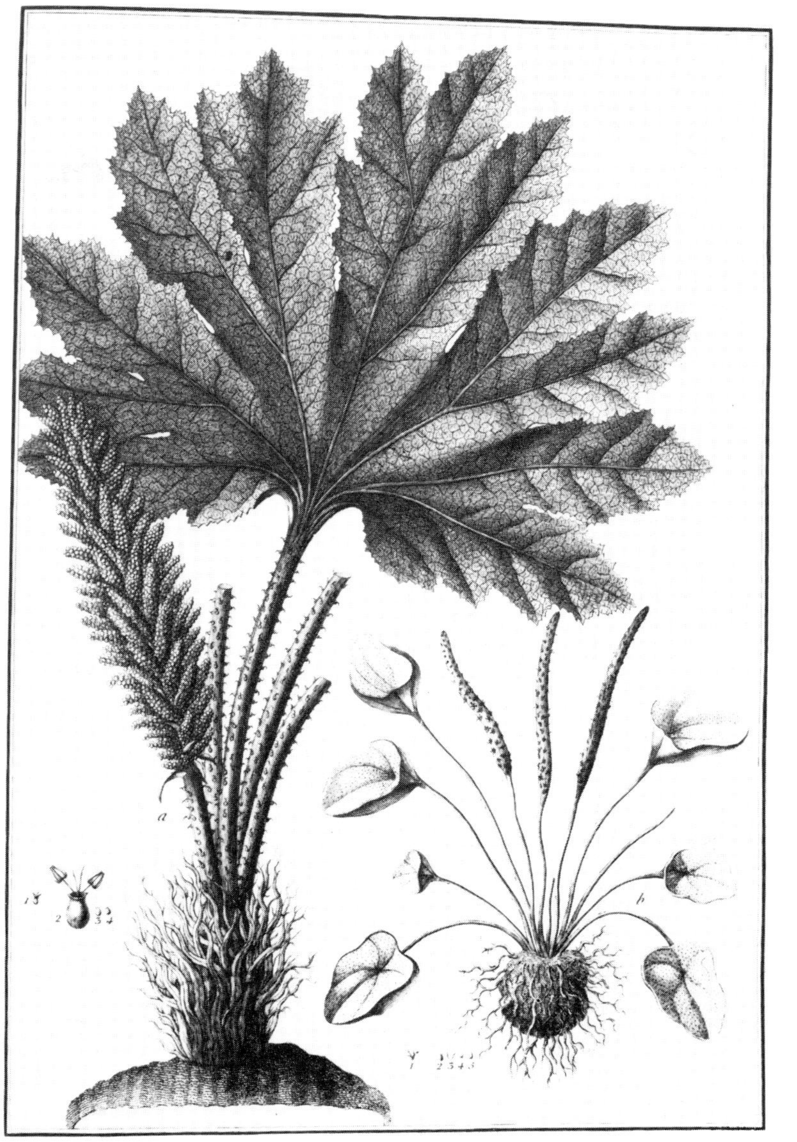

Three engravings after drawings by Isidor Galvez from Ruiz and Pavon's *Flora Peruviana et Chilensis* (1798–1802) *Left:* a wild ginger, *Costus argenteus*, found in woods in Peru. *Above: Calceolaria pinnata* and *Calceolaria salicifolia*. *Right: Gunnera chilensis*, a medium-sized species, with leaves up to five feet in diameter, of a genus named after Bishop J. E. Gunnerus (1718–73), a Norwegian botanist, with *Peperomia scutellatifolia*.

Overleaf:
Claytonia perfoliata, a native of western America, from Alaska to Mexico, and of Cuba; it is also found as a weed on sandy soils in many places in England. Engraving by Sellier after a drawing by Pierre-Jean-Francois Turpin from *Plantae Aequinoctiales* (1805–18) by Alexander von Humboldt and Aimé Bonpland.

offered a post as physician at a salary of 10,000 livres a year (nearly twice his original salary) and the hand of a beautiful heiress, but he refused in order to return to France.

He landed with his share of the collections at Cadiz in February 1785, having narrowly escaped shipwreck while rounding Cape Horn. But his troubles were not over: the first cases of his collections to be sent home had been captured by the British and sold at auction in Lisbon; the French consul had been able to ransom only some of the less valuable. The collection he brought with him was impounded by the Spanish, who demanded half, because their share of the expedition's collection, sent back by Ruiz and Pavon, had been completely lost in a shipwreck. When Dombey got to Paris in October 1785, he had only half of his collection, and much of that was damaged through careless handling. He is said to have become a 'misanthrope and a recluse', and the French Revolution completed his misery. He managed to get permission to go and buy corn for France in America. In Guadeloupe in the West Indies, however, he ran into another revolution, became sick with fever, was imprisoned and expelled. His ship was attacked and he was captured, 'disguised as a Spanish sailor', by the British; he died on the island of Montserrat, a prisoner of the British, in 1796.

Dombey's companions, Ruiz and Pavon had fared little better. They remained in Peru for two years, but lost many of their later collections and notes in a disastrous fire which destroyed their whole camp. Although they got back to Spain in 1788, it was not until 1794 that the results of their expedition began to be published. There is scant recognition of the important contributions made to their work by the unfortunate Dombey in *Flora Peruviana et Chilensis Prodromus* (1794) or the four-volume folio *Flora Peruviana et Chilensis* (1798–1802), which were published in Madrid. The 325 engravings are by various artists and show for the first time many genera which have since become garden plants: the tropical *Calceolaria*, *Schizanthus* and *Peperomia* and the temperate *Embothrium*.

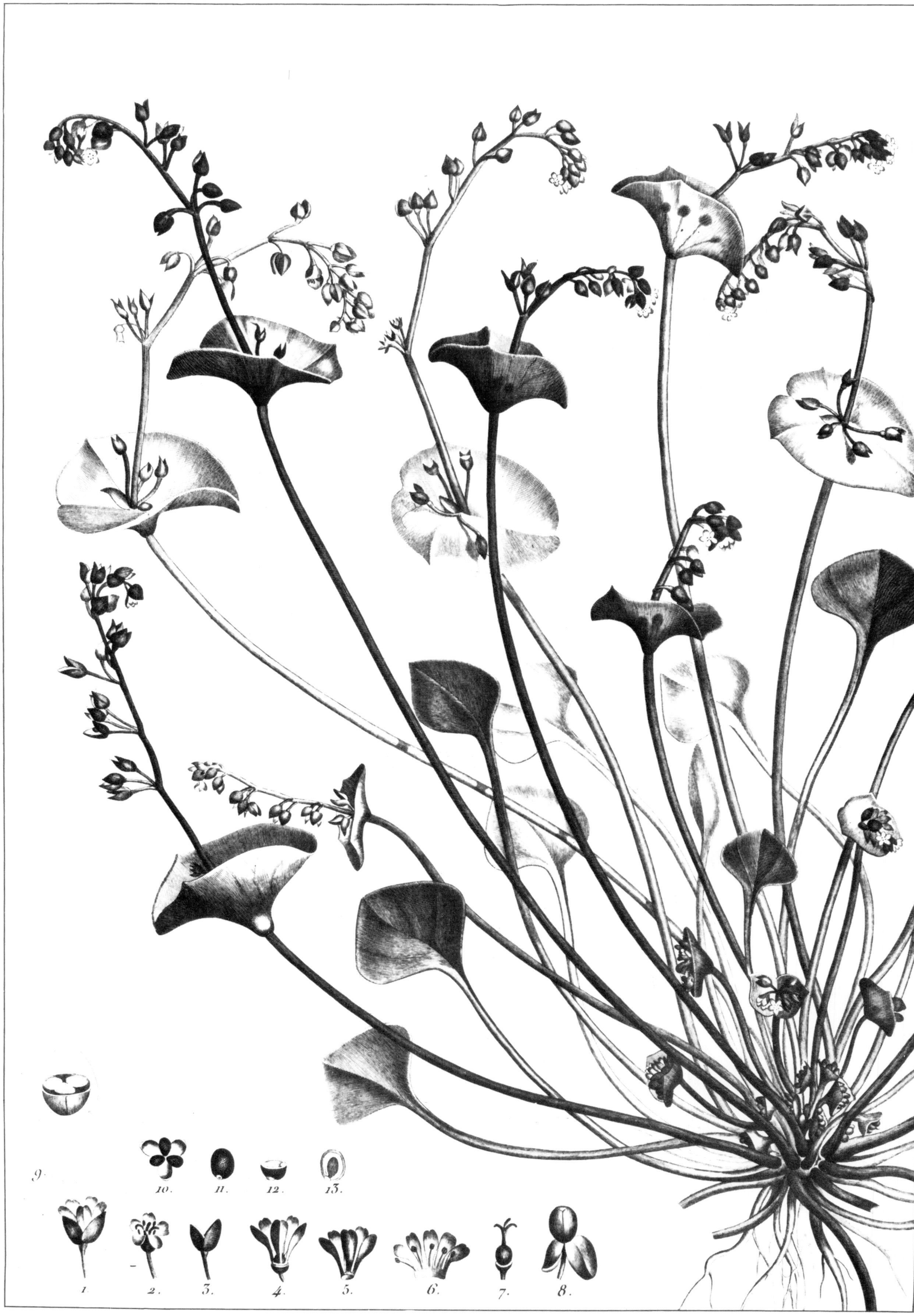
9.
10.
11.
12.
13.
1
2.
3.
4.
5.
6.
7.
8.

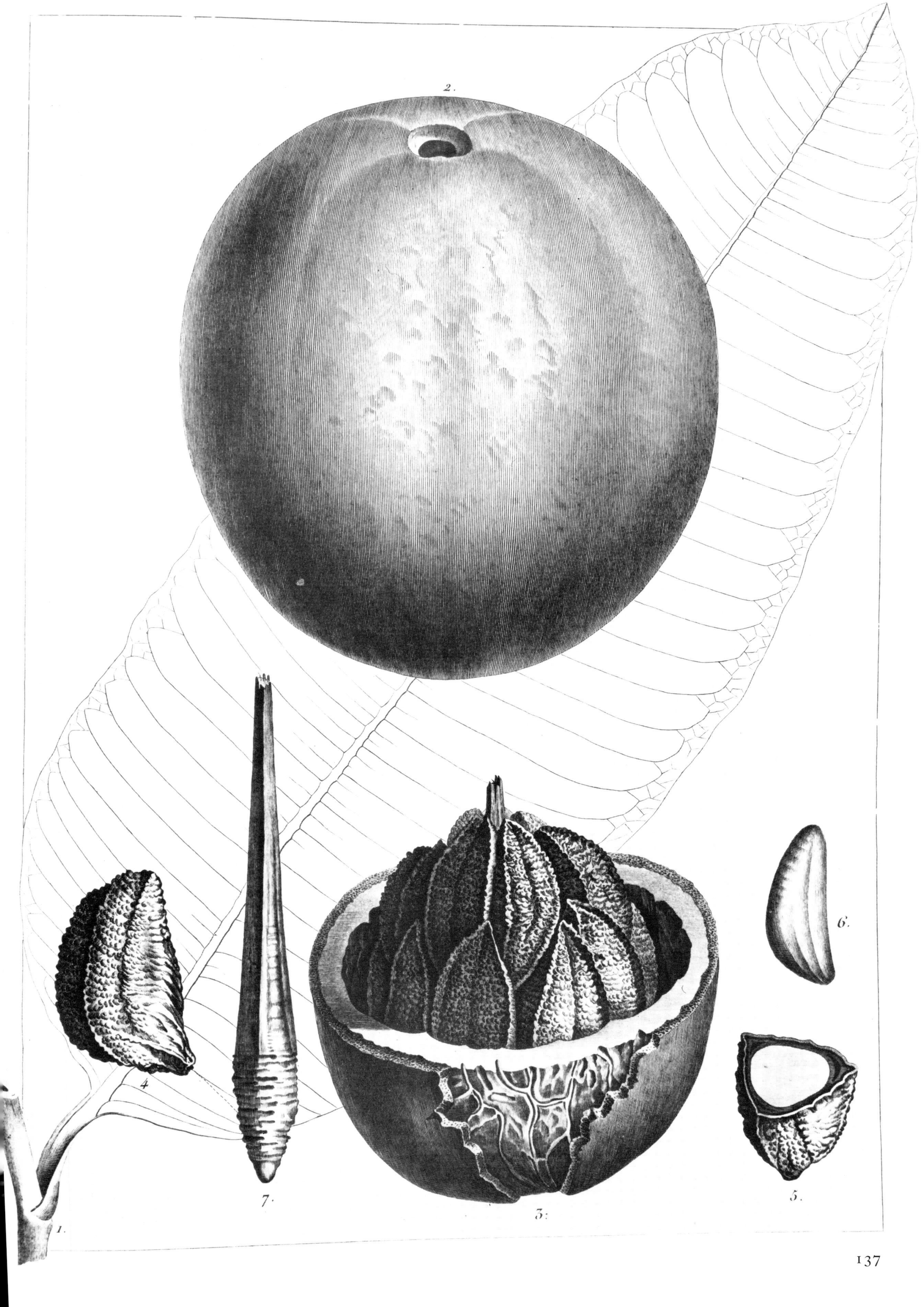
2.
4
7.
3:
6.
5.
1.

Previous page:
Brazil nut (*Bertholletia excelsa*), a large forest tree from South America with fruits that weigh up to four pounds and contain 12–24 nuts. Engraving by Sellier after a drawing by Turpin from *Plantae Aequinoctiales* (1805–18) by Alexander von Humboldt and Aimé Bonpland.

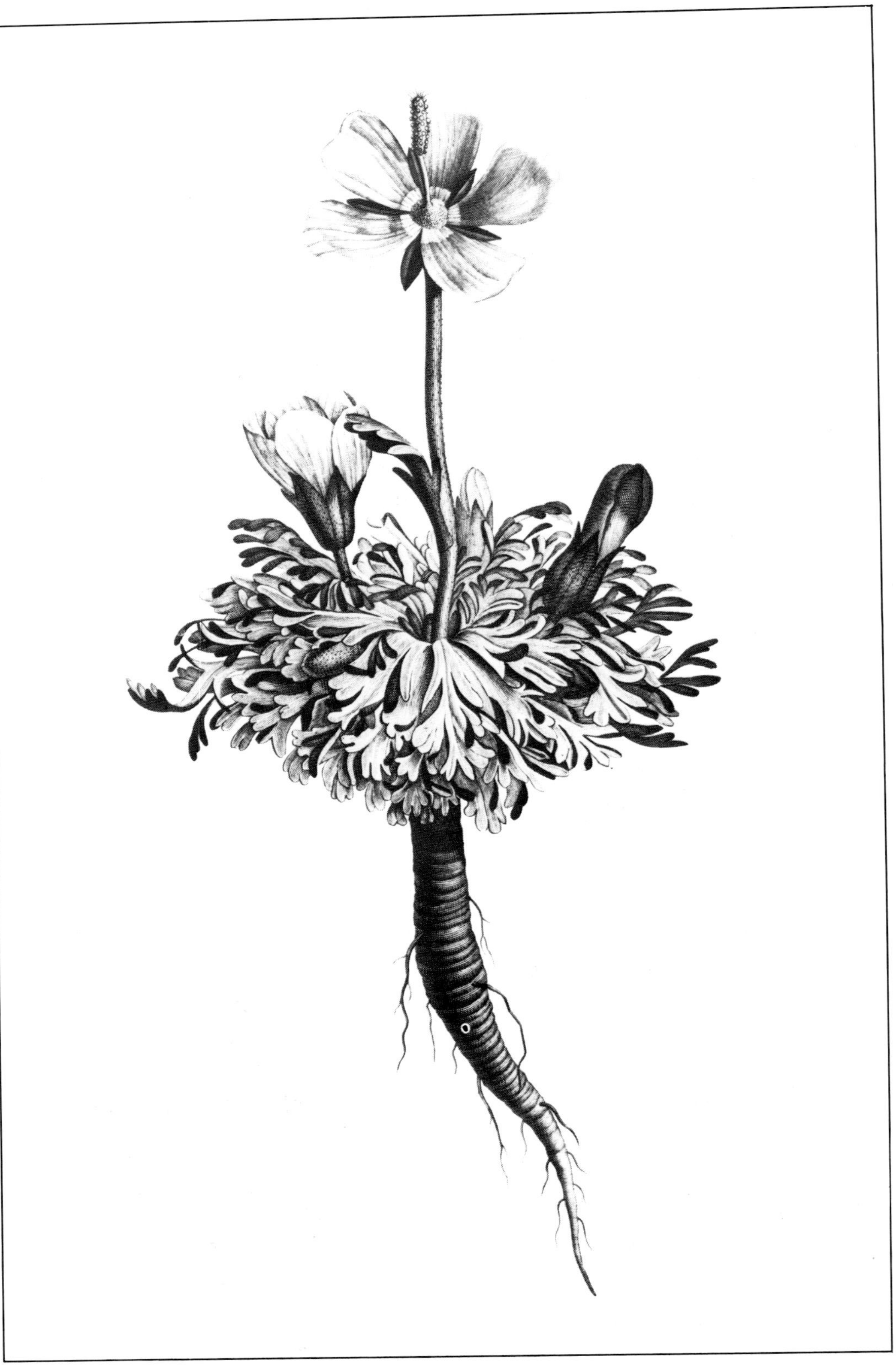

Two engravings by Sellier after drawings by Pierre-Antoine Poiteau from *Plantae Aequinoctiales* (1805–18) by Alexander von Humboldt and Aimé Bonpland. *Left: Sida pichinchensis. Right: Loasa ranunculifolia.*

The South American and Mexican journeys of Alexander von Humboldt (1769–1859) were, in terms of their published scientific results, the most important ever undertaken to the Americas. Humboldt was German and a mining engineer by training, but he made very important contributions to all branches of natural history, geography and geology. His companion was Aimé Bonpland (1773–1858), a surgeon and botanist from Paris. The two met in 1798 in Paris and decided to travel together, but political difficulties thwarted their plans to visit north Africa and Egypt. Baudin's proposed expedition to Australia, on which Bonpland had already been appointed botanist, was indefinitely postponed because of cuts in French government spending caused by the outbreak of war with Austria. The two decided instead to visit Spain, which was then still little known to botanists; in January 1799, they hired mules and set off. Humboldt soon discovered that an old family friend was the ambassador of Saxony to Spain and through him obtained an interview with the king, Charles IV. In an uncharacteristic moment of helpfulness, Charles gave Humboldt permission to visit the Spanish possessions in South America and provided him with a passport and a royal letter of recommendation.

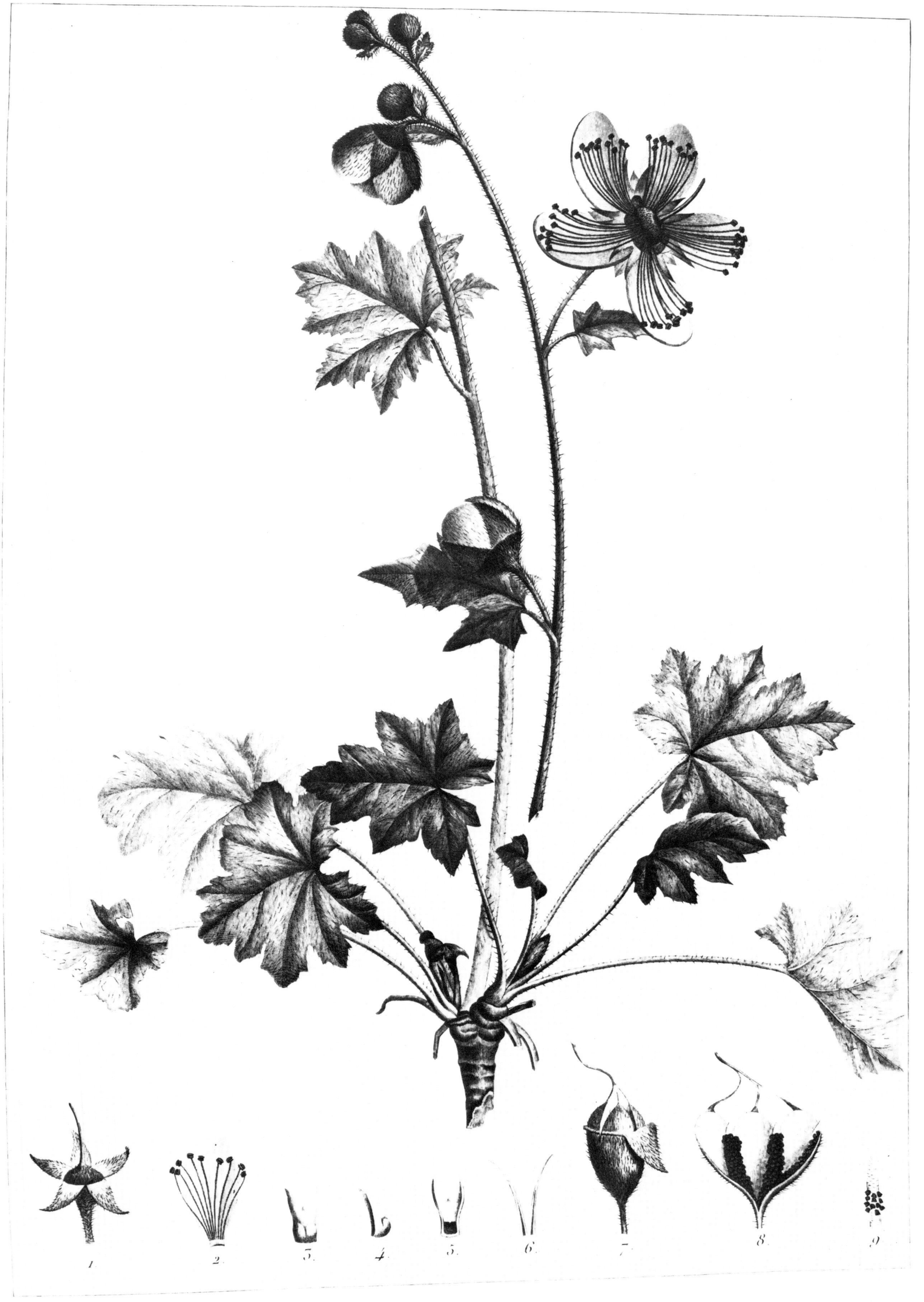
1.
2.
3.
4.
5.
6.
7.
8.
9.

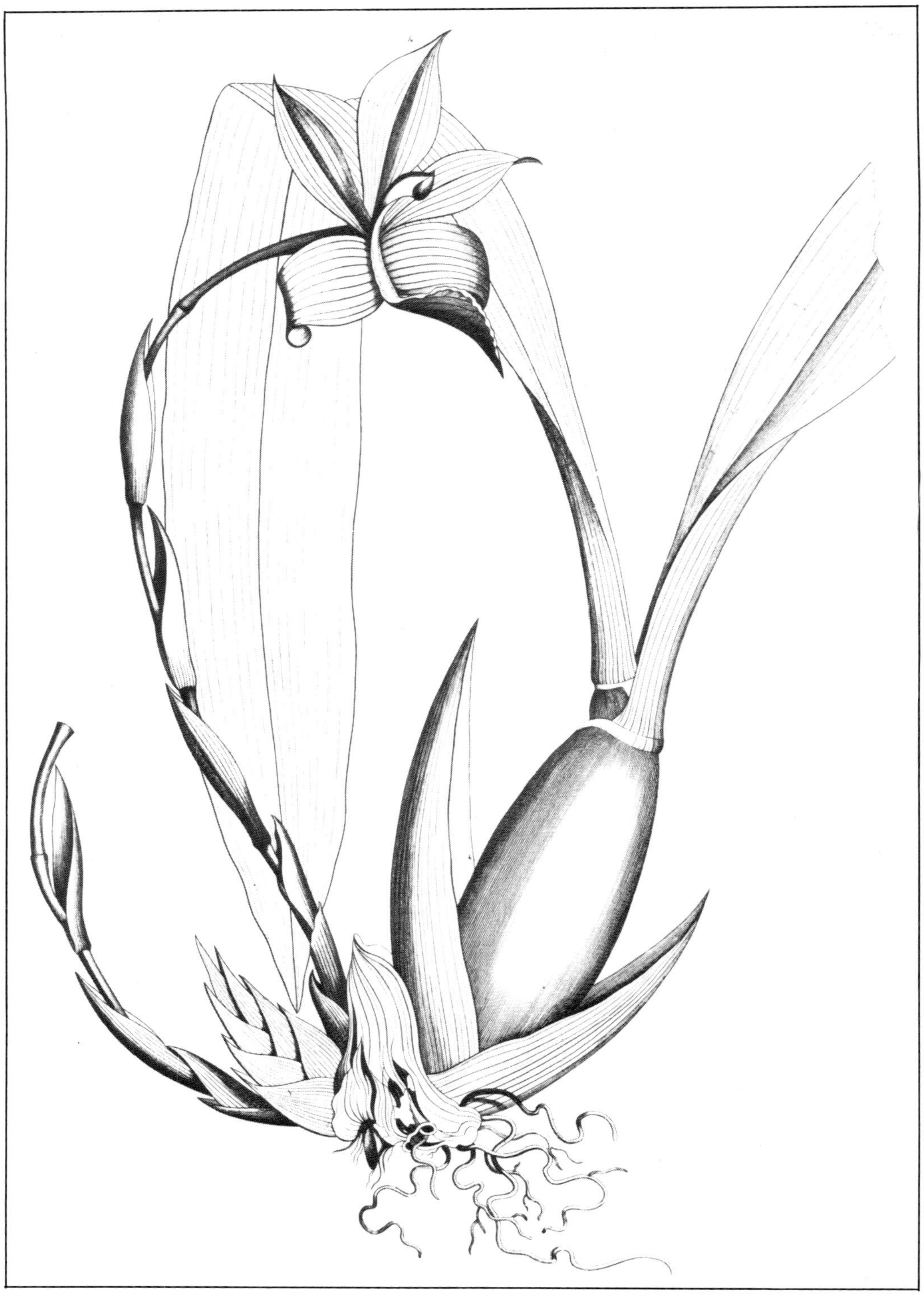

'*Dendrobium grandiflorum*', an orchid. Engraving after a drawing by Pierre-Jean-Francois Turpin from *Nova Genera et Species Plantarum* (1815–25) by Alexander von Humboldt, Aimé Bonpland and C. S. Kunth.

The two travellers left Spain in June 1799 and travelled around Central and South America for the next five years. In 1800 they explored the area of the Upper Orinoco and Rio Negro rivers on the borders of Venezuela and Brazil. In 1801, they visited Mutis in Bogotá. In 1802, they spent some months in Ecuador, where Humboldt made a special study of the volcanoes and climbed to 19,286ft. on Chimborazo (the top is 21,424ft.), then the highest recorded altitude climbed by man. They travelled southward down the coast of Peru, and visited the Chincha islands from which the guano had been used as fertiliser by the Indians for centuries: it was Humboldt who introduced it as a phosphate fertiliser to Europe. He also took measurements of the water temperature along the coast and deduced the presence of the cold northward-flowing current which is now named after him. In February 1803, Humboldt and Bonpland sailed to Acapulco in Mexico and spent the year exploring the country and investigating more volcanoes. Finally, in 1804, they left Mexico, visiting Jefferson in Philadelphia, before sailing from New York to France and arriving at Bordeaux on 4th August 1804.

The publications which resulted from these travels were among the most sumptuous produced by any expedition, with the inevitable exception of the *Flora Graeca*. Of the *Voyage aux Regions Equinoxiales du Nouveli Continent, fait en 1799–1804*, only the sixth part was devoted to botany. This itself consisted of six sections, published between 1805 and 1829.

Bonpland contributed the text of the earlier sections, but the later ones were mainly the work of C. S. Kunth. The illustrations in the earlier volumes, *Plantae Aequinoctiales* (1805–18) and *Monographie des Mélastomacées* (1806–23), were by Pierre-Antoine Poiteau (1766–1854) and Pierre-Jean-François Turpin (1775–1840); those in *Nova Genera et Species Plantarum* (1815–25) and *Mimoses* (1819) were by Turpin alone.

Turpin was the son of a poor craftsman from Vire in Normandy and became one of the foremost botanical painters of his day. At the age of fourteen, he joined the army and was sent to San Domingo where he met the young and enthusiastic botanist Poiteau. In 1801, Turpin got to know Humboldt in New York, and so began a collaboration that lasted for about twenty years. After about 1820, though, Turpin seems to have devoted his energies to the study of microscopic plants and plant organs, such as the internal tissues of the edible truffle (1827) or the cause of alcoholic and acetic fermentation (1838).

Turpin's work is very delicate and detailed, and that of Poiteau only slightly less clear and crisp in line. Compared with the originals of the *Monographie des Mélastomacées*, which are in the Fairhaven collection of the Fitzwilliam Museum, Cambridge, there is very little loss of detail or delicacy in the engravings by Bouquet. The *Plantae Aequinoctiales* was engraved by Sellier and the *Mimoses* by various engravers including Dien. All these engravers used a combination of line and stipple engraving to produce wonderfully subtle work for the many botanical painters working in France during the early years of the nineteenth century.

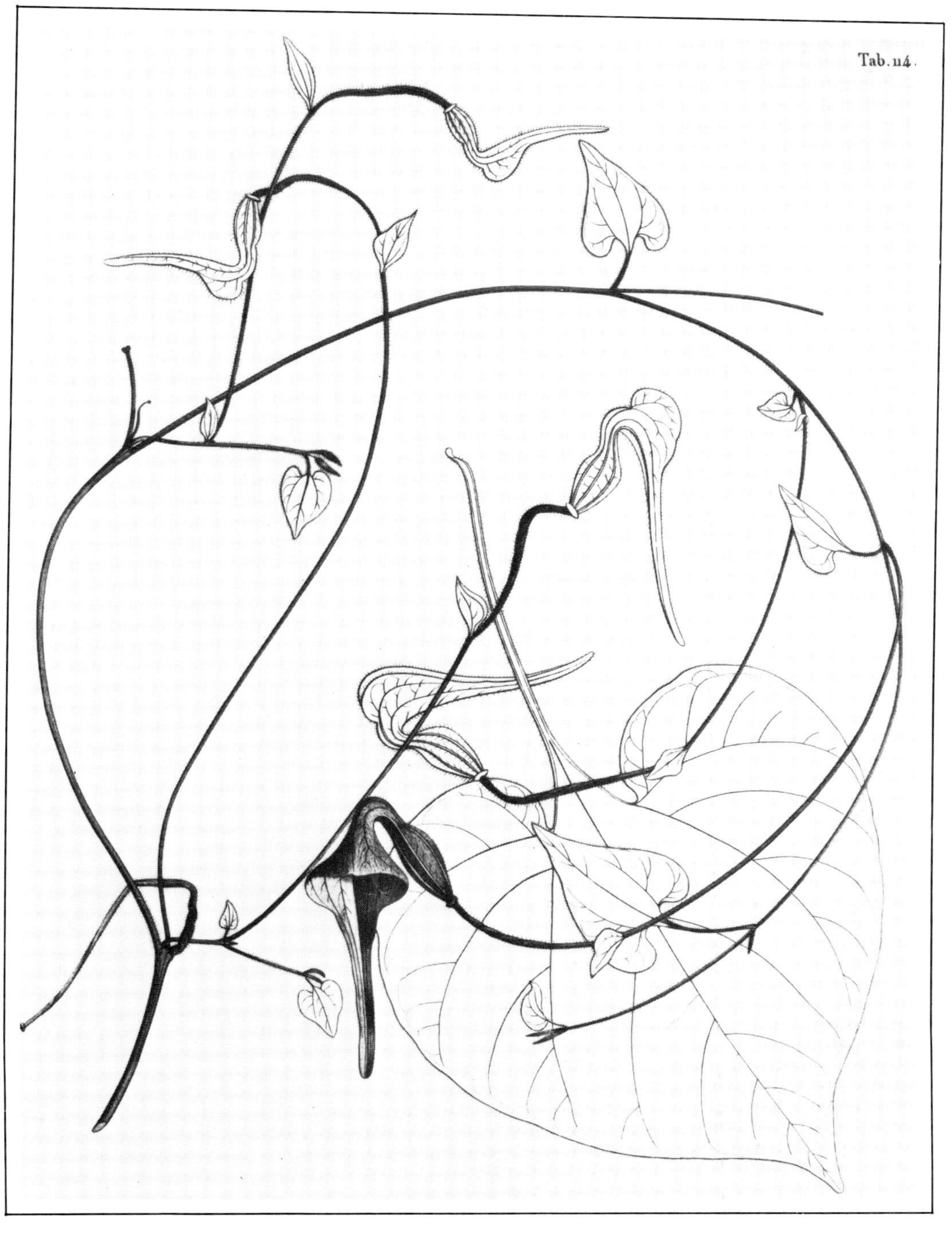

Aristolochia foetida, a pitcher plant. Engraving after a drawing by Turpin from *Nova Genera et Species Plantarum* (1815–25) by Alexander von Humboldt, Aimé Bonpland and C. S. Kunth.

CHAPTER X

Floras and Florilegia

In England, the first more or less complete illustrated floras began to appear in the late eighteenth century. The earliest of these was William Curtis's *Flora Londinensis* (1775–87). Sowerby's *English Botany*, which was to be especially influential, started to appear in 1790 and continued through various revisions until the final edition was published in 1902. These two were the largest and finest English illustrated floras of the period, but some other, smaller illustrated floras appeared in the early nineteenth century. Among these were the works of James Bolton of Halifax, which were published in the North of England. His *Filices Britanniae* (Leeds, 1785) and *History of Fungusses Growing about Halifax* (Huddersfield, 1788–91) were drawn and engraved from the living plants.

W. Baxter, Curator of the Oxford Botanic Garden, himself published *British Phaenerogamous Botany* in a six-volume octavo edition between 1834 and 1843. Its 509 plates are delicately engraved, with beautiful, if slightly amateurish hand-colouring by various painters. The text by Baxter gave detailed localities in Oxfordshire, the west of England and Scotland.

A larger work, intended to be complementary to Sowerby's *English Botany* was Robert Greville's *Scottish Cryptogamic Flora* published in six volumes between 1823 and 1828. The 360 coloured plates containing many algae and fungi were by the author. Greville was an interesting character, a wide-ranging naturalist, a friend of the great botanist Sir Joseph Hooker and a keen campaigner against slavery and drink.

William Curtis (1746–1799) was the son of a tanner from Alton in Hampshire and had an early training as an apothecary. His interest in plants was aroused by a botanically-minded ostler at the Crown Inn in Alton with whom he explored the Hampshire countryside. In 1772 he became *praefectus horti* at the Chelsea Physic Garden in London.

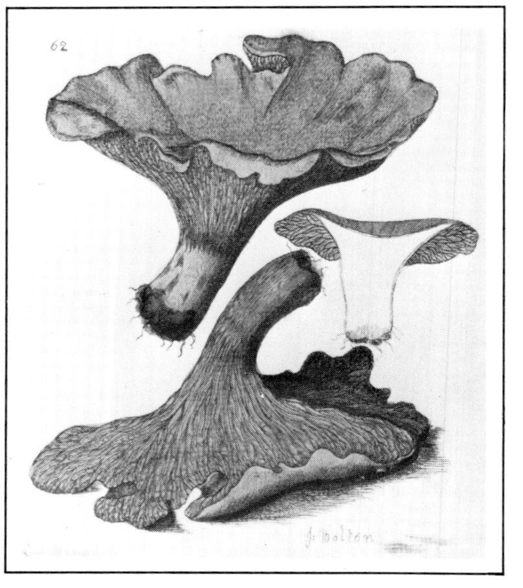

Chanterelle (*Cantharellus cibarius*). Hand-coloured engraving from his own drawing by James Bolton of Halifax from his book, *An History of the Fungusses Growing about Halifax* (1788).

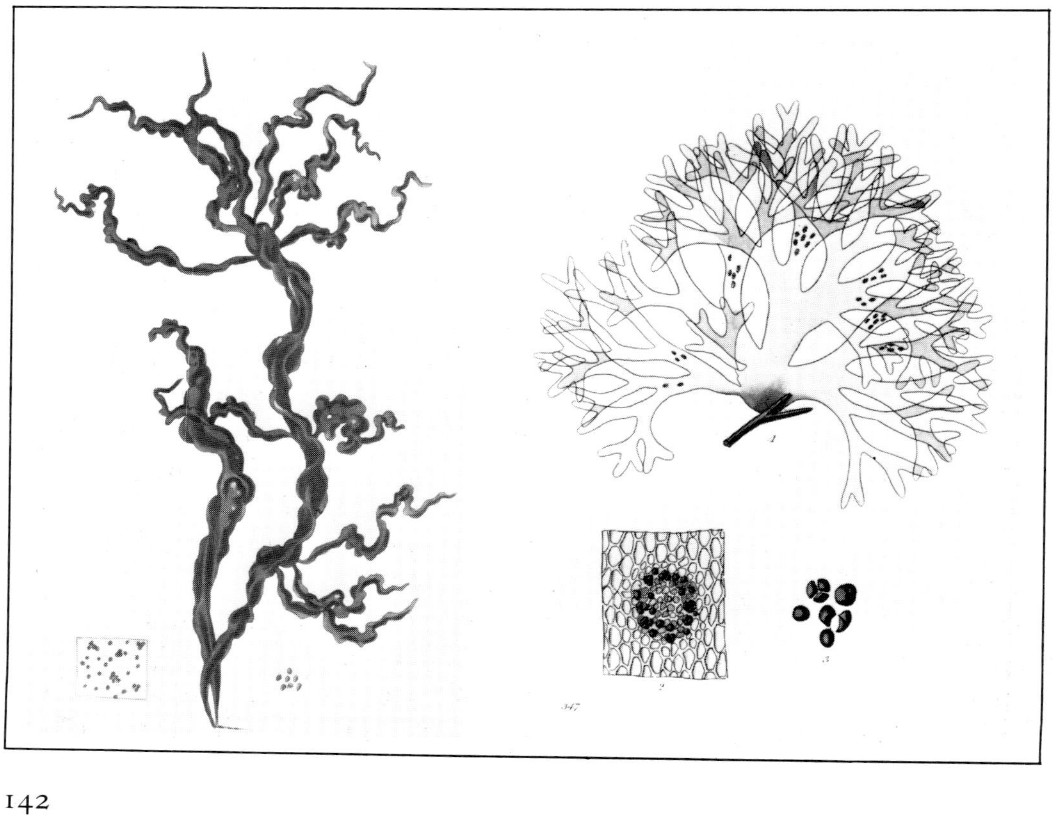

Two seaweeds. Two colour plates by the author from Robert Greville's *Scottish Cryptogamic Flora* (1823–28).

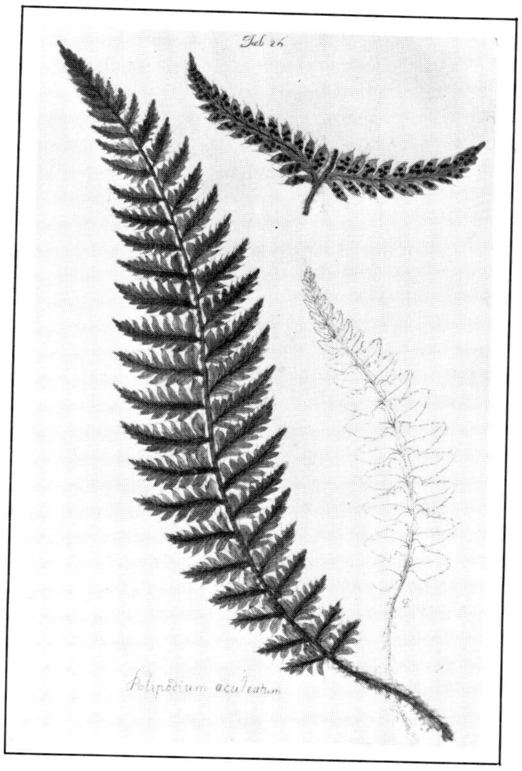

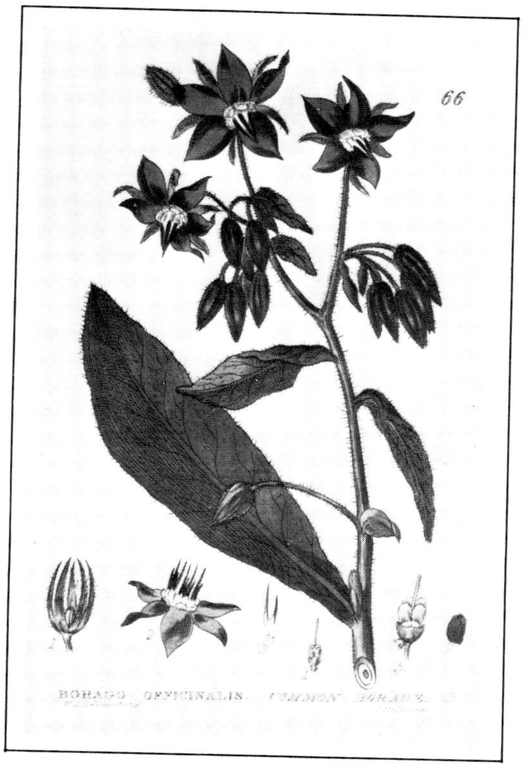

'Polipodium aculeatum', now *Polystichum aculeatum*, a common fern of damp woods throughout the British Isles. Hand-coloured engraving by James Bolton of Halifax from his *Filices Britannicae; an History of the British Proper Ferns* (1785).

Far right:
Borage (*Borago officinalis*). Hand-coloured engraving by C. Mathews from a drawing by W. A. Delamotte from *British Phaenerogamous Botany* (1834–43) by William Baxter.

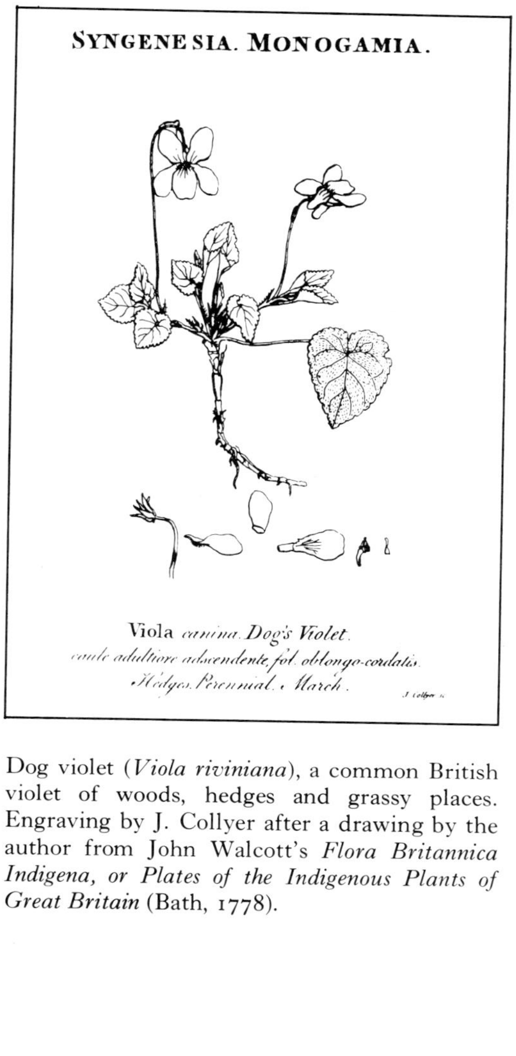

Dog violet (*Viola riviniana*), a common British violet of woods, hedges and grassy places. Engraving by J. Collyer after a drawing by the author from John Walcott's *Flora Britannica Indigena, or Plates of the Indigenous Plants of Great Britain* (Bath, 1778).

Next page:
Three hand-coloured engravings from Sowerby's *English Botany*. *Above:* toothwort (*Lathraea squamaria*), a parasite on the roots of hazel and other trees, Plate 50, from 1790. *Left: Chara aspera*, an alga found in fresh water where it sometimes forms a carpet on the beds of clear lakes, Plate 2738, dated April 1st 1832. *Below: Astragalus alpinus*, Plate 2717, dated November 1st 1831, from a drawing by R. K. Greville; this plate commemorates the discovery in Scotland of this species, which is now known from about three places in the eastern Highlands.

Curtis's *Flora Londinensis* was intended to include all the wild flowers growing within ten miles of London. The first edition is a folio with 432 plates showing most of the common flowers of the South of England, as well as some mosses and fungi. Some interesting plants from other parts of England, such as the northern lungwort (*Mertensia maritima*), and the birdseye primrose (*Primula farinosa*) also appear. Many of the illustrations in the early volumes were painted by William Kilburn (1745–1818). After Kilburn left Curtis to devise designs for calico printing, the main artists were James Sowerby and Sydenham Edwards; it is not always possible to be certain who painted any particular plate, as many are unsigned. The engraving seems to have been done by the painters themselves or by S. Samson.

The book contains a long list of subscribers, who committed themselves in advance to buying it. Nevertheless, the subscriptions were not sufficient to cover all the costs, and Curtis had the financial support of Lord Bute, 'the Maecenas of the Present Age', who had been Prime Minister for a few years in the 1760s and to whom the flora is dedicated. Three hundred copies were printed, and the parts, each containing about six plates, began to be issued in 1775. Most of the plates were un-numbered so that they could be bound in any order that the owner saw fit. The text contains a synonymy, a description of the plant, an account of its habitat and a list of the localities where Curtis had seen it. The first edition was completed in 1787. An enlarged edition with 647 plates and a text by William J. Hooker was published between 1817 and 1828. Among the new plates are many of plants from other areas of the British Isles, including the first description and picture of the Scottish primrose (*Primula scotica*), which is confined to the north coast of Scotland and the Orkneys.

The plates in *Flora Londinensis* are hand-coloured engravings usually life size. While they lack the ultimate elegance of the plates by Poiteau and Turpin in *Flora Parisiensis*, which was published a few years later, they are simple and beautiful and give a good impression of each plant.

The majority of the plates for *English Botany* were painted by James Sowerby (1757–1822), a pupil of the marine painter Richard Wright. Sowerby was followed as a natural history artist by three more generations, including George Brettingham Sowerby, *primus*, *secundus* and *tertius*, all of whom specialised in painting shells. James's eldest son, James de Carle Sowerby (1787–1871), was also a botanical artist and contributed many paintings for *English Botany*, particularly for the supplements. The plates are small and delicately engraved and hand-coloured; they depict mosses, liverworts, algae and lichens as well as flowering plants. The first edition plates were not issued in any particular order, but the second edition, with shortened descriptions, followed the Linnean classification. The same plates were used for this and are scarcely, if at all, inferior to the original. For the third and later editions, the plates were lithographed and have suffered in the process, although more botanical details have been added.

The text of the first edition was by Sir J. E. Smith and that of later supplements by W. J. Hooker and other botanists with illustrations by James de Carle Sowerby. For subsequent editions, the text was extended by J. T. Boswell Syme; popular material, with details of medicinal uses and folklore was contributed by Mrs Lankester.

The immensely industrious James Sowerby also published *Coloured Figures of British Fungi*, with four hundred illustrations, between 1795 and 1803. He worked with many other botanists, including Curtis on *The Botanical Magazine* and Thornton on the *Temple of Flora*, as well as illustrating many of the books by Sir J. E. Smith. Perhaps the finest example of Sowerby's skill as an engraver can be seen in another work that involved Smith, Sibthorp's *Flora Graeca*, for which he engraved many of the plates; the others were engraved by his son James de Carle Sowerby and by Ferdinand Bauer himself.

France with its much greater range of climate and therefore much larger number of species than Britain was slower to produce a complete flora. In 1808, Poiteau and Turpin produced the first volume of *Flora Parisiensis*, an account of plants growing in the region of Paris. This is in folio format with very beautiful and delicate colour stipple engravings in a style characteristic of the period.

Local floras and accounts of rarer species had been appearing in France since the latter part of the eighteenth century. Thus Dominique Villars's *Histoire des Plantes de Dauphiné*, published in Lyon and Paris between 1786

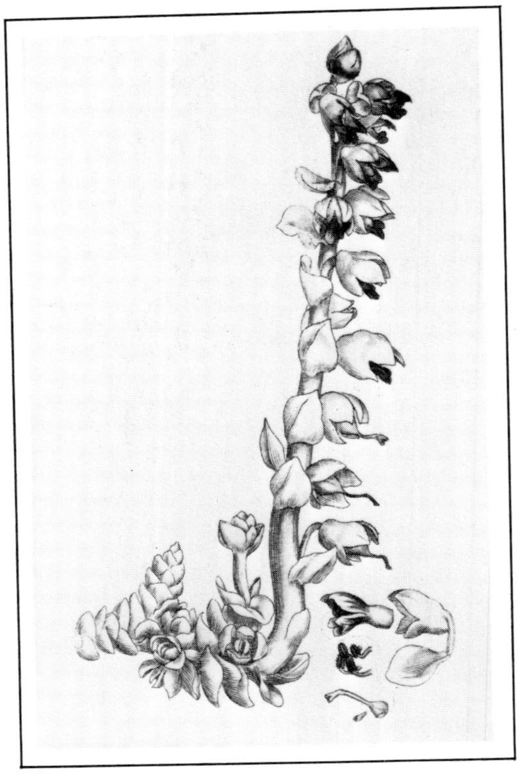

Opposite page :

Large bindweed (*Calystegia sepium*), a beautiful but noxious weed. Hand-coloured engraving by Michael Rössler after a drawing by Martin Rössler from an early volume of *Flora Danica*.

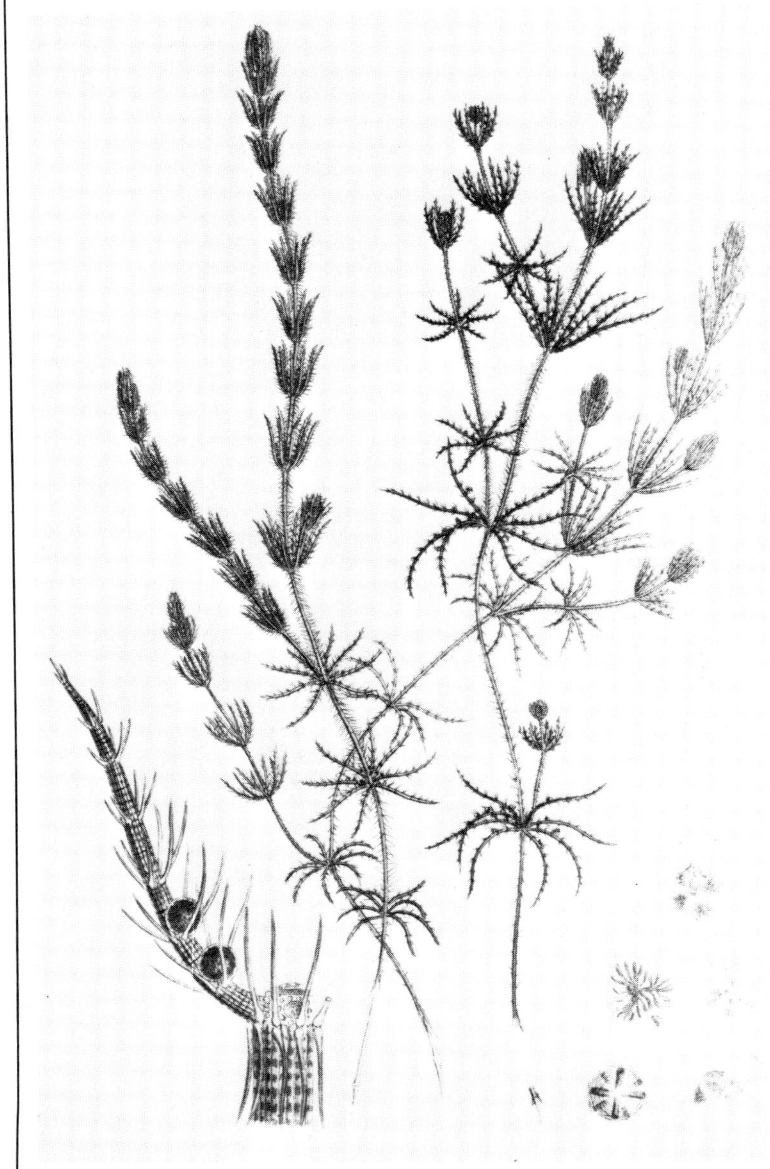

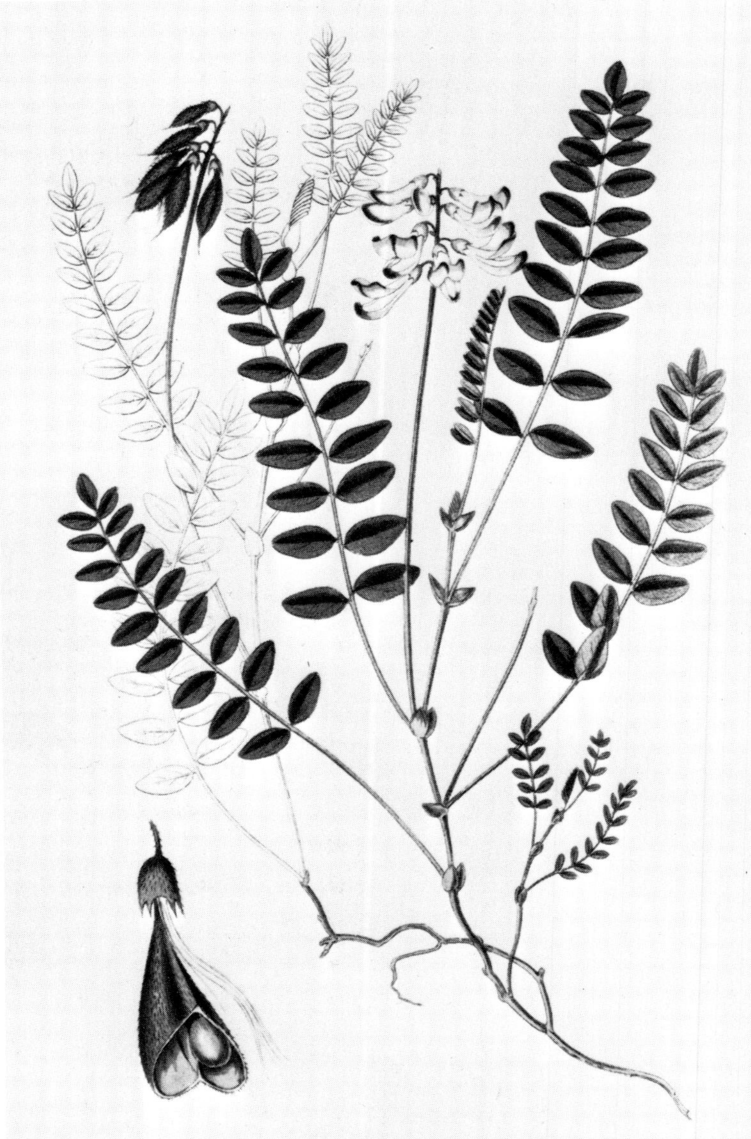

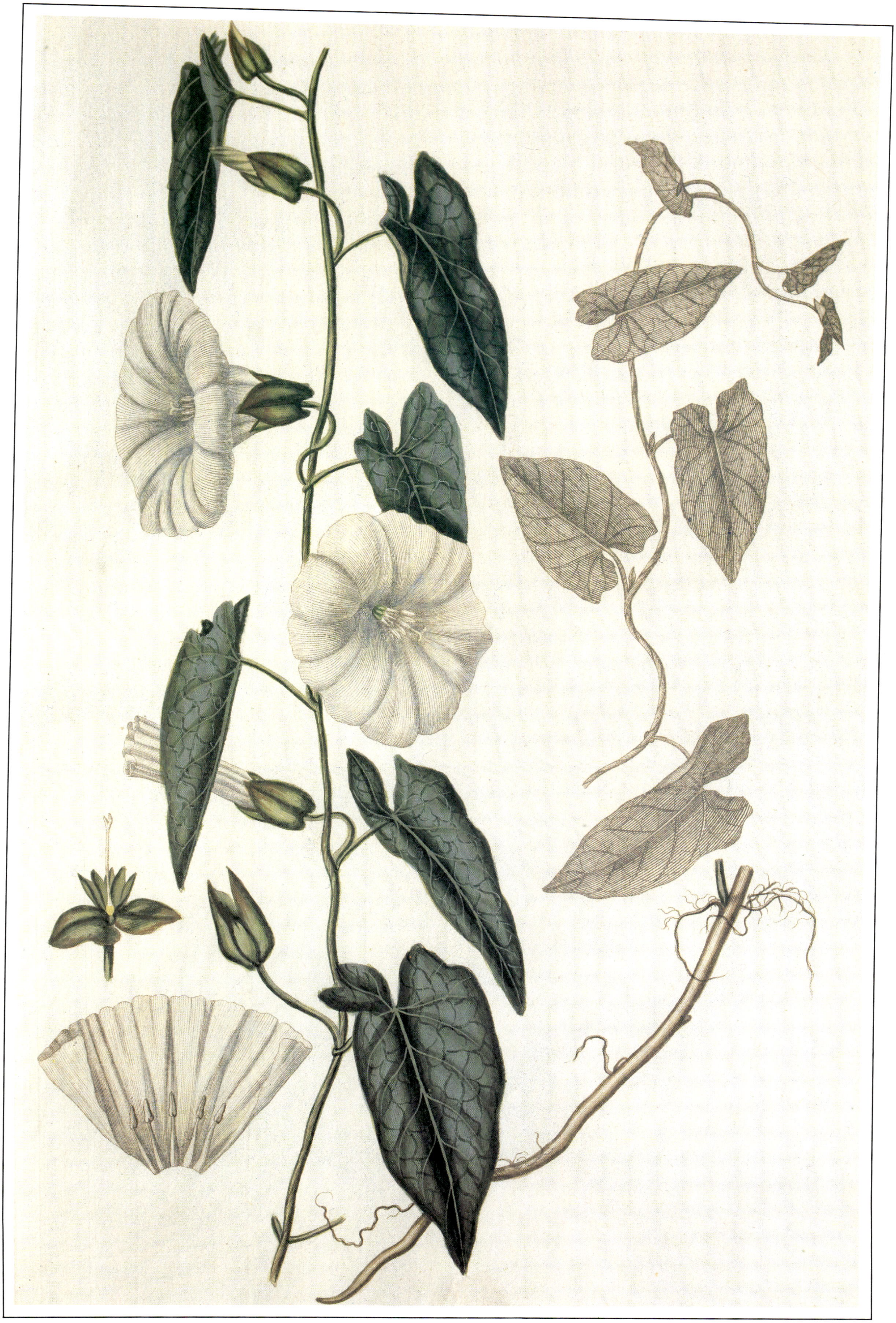

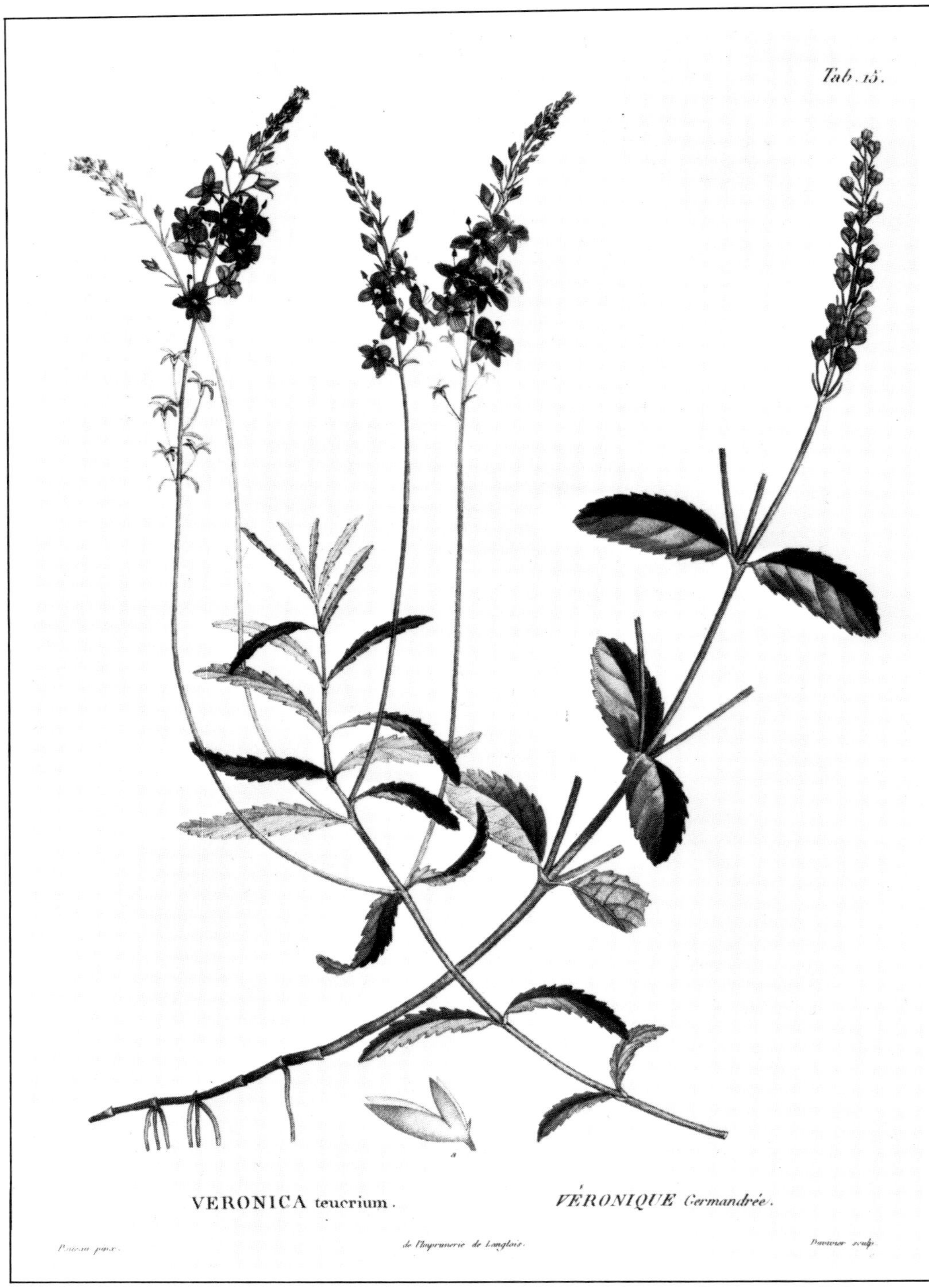

A speedwell, *Veronica teucrium*, which is found throughout continental Europe. Coloured stipple engraving by Duvivier after a painting by Pierre-Antoine Poiteau from *Flora Parisiensis* (1808) by Poiteau and Pierre-Jean-François Turpin.

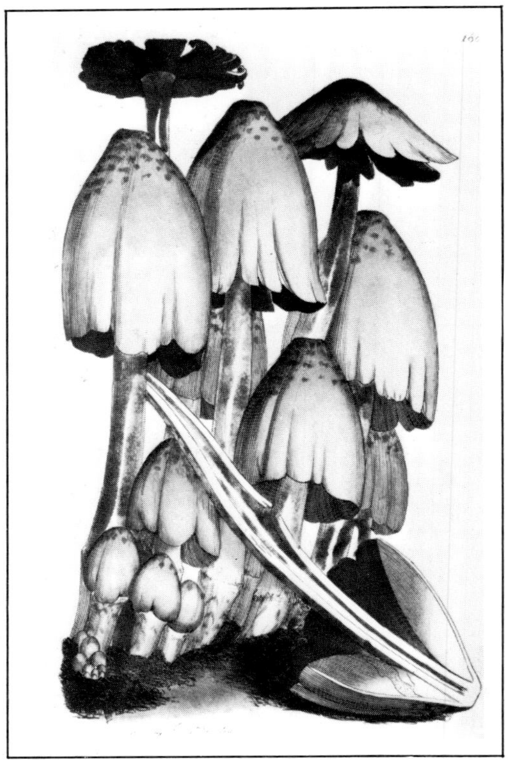

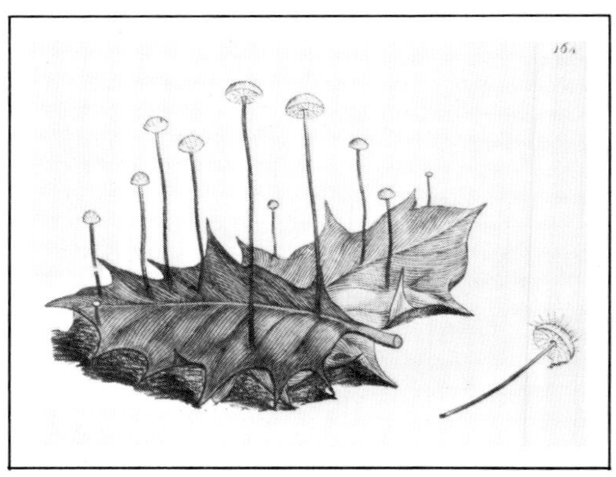

Two hand-coloured engravings from James Sowerby's *Coloured Figures of English Fungi* (1797–1803). *Top: Coprinus atramentarius*, a species of ink cap that produces nausea if consumed with alcohol. *Above: Marasmius hudsonii*, a minute fungus specific to fallen holly leaves.

and 1789, contains lists of plants found in particular localities together with simple and rather crude engravings of many of them.

Augustin Pyramus de Candolle (1778–1841) and his equally famous son Alphonse-Louis-Pierre Pyramus de Candolle, were almost as prolific and just as influential in France as the Hookers were to be slightly later in England. Augustin de Candolle, who was Swiss by birth, devised a natural system of plant classification and first published it in his *Théorie Elémentaire de la Botanique* (1813). Before that, in 1808, his *Icones Plantarum Galliae Rariorum* had appeared. The first and only part to be published contained fifty simple and elegant line engravings after Poiteau and Turpin showing plants from the whole of France, including the strange and aberrant *Cyclamen linearifolium* from near Draguignan in Provence, and several of the cushion-like *Androsace* species from the high Alps.

Other countries in Europe also began to produce large illustrated floras towards the end of the eighteenth century, some of them similar in scale and concept to Sowerby's *English Botany*. The earliest was *Flora Danica* which began publication in 1761 with text by Georg Christian von Oeder. It was over a century before the enterprise reached completion, by which time it contained 3060 engravings. It provided the source material for the famous *Flora Danica* service of Copenhagen porcelain. Jacquin's *Flora Austriaca*, with a mere five hundred plates, appeared between 1773 and 1778. In Italy, four illustrated floras were published: Carlo Allioni's *Flora Pedemontana*

Cyclamen linearifolium. Engraving by Plée after a drawing by Turpin from *Icones Plantarum Galliae Rariorum* (1808) by Augustin-Pyramus de Candolle.

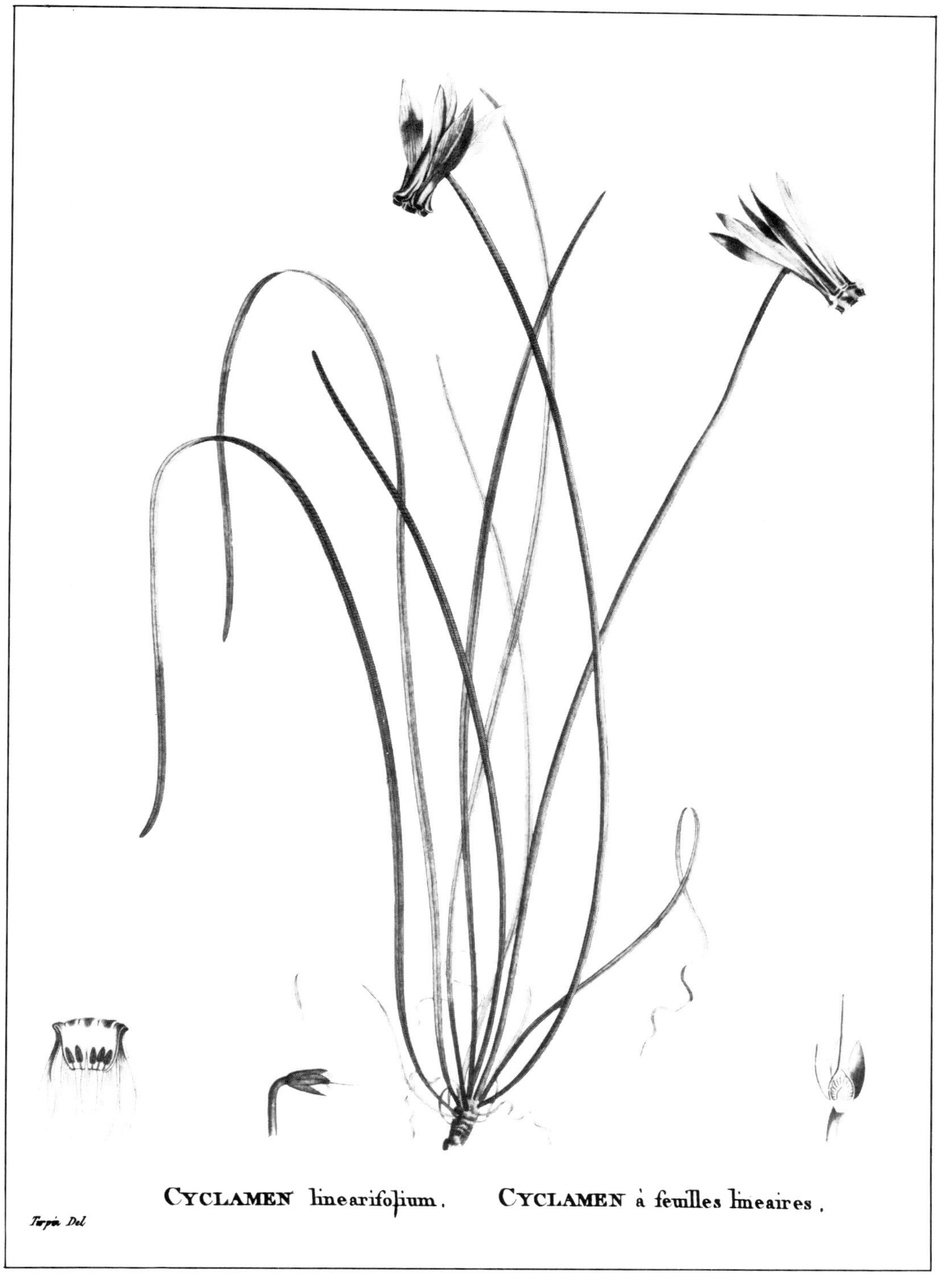

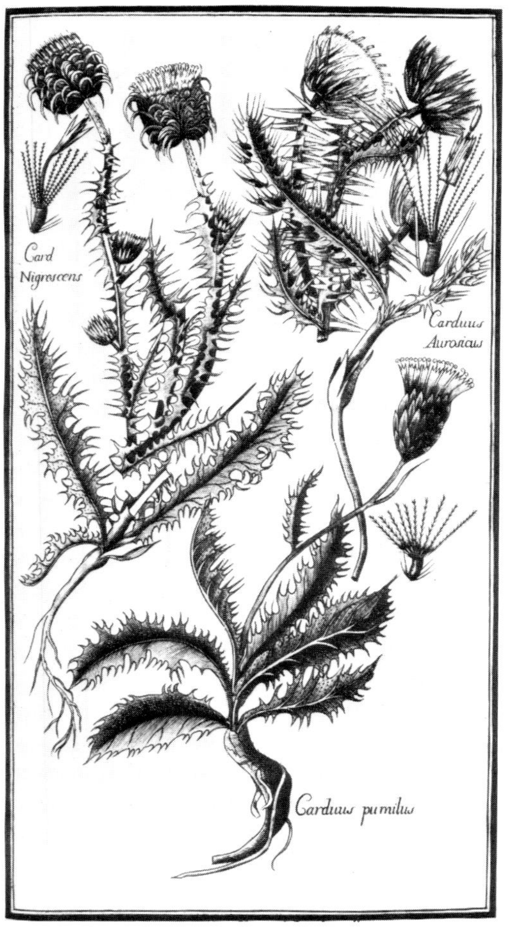

Various thistles. Engraving by Chauvin of Grenoble from *Histoire des Plantes de Dauphiné* (1786–89) by Dominique Villars.

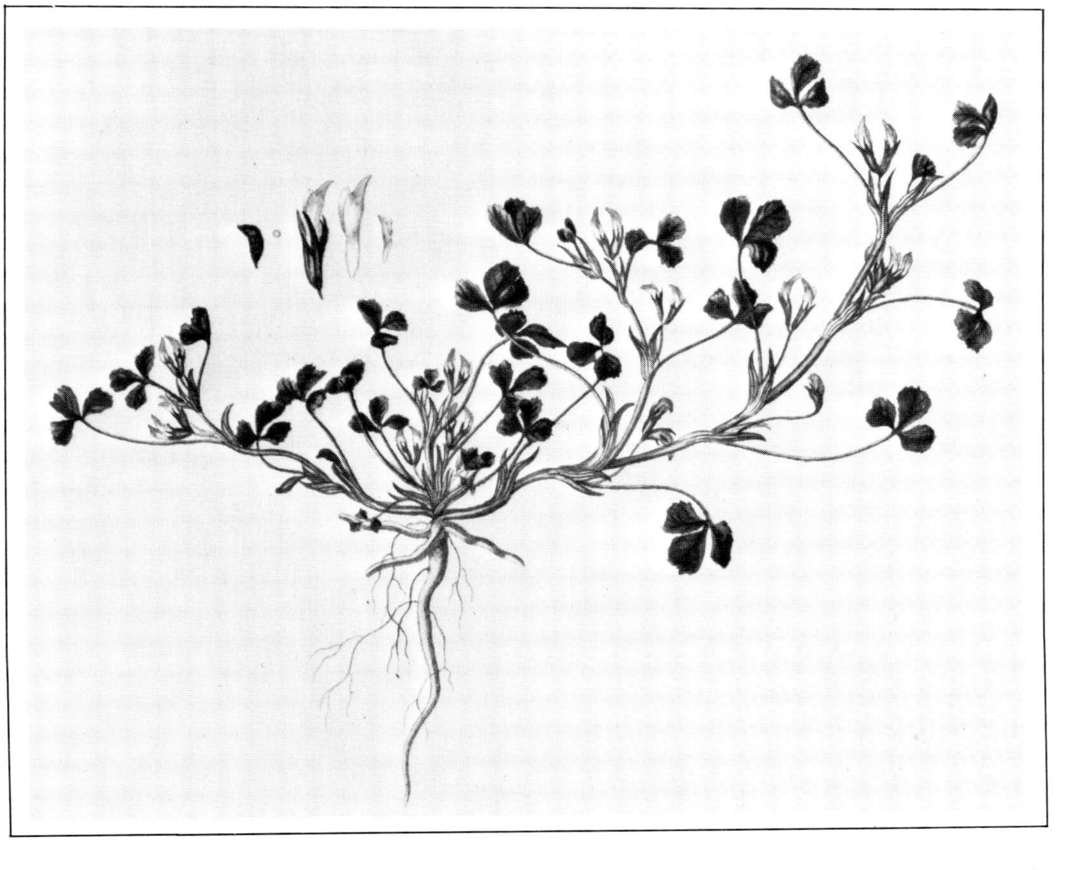

A small clover, *Trifolium ornithopodioides*, which reaches its northern limit in Denmark. Hand-coloured engraving from *Flora Danica* (1761 onwards).

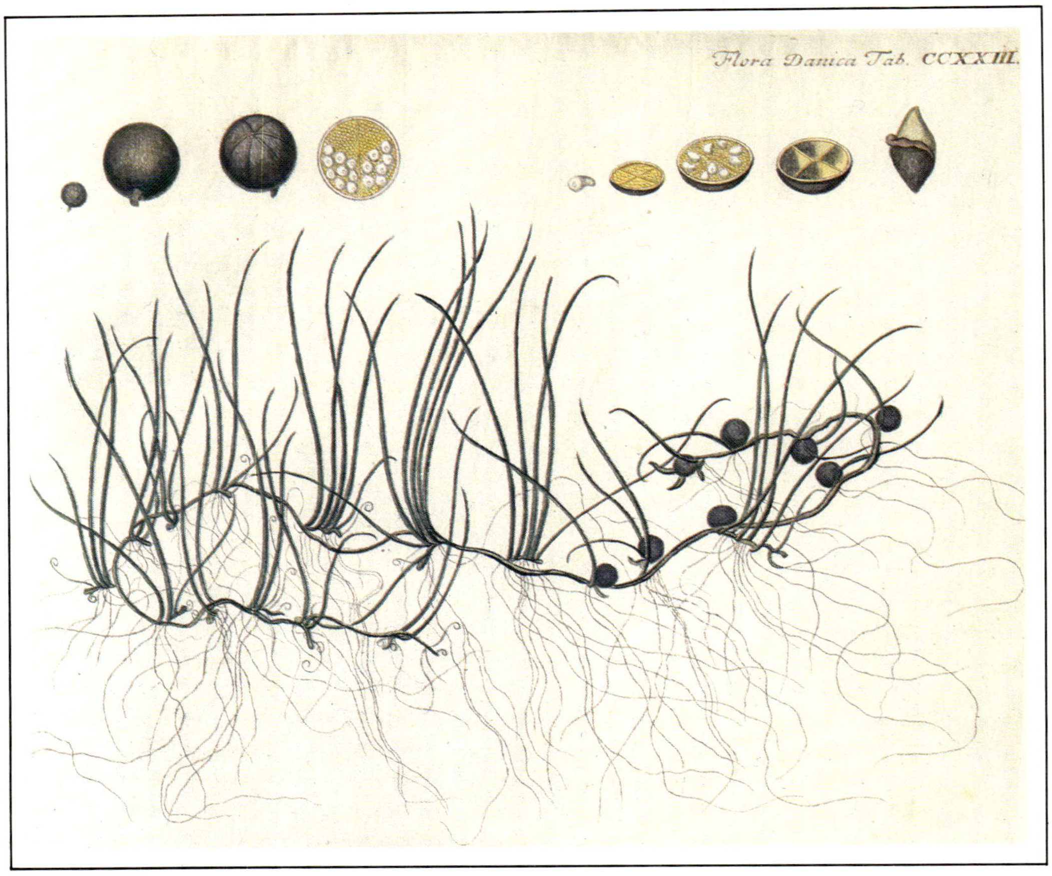

Pillwort (*Pilularia globulifera*), a tiny fern-like plant that grows in the edges of ponds and lakes throughout northern Europe; the pills contain sporangia of two sizes which in turn contain the spores. Hand-coloured engraving by Michael Rössler after a drawing by Martin Rössler from an early volume of *Flora Danica*, which commenced publication in 1761.

(1785), which has unusually bold engravings, Michele Tenore's *Flora Napolitana* (1811–38), Gaetano Savi's *Flora Italiana* (1818–24) and Giovanni Gussone's *Flora Sicula* (1826). Between 1798 and 1855, Jacob Sturm produced the miniature *Deutschlands Flora*, its tiny but beautiful plates embellished with particularly good floral details. The hand-coloured engravings, which show non-flowering as well as flowering plants, are by various artists, including Sturm and his son and H. G. L. Reichenbach. Sturm himself contributed many drawings to other books, notably the supplement to Caspar Sternberg's monograph on *Saxifraga* (1810), and to works on fungi, mosses and lichens.

Although the post-Linnean era of botany was characterised by the production of comprehensive and often systematic floras, the florilegium, illustrating an unsystematic miscellany of cultivated plants, often in a very grand format, still found patrons in France and Austria. Its last two great exponents were Nikolaus von Jacquin and Pierre-Joseph Redouté.

The prolific Nikolaus von Jacquin (1727–1817), who was also the author of *Flora Austriaca*, went on to publish several florilegia, mainly of plants found in the gardens of Vienna. He was born in Holland of French parents and educated in Holland and France. He went to Vienna in 1752 to complete his medical studies and was soon involved in organising a botanical collecting expedition for the Emperor Francis I, husband of Empress Maria Theresa. This expedition lasted from 1754 to 1759 and sent back a very rich collection from the West Indies to the gardens of the Imperial Summer Palace at Schönbrunn. In 1768, Jacquin became Director of the University Gardens in Vienna and Professor of Botany and Chemistry at the University, posts he held until his retirement in 1797.

Jacquin was himself a skilled draughtsman, as is demonstrated by his illustrated letters to Joseph Dryander which are now in the Natural History Museum in London, but little of his drawing was used in his publications with the exception of his early *Observationes Botanicae* (1764–71) and his last work on *Stapelia* (1806–20). He employed several artists, notably Ferdinand and Francis Bauer, but also Francis Scheidl and Joseph Hofbauer, and finally Johannes Scharf.

Many of the plants illustrated in Jacquin's florilegia came either from the Royal Gardens at Schönbrunn or from the University Botanic Garden; a large number of them are of South African origin and were sent back by the collectors Georg Schall and Franz Boos in the late 1780s. The florilegia

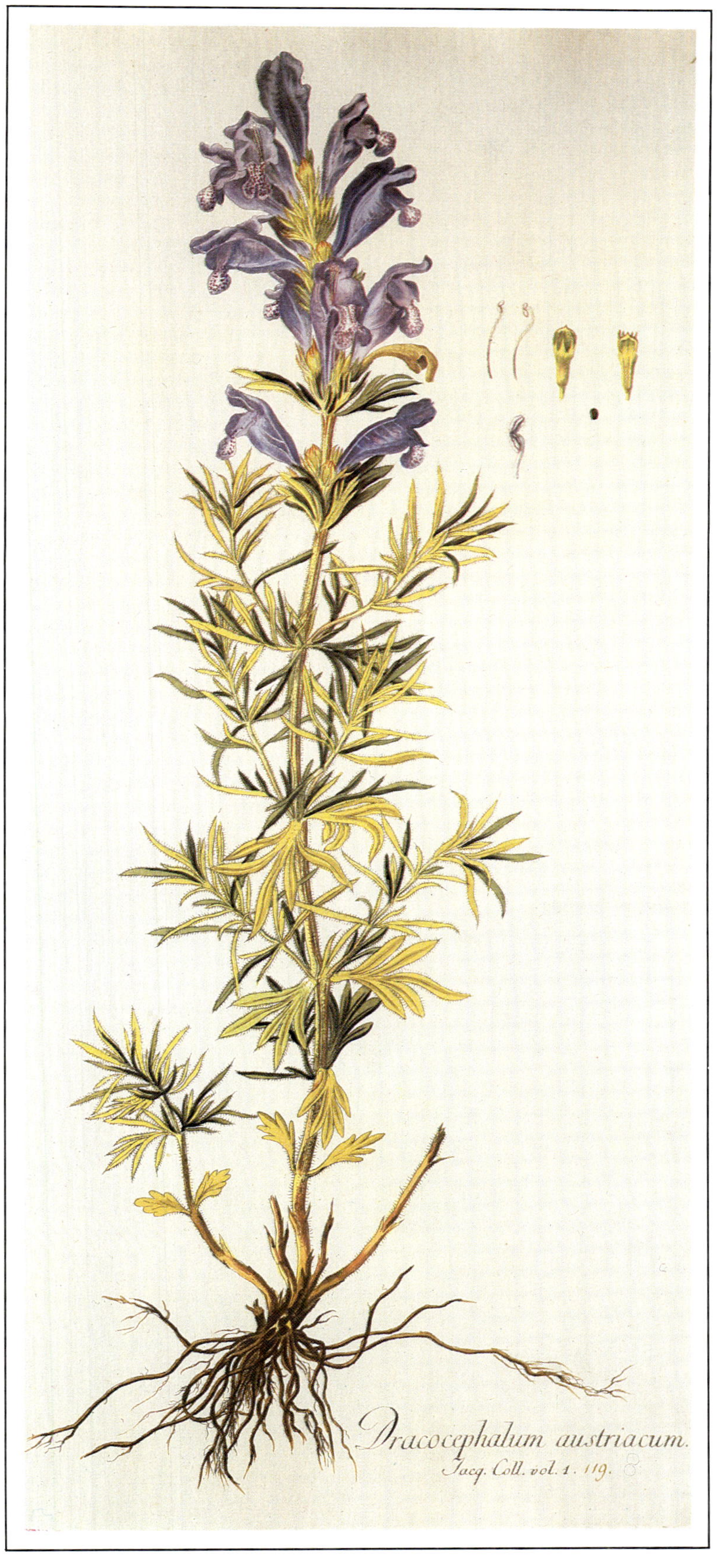

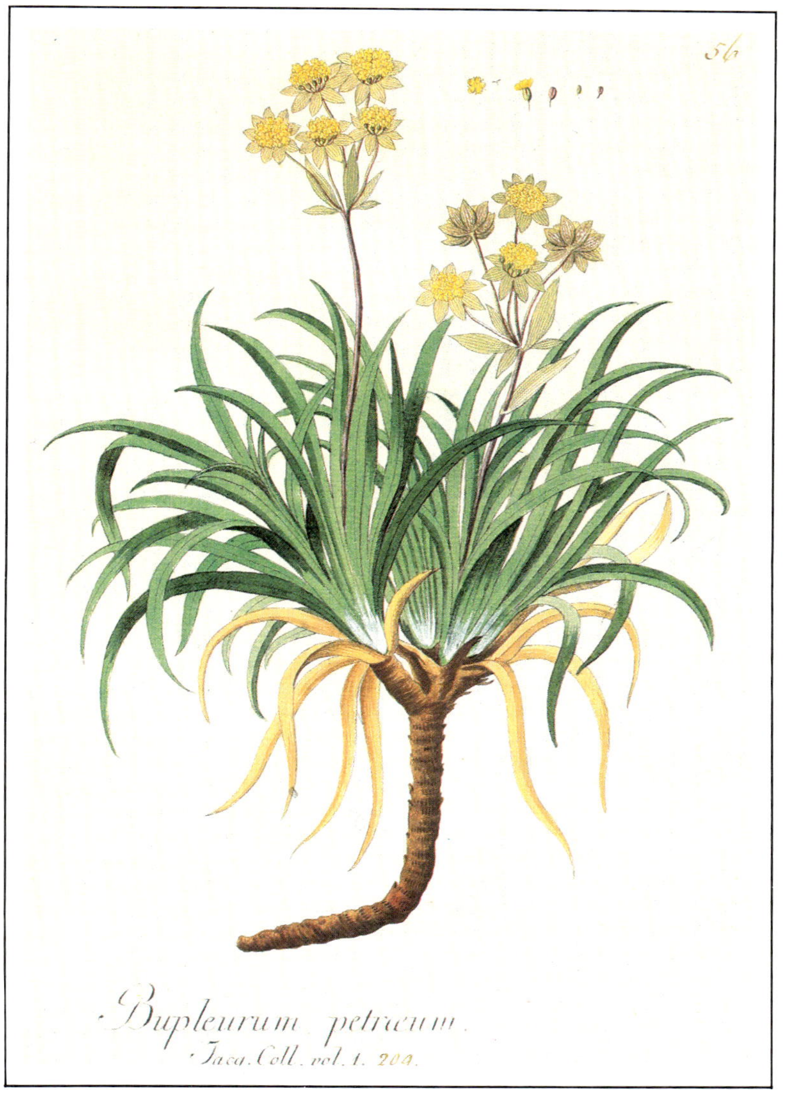

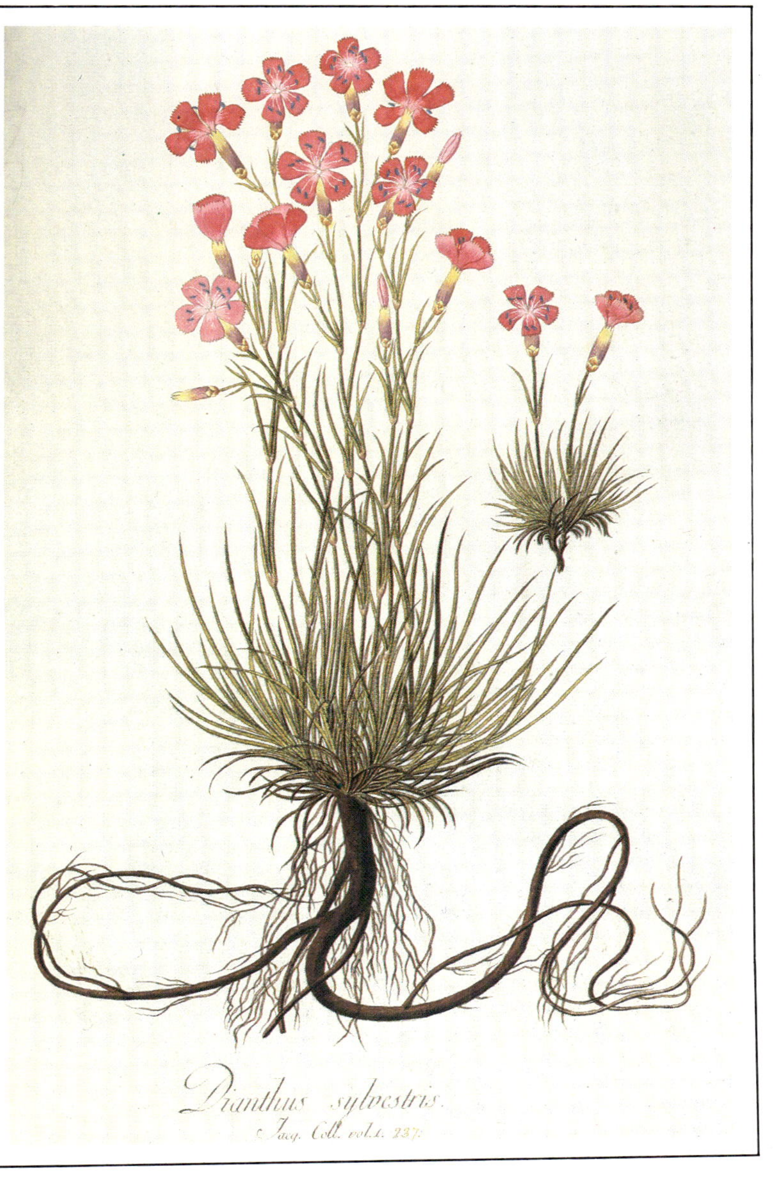

Three hand-coloured engravings by Joseph Hofbauer from *Icones Plantarum Rariorum* (1781–93) by Nikolaus von Jacquin. The pink, *Dianthus sylvestris*, grows wild in the southern Alps. The hare's-ear, *Bupleurum petraeum*, is found on cliffs and rocks in central Europe. *Dracocephalum austriacum*, a labiate, is a native of southern Europe and the Caucasus, where it grows on dry, grassy hillsides.

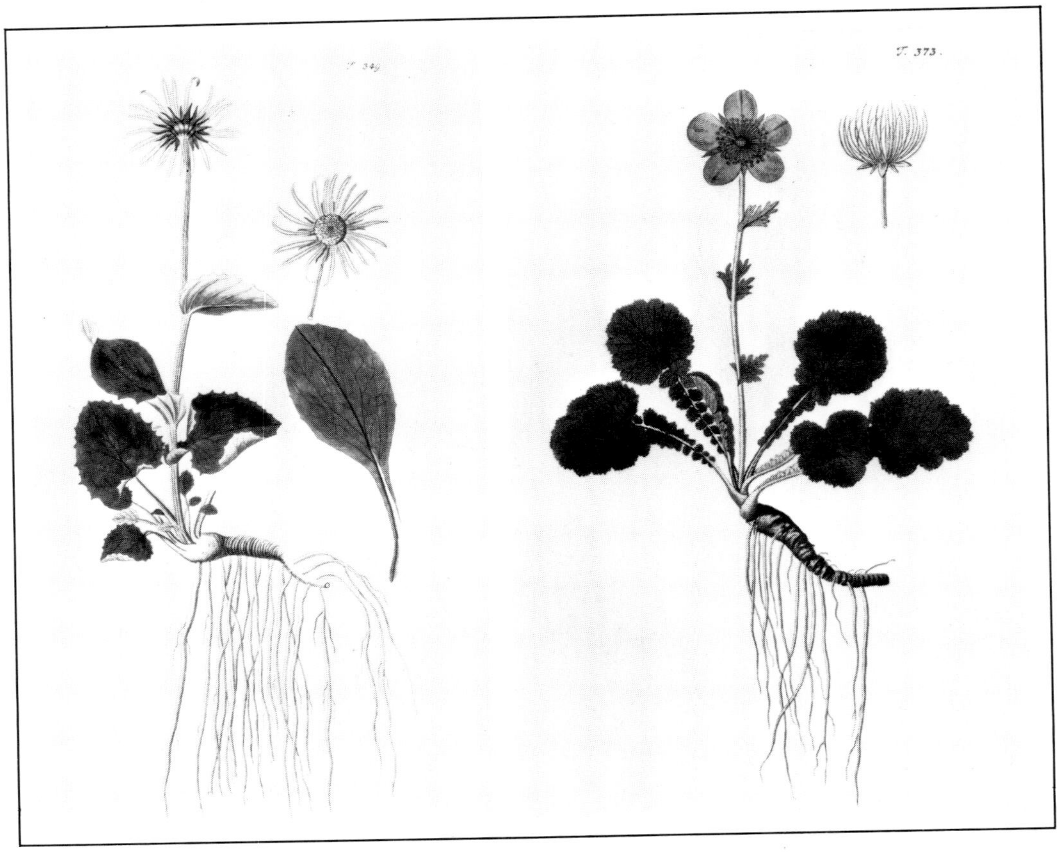

Left:
Doronicum grandiflorum and *Geum montanum*. Two hand-coloured engravings from *Florae Austriacae Icones* (1773–78) by Nikolaus von Jacquin.

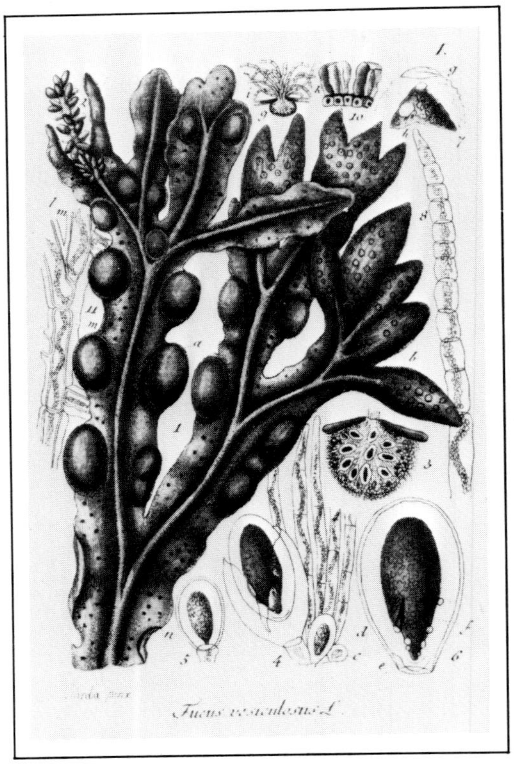

Bladder wrack (*Fucus vesiculosus*). Colour plate from an 1839 issue of *Deutschlands Flora* by Jacob Sturm.

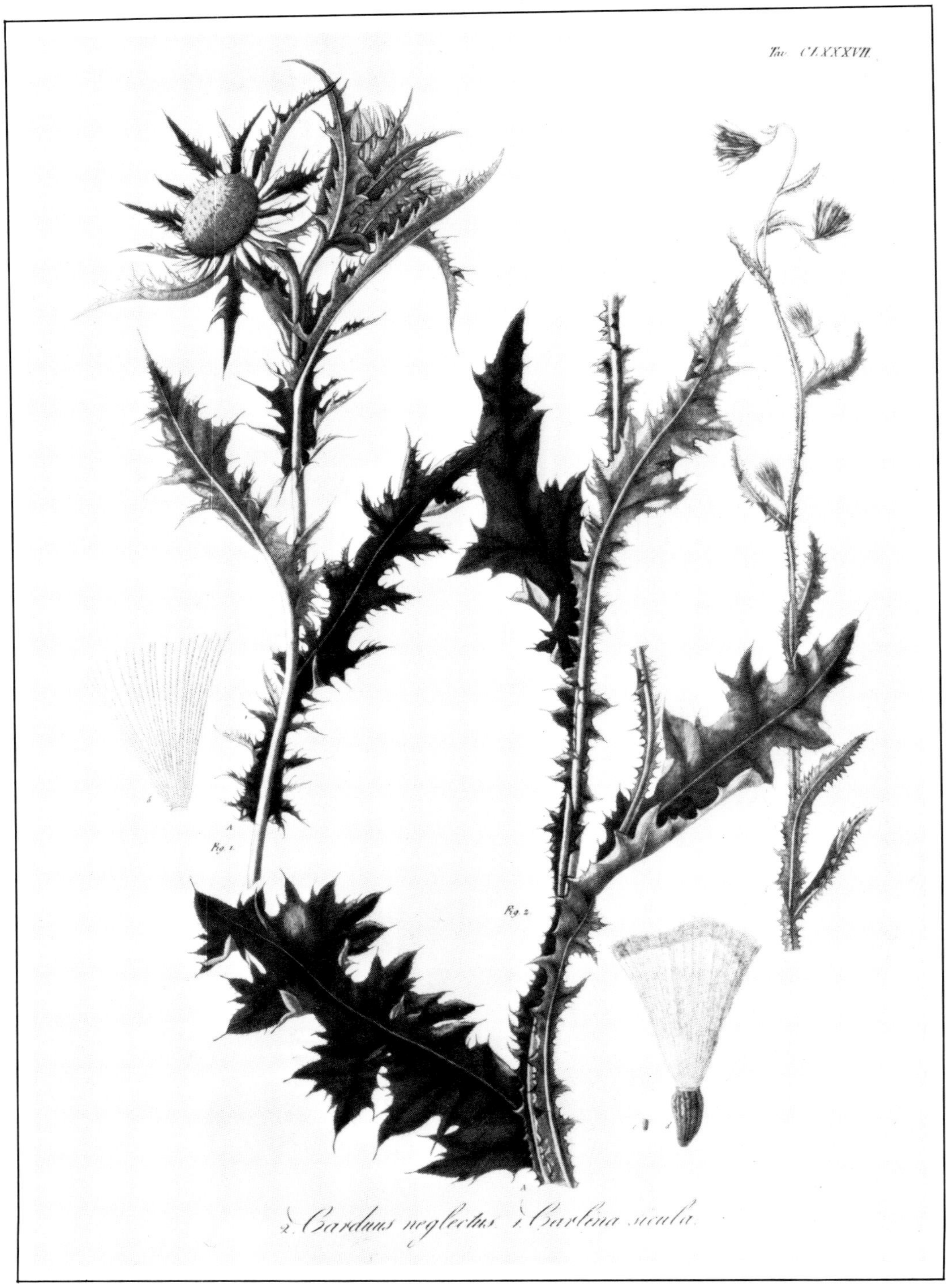

Left:
Two thistles from *Flora Napolitana* (1811–38) by Michele Tenore.

Right:
Cirsium ferox, a thistle found only in the southern Alps of France and Italy. Engraving by P. Pieroleri Jr after a drawing by F. Pieroleri from *Flora Pedemontana* (1785) by Carolo Allioni.

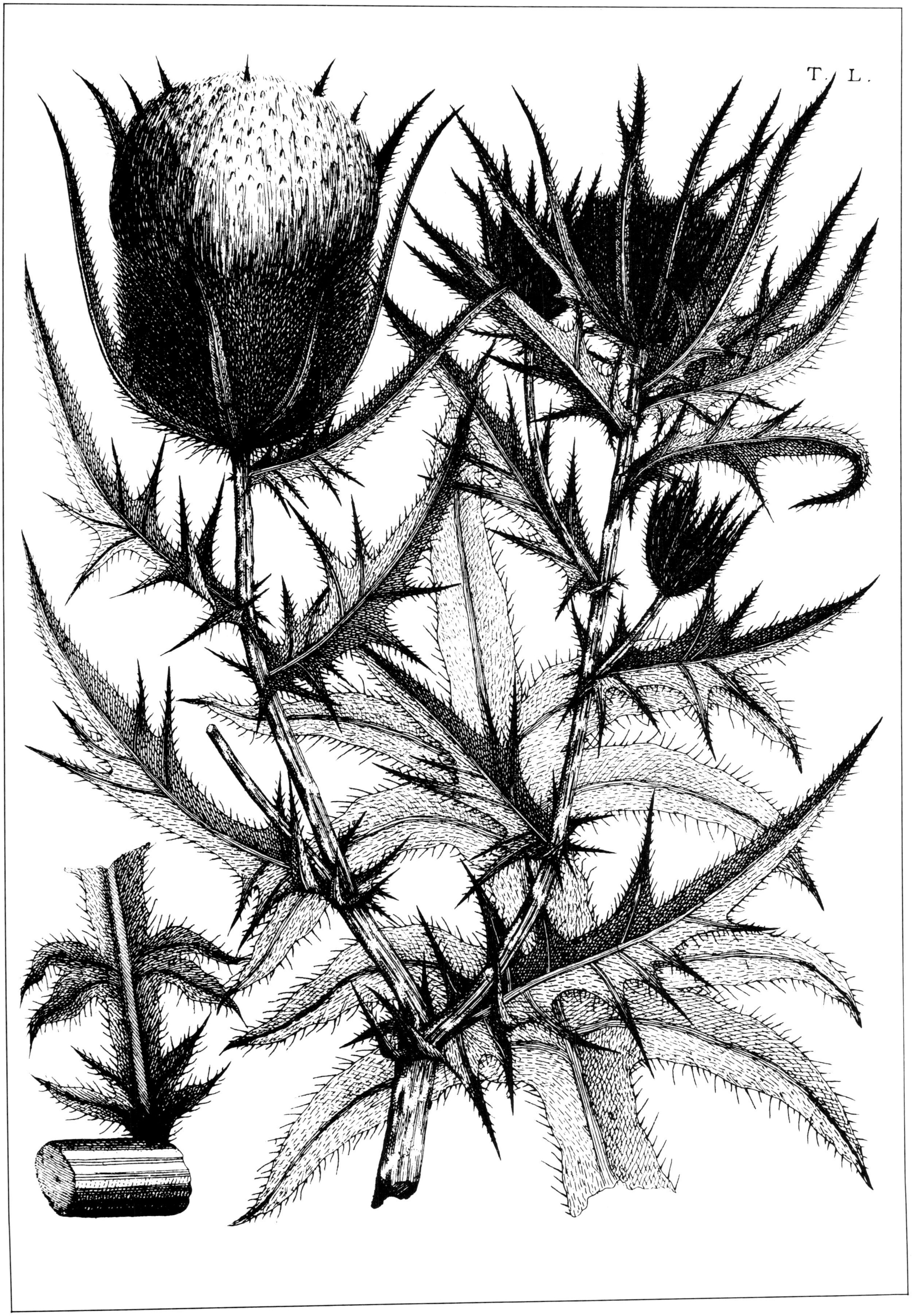
T. L.

were the *Hortus Botanicus Vindobonensis* with 300 engravings (1770–76), *Icones Plantarum Rariorum* with 648 engravings (1781–93) and *Plantarum Rariorum Horti Caesarei Schoenbrunnensis* (1797–1804) with 500 engravings. Among Jacquin's other books was the *Selectarum Stirpium Americanarum Historia* (1763), in which he published the results of his trips to the West Indies and Central America. All his books were illustrated with line engravings on copper, many with hand colouring. Although the technique is less subtle than the combination of line and stipple used by the French engravers, it is still of a very high standard, both scientifically and artistically; the plants are interesting as well as beautiful. As most of the plates are unsigned, it is not usually possible to tell who the artist was. Anything by the Bauers must date from before 1784 for Ferdinand and 1789 for Francis. Few of the localities in which the plants were originally collected are recorded – it is known that collectors in South Africa often gathered their seeds from gardens which had been established there.

While Jacquin was producing the great florilegia for the court in Vienna and the Bauer brothers were working first with him and later with Sibthorp and Banks in London, the most famous of all botanical painters, Pierre-Joseph Redouté (1759–1840) was beginning his career in Paris; he was to dominate botanical painting from 1790 to 1830 and has remained popular ever since. He was a native of St Hubert in the Belgian Ardennes where his father was a painter and interior decorator. His elder brother, Antoine-Ferdinand, went to Paris in 1776 and became very successful as an interior decorator, working on the house which is now the Elysée Palace, as well as on the château of Malmaison and the palace of Compiègne; he later worked as a stage designer and decorator, notably at the Italian Theatre in Paris. Pierre-Joseph was the second son: he left home at the age of thirteen and

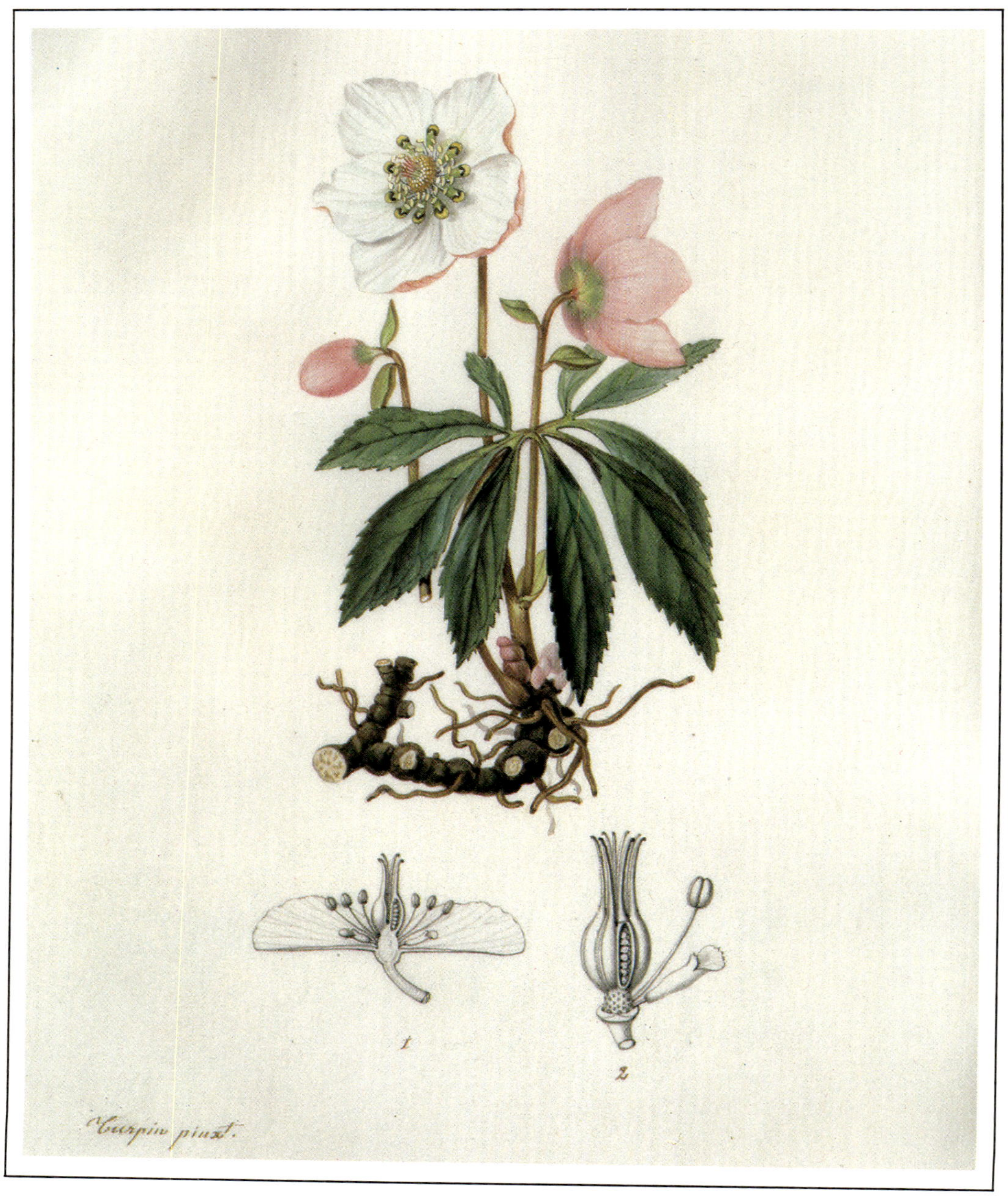

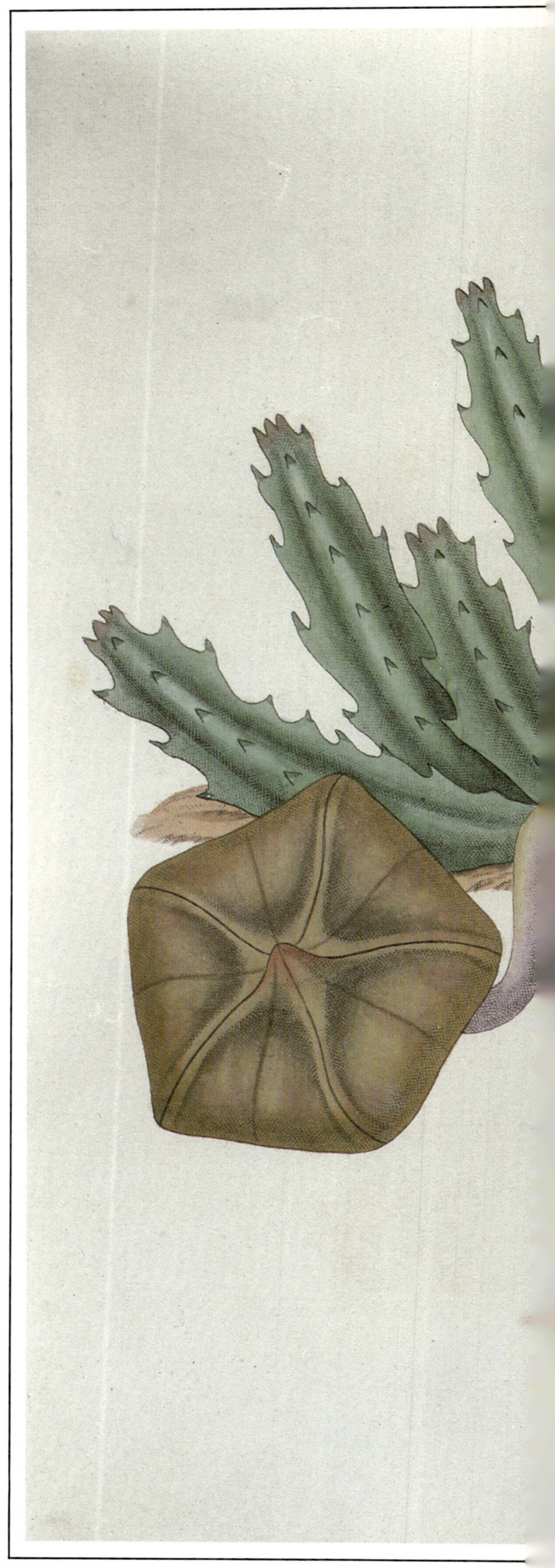

Christmas rose (*Helleborus niger*), a native of woods in the southern Alps which has been commonly grown in gardens since the sixteenth century. Painting in watercolour on vellum by Pierre-Jean-François Turpin from a miniature album of great beauty that is now in the Lindley Library of the Royal Horticultural Society, London.

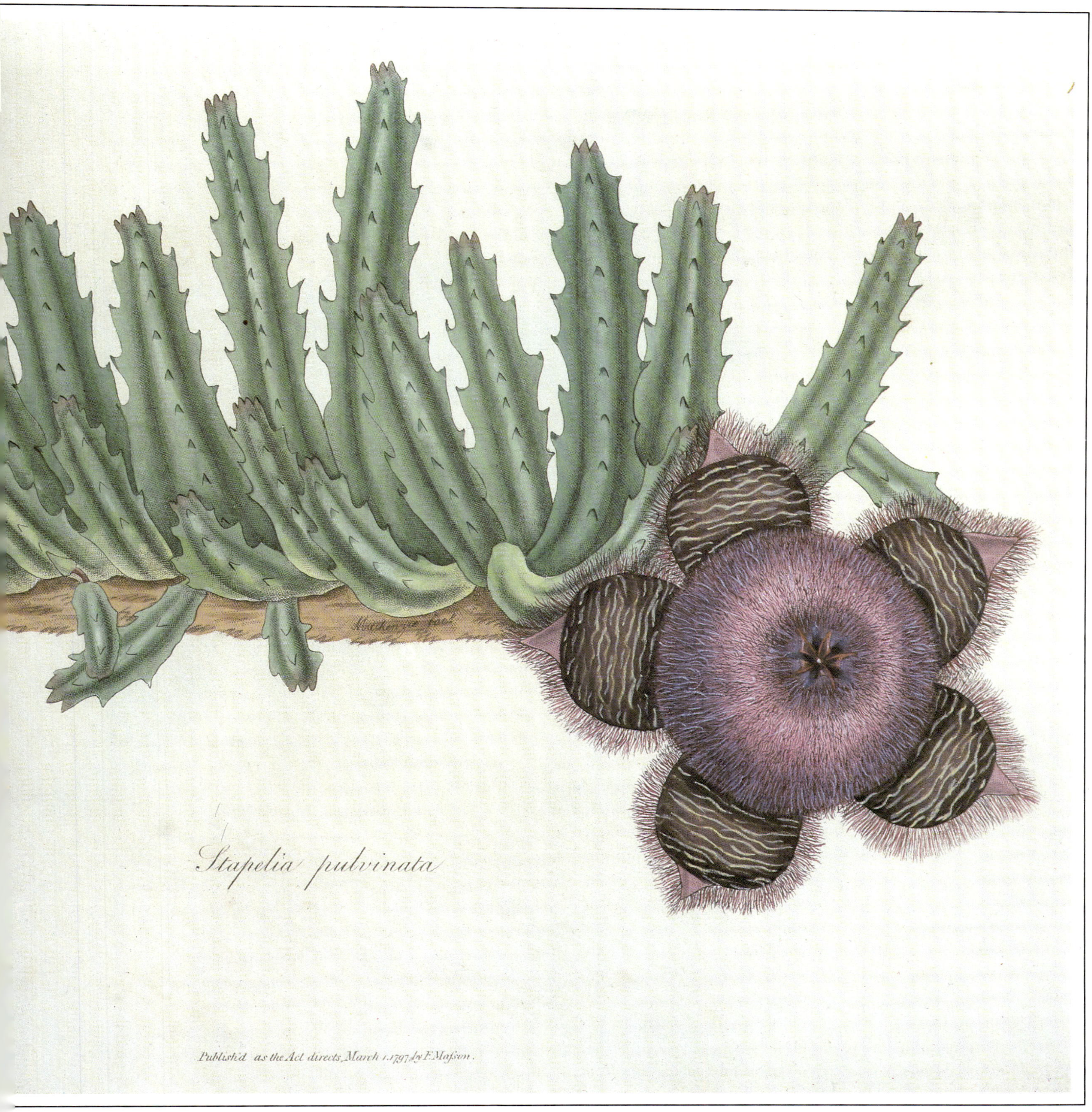

Stapelia pulvinata. Hand-coloured engraving by Mackenzie after a painting by Francis Masson for his *Stapeliae Novae* (1796–97). The paintings, which were done in Africa, are now in the Natural History Museum in London. This species of *Stapelia*, one of forty collected and introduced into cultivation by Masson, is a native of Cape Province.

became an itinerant decorator, travelling through Belgium and Holland for the next decade, apart from a year spent studying painting in Liège.

In 1782, at the age of 23, he went to join his brother in Paris. There he started to visit the Jardin du Roi and to paint flowers, at the same time attending the lectures of the royal professor of painting, the Dutchman Gerard van Spaendonck (1746–1822), whose personal pupil he became. Soon, he also came under the influence of the great French amateur botanist, Charles Louis l'Héritier de Brutelle (1746–1800). Under l'Héritier's guidance, Redouté produced 54 of the 91 illustrations in *Stirpes Novae aut Minus Cognitae* (1784–85). These appear to be the earliest published drawings by Redouté; they do not really bear comparison with his later work, though it is hard to say how far this is in the engraving rather than in the original drawing – the plates are engraved entirely in line, without any stipple; some of the copies were printed in black only, while others were in colour.

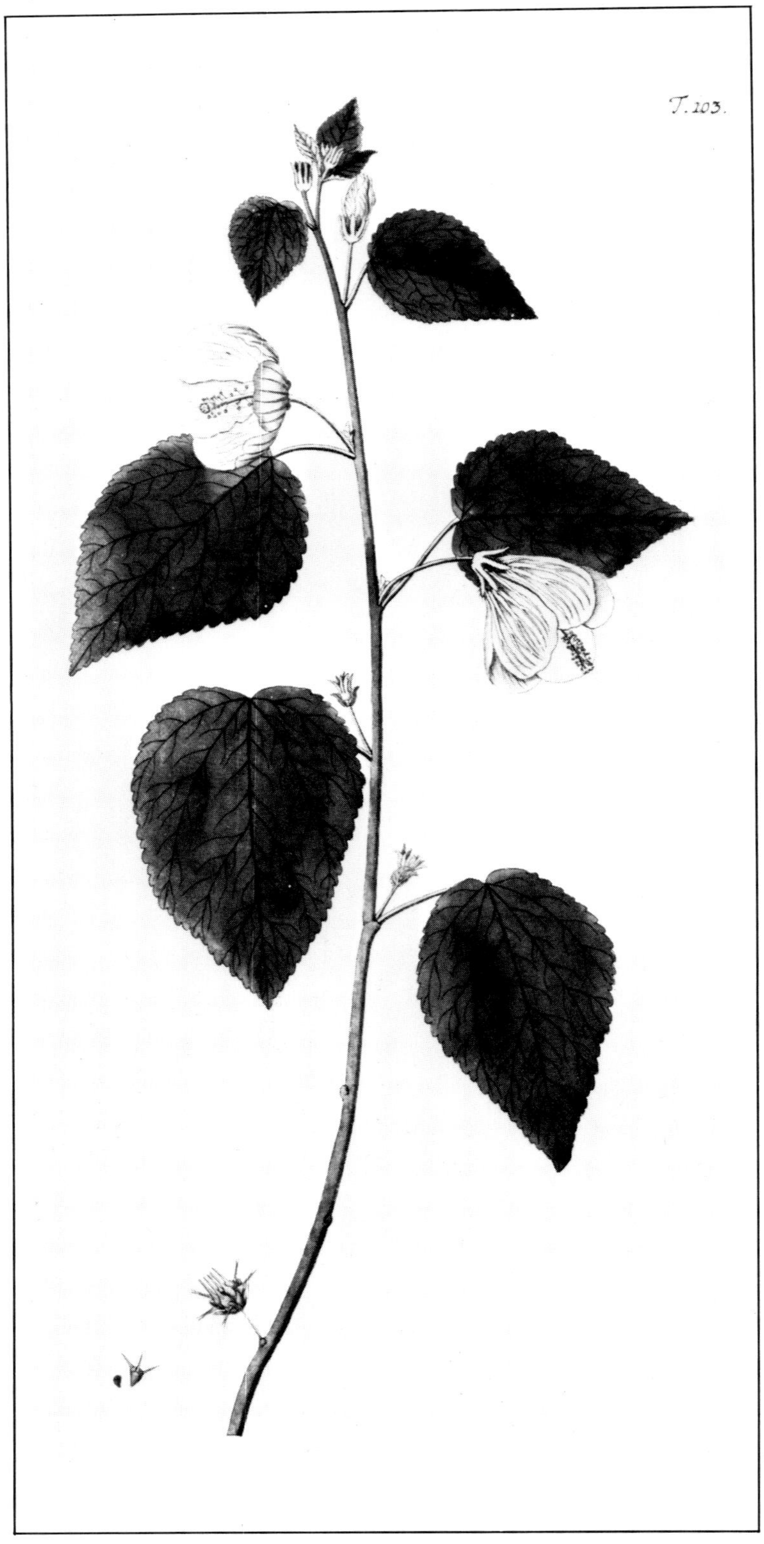

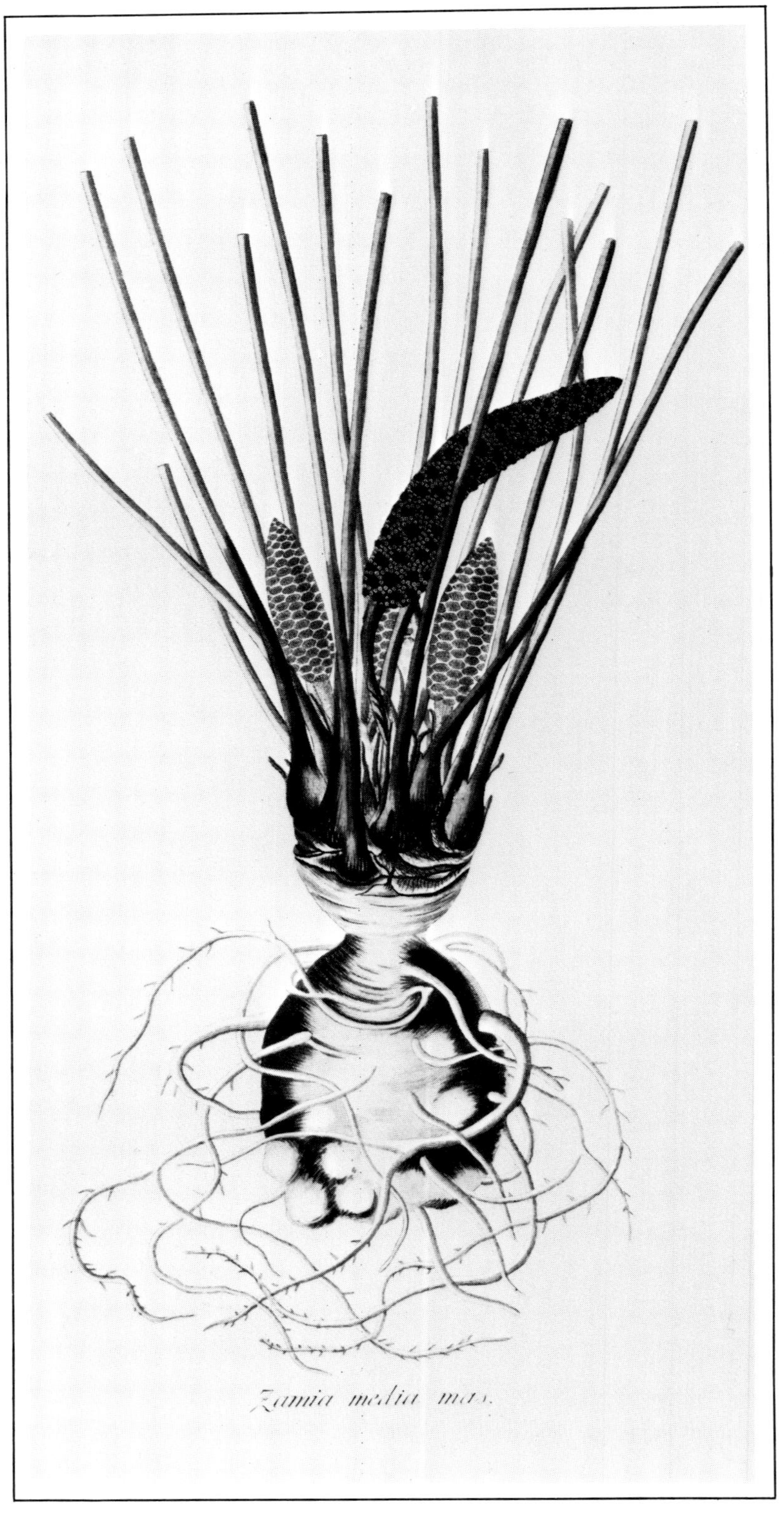

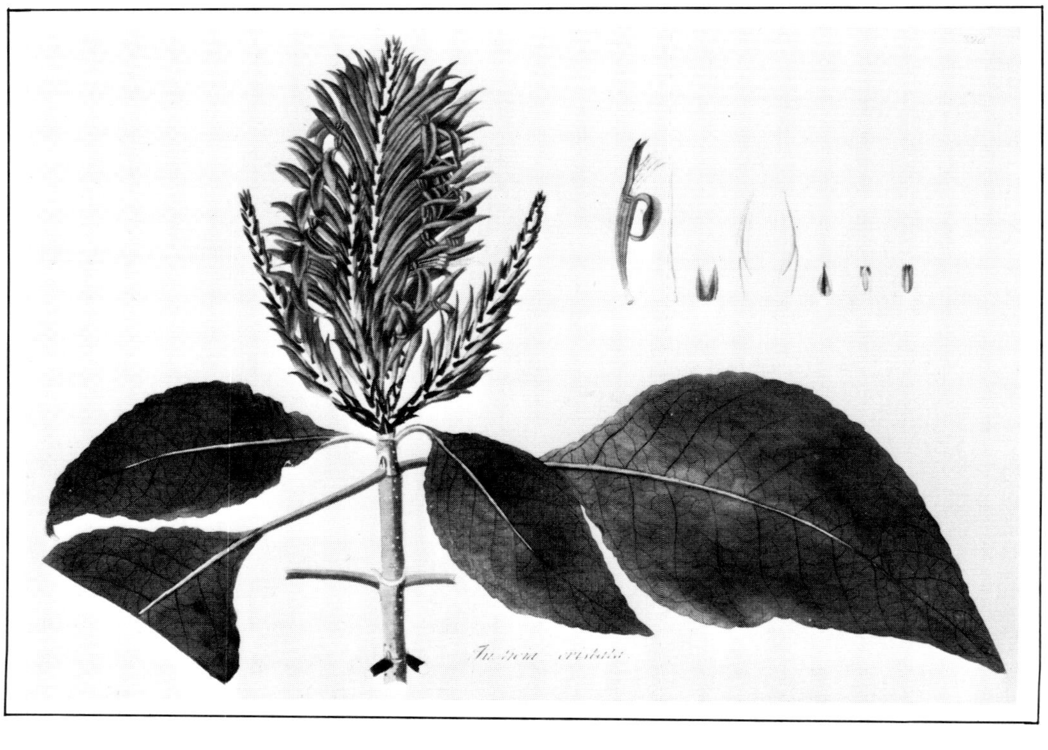

Hand-coloured engravings from *Plantarum Rariorum Horti Caesarei Schoenbrunnensis* (1797–1804) by Nikolaus von Jacquin.
Above: Zamia media, a cycad from Cuba; the male inflorescences are shown, together with the tuberous rootstock and the leaf bases.
Left: Justicia cristata, which has bright orange-red flowers, from Caracas.

Top left:
Hibiscus spinifex. Engraving after a drawing by Franz Scheidl from *Hortus Botanicus Vindobonensis* (1770–76) by Nikolaus von Jacquin.

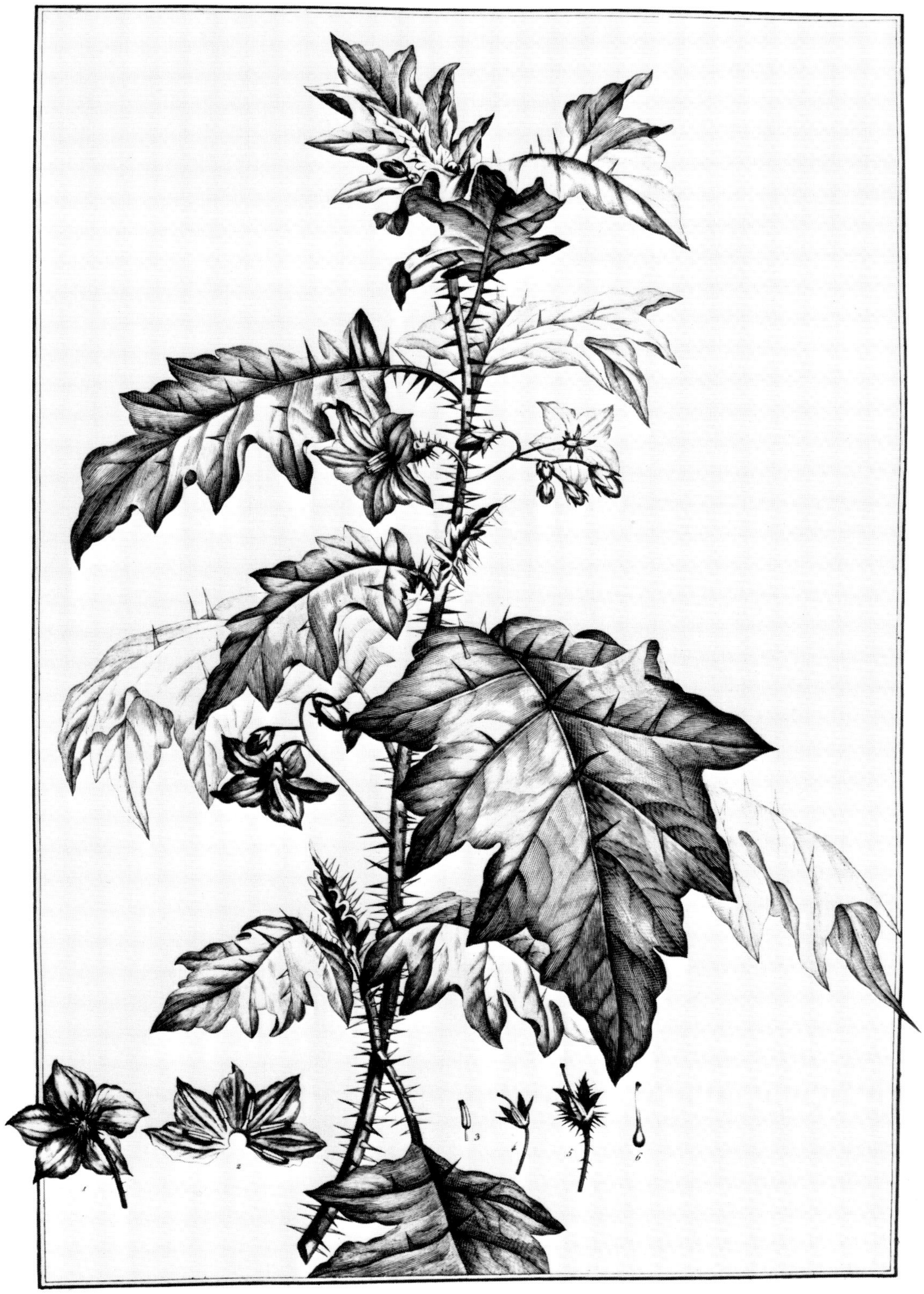

Solanum fuscatum. Engraving by Milsan after a drawing by Pierre-Joseph Redouté from *Stirpes Novae aut minus cognitae* (1784–85) by Charles-Louis l'Heritier de Brutelle.

If l'Héritier was the main botanical influence on Redouté, van Spaendonck was undoubtedly the greatest artistic influence on the young painter. Van Spaendonck had arrived in Paris in 1766; he became a fashionable painter of snuff boxes and was appointed as teacher of flower painting at the Jardin du Roi in 1780. He published only one book, *Fleurs Dessinées d'après Nature*, but also contributed to the royal collection of *vélins*. As Wilfrid Blunt discovered in his study of the *vélins*, van Spaendonck's paintings before 1782 were only in gouache; by 1784, his technique had changed and he was using pure watercolour, as Redouté did later. Blunt notes: 'Redouté was in fact the populariser and exploiter of van Spaendonck's technical discoveries.'

Redouté continued to work closely with l'Héritier. When Joseph Dombey arrived back in Paris from South America in 1785, his collections, the property of the French government, were lent to l'Héritier to study. In 1786, l'Héritier happened to hear that the Spanish ambassador was at court and had requested that the collections be returned to Spain. That same night, assisted by Redouté and the botanist Broussonet, l'Héritier packed up the collections and spirited them away by fast coach to Bologne and thence across the Channel to Sir Joseph Banks's house in Soho Square,

Overleaf:
Crinum giganteum, a native of Sierra Leone which was sent by Broussonet to M. Cels, in whose garden it flowered. Stipple engraving by Allais, colour-printed *à la poupée* and retouched by hand, after a painting by Pierre-Joseph Redouté from *Les Liliacées*, Volume IV, 1808.

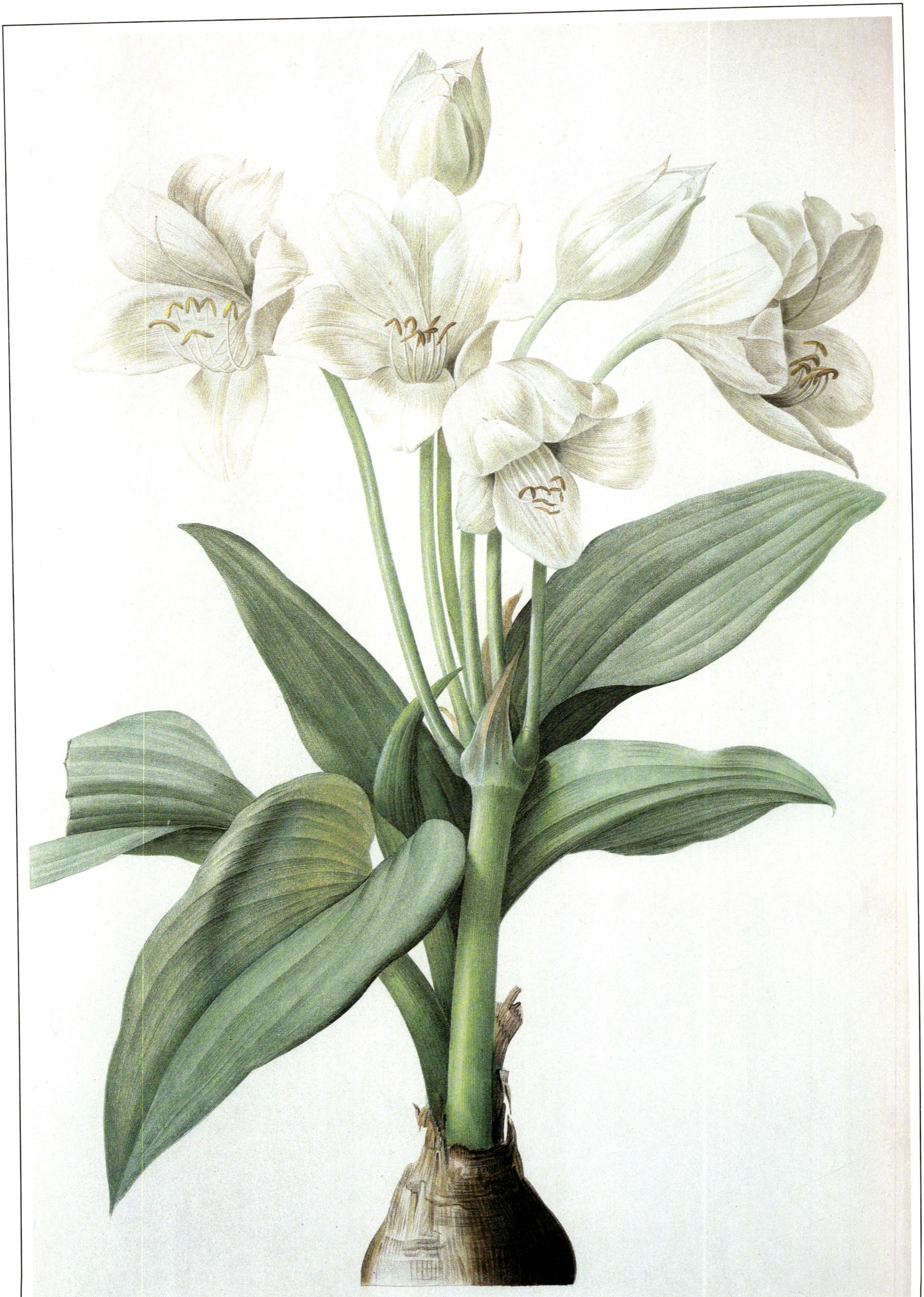

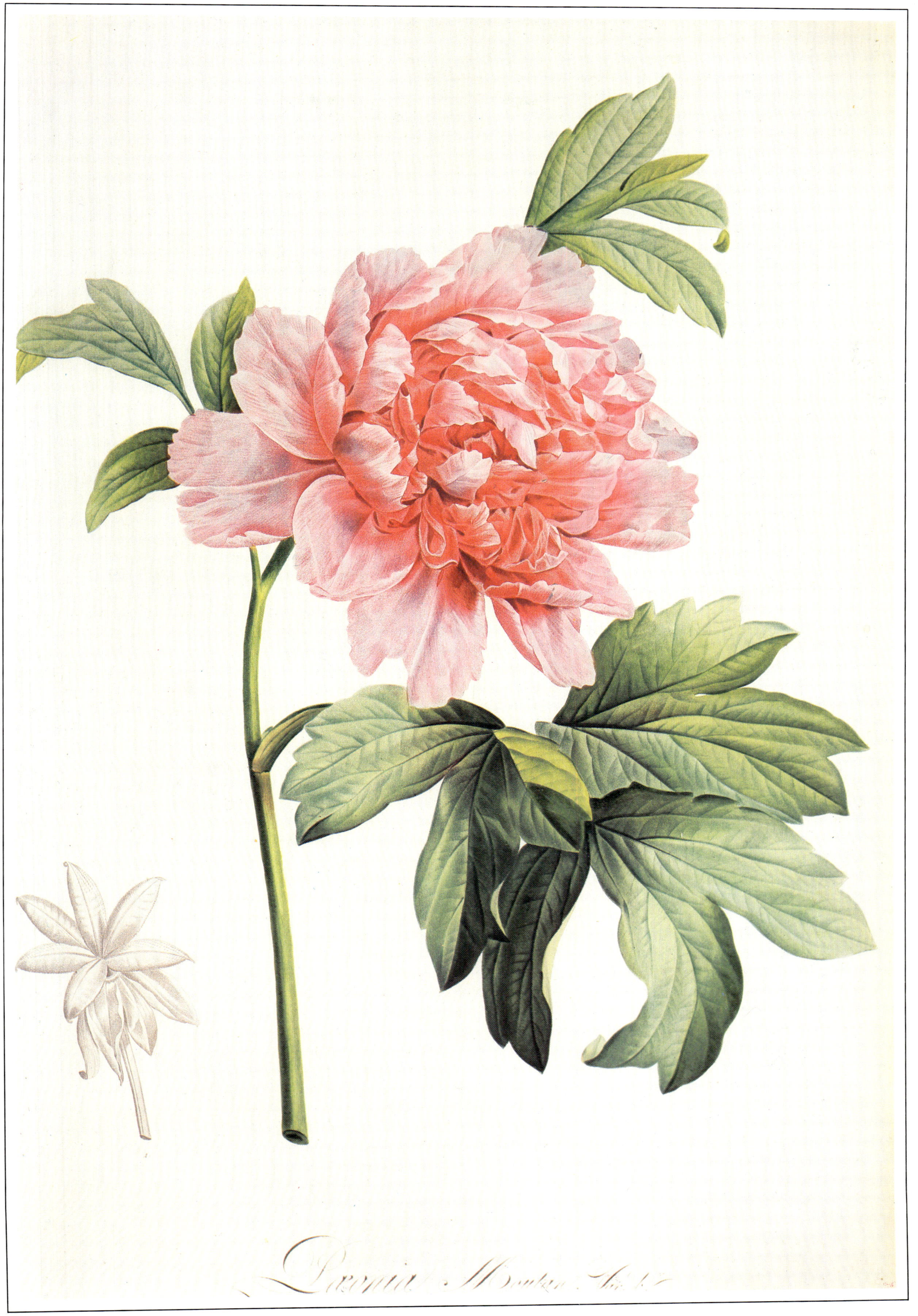
Pæonia Moutan

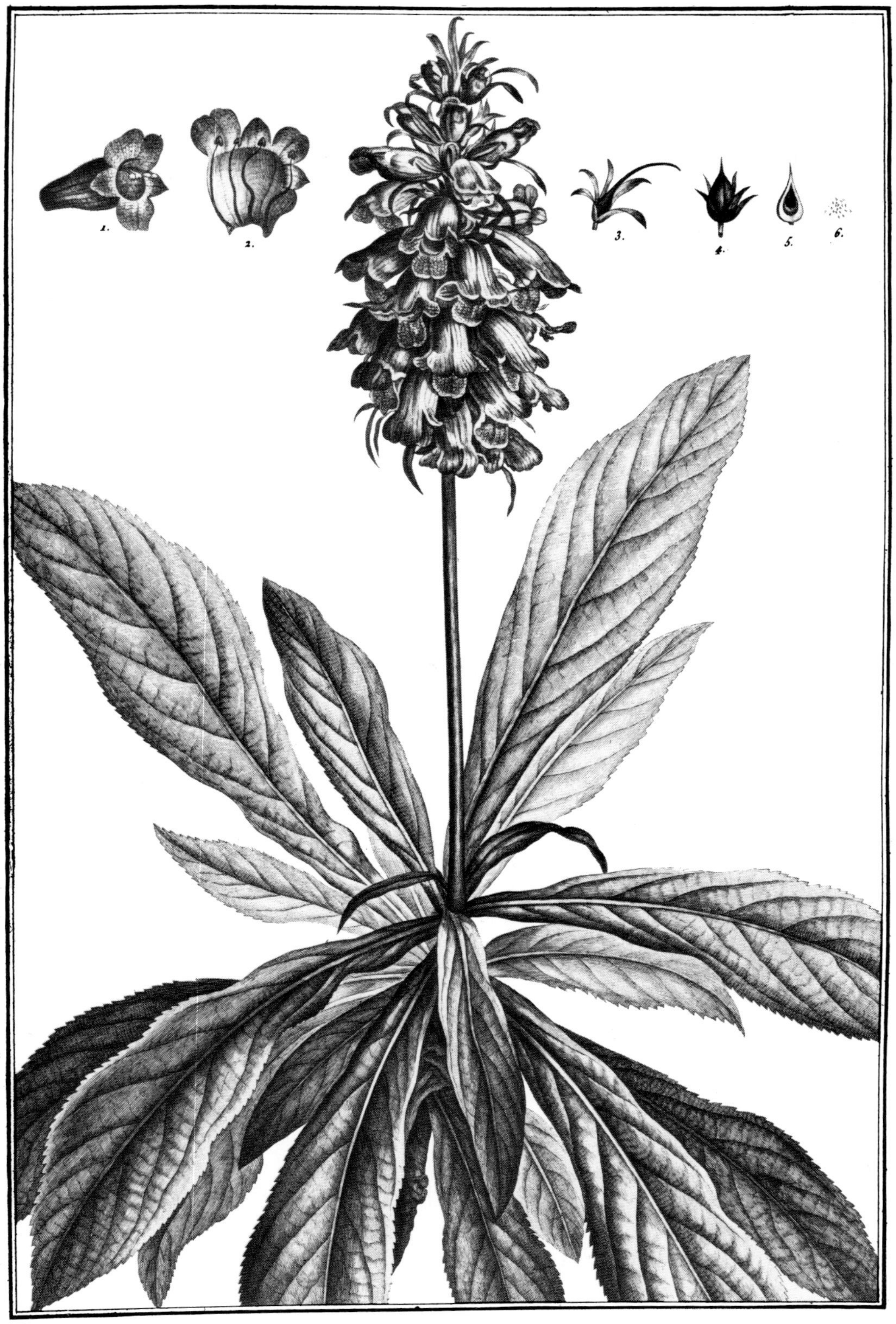

Previous page:
'Paeonia moutan', a tree peony (*Paeonia suffructicosa*). Stipple engraving by Bouquet, colour printed *à la poupée* and retouched by hand, after a painting by Pierre-Joseph Redouté from *Description des Plantes Rares Cultivées à Malmaison et à Navarre* (1813) by Aimé Bonpland. The tree peony, like the camellia and the chrysanthemum, was introduced from China, where it had been cultivated and developed since the seventh century AD.

'Digitalis sceptrum', now known as *Isoplexis sceptrum*, from the Canary Islands. Engraving by Milsan after a drawing by Pierre-Joseph Redouté from *Sertum Anglicum* (1788) by Charles-Louis l'Héritier de Brutelle.

Geranium phaeum variety *lividum*, a variety from southern Europe with strange greyish-mauve flowers. Engraving by Maleuvre after a drawing by Henri-Joseph Redouté from *Geraniologia* (1791 or 1792) by Charles-Louis l'Héritier de Brutelle.

London. But instead of working on Dombey's herbarium as he had intended, l'Héritier became more interested in the exciting new plants which were being grown at Kew and in private collections around London, particularly those sent back by Francis Masson. In 1787, l'Héritier invited his young protegé, Redouté, to join him in London. Here Redouté was able to meet such English botanical painters as James Sowerby and to learn some of the colour printing techniques that he later developed with such success. Many of his paintings from this period survive, notably the often-reproduced Canterbury Bell, which he gave to James Lee, a botanist with whom he stayed in Hammersmith, and a wonderful monochrome painting of the Peruvian *Cantua buxifolia*, now in the Fitzwilliam Museum, Cambridge. L'Héritier's *Sertum Anglicum* (1788), designed as a compliment to his English hosts, is a collection of plates of rare plants that he had seen in London. Redouté was responsible for much of it, but some of the paintings had been commissioned by l'Héritier from James Sowerby.

On his return to Paris, Redouté was appointed draughtsman and painter to the cabinet of Marie Antoinette. He continued to learn from van Spaendonck and from the royal engraver, Demarteau, who taught him the techniques of stipple engraving and printing, which he perfected in his later

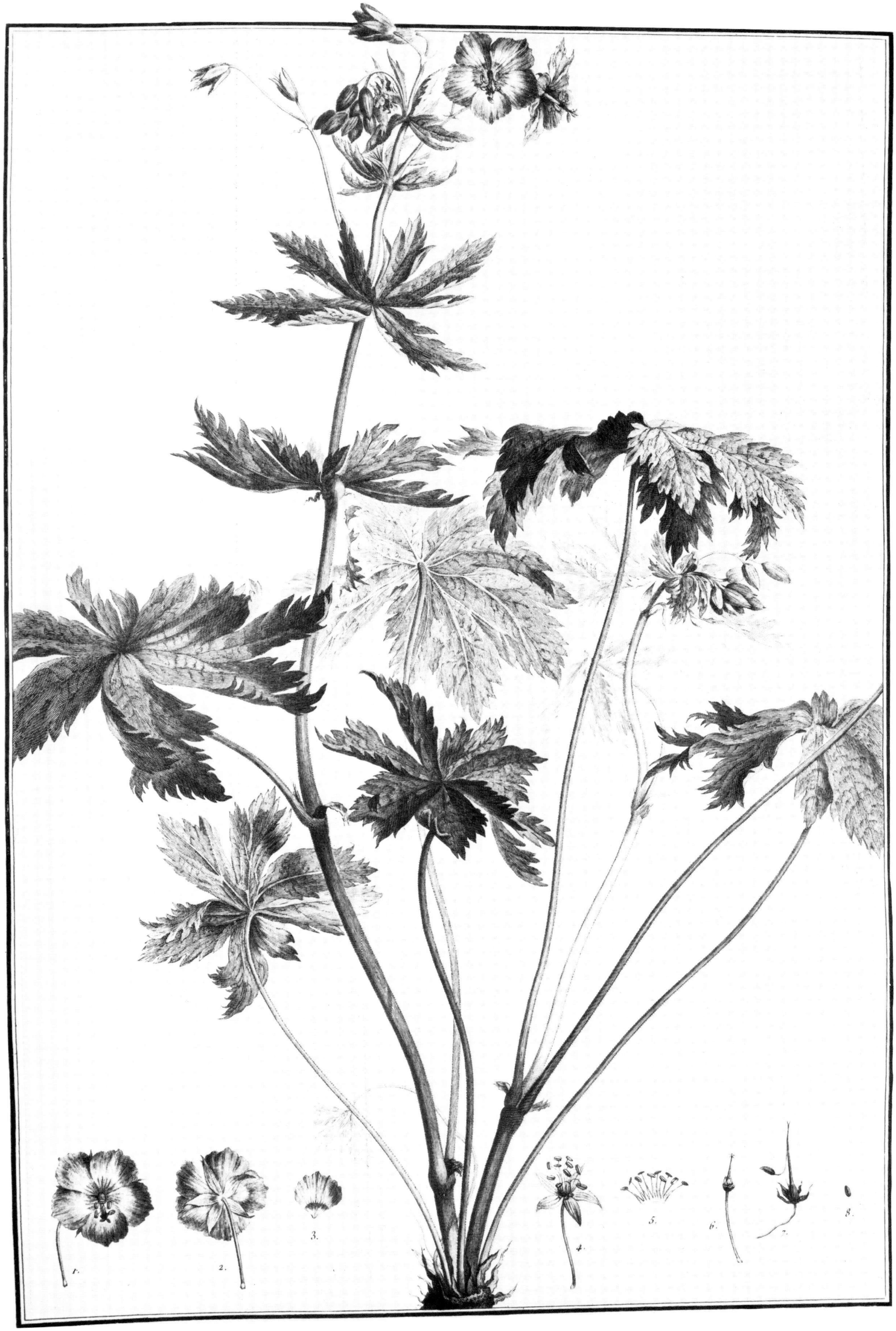
1.
2.
3.
4.
5.
6.
7.
8.

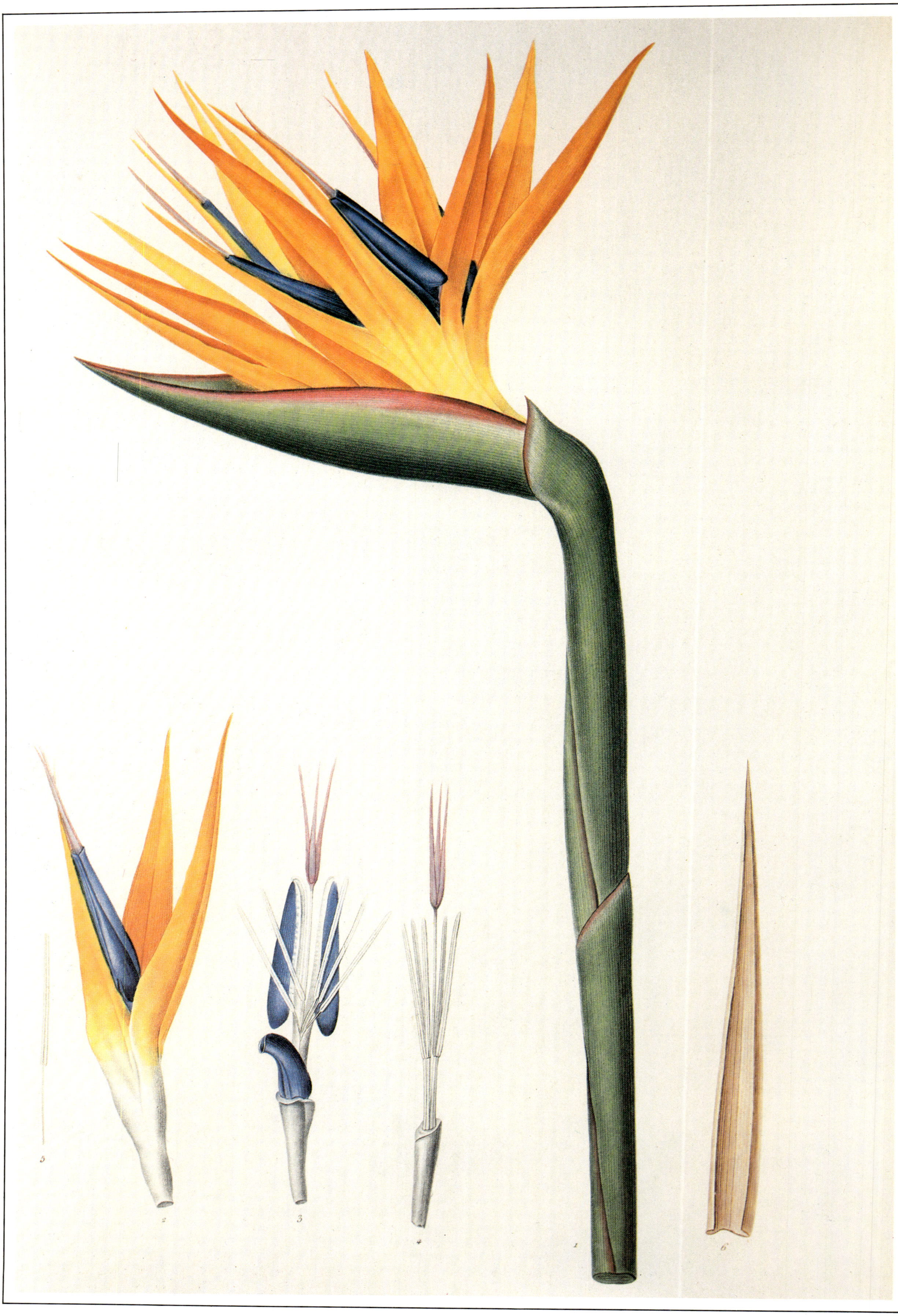
5
2
3
4
1
6

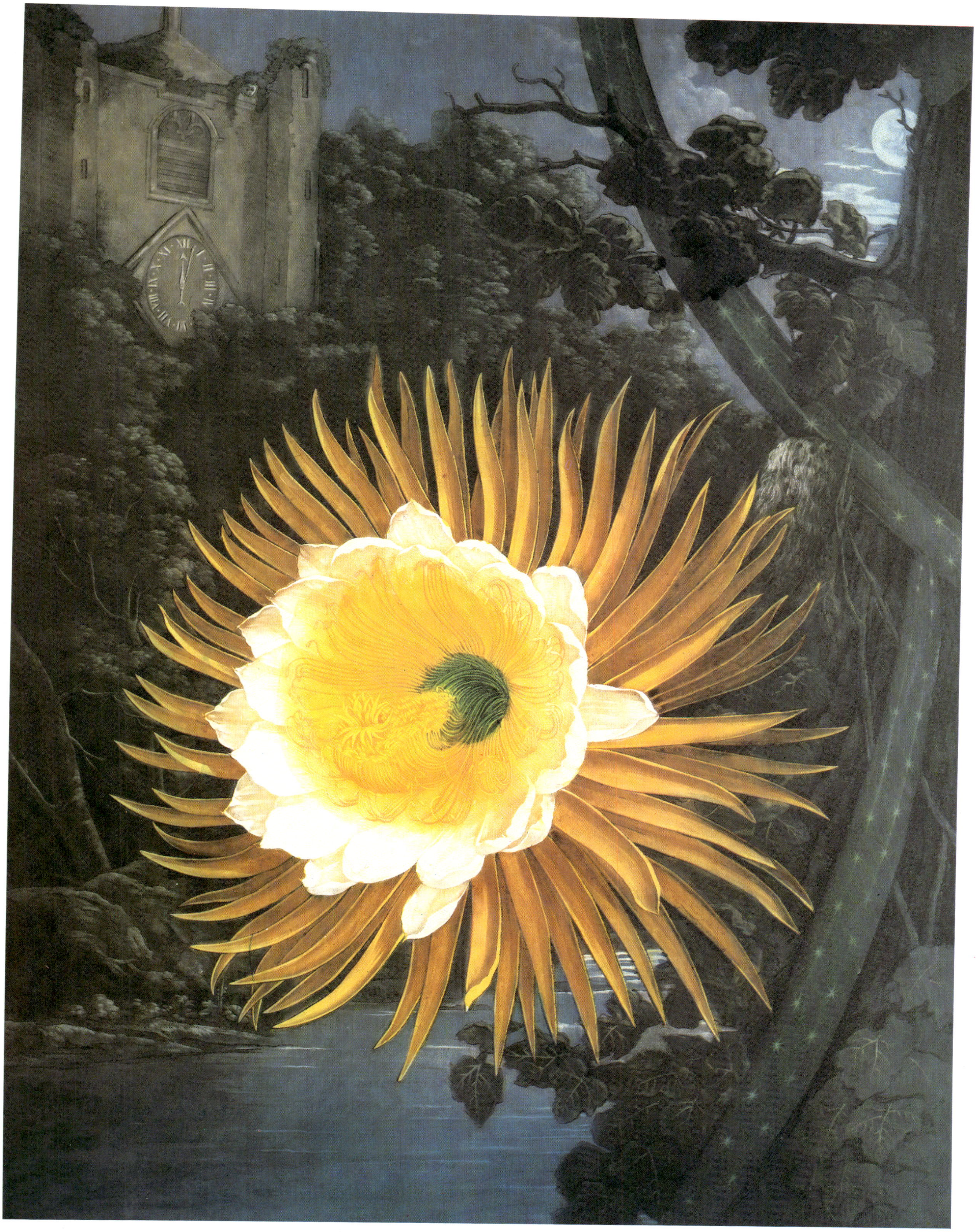

Left:
Strelitzia reginae. Stipple engraving by Phelippeaux, colour-printed *à la poupée* and hand-coloured, after a painting by Pierre-Joseph Redouté from *Les Liliacées*, Volume II (1804). *Strelitzia*, a native of the Cape region of South Africa, first flowered in cultivation at Kew and was named in honour of Queen Charlotte, who was Princess of Mecklenburg-Strelitz. It is a member of the same family, Musaceae, as the banana.

'Night-blowing Cereus or Queen of the Night' (*Selenicereus grandiflorus*). Mezzotint by Dunkerton from a painting, the flower by Philip Reinagle, 'moonlight by Pether', from *The Temple of Flora* (1799–1801) by Robert Thornton.

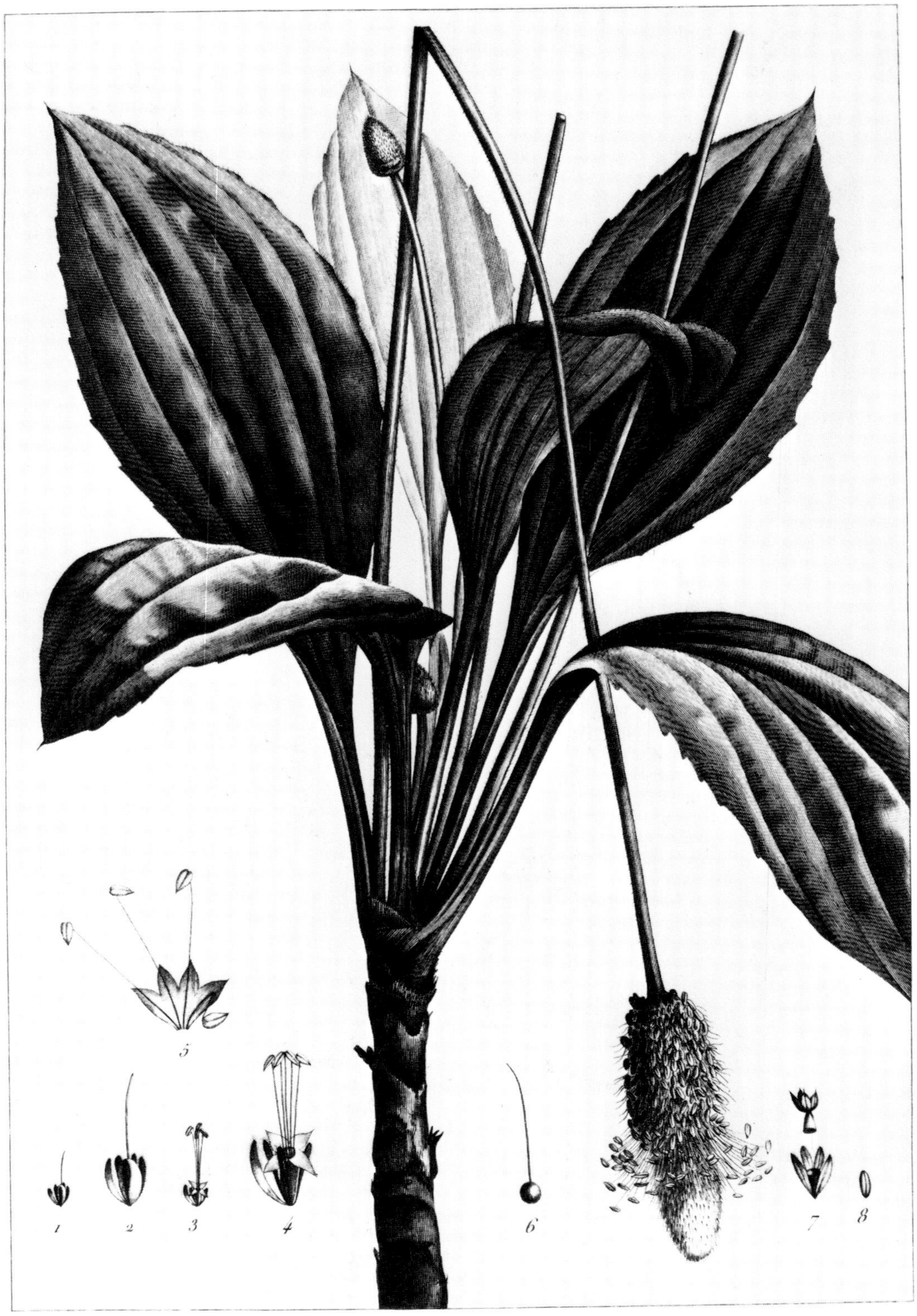

A plantain, *'Plantago vaginata'*. Engraving by Sellier after a drawing by Pierre-Joseph Redouté from *Description des Plantes nouvelles et peu connues cultivées dans le Jardin de J.M. Cels* (1800) by Etienne-Pierre Ventenat.

works. In spite of his royal connections, the French Revolution does not seem greatly to have affected him; his last work for Marie Antoinette was of a night-flowering cactus and was painted while she was in prison in 1792. At this time he worked also on l'Héritier's monograph of the geranium family, *Geraniologia*, which was published in 1792. L'Héritier himself was less fortunate. His correspondence with Sir J. E. Smith and Sir Joseph Banks in England describes many of the political difficulties of the time. He had been a magistrate before the Revolution, but a fair one, and was on the whole in favour of change. Although he was imprisoned for a short time and lost most of his money, he retained many of his public responsibilities and just survived the reign of terror (1792–94), only to be murdered without apparent motive on the streets of Paris in 1800, while he was on his way home from the Botanical Institute.

After the Revolution, Redouté and his younger brother, Henri-Joseph, were appointed to the staff of the Musée d'Histoire Naturelle and continued to produce watercolours to add to the collection of *vélins* as well as making drawings for numerous scientific botanical publications, notably Desfontaines's *Flora Atlantica* of 1798.

From that year onwards, Redouté worked mostly on colour-printed

'Rudbeckia pinnata'. Engraving by Sellier after a drawing by 'J.P. Redouté' from Ventenat's ... *Jardin de J.M. Cels* (1800).

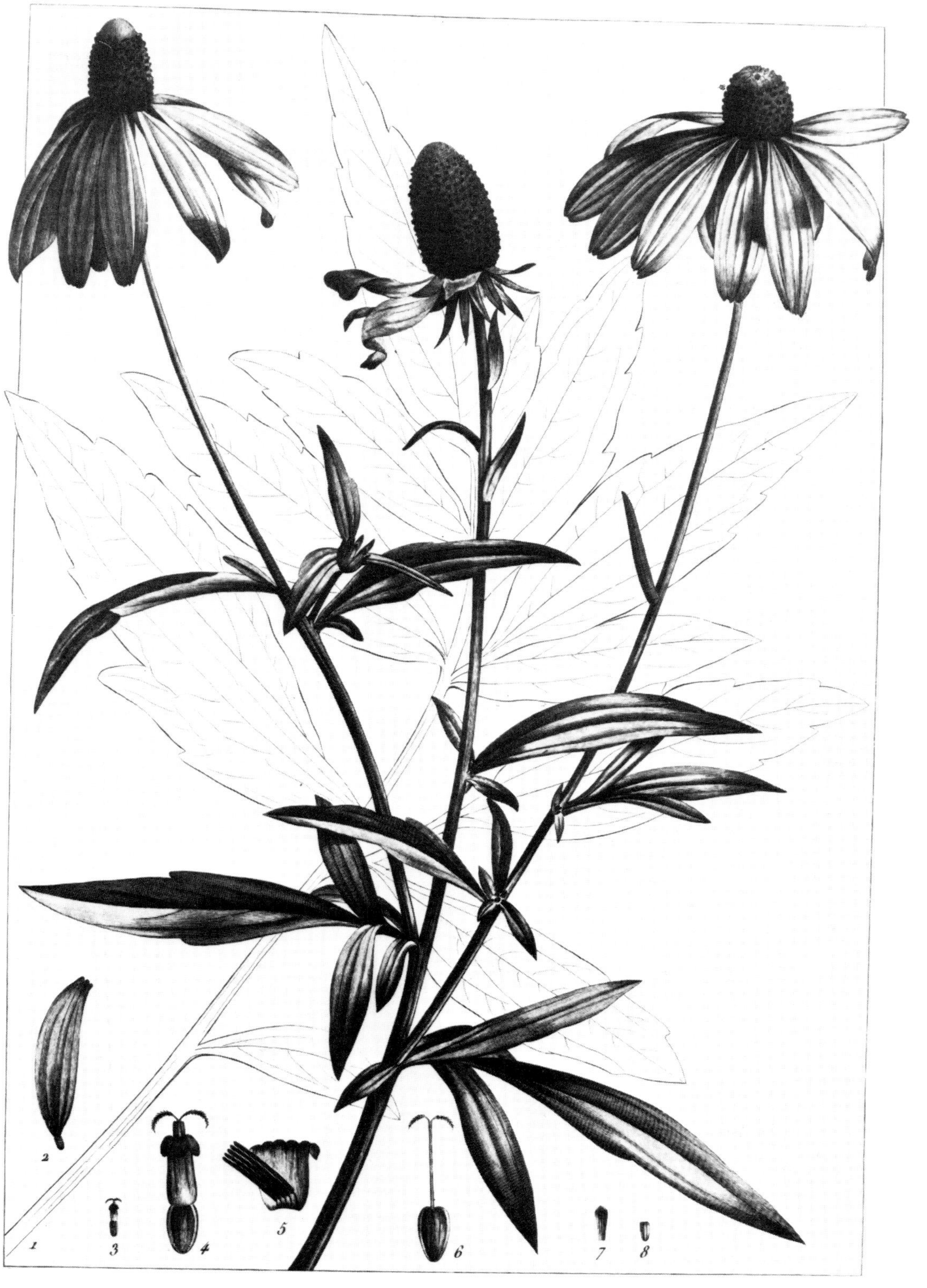

books, the first of which was the great work on succulents and cacti, *Plantes Grasses*. The idea for this was l'Héritier's, but Desfontaines played a great part in its production, and the text was written by the young Augustin de Candolle. *Plantes Grasses* was published between 1799 and 1831. It contains around 185 plates, all engraved, as well as painted, by Redouté himself.

The majority of the strictly botanical publications to which Redouté and his brother contributed appeared around this time, notably Augustin de Candolle's *Astragalogia* (1802), a monograph on the genus *Astragalus*, André Michaux's *Histoire des Chênes de l'Amérique* (1801), Philippe La Peyrouse's *Figures de la Flore des Pyrénées* (1795–1801) and a new edition of Henri-Louis Duhamel du Monceau's *Traité des Arbres et Arbustes*. This last had about five hundred plates, printed in colour and finished by hand, mainly by Redouté, but in later volumes by Pancrace Bessa.

Having survived the Revolution, Redouté continued to add to his reputation and reached the apex of his fame under the patronage of Josephine, first wife of Napoleon Bonaparte. In 1798, she acquired the property at Malmaison, and with great energy set about making a garden and acquiring a large collection of plants, especially roses. This garden was

Next page :
Cattleya labiata. Hand-coloured engraving by C. Fox after a painting by John Curtis from *Collectanea Botanica; or, figures and botanical illustrations of rare and curious exotic plants, chiefly cultivated in the gardens of Great Britain* (1821) by John Lindley. This orchid, one of the parents of the huge-flowered hybrids, was introduced to English hothouses from Brazil in 1818.

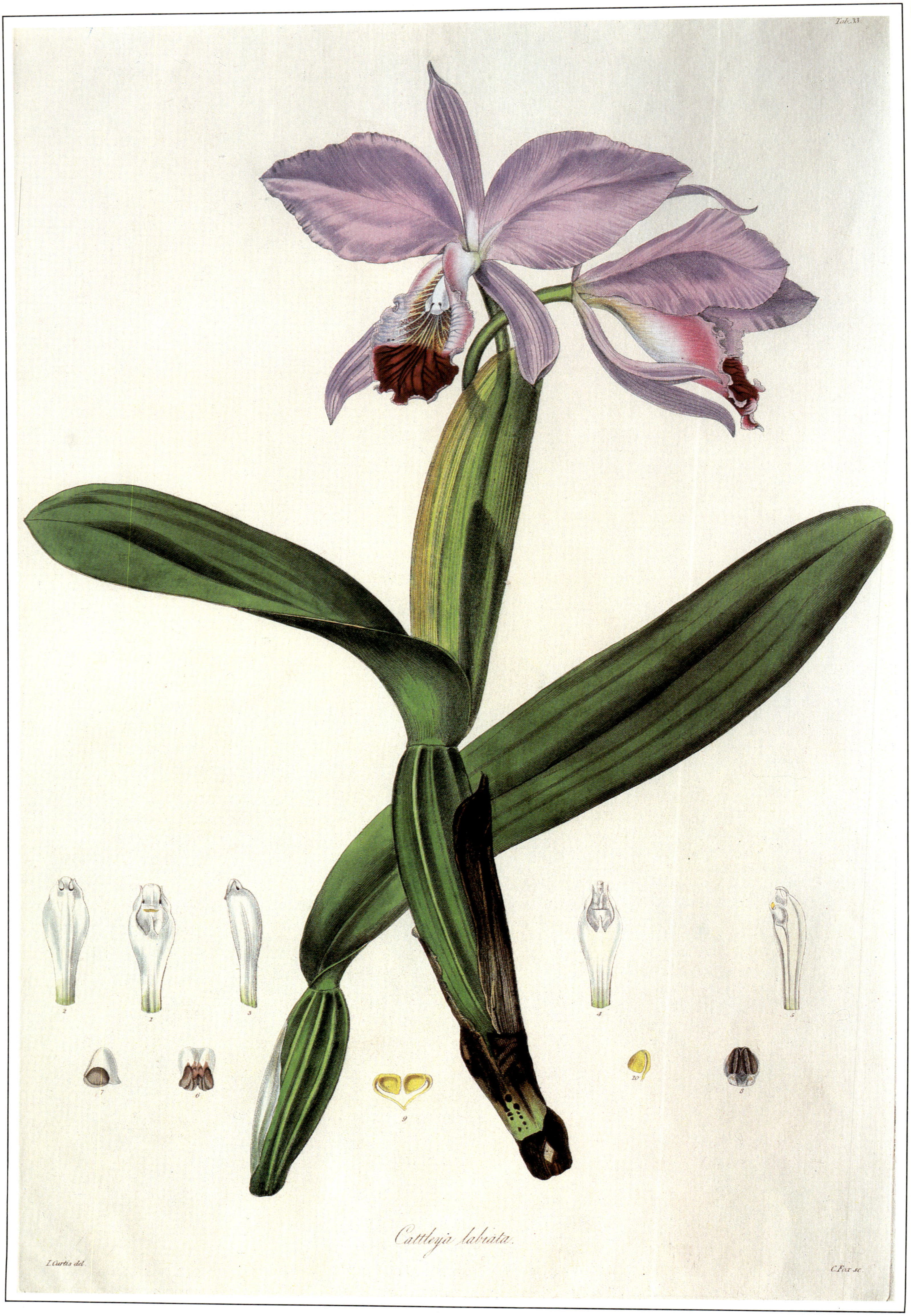
Tab. 33.
Cattleya labiata.
I. Curtis del.
C. Fox sc.

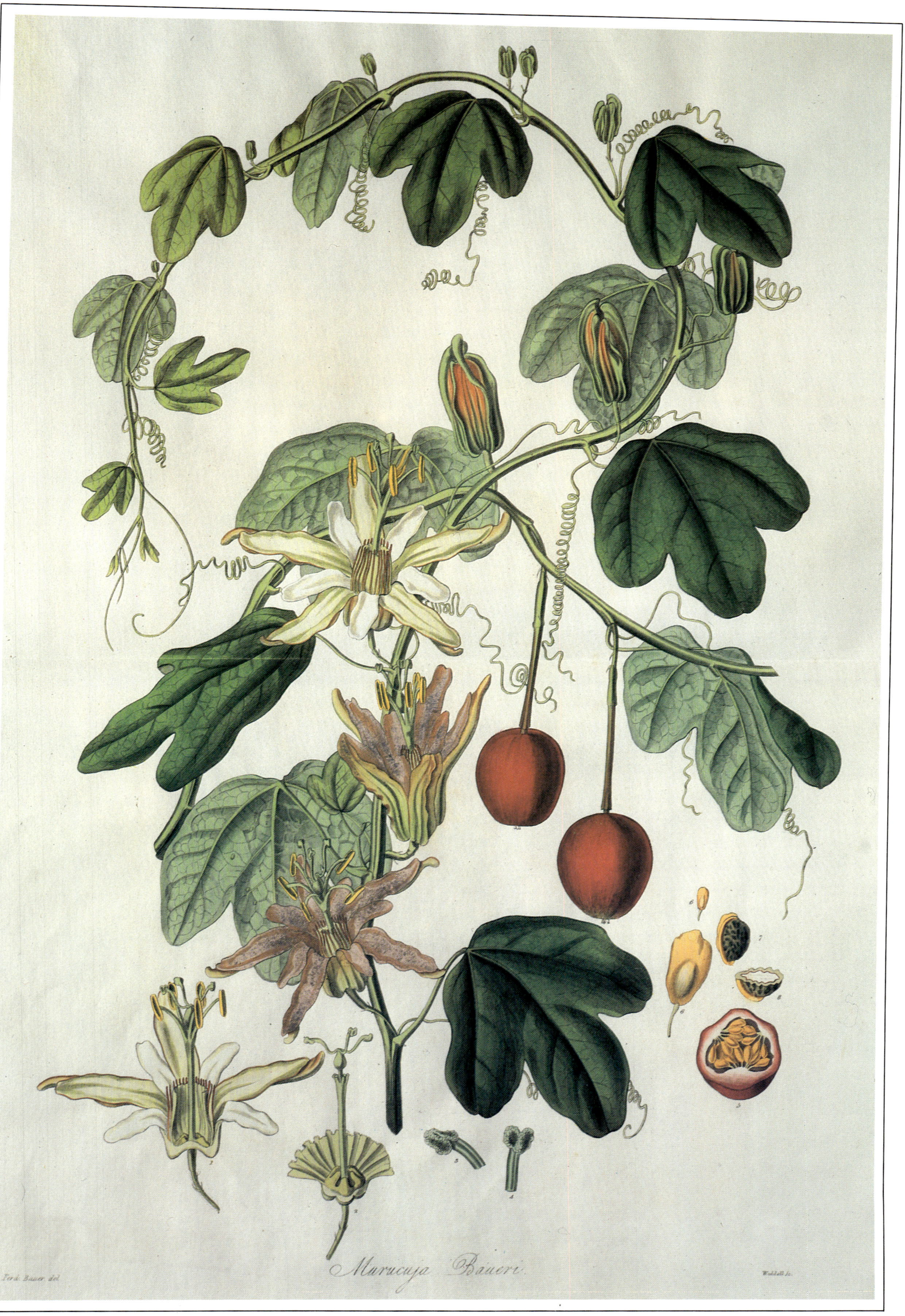
Murucuja Baueri

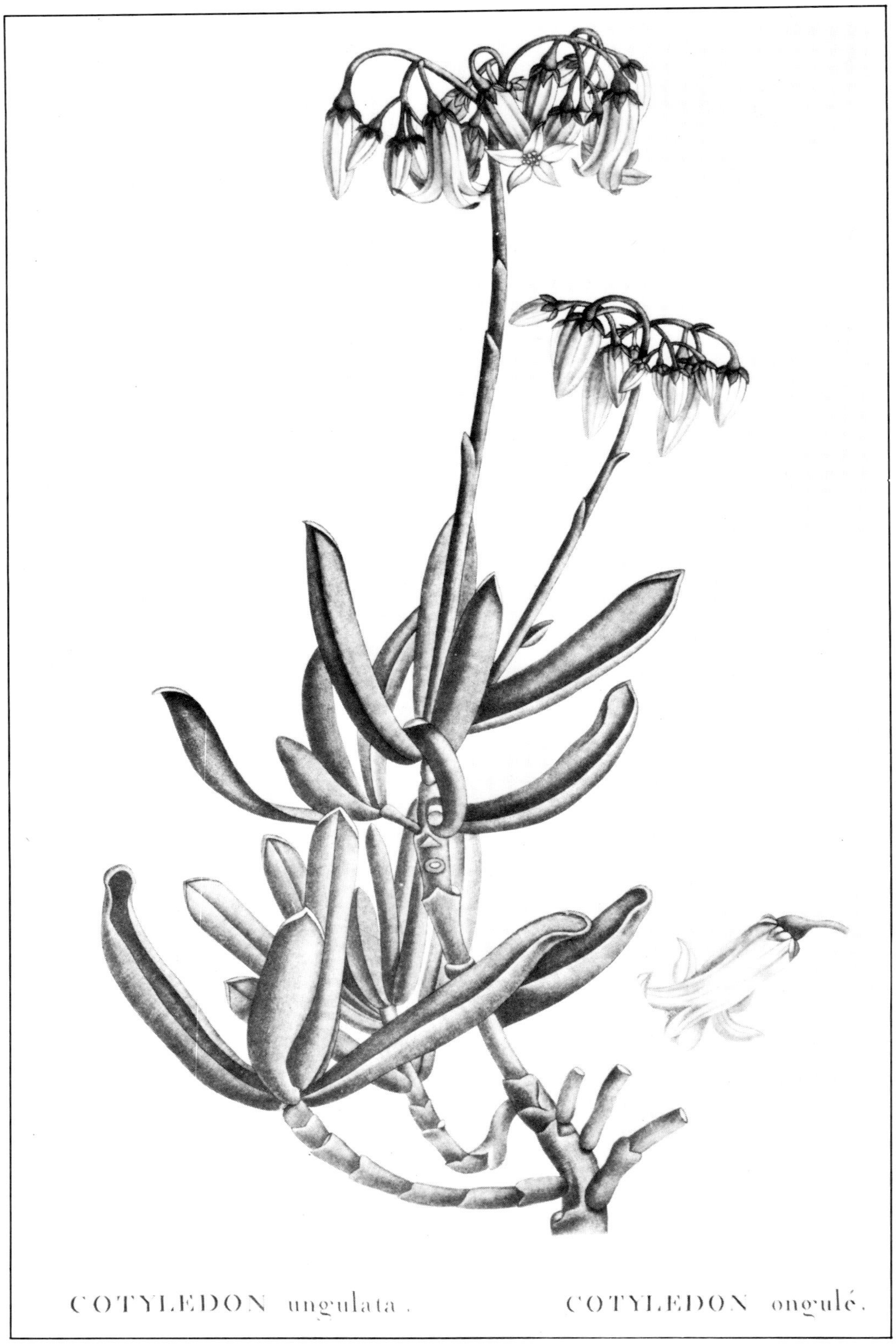

Right:
The Earl of Bute (1713–92), 'The Maecenas of Botany' and patron of several large-scale botanical works, notably Curtis's *Flora Londinensis*. Beneath him is 'An Allegorical Representation of the Prosperity of Great Britain under Lord Bute's Administration' – Bute was Prime Minister from 1761 to 1763; his administration was a complete disaster, surviving extreme unpopularity only through political corruption and the support of George III. One of many such plates, each celebrating a scientific or political figure, in *A New Illustration of the Sexual System of Linnaeus* (1799–1801) by Robert Thornton.

Cotyledon ungulata/ from South Africa. Stipple engraving, colour printed *à la poupée*, after a drawing by Pierre-Joseph Redouté from *Histoire Naturelle des Plantes Grasses* (1799–1831) by Augustin-Pyramus de Candolle.

the source and inspiration of Redouté's *Jardin de la Malmaison* (1803), with text by Etienne-Pierre Ventenat, and *Description des Plantes Rares Cultivées à Malmaison et à Navarre* (1813), with text by Bonpland; it also provided many of the plants for Redouté's masterpieces, *Les Liliacées* (1802–16) and *Les Roses* (1817–24). Napoleon divorced Josephine in 1809, but she continued to live at Malmaison until her death in 1814.

Of all Redouté's work, *Les Liliacées* is probably his finest achievement, containing 486 plates of lilies, irises and other monocotyledons. Redouté engraved and printed many of the plates himself, and the more sumptuous copies were also retouched by him. In honour of his patroness, the most magnificent plate is of *Amaryllis josephinae* (now known as *Brunsvigia josephinae*), one of the largest of all flowering bulbs. It had been brought to Holland from South Africa about thirty years before and did not begin to flower until it had been in cultivation for about 20 years. Hearing about this fabulous bulb, Josephine managed to acquire it, and it flowered in the garden at Malmaison. Another striking plate is that of *Strelitzia reginae*, which Redouté probably first saw at Kew. It was introduced into England in 1773 and was named in honour of Queen Charlotte, the wife of George III.

Previous page:
A passion flower (*Murucuja baueri*). Hand-coloured engraving by Weddell after a painting by Ferdinand Bauer from *Collectanea Botanica* (1821) by John Lindley.

Redouté was also patronised by Napoleon's second wife, Marie-Louise, and, after Waterloo, by other ladies of the court, many of whom had been his pupils. However, after the *Choix des Plus Belles Fleurs* (1827–33), his pictures lost much of their botanical significance, and he worked mainly on elegant paintings of garden flowers and on reissuing his earlier plates in an attempt to relieve the pressure of his debts. He died in poverty in 1840, but his drawings continued to be used and published after his death, notably in Jaubert and Spach's *Illustrationes Plantarum Orientalium* (1842–57).

Redouté had at least been extremely successful for most of his career. The same cannot be said of his English contemporary, Robert Thornton, who was responsible for one of the most famous British plant books, *The Temple of Flora*. Its story is one of the strangest in the history of botanical publishing.

Robert Thornton (*c* 1765–1837) came from a well-to-do family; his father, who died in 1768, was described by the young James Boswell as 'well-bred, agreeable, lively and odd'. As a boy, Thornton kept his own botanic garden and an aviary containing every species of British hawk. His mother sent him to Cambridge, hoping that he would be ordained, but he preferred medicine. After taking his MB degree at Cambridge, he went to Guy's Hospital in London and later continued his studies abroad. He started to practise in London in 1797. At about this time, both his mother and his elder brother died, leaving Thornton to inherit the family fortune. Soon, he was starting to plan his ambitious series of botanical publications which was finally entitled *New Illustration of the Sexual System of Carolus von Linnaeus; comprehending an illucidation of the several parts of the fructification; a prize dissertation on the sexes of plants; a full explanation of the*

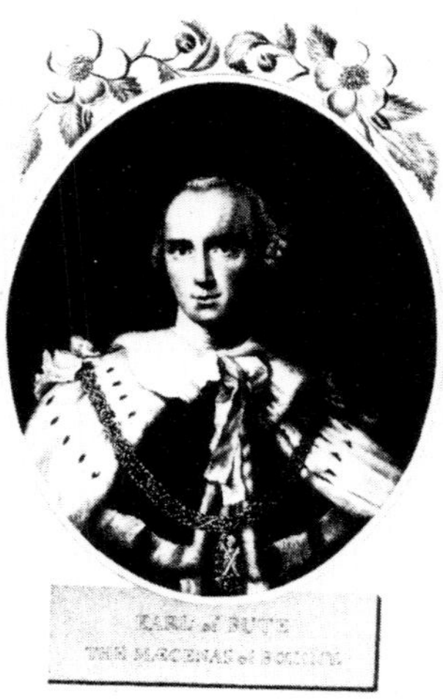

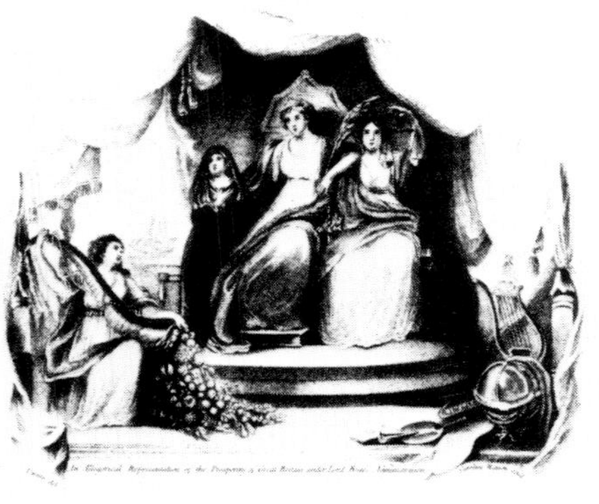

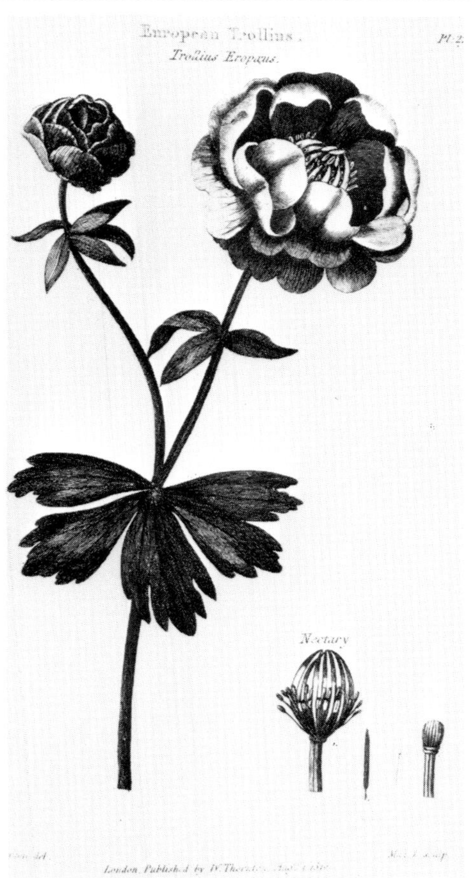

Globe flower (*Trollius europaeus*). Engraving by Mazel after a drawing by Henderson from *Elements of Botany, Part I: Classification* (1812) by Robert Thornton.

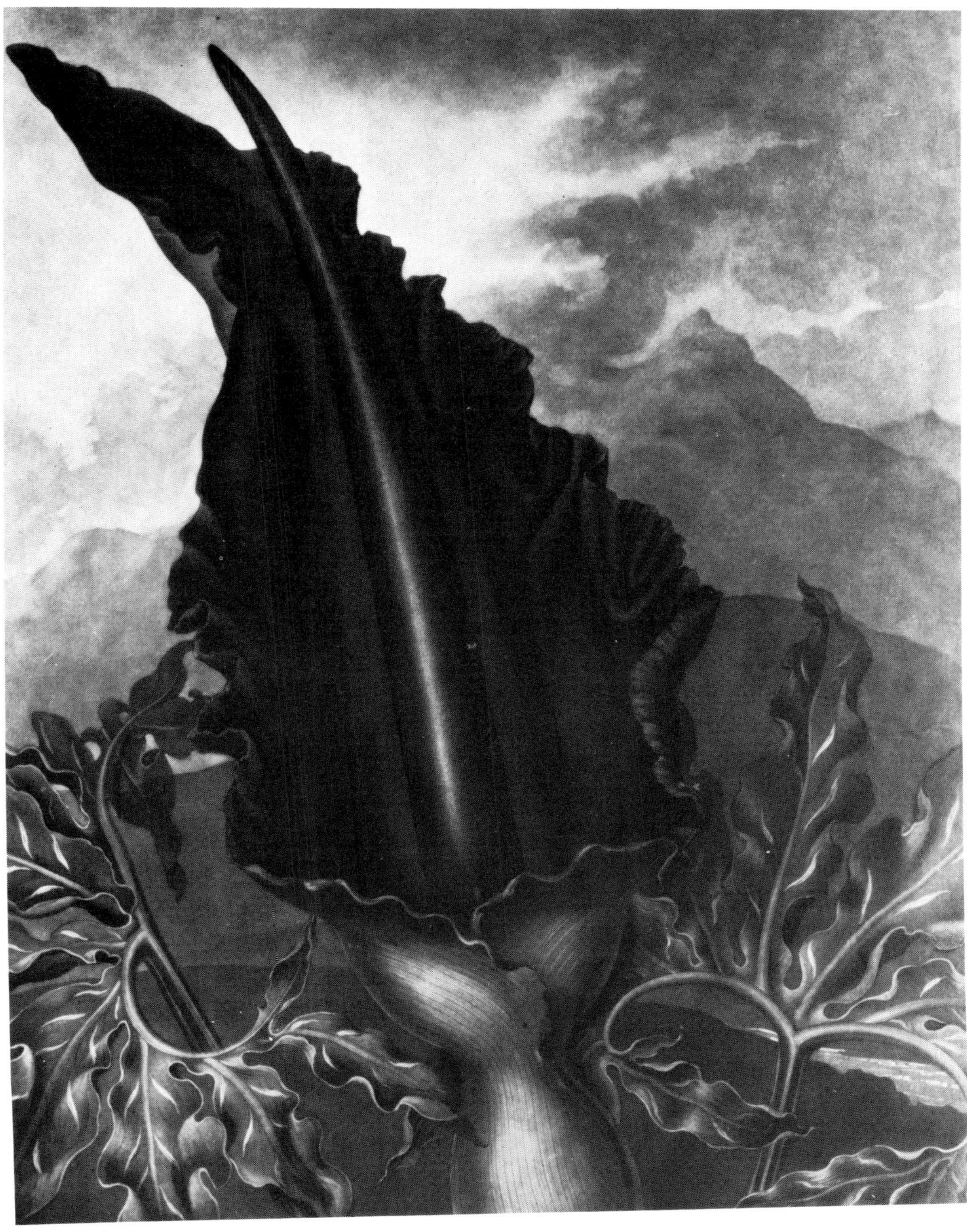

Dragon arum (*Dracunculus vulgaris*). Mezzotint with aquatint added by Ward after a painting by Peter Henderson from *The Temple of Flora* (1799–1801) by Robert Thornton. The dragon arum is an evil-smelling plant that can reach four feet in height and is found wild in thickets in the eastern Mediterranean area.

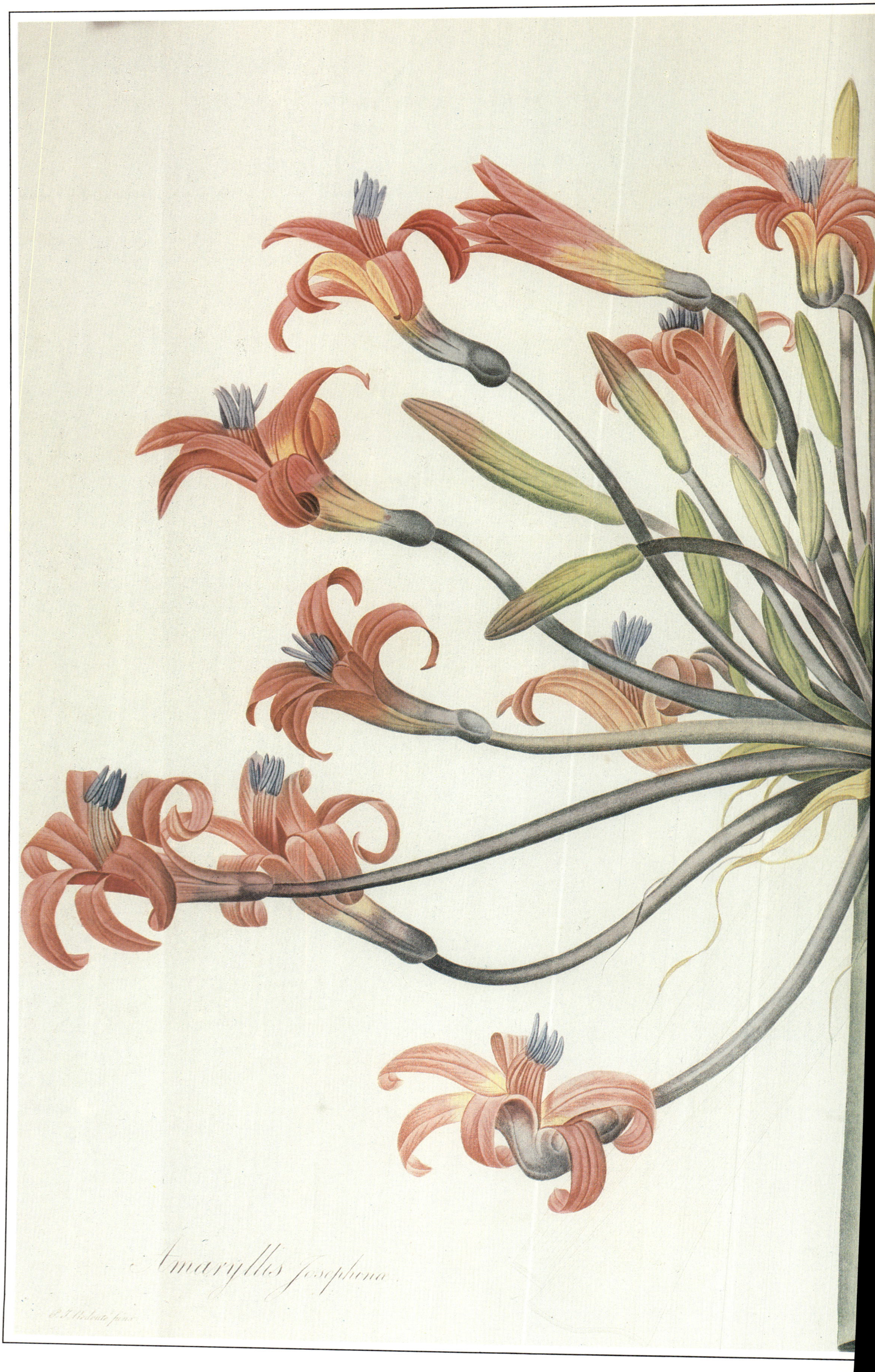
Amaryllis Josephinæ
P. J. Redouté pinx.

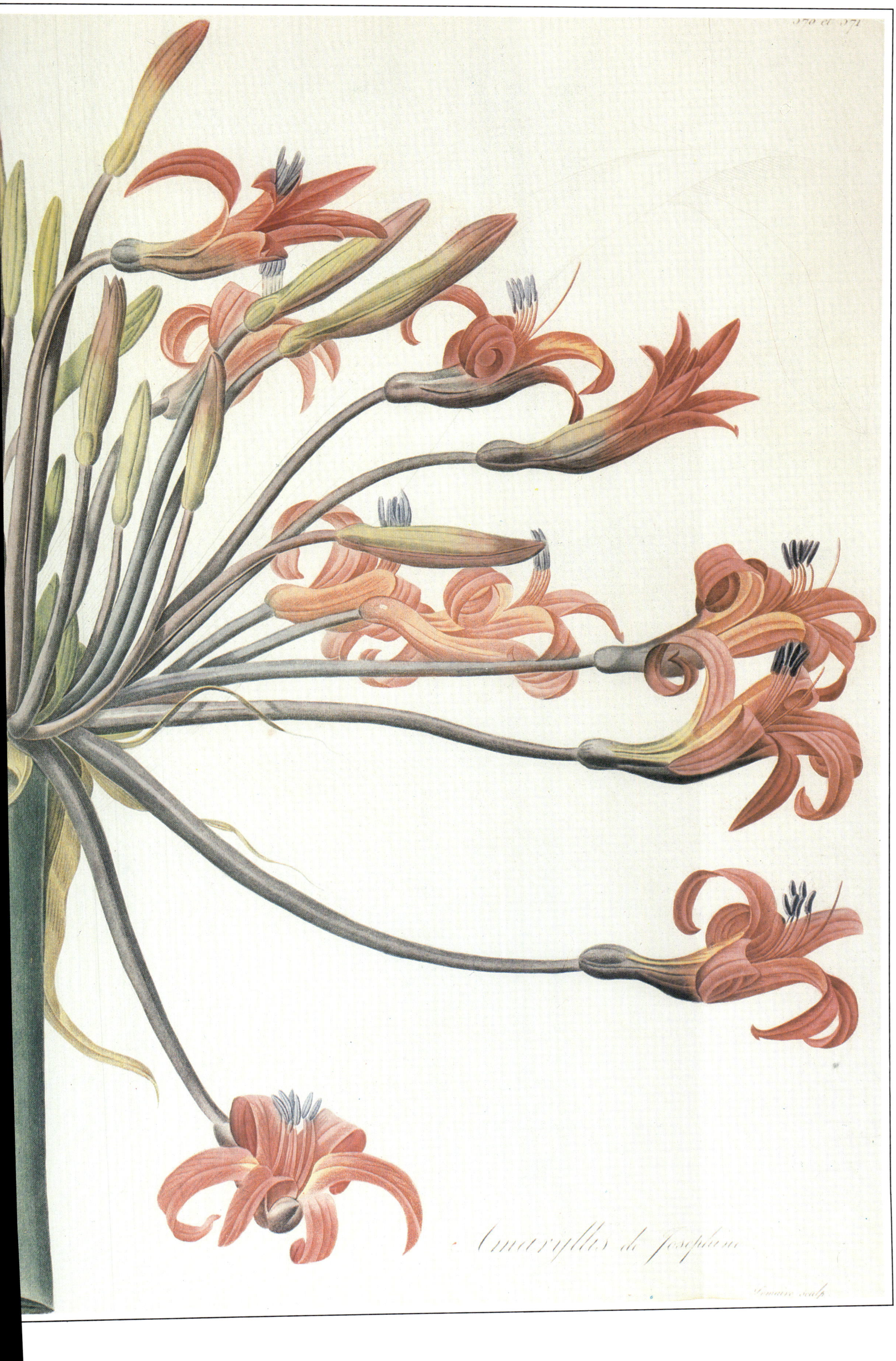
370 et 371
Amaryllis de Josephine

classes and orders, of the sexual system; and the Temple of Flora or Garden of Nature, being picturesque, botanical, coloured plates of select plants, illustrative of the same, with descriptions.

Thornton's intention was to popularise Linnaeus's comprehensive system of botanical classification, which had been published some forty years earlier, and to produce a botanical book that would surpass anything published in France or elsewhere.

> Shall Britons in the field
> Unconquered still, the better laurel lose?–
> In finer arts and public works shall they
> To Gallia yield?

The same theme was put even more forcefully in the dedication, which was to Queen Charlotte.

The Temple of Flora, which was only a part of the whole scheme, began to appear in 1799. For the huge plates, which combine floral foregrounds with landscape backgrounds that Thornton deemed to be appropriate, various painters were commissioned, but the majority of the plates were by two portrait painters, Peter Henderson and Philip Reinagle, with much smaller contributions by the botanical artists Sydenham Edwards and James Sowerby. Thornton himself painted one plate of a large vase of roses with a rather stiff bird nesting in the middle. Several engravers worked on the plates, which are variously in mezzotint, aquatint, stipple and line engraving, and in combinations of these.

In 1804, while the parts were being published, Thornton set up 'Dr Thornton's Linnean Gallery' at 49 New Bond Street. This was embellished with 'a bower about which are disposed foreign as well as curious English birds and butterflies.' Here the public could view the original paintings and buy a catalogue which included poetic compositions on various subjects. In spite of all this publicity, the venture was not a financial success. By 1811, Thornton was forced to petition Parliament for permission to set up a lottery; in May of that year, there was passed 'An Act to enable Dr Robert John Thornton to dispose of his collection of paintings, drawings and engravings, together with several copies of certain books therein mentioned, by way of chance.' A similar lottery had been held in 1805 to dispose of Boydell's Gallery of paintings for his massive folio, *The Dramatic Works of Shakespeare*, and had been a success, raising about £66,000.

Apart from the complete contents of the gallery which were to constitute the first prize, there were to be ten thousand other prizes, ranging from complete sets of the book, value £80, to copies of *The Elements of Botany* worth £3 apiece. Tickets were to cost two guineas each, and not more than twenty thousand were to be sold. Lots were to be drawn before 4th June 1812, 'provided that one third of the tickets had been sold by that time.' But they had not, and the draw did not finally take place until May 1813. It is not known how much was raised, but Thornton was ruined. Although he continued in medical practice and tried one or two minor publishing ventures including a school edition of Virgil with wood engravings by William Blake and a new translation of The Lord's Prayer, he died a poor man.

The plates in *The Temple of Flora*, though magnificent in a curious way as works of art, are without scientific merit, in spite of the grandiose title of Thornton's proposal. The landscape settings are appropriate to the romantic spirit of the plant rather than to its native habitat. Thus, the tropical 'Night-blowing Cereus' from Jamaica has as its background an English church tower, lit by moonlight, with the hands of the clock at five past twelve. Reinagle, the painter of the flower was helped here by Abraham Pether, who contributed his own speciality, the moonlight. The text is in a similar vein and owes much to the spirit of Erasmus Darwin's long poem, *The Botanic Garden*, the first part of which, *The Loves of Plants*, was published in 1789; the second, *The Economy of Vegetation*, appeared in 1799, the same year as the first parts of Thornton's ill-fated enterprise.

Previous page:
'*Amaryllis josephinae*', now known as *Brunvigia josephinae*, which was brought to Holland from South Africa and then acquired by the Empress Josephine for her garden at Malmaison. Stipple engraving by Lemaire, colour-printed *à la poupée* and retouched by hand, after a painting by Pierre-Joseph Redouté from *Les Liliacées* (1802–16).

Overleaf:
'*Aster Virginianus*'. Colour mezzotint by Kirkall after a painting by Jan van Huysum from *Historia Plantarum Rariorum* (1728) by John Martyn.

A poppy, *Meconopsis paniculata*, from Nepal; it has beautiful rosettes of hairy leaves in winter and a tall stem with yellow flowers. Hand-coloured lithograph by W.H. Fitch after a drawing by the author from J.D. Hooker's *Illustrations of Himalayan Plants* (1855).

PART THREE

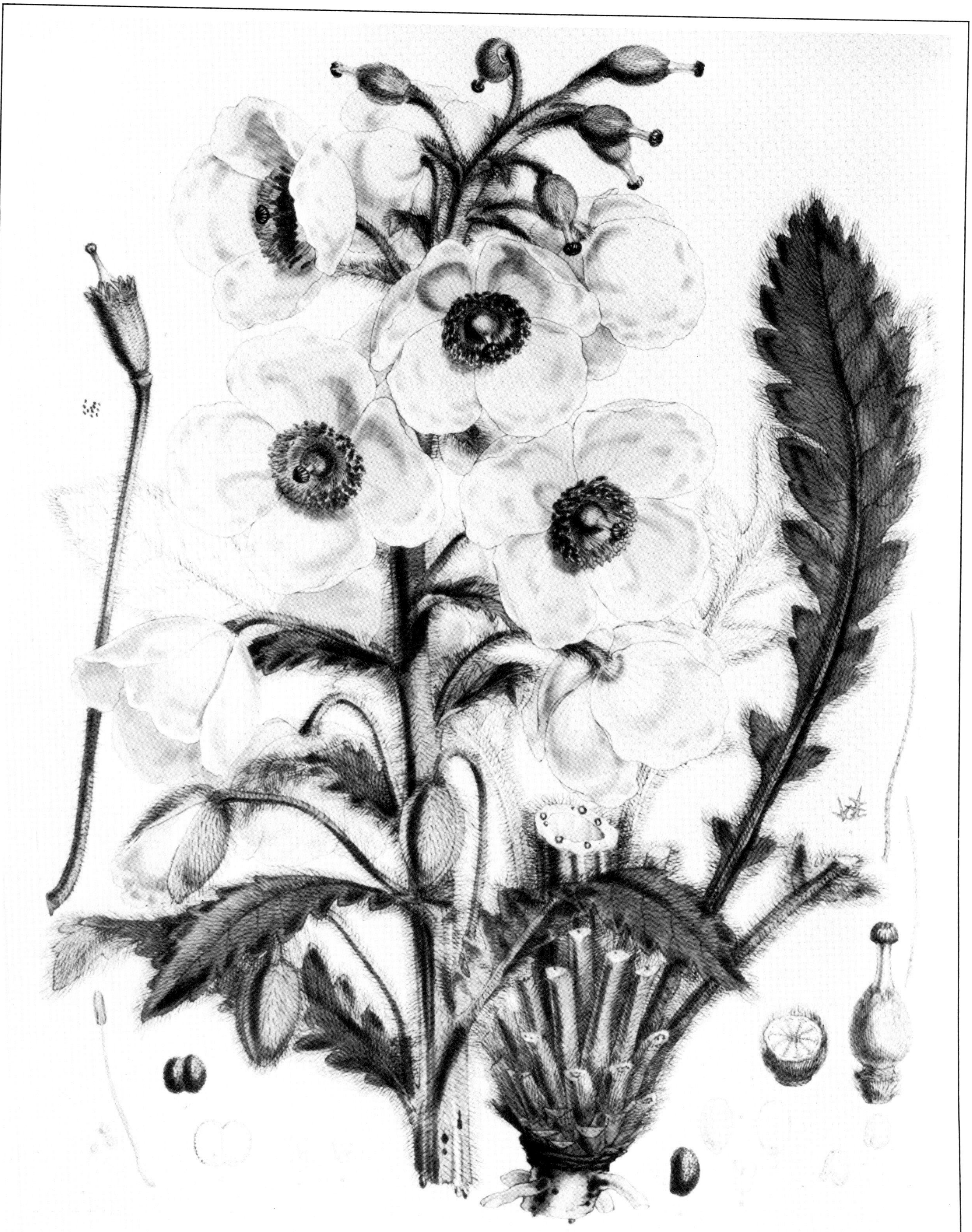

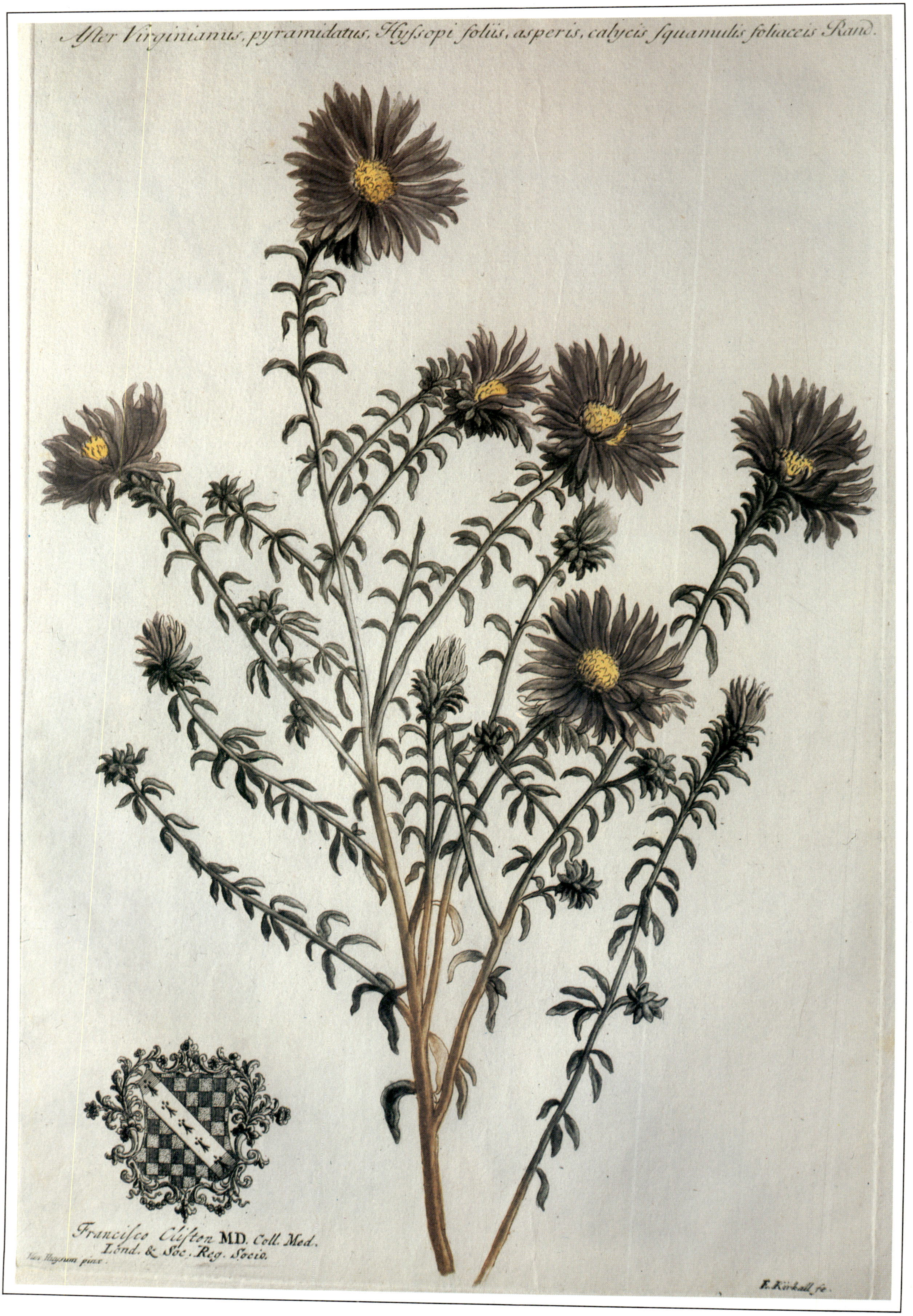
Aster Virginianus, pyramidatus, Hyssopi foliis, asperis, calycis squamulis foliaceis Rand.
Francisco Clifton M.D. Coll. Med.
Lond. & Soc. Reg. Socio.
pinx.
E. Kirkall fe.

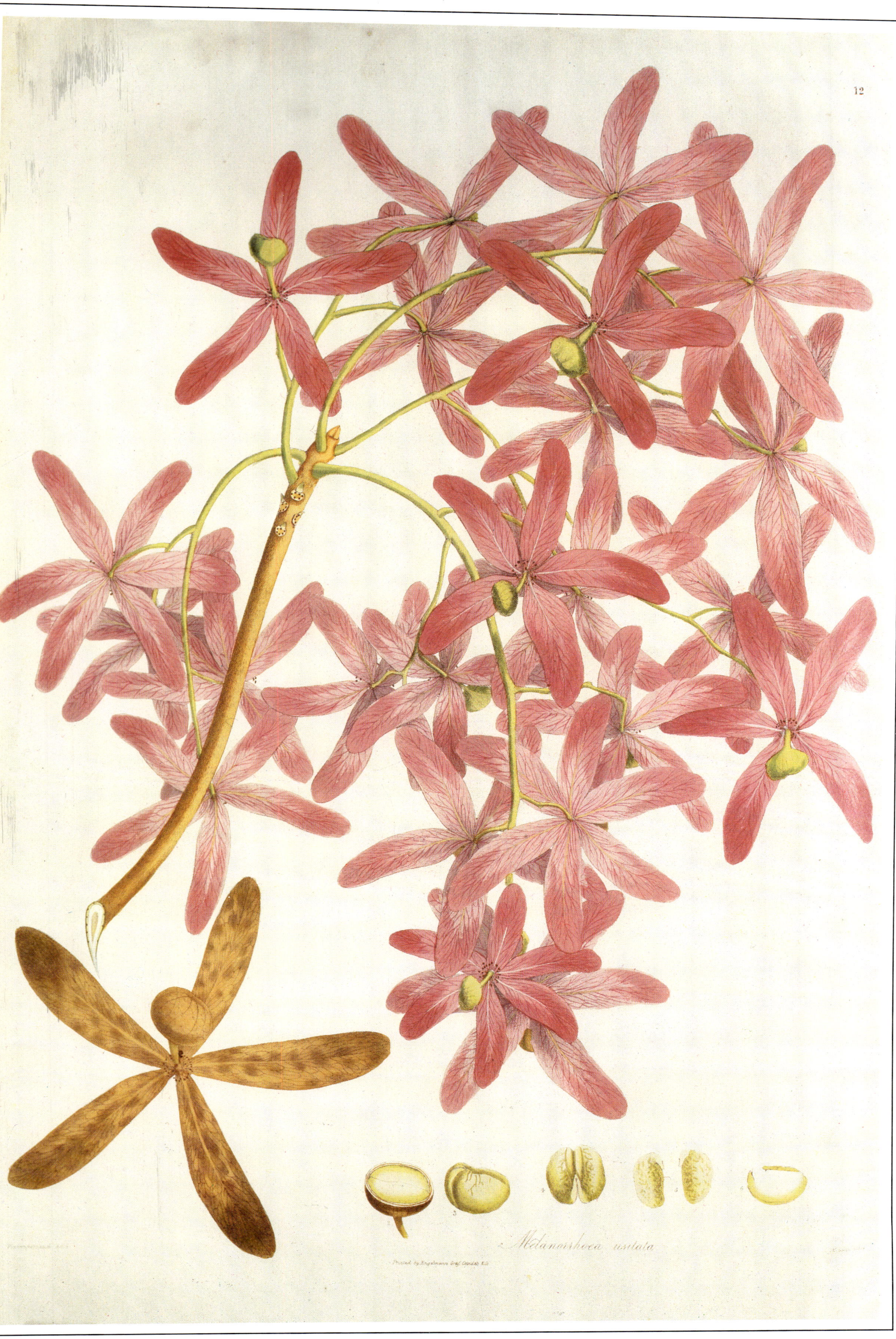
12
2
3
4
5
6
Melanorrhoea usitata
Printed by Engelmann Graf. Coindet & Co

CHAPTER XI

The Coming of Lithography

The number of well-illustrated botanical books published in the nineteenth century was enormous, probably far larger than the total for all the previous centuries since the introduction of printing. European floras, both scientific and popular, were joined by similar volumes for other parts of the world. From the late eighteenth century, too, knowledge of the world-wide distribution of many plant groups had reached a high enough level to allow the compilation of monographs devoted to a single family, genus or even species. In the face of the vast array of nineteenth-century botanical literature, the present selection concentrates on some of the most beautiful and on those which proved particularly useful or influential among botanists.

The nineteenth century saw what S. Peter Dance has called the Lithographic Revolution in natural history books, a revolution which produced the great animal illustrations of Edward Lear, John Gould and Joseph Wolf, as well as some botanical work of equal merit but lesser fame.

Over the previous centuries, the methods of reproducing illustrations had developed greatly in sophistication. Indeed, it is probably true to say that progress in botanical art between the sixteenth and nineteenth centuries is

Previous page:
Varnish tree (*Melanorrhoea usitata*). Hand-coloured lithograph by M. Gauci after a painting by Vishnupersaud from *Plantae Asiaticae Rariores* (1830–32) by Nathaniel Wallich. Waterproof varnish was extracted from this large tree, which is a native of Manipur and Burma. After the leaves have fallen in the dry season, the striking red bracts remain around the fruits on the otherwise bare tree.

A double-page spread from Fabio Colonna's *Phytobasanos sive Plantarum Aliquot Historia* (1592), the first botanical book illustrated with engravings on metal. The plants are a *Campanula* and a crucifer, possibly watercress.

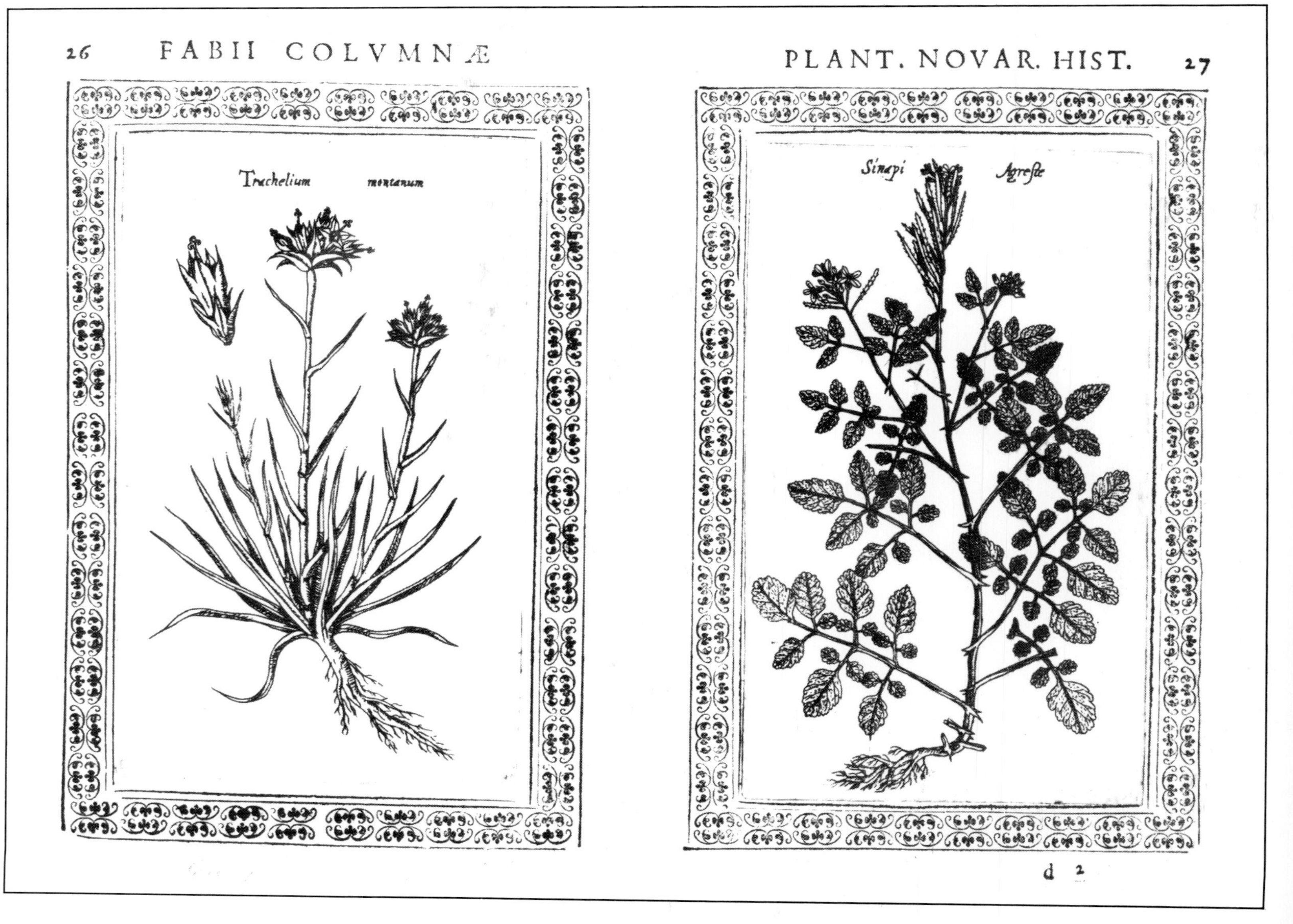

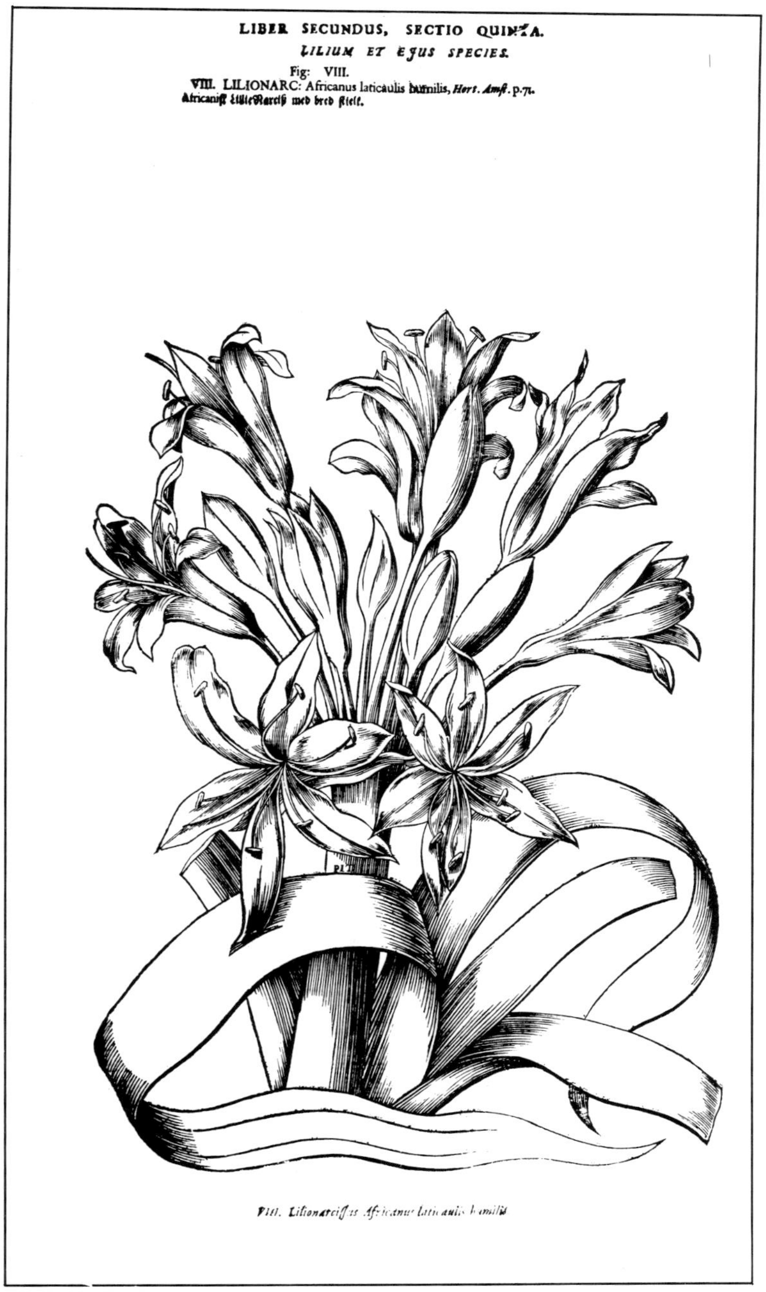

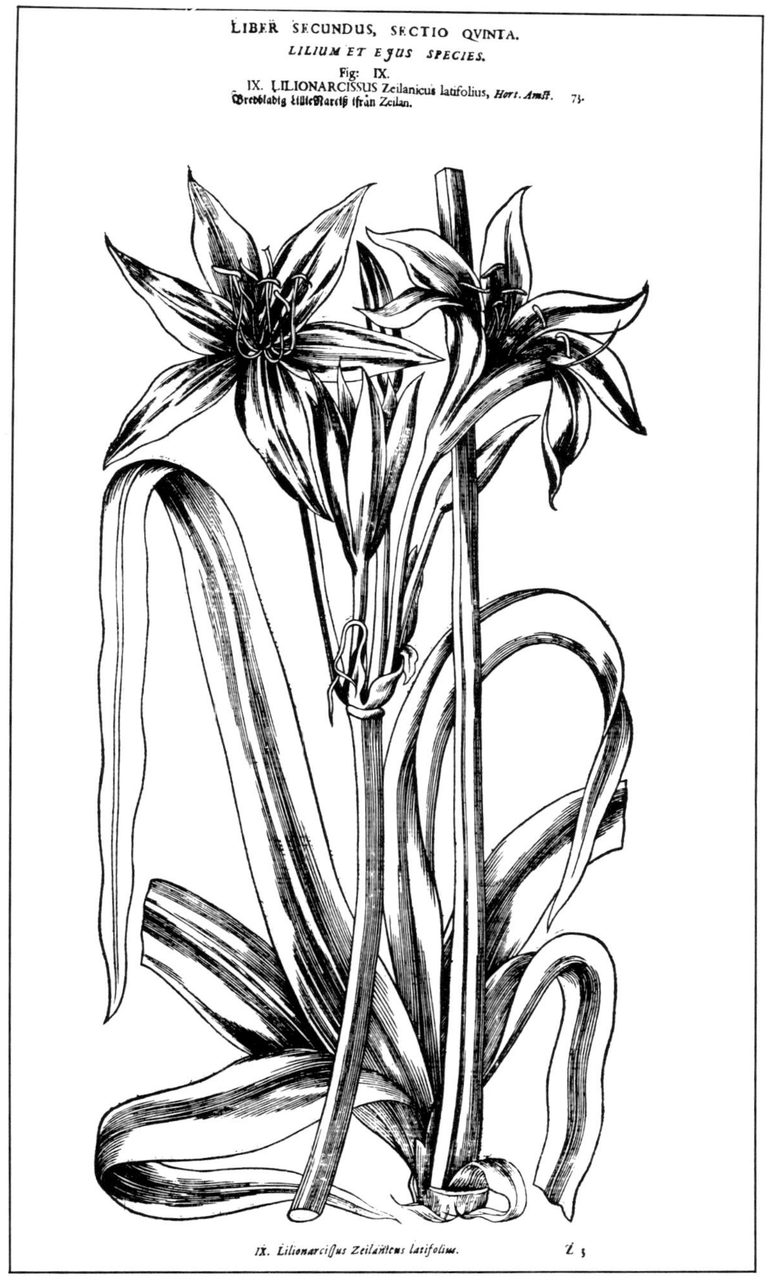

Two pages from a very late example of woodblock printing, *Campi Elysii Liber Secundus* by Olof Rudbeck, father and son, printed in Uppsala in 1701. The book was an ambitious undertaking and was to comprise twelve large volumes, of which the second was the first to appear, but the material for the book, including eleven thousand drawings and all except two copies of the first volume, were destroyed in a fire in 1702. The plants illustrated here are members of the genus *Crinum* in the family Amaryllidaceae.

An early plant engraving, showing roses and onions to illustrate the idea of 'Opposites', from *Symbolorum et Emblematum* (1590) by Joachim Camerarius the Younger.

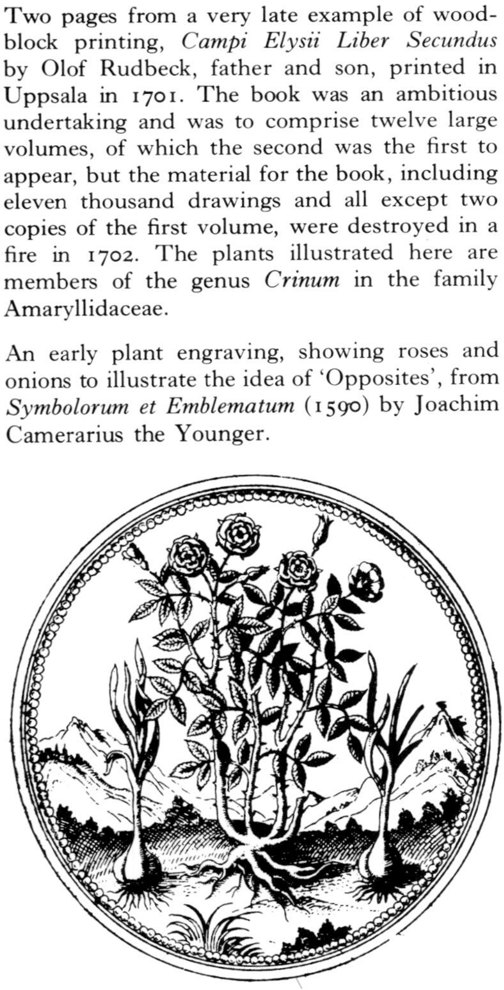

to be found in the techniques of reproduction rather than in the skill of the artists. After all, the plant studies of Dürer were produced at a time when the pictures in printed herbals were so crude as to be largely unrecognisable. Even eighty years later, when Ligozzi was producing botanical paintings that were the equal of the finest work of the eighteenth and nineteenth centuries, the best woodcuts in books could convey only a fraction of what an artist could put into his original. The first books with plant illustrations engraved on metal appeared in the 1590s, forty years later than the engravings of fish in Hippolyte Salviani's *Aquatilium Animalium* (1554–57). Engravings offer the possibility of finer line and greater detail than woodcuts, because the ink is held in indentations cut or etched with acid in the surface of the metal, where most of the surface of a woodblock has to be cut away, leaving the image standing out in relief – there is obviously a limit to the fineness of line which can be achieved in this way. Wood engraving, which was popular for much of the nineteenth century, particularly for popular books produced on long runs with small black and white illustrations, is a refinement of the woodcut using a surface cut across, rather than along, the grain of a very fine, hard wood such as box.

Throughout the seventeenth and eighteenth centuries, the techniques of engraving developed in skill and accuracy. An important innovation was the use of stipple engraving for shading; the stipple was made up of dots bitten into the surface of the plate with acid and could be manipulated to produce quite subtle tonal differences. To begin with, stipple was used in combination with line engraving, but some of the most exquisite plant engravings from late eighteenth and early nineteenth century France are in stipple alone.

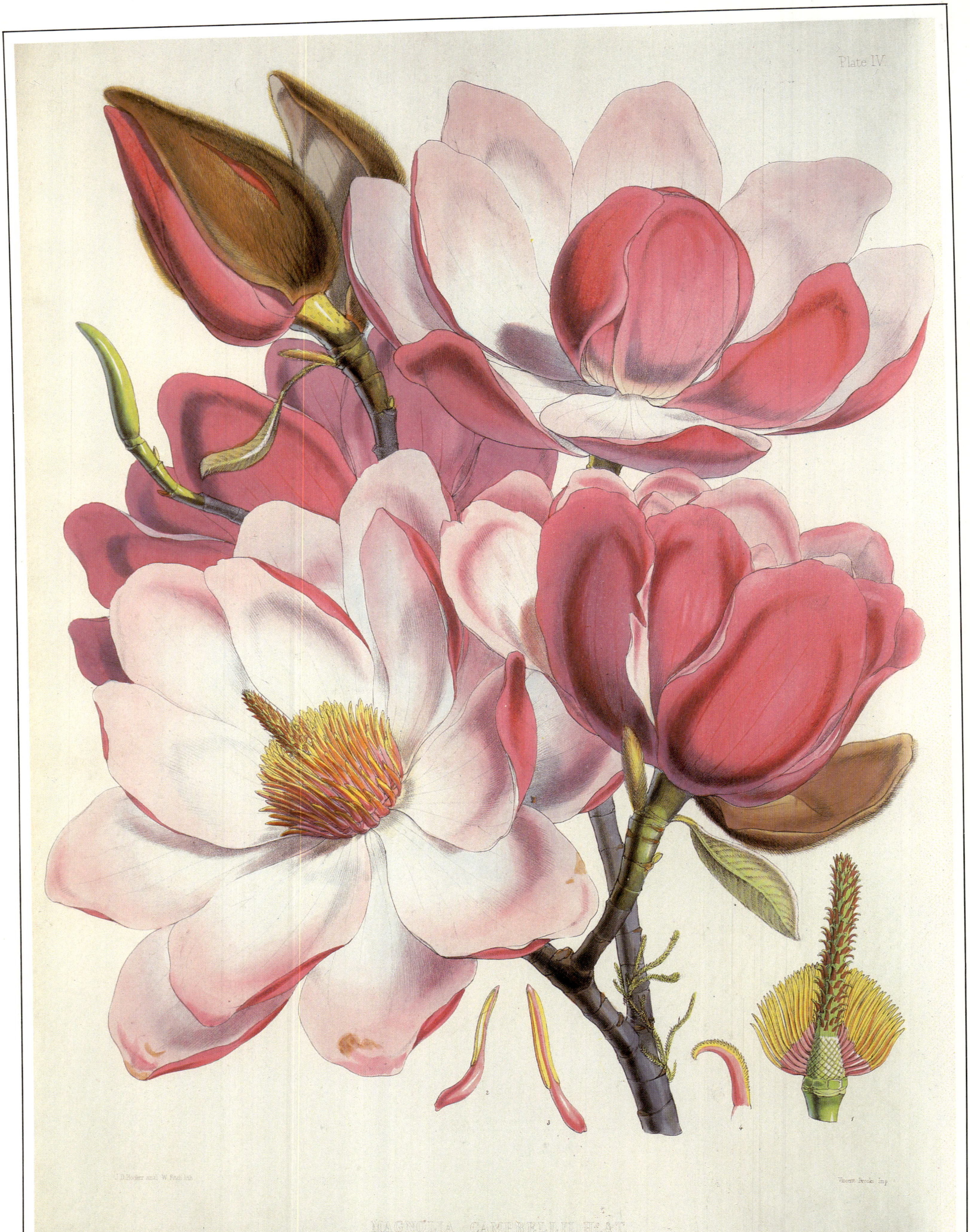

Magnolia campbellii, one of the most magnificent of all flowering trees. It has reached a height of seventy feet in suitable climates such as that of Cornwall and may be covered in flowers about ten inches across. Hand-coloured lithograph by W. H. Fitch after a painting made for J. F. Cathcart from *Illustrations of Himalayan Plants* (1855) by J. D. Hooker.

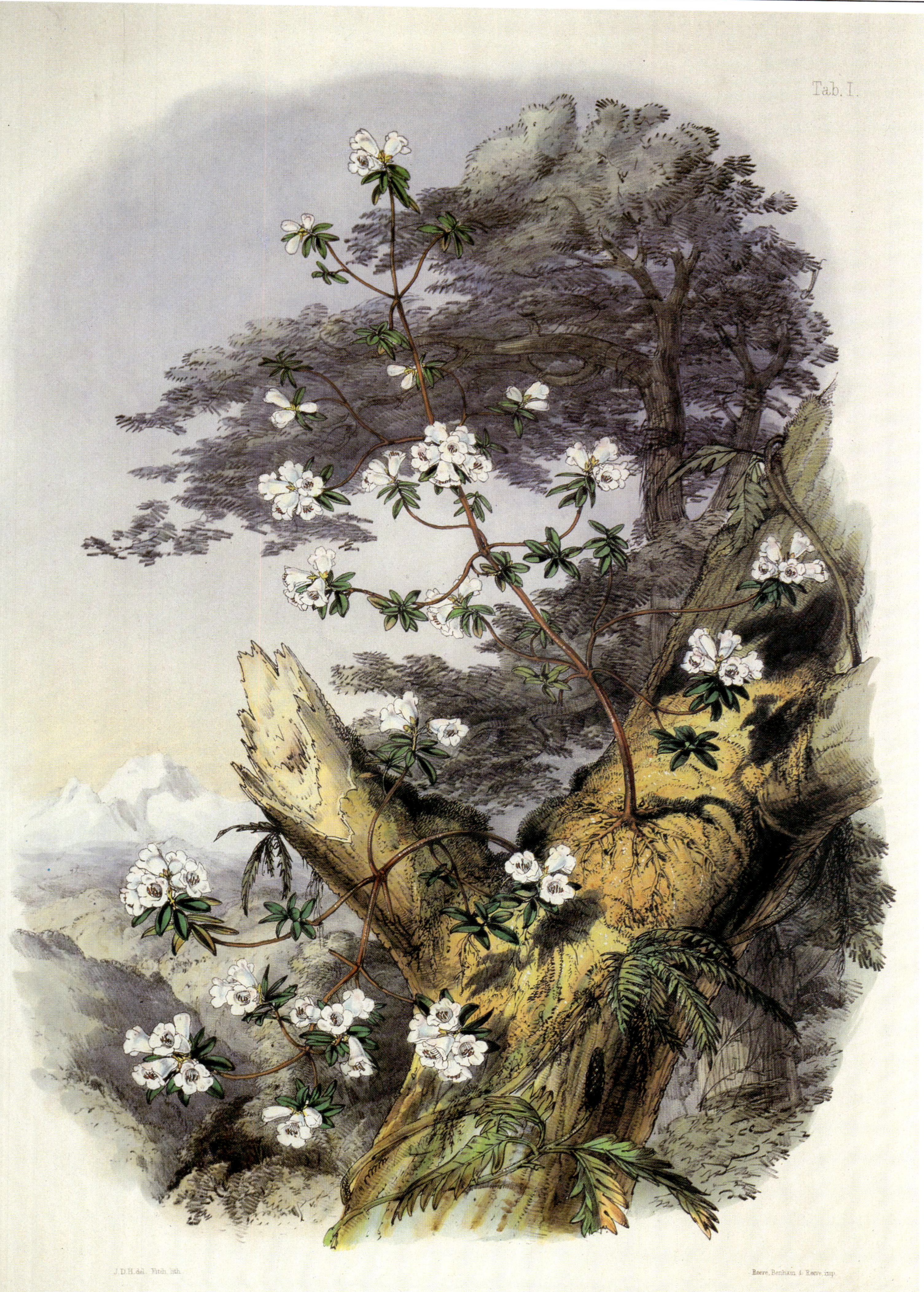

RHODODENDRON DALHOUSIÆ, Hook. fil.

(in its native locality)

This was the period when engraving reached its highest degree of perfection with the achievements of Ferdinand Bauer in England and of Sellier, Dien and their co-workers in France. A method of producing stipple engravings in colour was devised by Bartolozzi in England in the late eighteenth century. For this, coloured inks were applied directly to a single plate with a rag stump called a *poupée*. The actual dots of the print are thus in colour with white paper in between them, where hand-coloured engravings have watercolour washes applied over a monochrome image. In most cases, the two colouring techniques were combined, with watercolour being used to retouch and add to an already coloured print. Redouté, who had learned about colour printing in England, was very successful with it, particularly in some of the plates for the *Choix des Plus Belles Fleurs*.

The earliest colour-printed pictures of flowers, though, had appeared much earlier: a form of coloured mezzotint printed *à la poupée* was used in the *Historia Plantarum Rariorum* by John Martyn in 1728 and in the *Catalogus Plantarum* by a Society of Gardeners in 1730; mezzotint combined with hand-colouring is also one of the techniques that can be seen in *The Temple of Flora*. In the mezzotint process, the whole plate is roughened and areas are smoothed to various degrees, so that they hold less ink and print paler, or completely smoothed to produce white. Other plates in *The Temple of Flora* were printed by the aquatint process. In this, the plate is covered in powdered resin and heated so that, when it cools, its resin coating has a fine network of cracks through which acid will etch the plate; gradations of tone are produced by stopping the etching at different stages, using varnish to protect the pale areas from acid while the darker areas are further etched. The printed result is a very fine network of irregular lines building up various depths of tone. Neither mezzotint nor aquatint became popular for natural history illustration as both were more suitable where large areas of tone were required, for example in landscapes. However, aquatint was used very strikingly for John James Audubon's *Birds of America* (1827–38).

Lithography was discovered by a German, Alois Senefelder (1771–1874) in 1797. The image is drawn with a greasy crayon on a slab of dry, fine-grained porous limestone. If the stone is then wetted and inked with a roller, the ink adheres only to the greasy image and can be printed on to a sheet of paper. Very fine gradations of tone could be produced, and very delicate, pale prints could be made for hand-colouring. The first lithographs to be used in botanical illustration were by Rudolph Ackermann in *A Series of Thirty Studies from Nature* (1812). The first purely botanical book in the process was William Roscoe's *Monandrian Plants*, which was published in Liverpool between 1824 and 1828, rather later than the first lithographed zoological book, Karl Schmidt's *Beschreibung der Vögel* (1818). Lithography proved much more flexible than engraving, and a skilled lithographer could work very much faster than an engraver. In England, M. Gauci produced some of the earliest and most magnificent lithographed plates, notably for Nathaniel Wallich's *Plantae Asiaticae Rariores* (1830–32) and James Bateman's *Orchidaceae of Mexico and Guatemala* (1837–43). The most productive artist and lithographer of the period was W. H. Fitch. It is said that he often drew directly on the stone, straight from the plant, without making an intermediate drawing on paper. Certainly his output for W. J. and J. D. Hooker and for the *Botanical Magazine* and many other publications was stupendous.

Another process, nature-printing, had a brief period of popularity in the mid-nineteenth century. It used the actual plant in making the print, but was never really successful, either artistically or scientifically except in the case of ferns.

Nature-printing was first attempted in the seventeenth century, and in 1733 Johannes Hieronymus Kniphof published a folio of nature-printed plants. Printer's ink was applied to the plant itself, which was passed through a press with the paper. Thirty years later, he published a vast work with 1200 similar plates entitled *Botanica in Originale seu Herbarium Vivum*. The finest nature-printing, by Henry Bradbury for Thomas Moore's *Ferns*

Previous page:
Rhododendron dalhousiae, which grows epiphytically in its native localities in the Himalayas; it was named by J. D. Hooker in honour of Christina, Countess of Dalhousie (1786–1839). Hand-coloured lithograph by W. H. Fitch from a drawing by J. D. Hooker, the frontispiece for Hooker's *The Rhododendrons of the Sikkim Himalaya* (1849–51). Fitch produced the magnificent plates for this book from partly-coloured field sketches by Hooker which are now kept at the Royal Botanic Gardens, Kew.

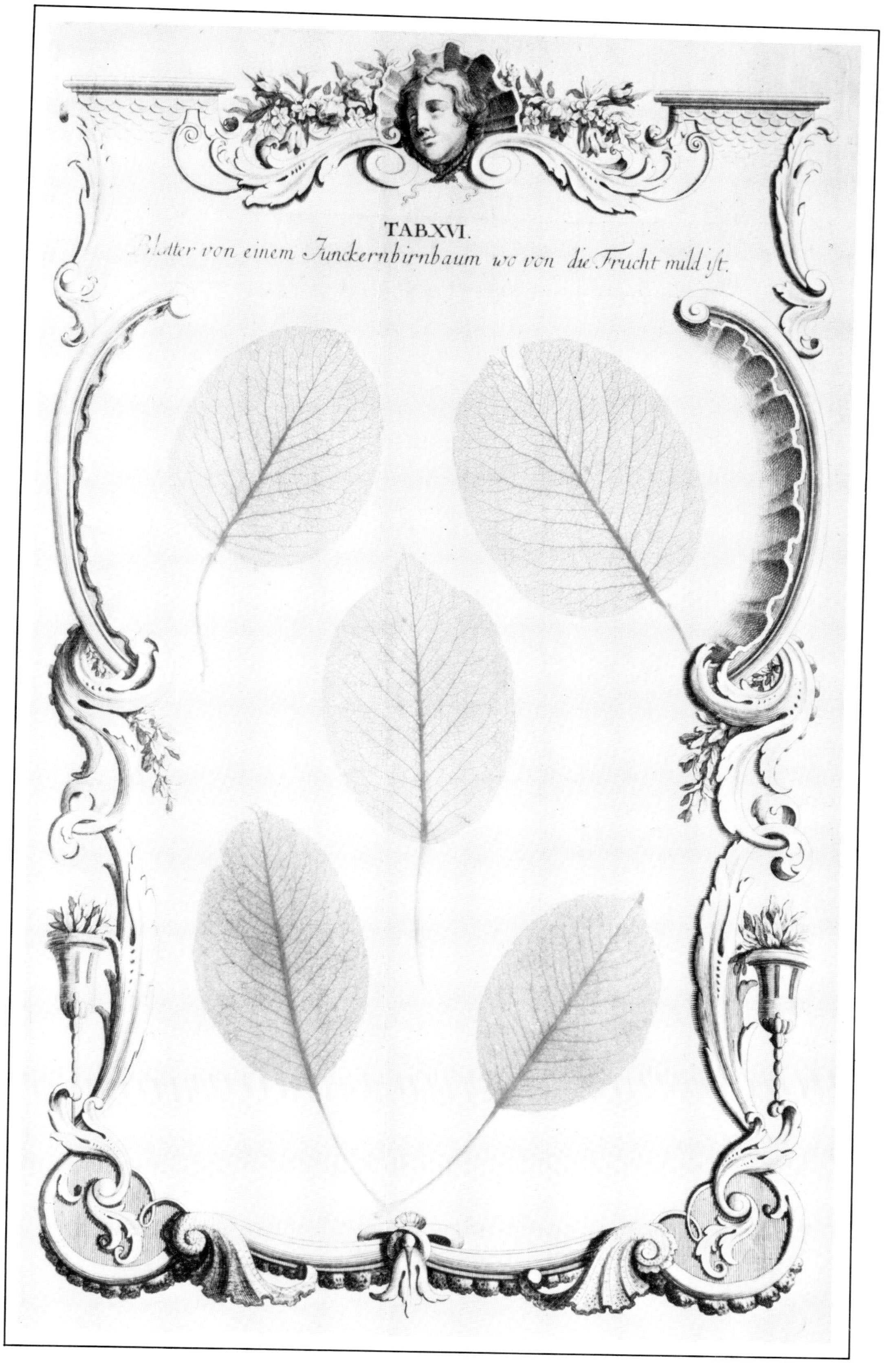

Early example of nature-printing from leaves in *Die Nahrungs-Gefässe in den Blättern der Bäume* (1748) by Johann Michael Seligmann. The leaves are printed in red and the engraved borders are in black.

of Great Britain and Ireland (1855), was done by taking an impression of the specimen on a soft lead plate, which was then copied by the electrotype process to produce a block for printing. Later, Bradbury produced *Nature-printed British Seaweeds* (1859) by the same method. It was ideal for showing the thin two-dimensional fronds of ferns and seaweeds, but its inflexibility prevented its being used more extensively, and the pressing required tended to distort the tissues, especially of more fleshy plants.

A somewhat similar process was used in the monumental work *Herbier de la Flore Française* by Louis-Antoine Cusin, published between 1867 and 1876. It consists of 26 volumes with about five thousand plates. The reproduction was *phytoxygraphique*: actual herbarium specimens were used to make a lithographic image, the enlarged details of the plants were drawn on stone, and the result hand coloured after printing. This process has most of the limitations of nature printing, but nevertheless the results are in some cases quite beautiful, for instance in Cusin's plate of butcher's broom (*Ruscus aculeatus*).

Next page:
Durian (*Durio zibethinus*), a tropical fruit that is well known for its vile smell but delicious taste; it is mainly found in Malaysia and Indonesia. Chromolithograph by G. Severeyne from *Fleurs, Fruits et Feuillages Choisis de la Fleur et de la Pomone de l'Ile de Java, peints d'après Nature par Madame Berthe Hoola van Nooten* (1863–64).

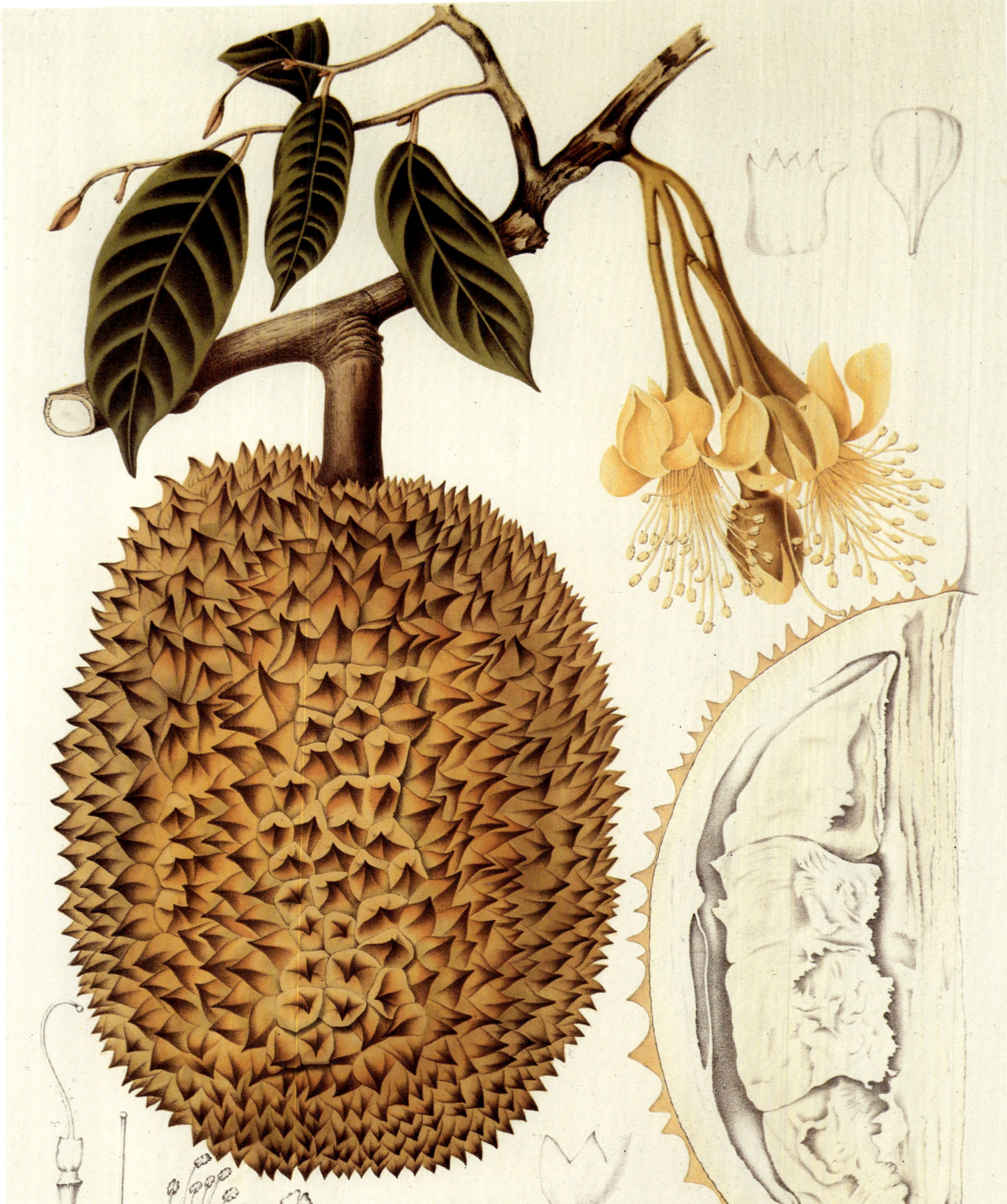

Peint d'après nature par Mme Berthe Hoola van Nooten, à Batavia.

Chromolith par G. Severeyns lith. de l'Acad. Roy. de Belgique.

DURIO ZIBETHINUS. L.

Emile Tarlier éditeur, à Bruxelles.

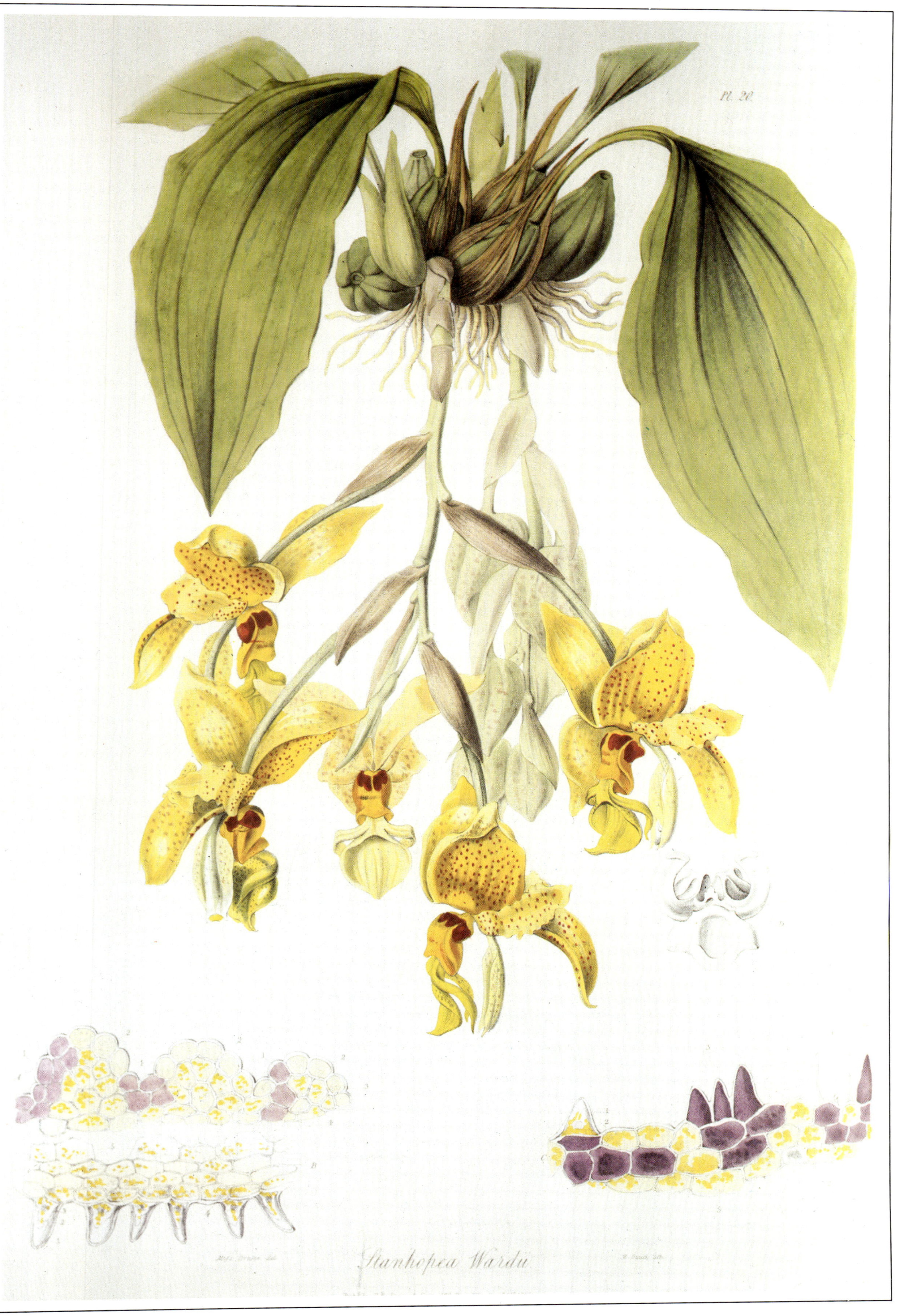
Pl. 20
Stanhopea Wardii

CHAPTER XII

Victorian India

The early European botanists in India had concentrated on the tropical flora of the lowlands and the south; they did not penetrate into the Himalayas. It was Nathaniel Wallich (1788–1854) who first got some idea of the unparalleled floral richness of Nepal and the countries on the south side of the Himalayan range. Even then, most of the plants he illustrated were from the sub-tropical Tarai jungle. His work on the Nepalese flora, *Tentamen Florae Napalensis Illustratae* was never completed – only two parts containing fifty lithographs were published in 1824 and 1826. Wallich was a Dane who went to India in 1807 as surgeon to the Danish settlement of Seranpur in Bengal. In 1815, on the retirement of William Roxburgh's successor, Dr Hamilton, he was put in charge of the East India Company's botanic garden in Calcutta, a position he held for the next thirteen years. He spent 1820 in Nepal, 1825 in the Oudh region on the Nepalese border and 1826–27 on a long sea voyage for the sake of his health, during which he visited Penang and Singapore and went on a trip up the Irawaddy in Burma. In 1828, he retired to London to study his collections and produce his *magnum opus*, the *Plantae Asiaticae Rariores* of 1830–32, a large folio with three hundred beautiful hand coloured lithographs by Gauci after paintings by C. M. Curtis, Miss Drake,

Previous page:
Stanhopea wardii. Hand-coloured lithograph by M. Gauci after a painting by Mrs Augusta Withers from James Bateman's *The Orchidaceae of Mexico and Guatemala* (1837–41). The orchid, a native of Venezuela and Guatemala, was named in honour of Earl Stanhope (1781–1855).

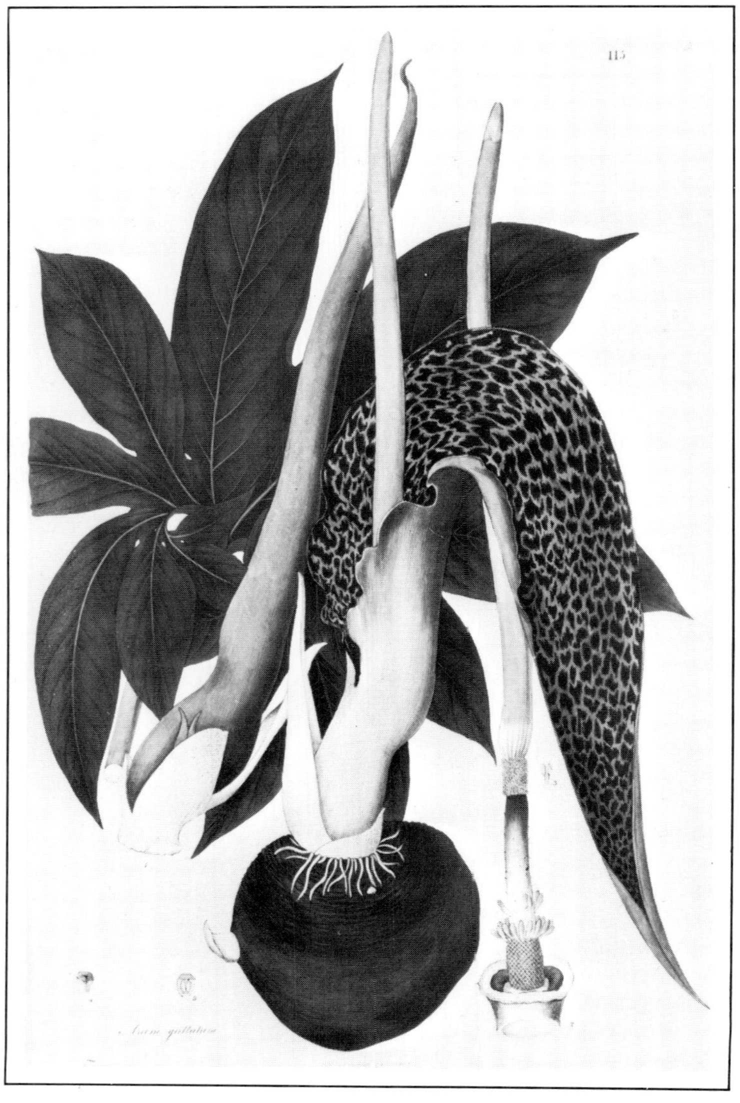

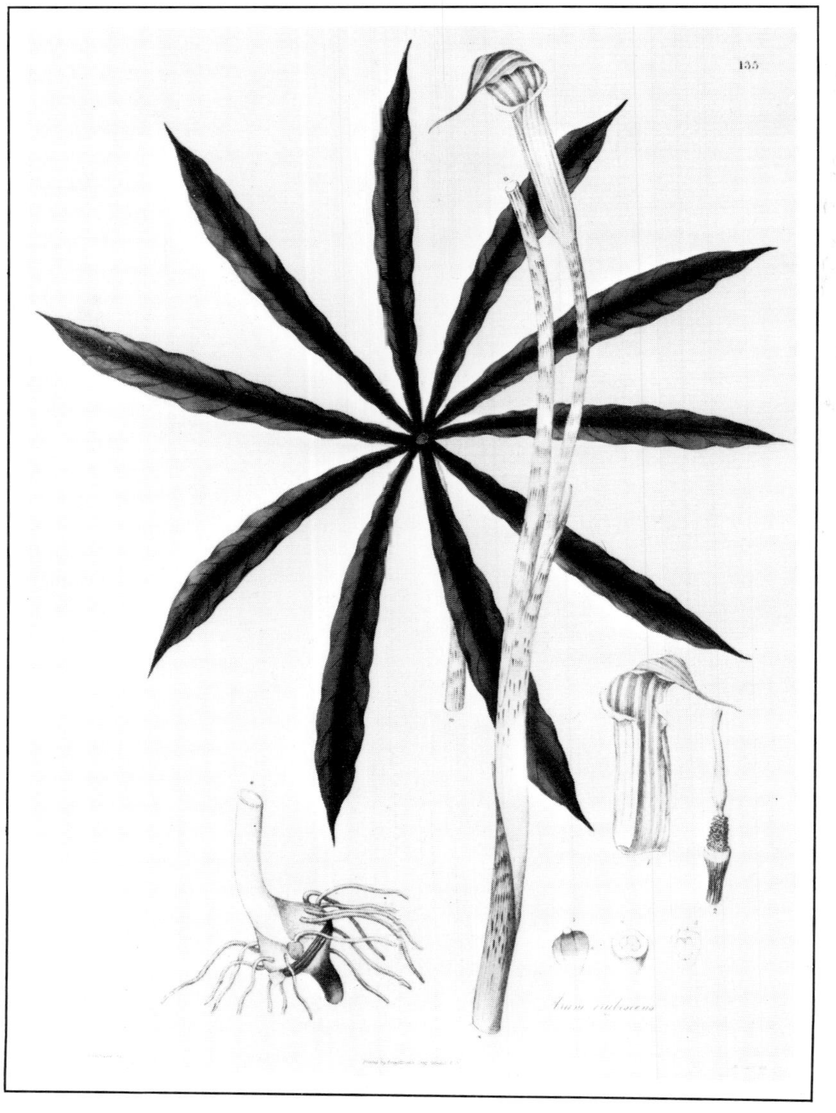

'*Arum guttatum*' and '*Arum erubescens*', both of which are now referred to the genus *Arisaema*. Hand-coloured lithographs by M. Gauci after paintings by, respectively, Vishnupersaud and Gorachand from *Plantae Asiaticae Rariores* (1830–32) by Nathaniel Wallich.

and the Indian artists Gorachand and Vishnupersaud. The first plate is the flowering tree *Amherstia nobilis*, which had just been discovered by a Mr Crawford near Martaban in Burma and had been introduced into cultivation by Wallich himself. The name commemorates the wife and daughter of Lord Amherst, Viceroy of India, who were both keen and energetic naturalists and collectors.

A contemporary of Wallich was John Forbes Royle, who was born in Cawnpore, studied medicine at Edinburgh University and became surgeon to the East India Company in Bengal in 1820. Three years later, he was made Curator of the Botanic Garden at Saharanpur, north of Delhi, and was able from there to undertake a special study of the Himalayas. *Illustrations of the Botany and other Branches of the Natural History of the Himalayan Mountains and of the Flora of Cashmere* (1833–39) was published after Royle's return to London. It is a pioneering ecological study, with descriptions of the different types of vegetation and notes on the distribution of families and genera in the region. There are a hundred illustrations of the mountains and of plants lithographed by Gauci after paintings by Vishnupersaud, Curtis, Miss Drake and James de Carle Sowerby. Royle remained in London until 1856 as Professor of Materia Medica at King's College, as well as being for a time Secretary of the Horticultural Society.

While Wallich and Royle were working mainly in northern India, Robert Wight was based in Madras. His six-volume flora, *Icones Plantarum Indiae Orientalis; or Figures of Indian Plants*, was published between 1840 and 1856. It was intended that it would appear in monthly numbers of 20 plates each, at the very cheap price of 2 rupees. 'My wish,' wrote Wight, 'is to diffuse as quickly and as extensively as possible a knowledge of Indian plants, by publishing as many as possible in the shortest period of time, and at the lowest charge.' This and smaller publications that make up his *Spicilegium Neilgherrense; or A Selection of Neilgherry Plants* are still the most useful sources of illustrations of south Indian plants. The lithographs are clear and simple, after drawings by Wight himself.

Bombax insigne (after a painting by Vishnupersaud) and *Dischidia rafflesiana*, a climber found in hot, wet forests; it is similar to *Hoya* but has both normal leaves and strange, hollow ones which make nests for ants. Hand-coloured lithographs by M. Gauci from *Plantae Asiaticae Rariores* (1830–32) by Nathaniel Wallich.

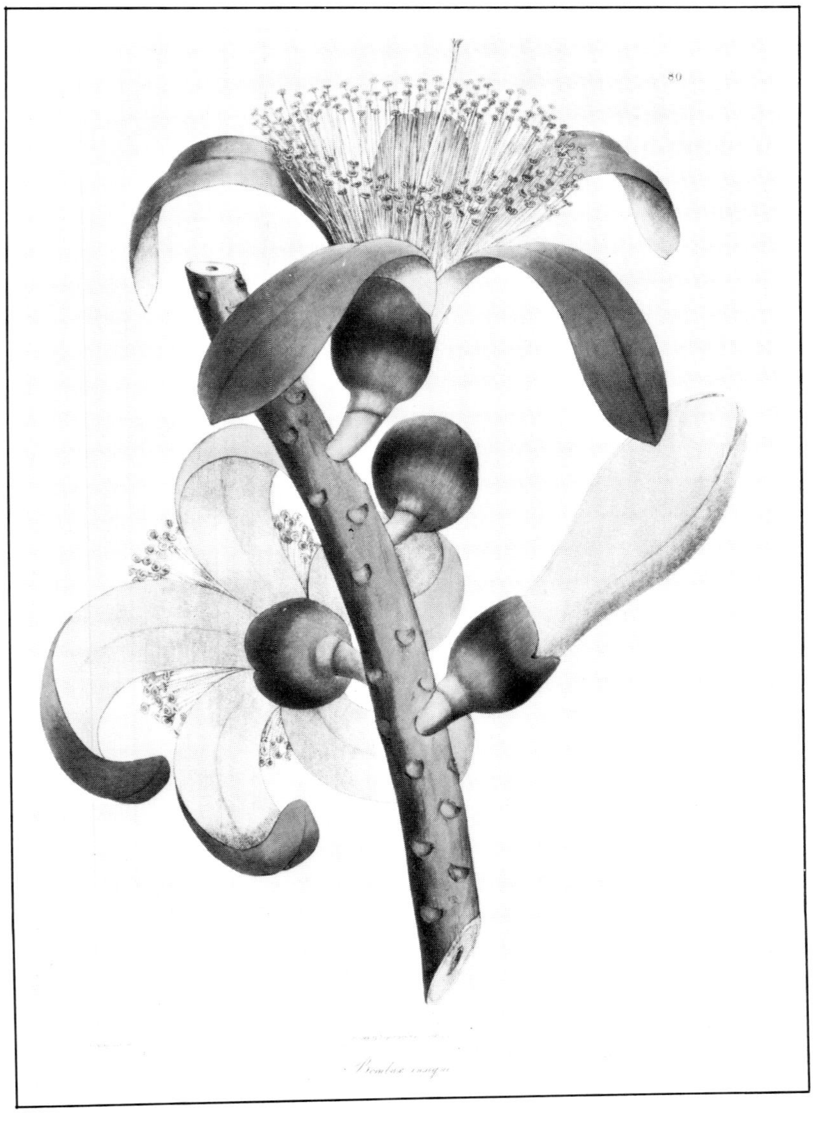

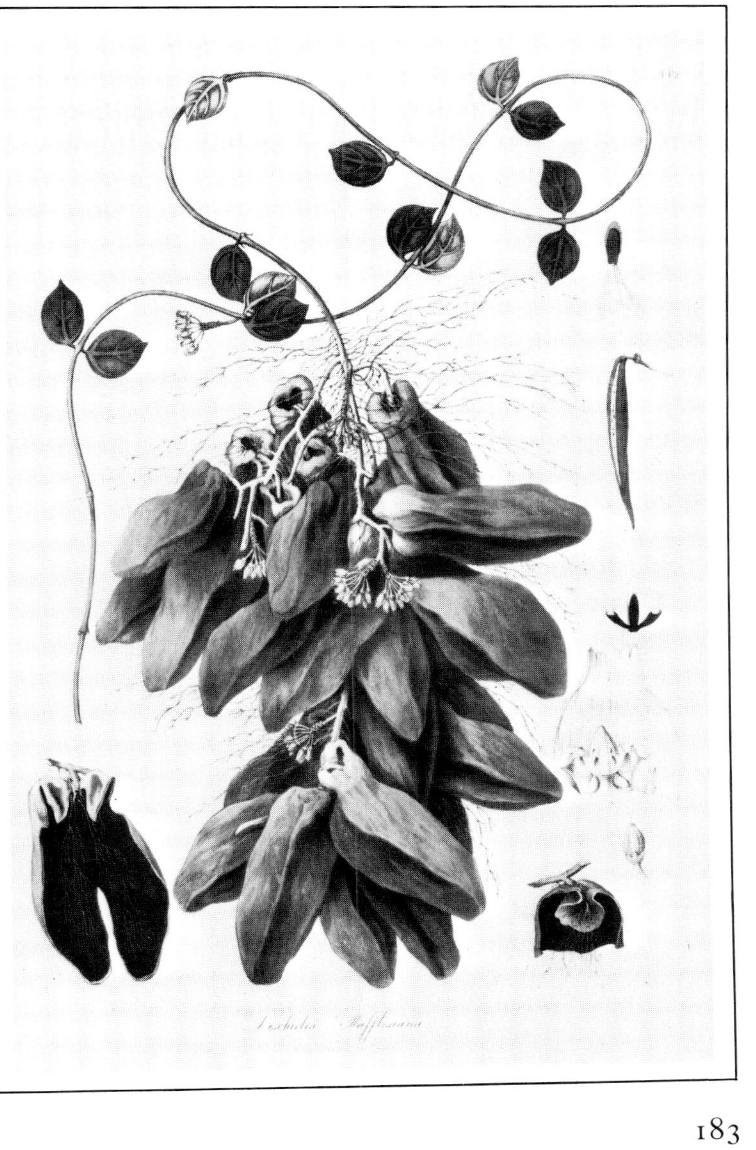

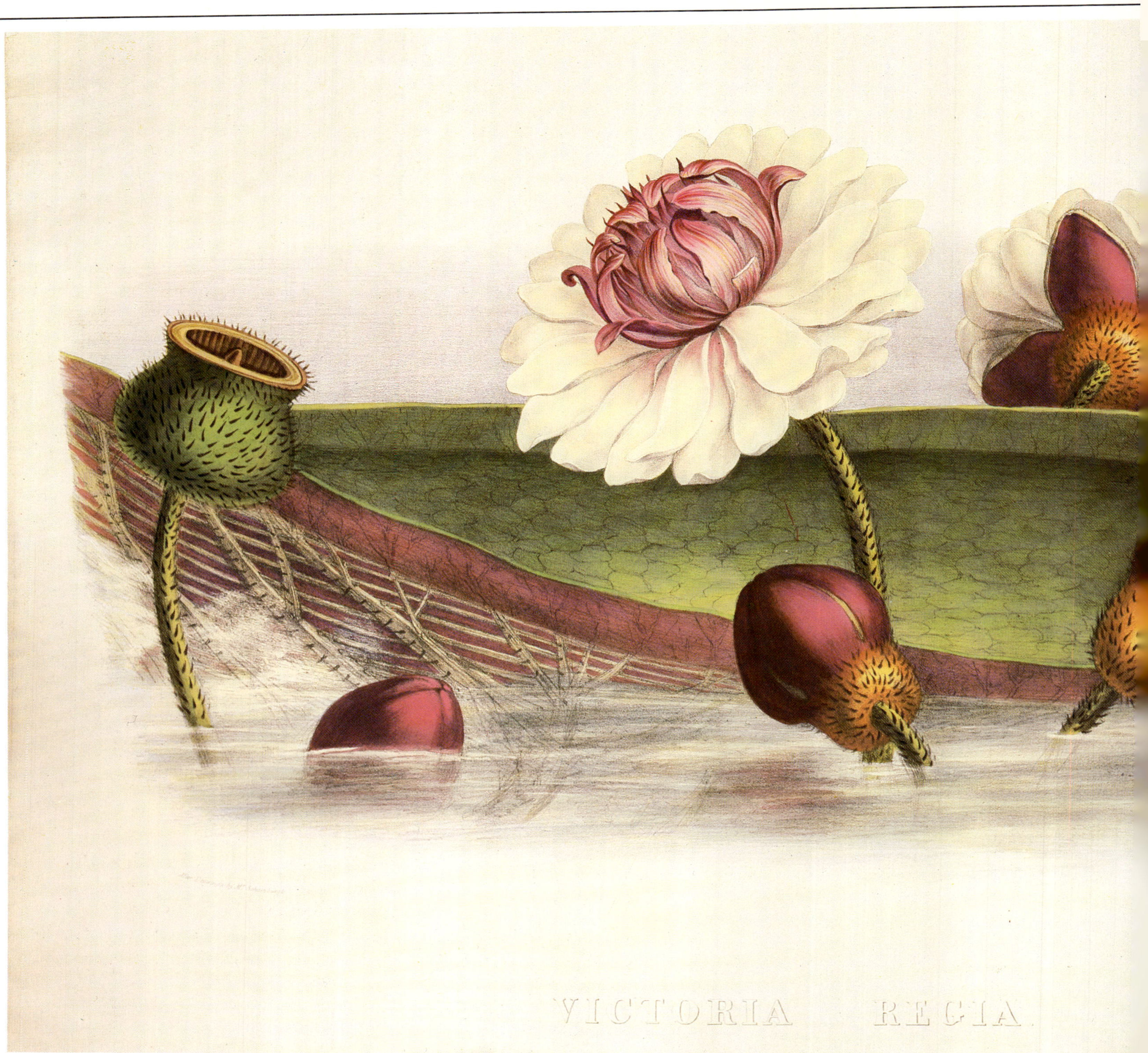

Although India was at this time becoming healthier, not all botanists survived to write their own books. Among them was the young Frenchman, Victor Jacquemont. He had left France to escape from a sad love affair, and was soon invited by the Jardin des Plantes to collect for them in a country of his own choice. He chose India and especially the Himalayas and, arriving in 1828, visited Royle in Saharanpur before going to Simla and up the Sutlej valley into Tibet. He returned to Simla for the winter and there obtained an invitation from the Maharajah of Kashmir to visit his country. He spent 1831 in Kashmir, visiting many places previously unexplored. He had discovered many new species before he died of cholera in Bombay in 1832, aged only 31. His collections were returned to France and the results were published under the title *Voyage dans l'Inde* between 1835 and 1844, with very fine elegant line engravings by Riocreux and E. Deline. Jacquemont's journey had an important influence on J. D. Hooker, who followed him ten years later and named several plants after him.

Another posthumous publication was that based on the plants collected by Dr Werner Hoffmeister, who accompanied Prince Waldemar von Preussen to India and Ceylon. The party arrived early in 1845, and their travels took them to Katmandu; they went on to visit the source of the

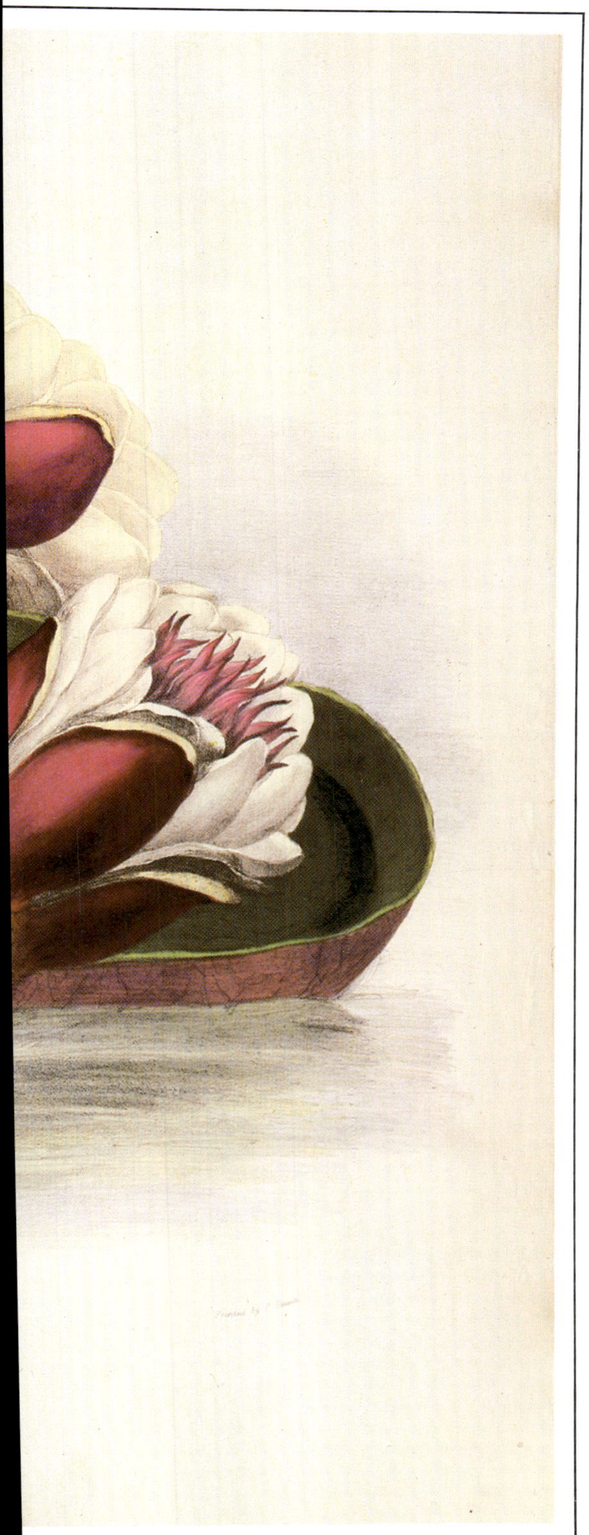

Victoria regia. Hand-coloured lithograph by P. Gauci after a painting by Robert Schomburgk from *Notice of Victoria regia* (1837) by John Lindley.

Ganges and the border of Tibet, returning through Simla. They arrived back in the midst of a Sikh war and were observing the battle of Ferozeshalir in the Punjab on 21st December, when the unfortunate Dr Hoffmeister was killed by grapeshot as he rode at the Prince's side. Many of the plants he found were described as new species, but few of them are upheld today. However, the lithographs by C. F. Schmidt which illustrate the account of his journey are excellent, both artistically and in botanical detail and accuracy.

The most important botanical traveller to Victorian India was J. D. Hooker, whose journeys resulted in the two most beautiful and famous books on the Himalayan flora. Joseph Dalton Hooker (1817–1911) was the son of William Jackson Hooker (1785–1865), an expert on ferns who was Professor of Botany at Glasgow University and later first Director of Kew Gardens. Joseph Hooker was brought up in an atmosphere of botany in Glasgow and Kew. At the age of 22, he sailed on the *Erebus* on an expedition to the Antarctic, under the command of Sir James Ross, which lasted from 1839 to 1843. This voyage had left Hooker with a particular interest in plant geography. He was now anxious to compare the floras of the tropics and especially the mountains with those of Europe and the Antarctic. His choice lay between the Andes and the Himalayas, and finally he was persuaded to visit the Himalayas by Dr Falconer, superintendent of the Calcutta Botanic Garden, who recommended Sikkim as being almost unknown and likely to be botanically very rewarding. Hooker obtained a government grant of £400 per annum for two years and this was extended for a third. He arrived in Calcutta in January 1848 on the same ship as the Marquess of Dalhousie, Governor General of India (1812–60), whose friendship was later very helpful to him. His first trip was into central India, and in April he went to Darjeeling, which was then becoming popular with the British as a hill station. There Hooker met J. F. Cathcart of the Bengal Civil Service, a keen amateur botanist, and Dr Archibald Campbell, superintendent of the sanatorium and later the Political Agent. Cathcart, a semi-invalid, could not go collecting in the hills himself, but employed local collectors to bring in plants, and artists to draw them. As Hooker wrote to his father in April 1850, 'Cathcart's drawings, I have surely explained, are all now going on with being made at or near Darjeeling, of Darjeeling plants. He has five artists at work, who turn out together about three plants a week – it costs him more than all my pay together.' These drawings eventually numbered more than a thousand. Hooker promised to have some of them published, which he did, but not before Cathcart had died suddenly in Switzerland on his way home in 1851. *Illustrations of Himalayan Plants, selected from drawings made for the late J. F. Cathcart Esq. of the Bengal Civil Service*, which appeared in 1855, is one of the most beautiful of all botanical books. The hand-coloured lithographs are by W. H. Fitch, who also redrew most of them, 'correcting the stiffness, and adding botanical details.' Some of the plates were by Hooker, himself no mean botanical artist. Among the interesting plants shown are *Buddleia colvilei*, the largest-flowered of the buddleias and *Rheum nobile*, a large rhubarb with strange hooded bracts.

The other book that came out of Hooker's journey was *The Rhododendrons of the Sikkim Himalaya*, published between 1849 and 1851; the first part appeared while Hooker was still in Sikkim, the proofs having been sent out to him in Darjeeling. It contains 30 folio plates of rhododendrons, drawn and lithographed by Fitch after field sketches by Hooker himself.

Before Hooker visited the Himalayas, only 32 rhododendron species were known from the whole world. By the time Hooker's monograph was complete, 43 species were known from India alone – and in a single day Hooker recorded collecting seed of 24 species. His first journey into the mountains was a short one, between October 1848 and January 1849. As permission to visit Sikkim had proved difficult to obtain, he started westwards from Darjeeling, and then north into Nepal along the Tamur river, as far north as the watershed at the Wallanchoon Pass on the border with Tibet. Here he found dwarf rhododendrons and *Cassiope*, which he compared to the heather

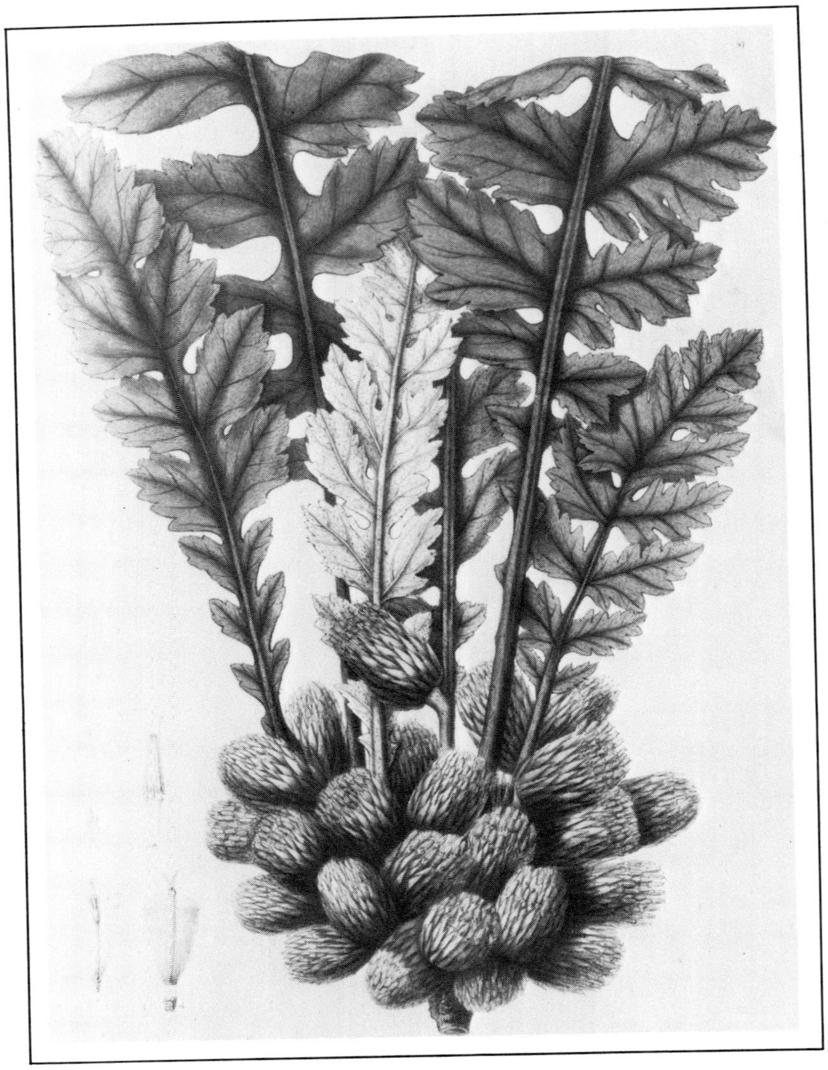

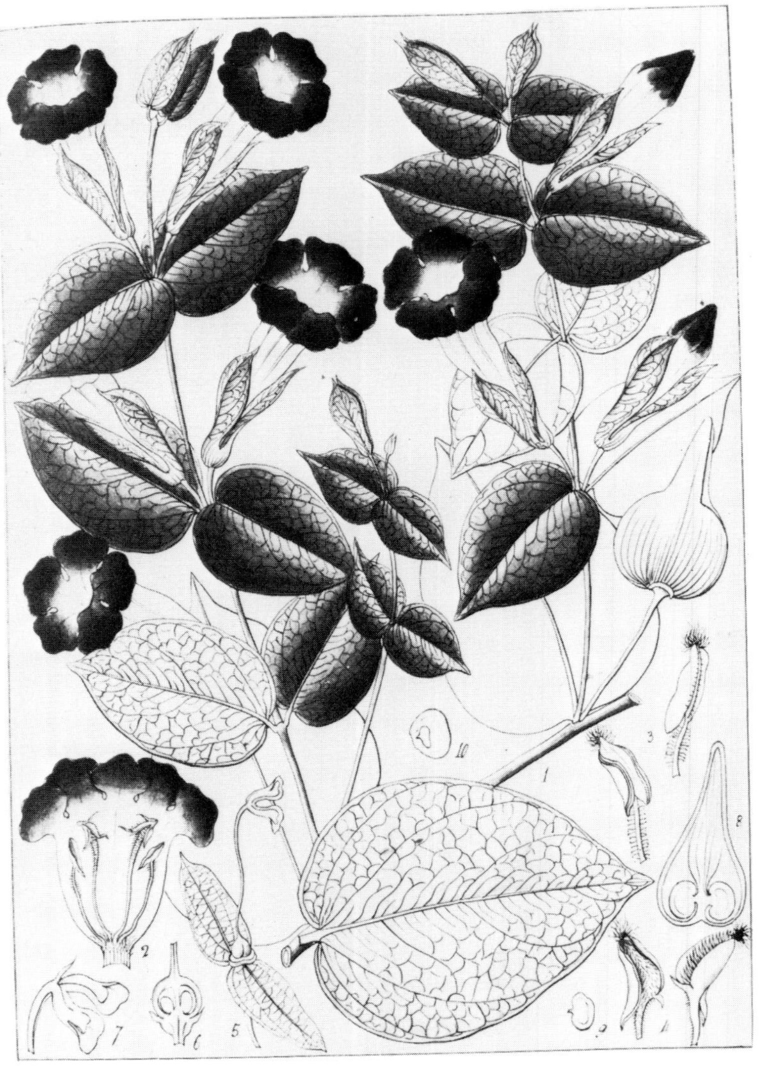

on the moors of Scotland, together with an assemblage of familiar European weeds, the grass *Poa annua*, shepherd's purse (*Capsella*), a bitter cress (*Cardamine hirsuta*) and the toad rush (*Juncus bufonius*), all 'evidently imported by man and yaks, and, as they do not occur in India, I could not but regard these little wanderers from the north with the greatest interest.' He also found ancient lake beds, moraines and signs of much more intensive glaciation in the past, and large herds of the central Asian wild sheep (*Ovis ammon*), which are about the size of a small pony.

Hooker then returned southwards before crossing into central Sikkim, where he met Dr Campbell. They managed to obtain an interview with the Rajah despite difficulties put in their way by the Dewan (or Regent).

The source of many of Hooker's troubles, the Dewan was Tibetan and strongly anti-European. He hoped to enrich himself by monopolising trade in the country and keeping out the British. Relations between the Rajah and the colonial government had at first been good – and he had sold them the land on which to build Darjeeling – but he was now old and dominated by the Dewan.

On the second year of his visit, Hooker started for the high parts of Sikkim on 3rd May, having received permission from the Rajah in spite of the Dewan's efforts to delay him. He had been travelling only a week when the rains began, greatly increasing the discomfort of the journey. Leeches crawled into his hair, hung on his eyelids, and climbed up his legs and down his back; he used snuff as a remedy, and rolled tobacco leaves around his feet. He also had to suffer the stinging flies, the humid heat in the valleys and continuous difficulties put in his way by the Dewan's agents; the villagers had been forbidden to supply him with food, the roads had been left unrepaired and the story had been spread around that if he fired his gun it might bring down rain and ruin the crops (a crafty plan by the Dewan to prevent him supplementing his provisions with game).

As he travelled slowly up the Hachen valley, the flora included a gradually increasing number of temperate plants and fewer tropical ones, until at about 9000 feet the vegetation became entirely temperate. Many of the

Left:
Dolomioea macrocephala. Hand-coloured lithograph by M. Gauci after a painting by Vishnupersaud from *Illustrations of the Botany and other branches of Natural History of the Himalayan Mountains and of the Flora of Cashmere* (1833–39) by John Forbes Royle.

Above:
Meyenia hawtaniana. Hand-coloured lithograph by Dumphy after a painting by Govindoo from *Spicilegium Neilgherrense; or a selection of Neilgherry Plants* (1846–51) by Robert Wight, published in Madras.

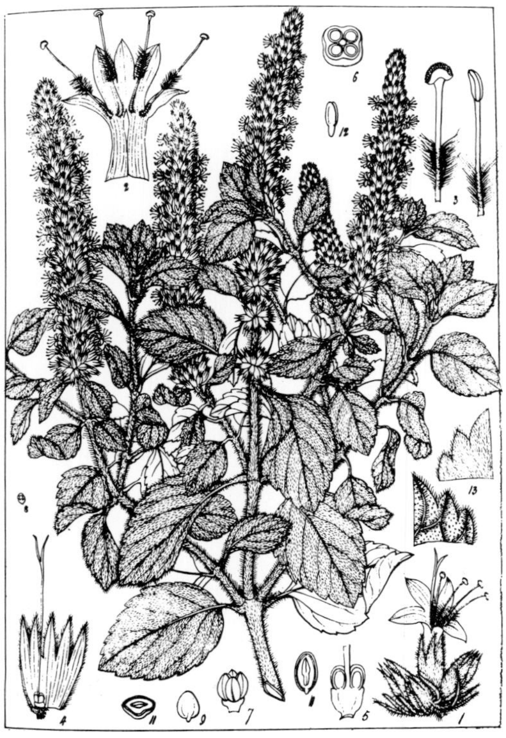

specimens he had collected were lost through damp, but the plants were wonderful and all were in flower. He did not succeed in reaching the Tibetan border that summer, but returned to central Sikkim, where he met Dr Campbell again on 2nd October. Together they reached Tibet on 16th October and negotiated permission to remain in the country for a day. Hooker found the rocks covered with an orange lichen, the species that he had seen colouring the rocks in the Antarctic. Returning to Sikkim by the Dongkya La pass, they went south to the capital, Gangtok; prevented from meeting the Rajah, they set out again eastward to visit the Yak La and Cho La passes.

It was on 7th November, when they were camped below the Cho La that Campbell was assaulted and captured by Sikkimese soldiers on the orders of the Dewan. He was taken to the capital to be used as a hostage in obtaining concessions from the British. Hooker himself was not molested, but followed behind quietly collecting seed. The negotiations for Campbell's release were delayed because the letter which contained the demands was put aside for Campbell to deal with on his return. When at last Hooker was allowed to write, he addressed his letter to Lord Dalhousie who was away from Calcutta at the time. His return produced swift action: an army was sent with a demand for Campbell's release. Hooker and Campbell finally arrived back in Darjeeling on Christmas Eve. The Dewan was disgraced and the Rajah lost the lower part of Sikkim, which was annexed by the British.

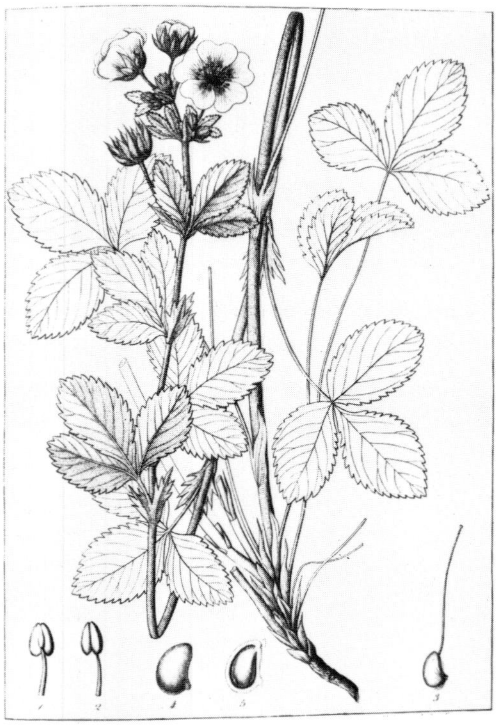

Potentilla jacquemontiana. Engraving after a drawing by E. Delile from *Voyage dans l'Inde, Atlas* (1844) by Victor Jacquemont.

Hooker could not visit Bhutan, which was closed to Europeans; his next journey was to the Cachar Hills with Dr John Thomson, a friend from his Glasgow days. Hooker's best find here was the blue orchid, *Vanda caerulea*, which grew in such quantity that he needed seven porters to carry the plants he collected. Hooker and Thomson went as far as the border of Manipur in the southeast and then back to Calcutta in January 1851 for Hooker to return to England.

Of all Himalayan expeditions, Hooker's was perhaps the most important because of the extraordinary variety of flowers that he found, many of which were brought into cultivation in England. Many of his new rhododendrons became familiar features of gardens in mild temperate climates; some, like *Rhododendron thomsonii, falconeri* and *cinnabarinum*, were introduced on this trip. One of the subtropical species, *Rhododendron dalhousiae*, named in honour of Lady Dalhousie, Hooker found growing epiphytically on the trunks of oaks and of *Magnolia campbellii*. The account of his travels, *Himalayan Journals*, was published in 1854 and immediately became a best seller; his rhododendron introductions began the great vogue for the genus which still continues. However, it was not until the twentieth century and the journeys to western China of Wilson, Forrest and Rock that the huge size and diversity of the genus was fully recognised.

After Hooker's work, which culminated in his unillustrated seven-volume *Flora of British India* (1875–97), the centre of Indian botanical research was in Calcutta. Accounts of many genera were published as parts of the *Annals of the Royal Botanical Garden of Calcutta*, with many very fine illustrations by the team of Indian artists at the garden. Hooker himself was responsible for *A Century of Indian Orchids* published in 1895 with a hundred lithographs, partly hand-coloured, after paintings by Lutchman Singh, G. C. Bass and others from the collection in Calcutta Botanic Garden. Accounts of monkshoods (*Aconitum*), louseworts (*Pedicularis*), figs (*Ficus*), bamboos and many other groups were published at the end of the nineteenth century. The finest of the series is *The Orchids of the Sikkim Himalaya* by King and Pantling. Sir George King was Director of the Calcutta Botanic Garden and Pantling was superintendent of the Government's chinchona (quinine) plantation. 485 species were illustrated by hand-coloured lithographs after drawings by Pantling himself.

A name that is often mentioned in connection with botanical books on India and, indeed, with Victorian botany in general is that of Walter Hood Fitch (1817–92), who was responsible for the lithographs and many of the original drawings for books by the Hookers. When he was discovered by

Left:
Pogostemon rotundatum. Lithograph by Dumphy after a drawing by Govindoo from *Icones Plantarum Indiae Orientalis; or Figures of Indian Plants* (1840–56) by Robert Wight, published in Madras.

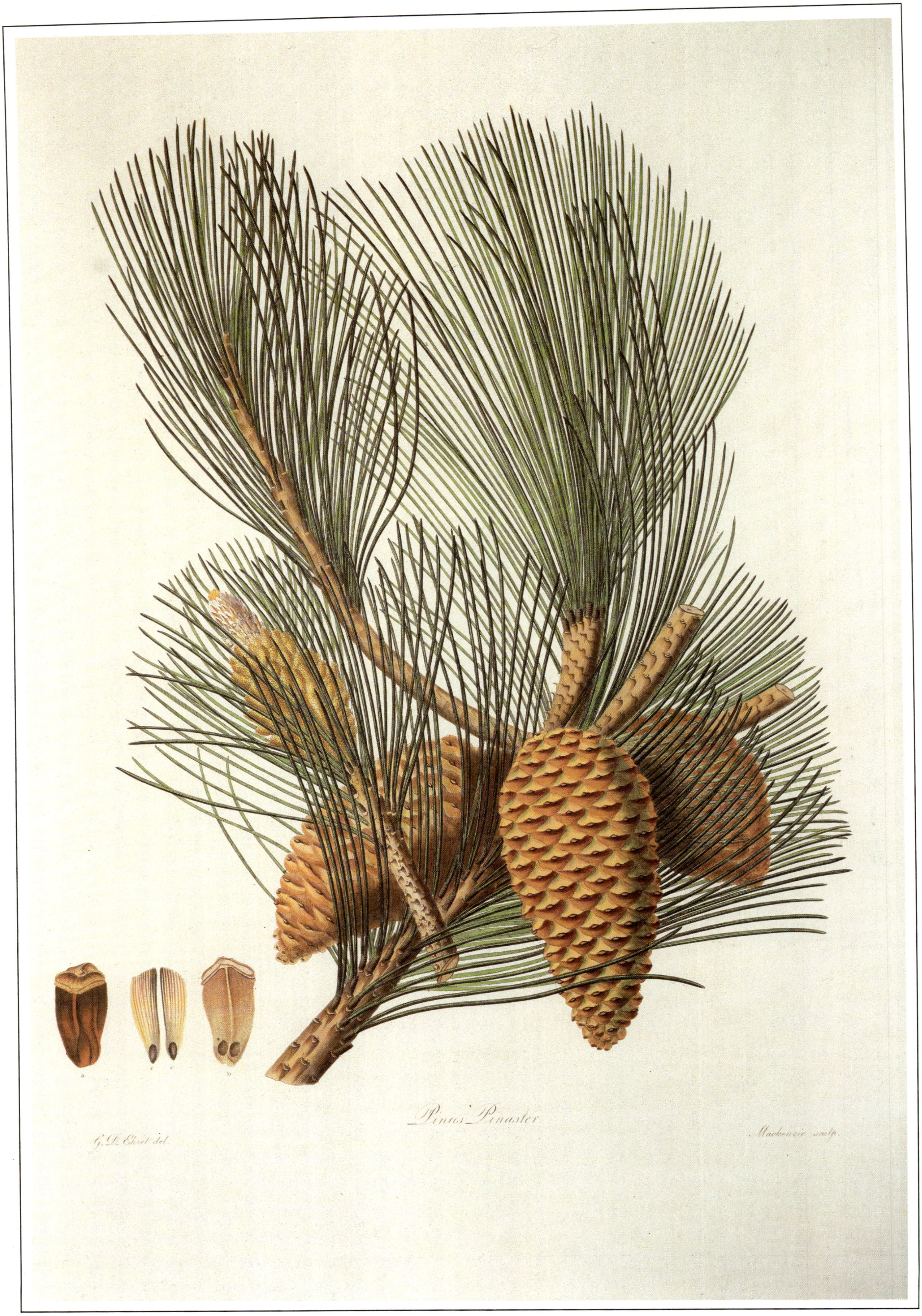
Pinus Pinaster
G.D. Ehret del.
Mackenzie sculp.

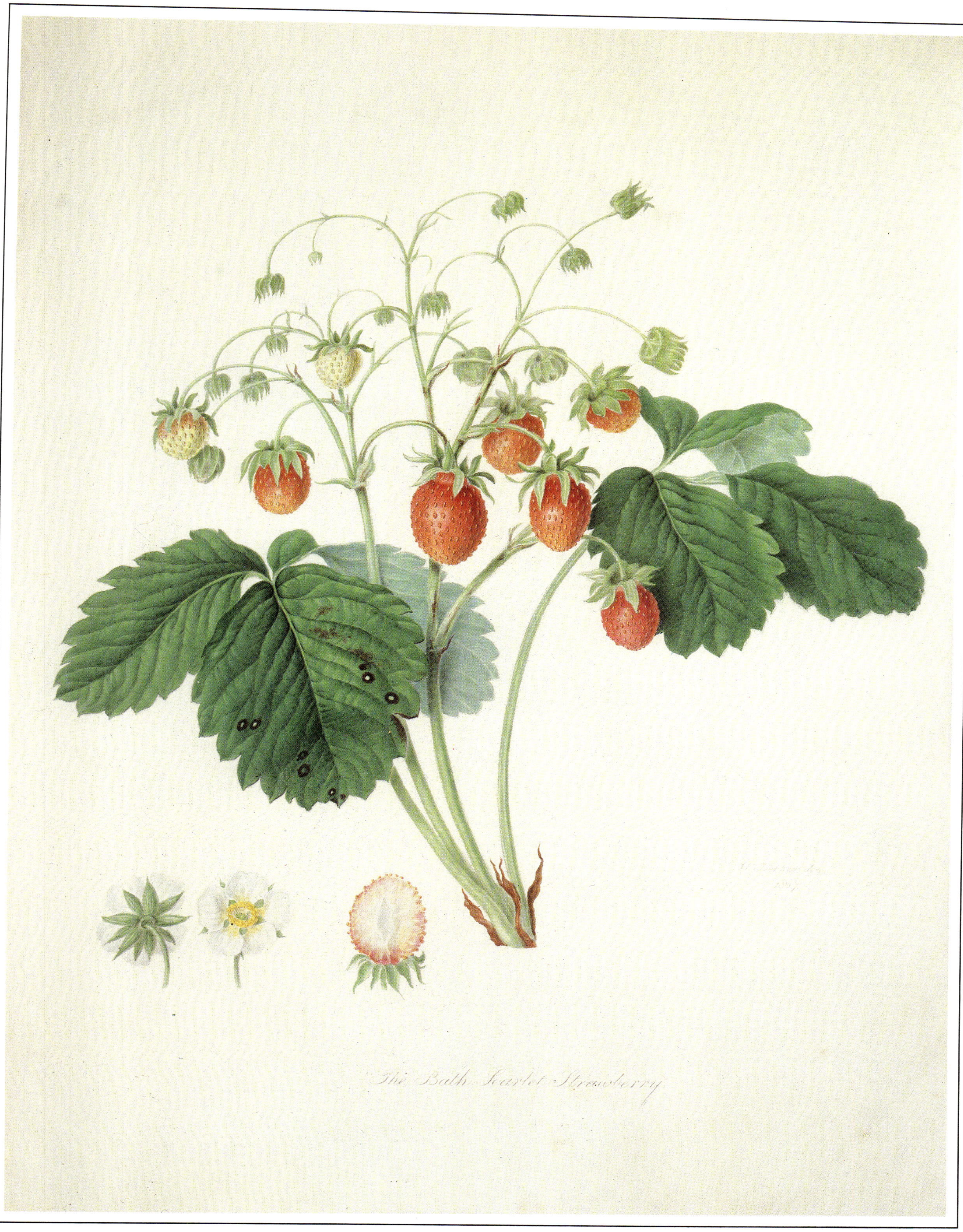

Maritime pine (*Pinus pinaster*). Engraving by Mackenzie, hand-coloured by William Hooker, after a drawing by Georg Dionysius Ehret from *Description of the Genus Pinus* (1803–24) by Aylmer Bourke Lambert. This was the only picture by Ehret that Lambert used, most of the plates being by Francis and Ferdinand Bauer. The maritime pine is a native of the Mediterranean region but has long been cultivated in northern Europe.

The Bath Scarlet strawberry, watercolour on paper, by William Hooker, 1817. Like the pictures of pears and grapes by Mrs Withers, this is one of the collection of paintings of fruit made for the Horticultural Society. Lindley Library, Royal Horticultural Society, London.

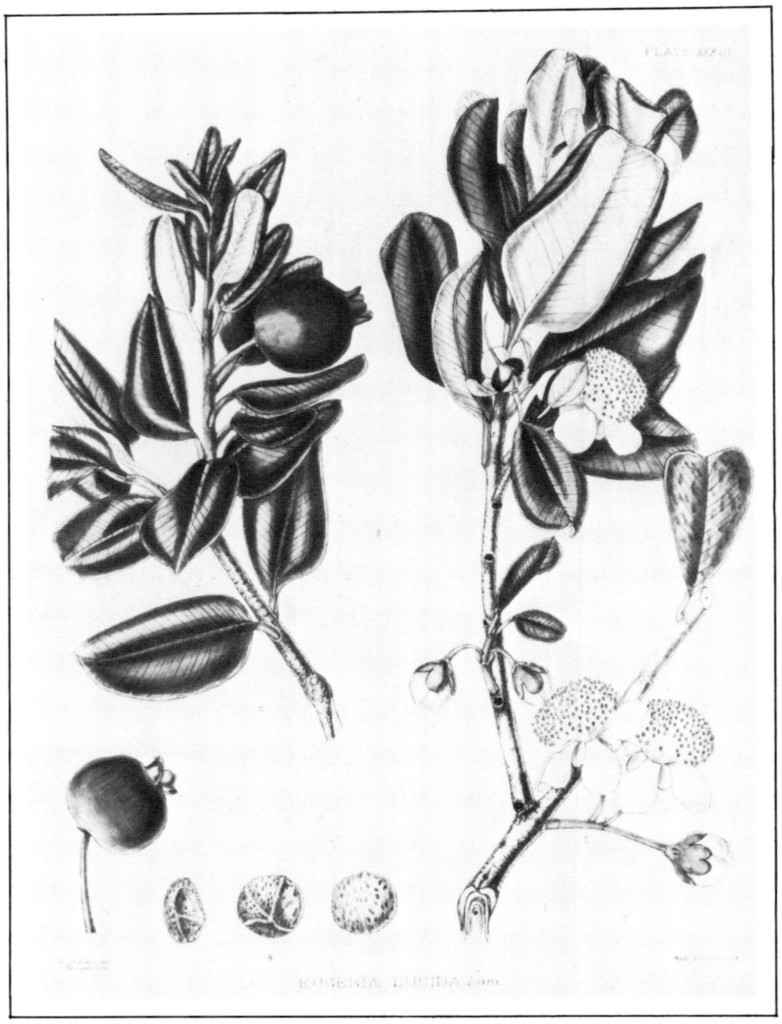

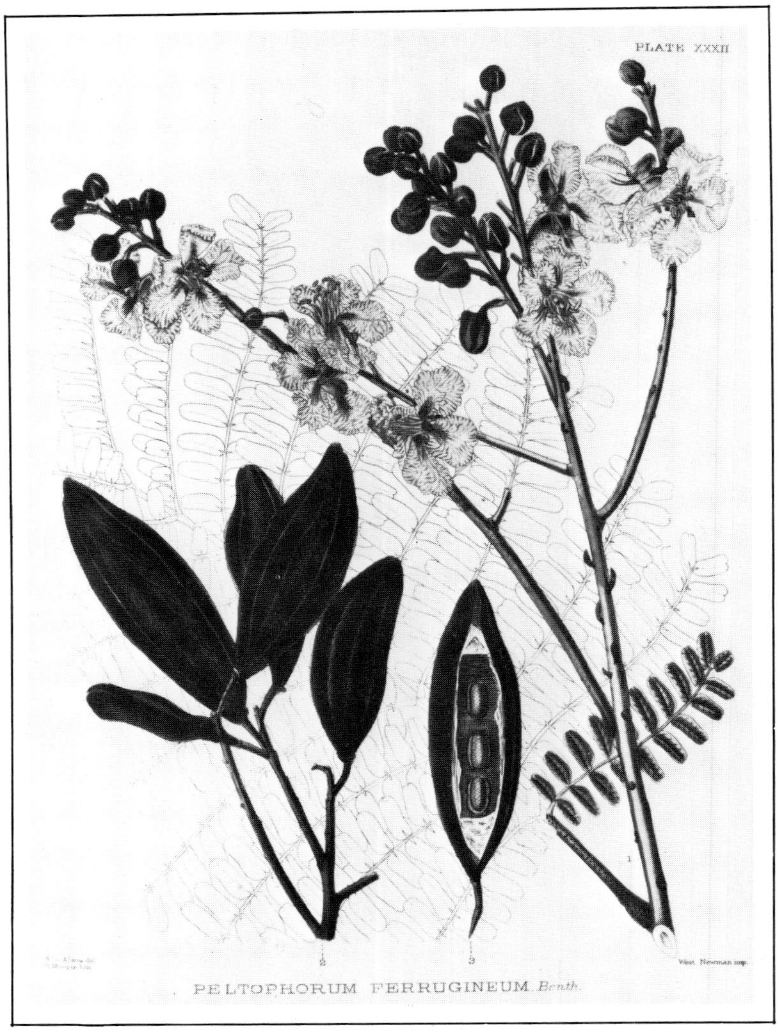

W. J. Hooker in Glasgow, he was apprenticed to a calico printer. Fitch went with Hooker to Kew and became certainly the most prolific botanical artist of the nineteenth century and probably of any century. His work ranged from the magnificent lithographs of Hooker's rhododendrons to the little drawings for Bentham's *Handbook of the British Flora* (1865) which were afterwards published separately as *Illustrations of the British Flora* (1880). As far as is known, he never travelled outside the British Isles. Instead, he was content to draw from plants cultivated in gardens and from dried specimens, which he had a knack of making appear alive. His first drawing to be published was in the *Botanical Magazine* of October 1834. He was official artist at Kew until 1877, when he retired through ill health, although he continued to draw; some of his greatest work was produced in this period, notably the plates for Elwes's monograph on *Lilium*, published in 1880. In all, he published over 10,000 drawings, of which over half were coloured, and his illustrations contributed much to the value of the Hookers' work (as they acknowledged at the time).

'Eugenia lucida' and *Peltophorum ferrugineum*. Lithographs by R. Morgan after drawings by W. De Alwis from *Plates in Illustration of A Handbook to the Flora of Ceylon* (1893) by Henry Trimen.

One illustrated entry and two other plates – stinking hellebore (*Helleborus foetidus*) and common columbine (*Aquilegia vulgaris*) – from George Bentham's *Handbook of the British Flora* (1865); the wood engravings are from drawings by W.H. Fitch.

1. Green Hellebore. Helleborus viridis, Linn. (Fig. 25.)

(Eng. Bot. t. 200.)

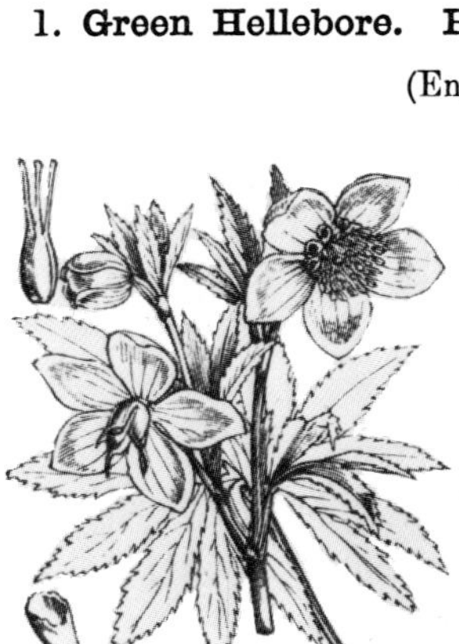

Fig. 25.

Radical leaves large, on long stalks, divided into 7 to 11 oblong, acute, toothed segments, 3 to 4 inches long, the central ones free, the lateral ones on each side connected together at the base so as to form a pedate leaf. Stem scarcely exceeding the leaves, bearing usually 2, 3, or 4 large, drooping flowers, of a pale yellowish green, and at each ramification a sessile leaf, much less divided than the radical ones, and the segments usually digitate.

In pastures and thickets, especially in calcareous soils, and about old walls and ruins in western and central Europe, but not extending to the eastern frontier, nor far to the north. Recorded from many parts of England, but in most cases introduced. It may however be really indigenous in some of the southern and eastern counties. *Fl. early spring.*

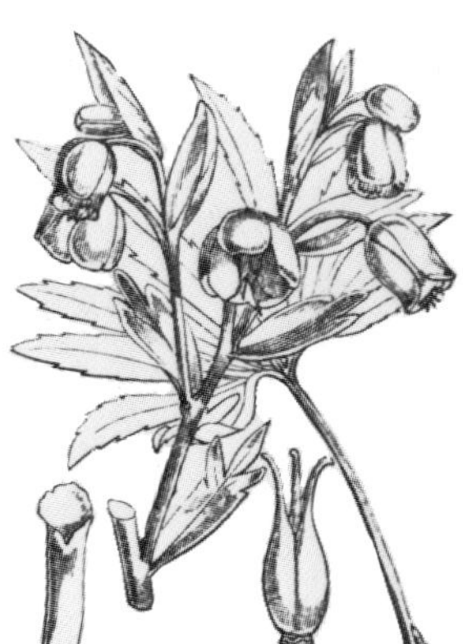

Fig. 26.

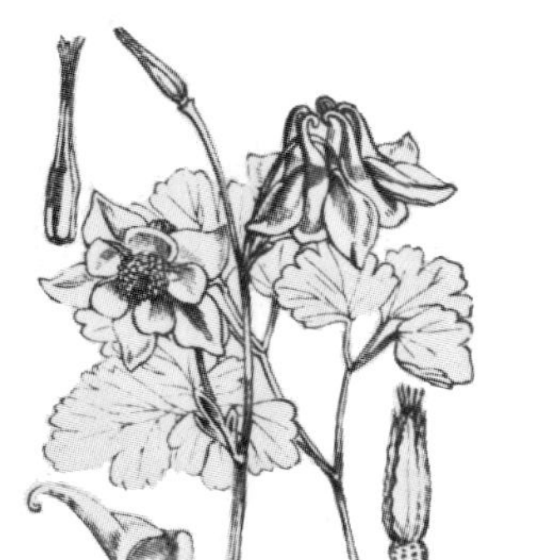

Fig. 27.

CHAPTER XIII
Floras from around the World

While India provided the material for the most beautiful nineteenth-century local floras, especially those of Wallich and J. D. Hooker, European botanists were producing fine illustrated works on other parts of the world. The most monumental of these was *Flora Brasiliensis*, the largest flora ever produced, which is fitting for an area of the world with an outstandingly rich variety of plants. The book was issued in folio format in 130 parts between 1840 and 1906. The editor under whom the project got started was Karl Frederick Philipp von Martius, Professor of Botany at the University of Munich, who was an expert on palms and author of the *Historia Naturalis Palmarum* (1823–50). *Flora Brasiliensis* contains nearly four thousand lithographs and

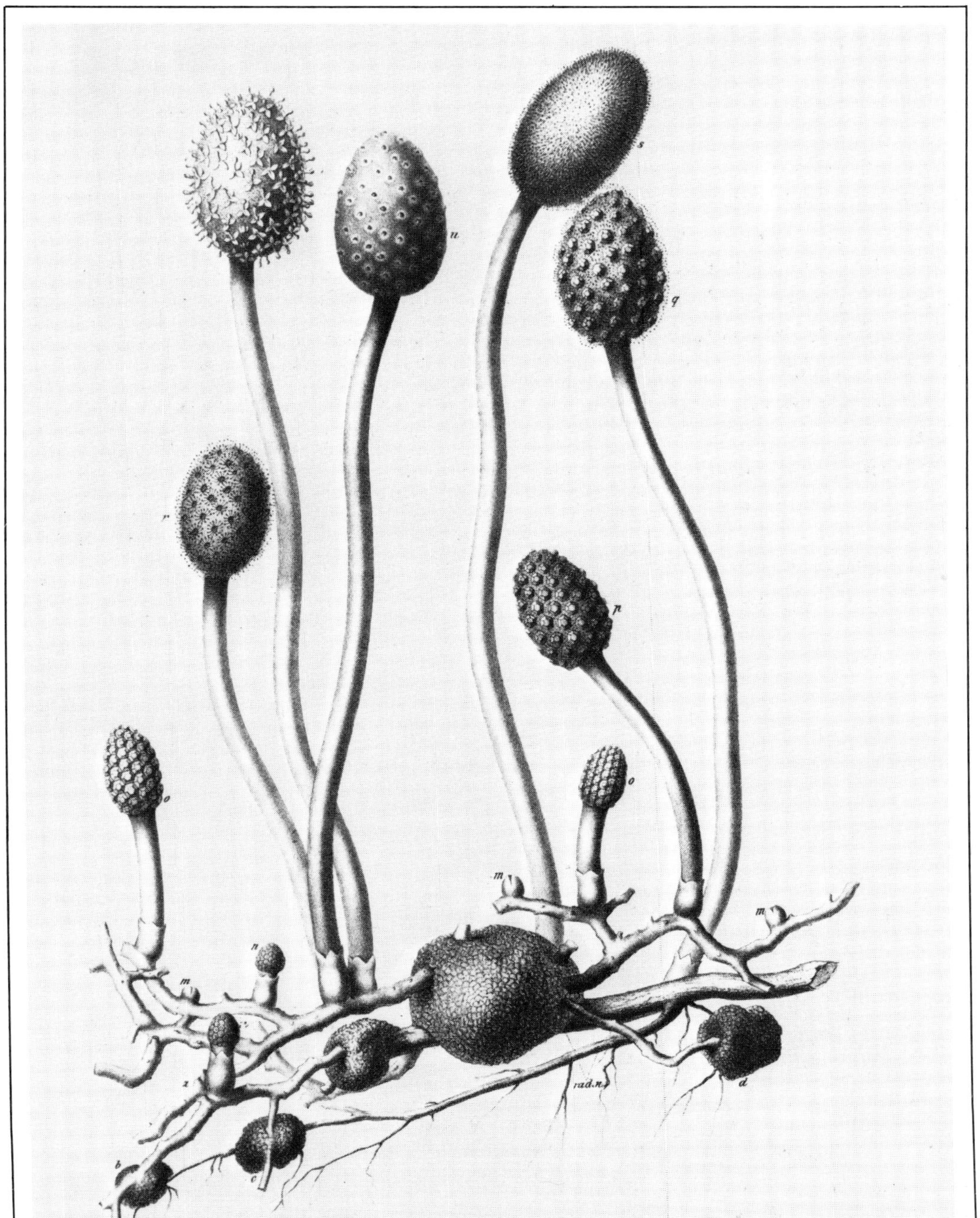

Helosis guyanensis, which in spite of first appearances is a flowering plant, a member of the family Balanophoraceae, which are parasitic on the roots of forest trees. Lithograph from the huge *Flora Brasiliensis* (1840–1906), edited by Karl Frederick Philipp von Martius.

Next page :
The Cannon Hall Muscat grape, watercolour on paper, by Mrs Augusta Withers, around 1825. Lindley Library, Royal Horticultural Society, London.

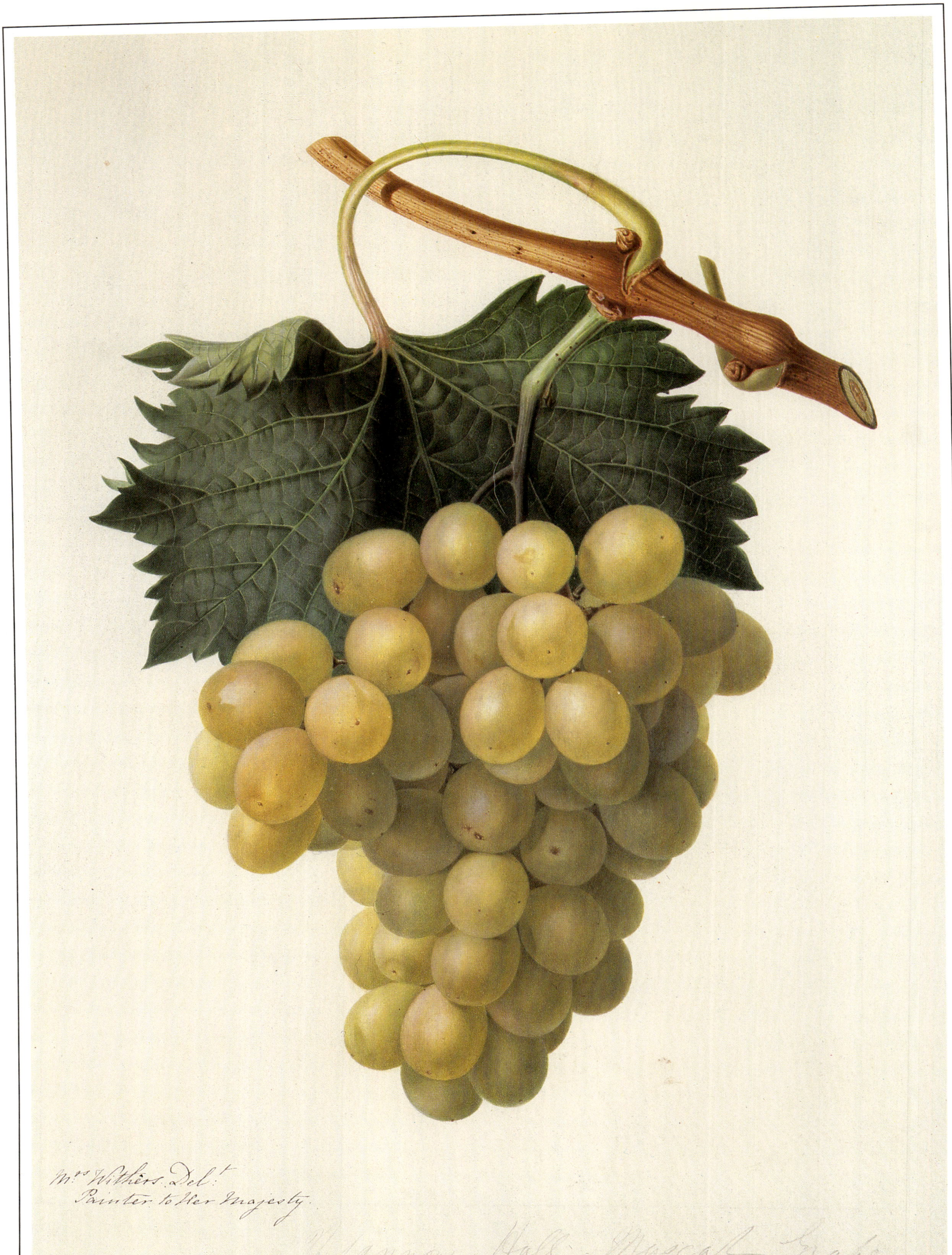
Mrs Withers Delt
Painter to Her Majesty
Cannon Hall Muscat Grape

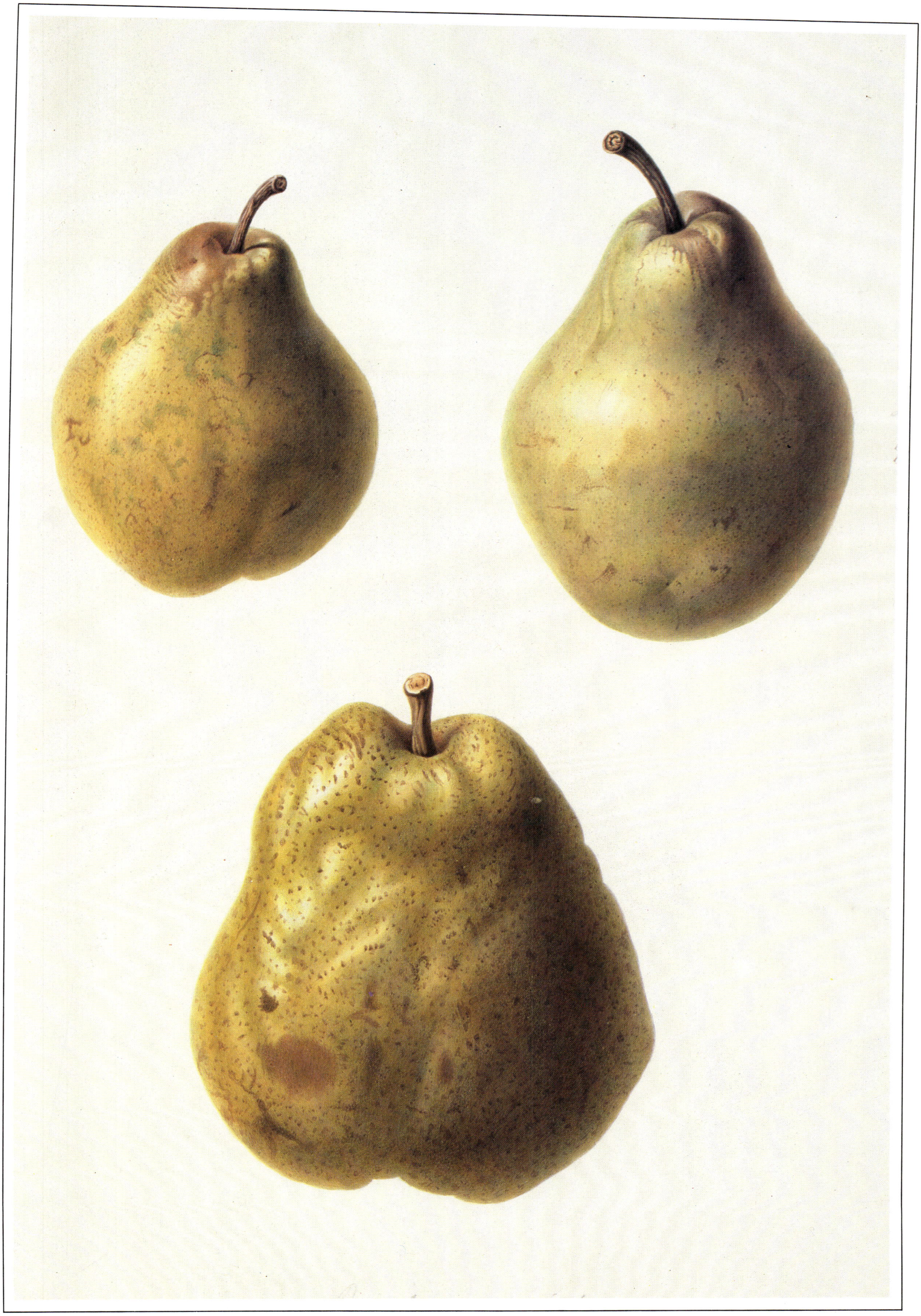

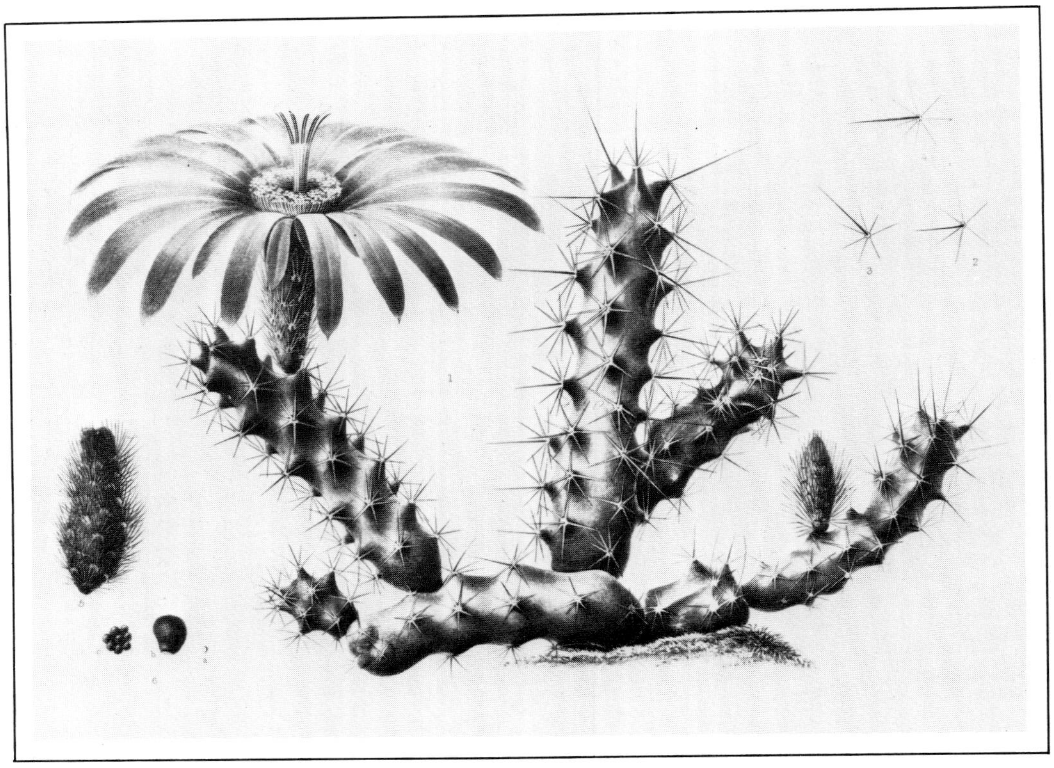

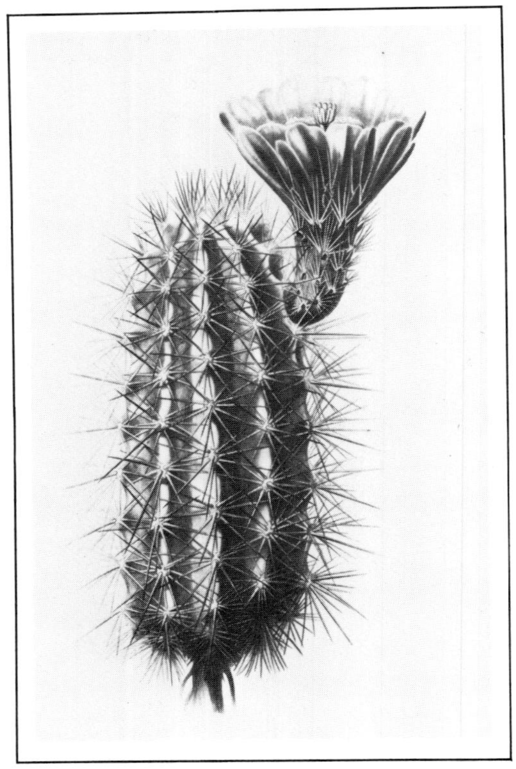

nature prints by at least 38 artists. In the first volume, there are plates of different types of vegetation, some by Rudolph Ackermann; each succeeding volume contained an account of a family or group of families. Particularly striking are the illustrations of the weird families Balanophoraceae, which are parasites on the roots of tropical forest trees, and Podostemaceae, which are alga-like flowering plants on rocks in tropical rivers.

The flora of North America inspired an ever-increasing number of books of different types. W. J. Hooker's *Flora Boreali-Americana* (1829–38) was subtitled *The Botany of the Northern Parts of British America, Compiled Principally from the Plants Collected by Dr Richardson and Mr Drummond on the late Northern Expeditions, under Command of Sir John Franklin R.N. to which are Added those of Mr Douglas from Northwest America.* John Richardson was primarily a zoologist and geologist, and Thomas Drummond was the main botanist to Franklin's second Arctic expedition. In 1825, they travelled from York Factory on the west coast of Hudson's Bay up the Saskatchewan River, then northwest past Lake Athabasca, Great Slave Lake and Great Bear Lake to the north coast, going west as far as Beechey Point on the north coast of Alaska and returning by Lake Winnipeg and the Great Lakes. David Douglas (1799–1834) did most of his collecting in the Pacific northwest, but did not take any part in Franklin's expedition. He had been one of Hooker's pupils in Glasgow and was first sent to America in 1823 to collect plants for the Horticultural Society. For the next eleven years, he travelled all over northwestern America and California, often suffering great hardship. He survived fevers and snowblindness, and once, when his canoe was wrecked on the Fraser River, is said to have spent an hour and forty minutes spinning in a whirlpool. At intervals, he sent back the seeds and specimens he collected to Hooker and the Horticultural Society in England. In 1834, on a trip to Hawaii, he was looking at a bullock caught in a pit trap, when he ventured too near the edge; the bank gave way, and he slipped into the pit to be gored to death beneath the animal's hoofs. Many of Douglas's plants grew well in England and were illustrated in the *Botanical Magazine* as they flowered. They include such familiar garden plants as the flowering currant (*Ribes sanguineum*), *Garrya elliptica* and the Californian poppy (*Eschscholzia californica*).

The illustrations in the *Flora Boreali-Americana* were very clear and simple line engravings by Swan after drawings by Hooker himself. They include *Acer circinnatum* and *Ceanothus velutinus*, which were discovered by Douglas in California.

The great American botanists John Torrey and Asa Gray were among those involved in producing the botanical sections of the early U.S. survey

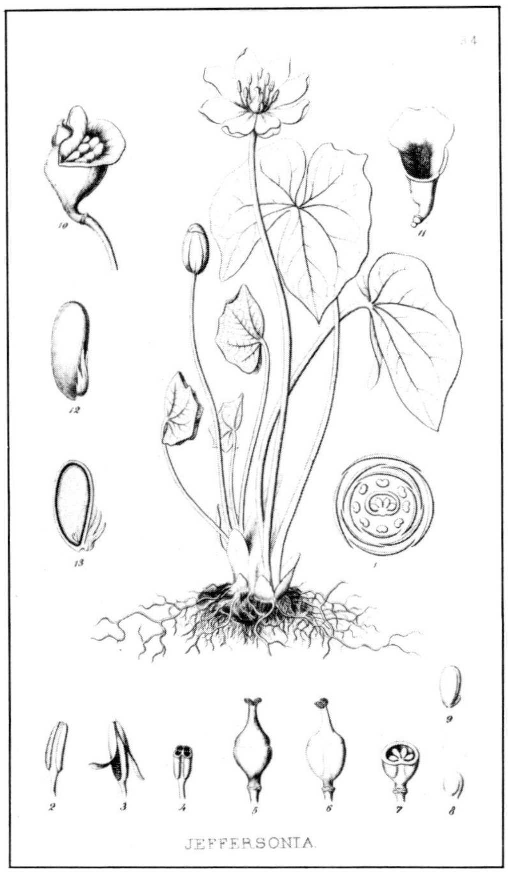

Jeffersonia diphylla, a native of woods in northeastern North America, named in honour of Thomas Jefferson, third President of the United States. Lithograph by Isaac Sprague from *The Genera of the Plants of the United States* (1848–49) by Asa Gray.

Previous page:
Three pears, Beurrée d'Aremberg, Gloria Morceau and Duchesse d'Angoulême, watercolour on paper, by Mrs Augusta Withers, 1826. Lindley Library, Royal Horticultural Society, London.

Two lithographs of cactuses by Cornelius Wenstell from *Report of the United States and Mexican Boundary Survey* (1857–59), botanical volume by C.C. Parry, J. Torrey and G. Engelmann under the overall editorship of William Hemsley Emory. *Far left: Cereus berlandieri. Left: Cereus polyacanthus.*

expedition reports, some of which have fine illustrations. The *Report of the Botany of the Exploration and Survey for a Railroad Route from the Mississippi River to the Pacific Ocean* (1855), based on specimens collected by various members of the expedition, has excellent lithographs by Isaac Sprague (1811–95), while those in the *Report of the United States and Mexican Boundary Survey* (1857–59) are by Cornelius Wenstell. Sprague also produced the illustrations for Gray's *The Genera of the Plants of the United States* (1848–49), of which ten volumes were planned, although only the first two appeared. Sprague had been taught by Gray and worked closely with him on the 186 very elegant lithographs,

Rather similar to Sprague's work is that of Charles Edward Faxon (1846–1918), who was an instructor in botany at Harvard. He made the drawings for the fourteen volumes of Charles Sprague Sargent's *Silva of North America* (1890–1902). Faxon often travelled with Sargent to study the trees for the book rather than working from specimens, so that the illustrations are admirable artistically as well as botanically. Faxon's drawings were engraved in Paris by Phillipert and Picart, under the direction of Alfred Riocreux. The 714 steel engravings used a fine mechanical stipple which produced great delicacy of detail. C. S. Sargent himself played much the same role at the Arnold Arboretum as W. J. Hooker had at Kew: he was director for 54 years, building it up from an inauspicious start to the greatest collection of hardy trees in the world.

Interest in the Antipodes continued to be maintained in the first half of the nineteenth century through the work of visiting European botanists. More of Ferdinand Bauer's paintings were published in 1833 by S. L. Endlicher as *Prodromus Florae Norfolkicae*, a catalogue of plants collected by Bauer on Norfolk Island. From 1839 to 1843, the young J. D. Hooker sailed as botanist and assistant surgeon on the expedition under Captain Sir J. C. Ross to Tasmania, New Zealand and the islands of the Antarctic. Hooker's *Botany of the Antarctic Voyage of HMS Erebus and Terror* was published in three parts, all with lithographs by W. H. Fitch, starting with *Flora Antarctica* (1844–47), which included the famous cabbage-like *Pringlea antiscorbutica*, discovered during Captain Cook's third voyage and named in honour of one of the pioneers in the study of scurvy, Sir John Pringle; it is restricted to Kerguelen and neighbouring small islands in the Southern Ocean. The next part of Hooker's work to appear was *Florae Novae Zelandiae* (1853–55), two quarto volumes with 130 lithographs, followed in 1855–60 by *Florae Tasmaniae* with 200 lithographs, showing ferns, mosses, lichens, algae and fungi as well as flowering plants.

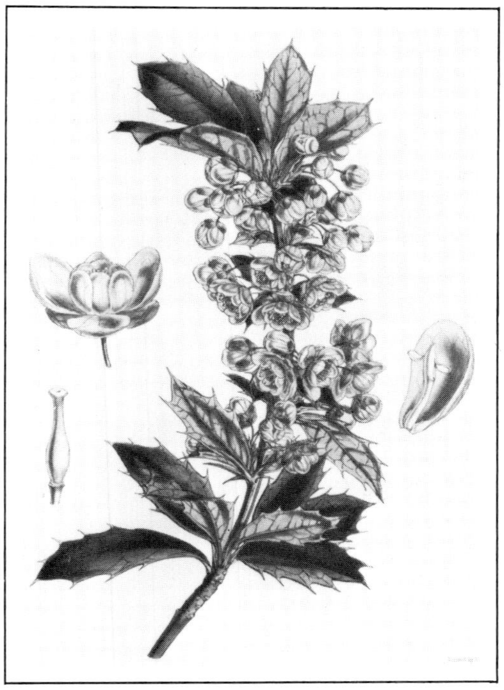

Berberis ilicifolia, which was found in Tierra del Fuego and first brought back by Ross's expedition. Hand-coloured lithograph by W.H. Fitch after a drawing by Joseph Dalton Hooker from *The Botany of the Antarctic Voyage: Flora Antarctica*, Volume II (1847) by Hooker.

At the same time as Hooker was in the southern hemisphere, a French botanist, M. E. Raoul, visited New Zealand as surgeon on the corvette *Aube*. His report, entitled *Choix de Plantes de la Nouvelle Zélande* was published in 1844 with thirty elegant engravings by Riocreux.

The first notable botanist to become a permanent resident in Australia was a German. Sir Ferdinand Jakob Heinrich von Mueller was the author of numerous works on the flora, trees and plant fossils of Australia which were published between 1854 and 1890. One of these, *The Plants Indigenous to the Colony of Victoria* (1860–65) contains the most delicate and detailed lithographs which have a real feeling of depth. The aesthetic effect, however, is spoiled by the mass of botanical details jammed into each plate; the lithographer, F. Schonfeld, does not seem to have done any other botanical work.

While the British were mainly responsible for the study of botany in India and the Antipodes, China was the scene of scientific exploration by three major European powers, the Russians and the French as well as the British. Travel remained difficult outside the major cities, with periodic wars and political disputes between the Chinese and the Europeans restricting the progress of botanical exploration.

When the First Opium War ended in 1842, the Treaty of Nanking gave the British the right to occupy Hong Kong, and it became much easier for foreigners to enter China. The council of the Horticultural Society was

Next page:
Iris iberica, watercolour on paper by F. H. Round for *The Genus Iris* (1913) by W. R. Dykes. *Iris iberica* is probably extinct in the wild, as its habitat has been destroyed by the growth of the city of Tblisi in Georgia. Lindley Library, Royal Horticultural Society, London.

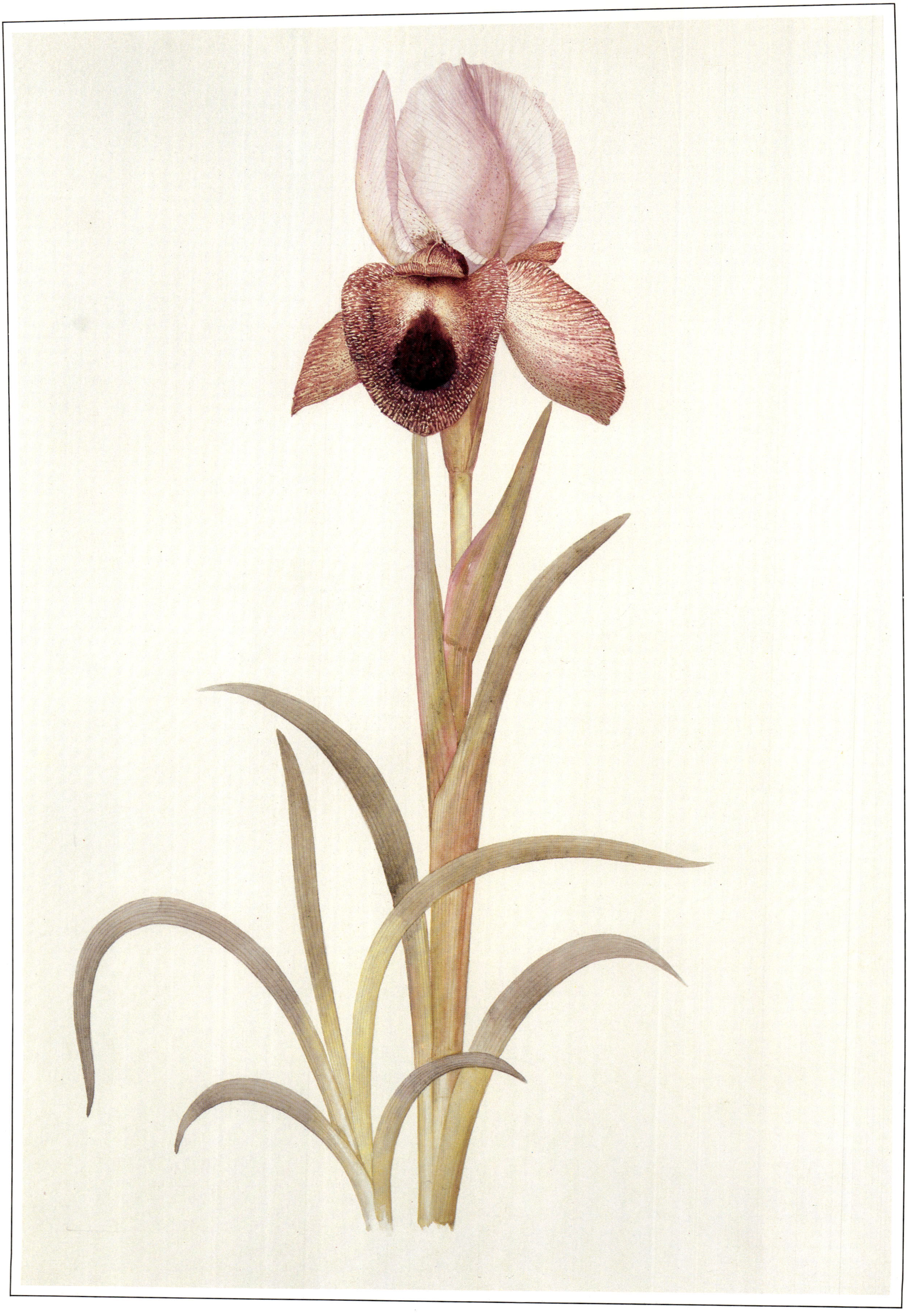

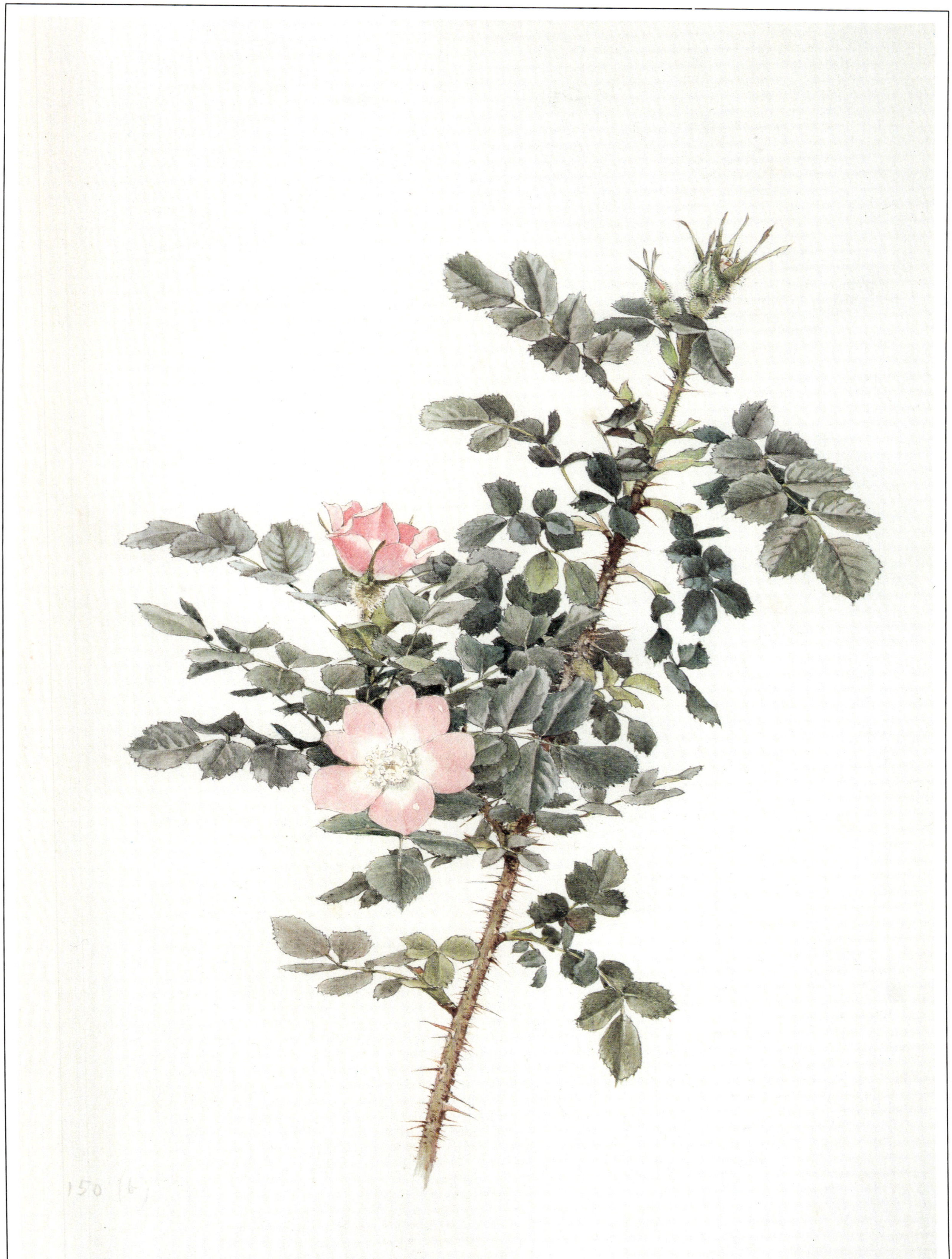

quick to react by sending a collector to China, with the idea of introducing to Europe more of the excellent garden plants that were known to be cultivated there, and of acquiring seeds of good varieties of tea to improve the plantations in India. They chose Robert Fortune, who reached Hong Kong in July 1843. He subsequently returned to China several times up to 1861 and brought back many new plants, which were often illustrated in journals such as Paxton's *Magazine of Botany*, Curtis's *Botanical Magazine* or the French *Flore des Serres*. No one book includes even a selection of Fortune's introductions, and his own accounts of his travels, *A Journey to the Tea Countries of China* (1852) and *A Residence Among the Chinese . . .* (1857), did not contain any botanical plates.

Other significant explorations were made as part of the Russian expansion in the Far East and the annexation of the Amur valley. Carl Maximovicz (1827–91) accompanied the Russian frigate *Diana* as botanical collector on an expedition which arrived in the Far East in 1854. In that year and the next, he botanised along the coast and up the valley of the Amur and Ussuri rivers, before returning to St Petersburg overland via Siberia in 1856. This was the beginning of his lifelong interest in the Far Eastern flora and resulted in the publication of *Primitiae Florae Amurensis* with ten clear and simple plates of new species. After his return, Maximovicz became chief botanist and keeper of the herbarium at the Imperial Botanic Garden, and worked on the collections of Przewalski and Potanin from northwestern China and Mongolia, as well as those of other lesser known travellers.

European travel in China again became easier after further wars in 1857–60 had ended in the defeat of the Chinese Imperial Army by a joint French, English, American and Russian force. Soon after this, several French missionaries began sending back to Paris plant collections from the borders of China and Tibet, which is probably the richest area for plants in the temperate world. The first was Armand David who joined the Lazarists in 1848, becoming attached to their mission in Peking in 1862. Père David was primarily a zoologist (a deer is named after him), but his botanical and geological collections were also very important. In 1866, he explored southern Mongolia, and, in 1868–70, he travelled in western China and eastern Tibet, spending a year in the semi-independent principality of Mupin, in western Szechuan, where there was a Catholic seminary. Here he remained for the summer, making a remarkable collection of animals and plants. It is the results of these journeys which are recorded in Adrien-René Franchet's *Plantae Davidianae*, of which the *Plantes de Mongolie* (1884) contains 44 good clear lithographs by Cuisin and d'Apreval and *Plantes du Thibet Oriental (Province de Moupine)* (1884–88) contains 39 lithographs by d'Apreval, mostly uncoloured but with an excellent coloured plate of *Davidia involucrata*. It includes other now familiar garden plants discovered by David, *Clematis armandii* and *Rhododendron moupinense*.

David's collections were exciting and contained a high percentage of new species, but were dwarfed by the huge numbers of specimens which arrived in Paris from another French missionary, Jean-Marie Delavay (1835–95). Père Delavay joined the Societé des Missions Étrangères and was sent to China in 1868. It was only in 1881, after Franchet had introduced him to David in Paris that he started to collect plants and send them back to the Muséum d'Histoire Naturelle. In the next fourteen years, the Muséum received over 200,000 specimens, beautifully preserved, including about 1,500 new species, mostly from Yunnan. Their study occupied Franchet for the rest of his life; he published the results in several journals and most notably in *Plantae Delavayanae*, which appeared in parts in 1889–90, but remained unfinished. Delavay died in a village in Yunnan in December 1895 after several years of ill health. Nevertheless he continued his botanical work to the end with the help of a Chinese collector.

There were no other botanical collectors in China during the nineteenth century to rival these great pioneers. Père Soulie and Père Farges are among the French missionaries who collected many new species, and Franchet gave their names to the well-known species of rhododendron that they

Previous page :
'Rosa glutinosa', now known as *Rosa pulverulenta*, watercolour painting by Alfred Parsons for *The Genus Rosa* by Ellen Willmott (1910–14). This rose is a dwarf shrub that is distributed from southern Italy to Iran and Afghanistan. Lindley Library, Royal Horticultural Society, London.

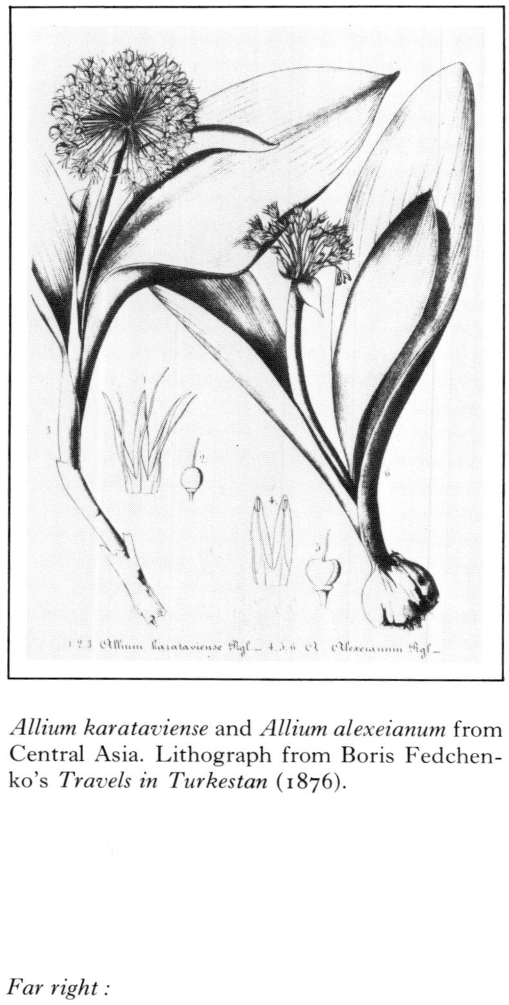

Allium karataviense and *Allium alexeianum* from Central Asia. Lithograph from Boris Fedchenko's *Travels in Turkestan* (1876).

Far right :
A scabious, *Cephalaria tchihatchewi*, from northern Turkey, around Trabzon. Engraving by Ph. Picart after a drawing by Heyland from *Asie Mineur, description physique de cette contrée* (botanical atlas volume, 1866), by Pierre de Tchihatcheff, a Russian count.

Overleaf :
'Paeonia cretica' (*Paeonia clusii*), watercolour on paper, dated 1931, by Lillian Snelling for *A Study of the Genus Paeonia* (1946) by F. C. Stern. This peony is a native of the Greek islands of Crete and Karpathos. Lindley Library.

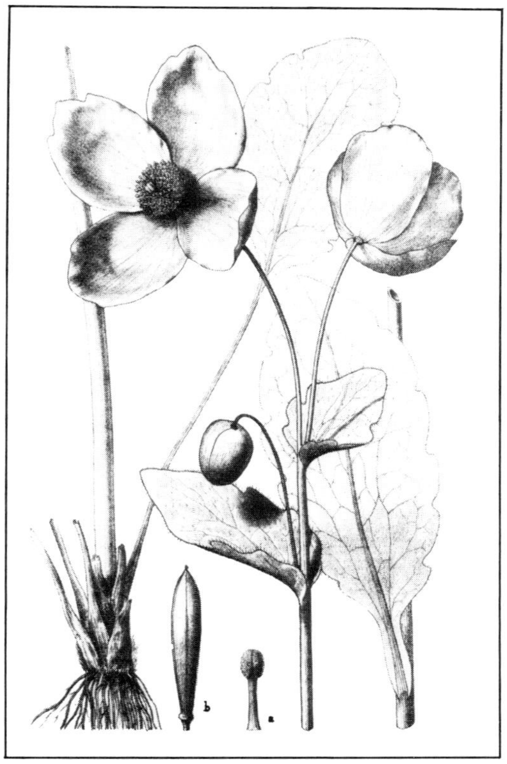

Himalayan blue poppy (*Meconopsis betonicifolia*), a plant that created a sensation among gardeners when it was introduced from Tibet by Kingdon-Ward in 1924 under the name *Meconopsis baileyi.* It later transpired that the same plant had been discovered in China by Jean-Marie Delavay in 1886 and described by Adrien-René Franchet in his *Plantae Delavayanae* (1889–90), from which this lithograph by d'Apreval after his own drawing is taken.

Below :
A broomrape, *Orobanche denudata*, that is found only in Sardinia and is parasitic on brambles. Engraving by A. Nizza after a drawing by Heyland from *Flora Sardoa* (1837–59) by Giuseppe Giacinto Moris.

discovered. An Irishman, Dr Augustine Henry, joined the Chinese Maritime Customs in 1881, and was stationed successively in Hupeh, Hainan and Yunnan. He made several collecting journeys in to the mountains, sending back a very large number of herbarium specimens. Many of his new species are described and illustrated in Hooker's *Icones Plantarum.* On his return from China, Henry lived near Kew, and collaborated with H. J. Elwes on the monumental *Trees of Great Britain and Ireland* (1906–13).

The French, Russians and English also showed an interest in the flora of the Middle East and Central Asia. Floras of particular areas appeared at intervals through the nineteenth century, but are more interesting for their content than for the quality of their illustration.

The flora of Afghanistan remained almost unknown, until Surgeon-Major Aitchison became botanist to a commission that surveyed the border between Afghanistan and the Russian and Persian territory to the north and west in 1884–85. He travelled from Quetta (now in Pakistan) through Baluchistan to Herat in northwestern Afghanistan and Meshed in Persia, then across the Paropamisus Mountains and into Russian Turkestan on the Kushka river. The spring flowers were magnificent, and among them were many new species including the giant species of seakale, *Crambe cordifolia*, which is now a popular herbaceous border plant. His report was published in 1888 and contained rather dull lithographs by J. Allen and Matilda Smith, W. H. Fitch's successor at Kew.

Eduard von Regel (1815–92), a German by birth, became director of the Botanic Garden at St Petersberg in 1855 after working at Göttingen, Bonn, Berlin and Zurich; he remained in the post until his death. His *Flora Turkestanica* has coloured lithographs of some particularly Asian genera such as *Eremurus* and *Allium.* In 1852, Regel founded the journal *Gartenflora*, of which he remained editor until 1884, and in 1871 he started the *Actae Horti Petropolitani.* It is in these two journals that illustrations of many of the new discoveries from central Asia are to be found. Foremost among the contributors was Regel's son, Albert, who was appointed district physician in Kuldja near Bukhara in 1875, and spent the years 1877–85 travelling and collecting in central Asia, sending the plants he collected back

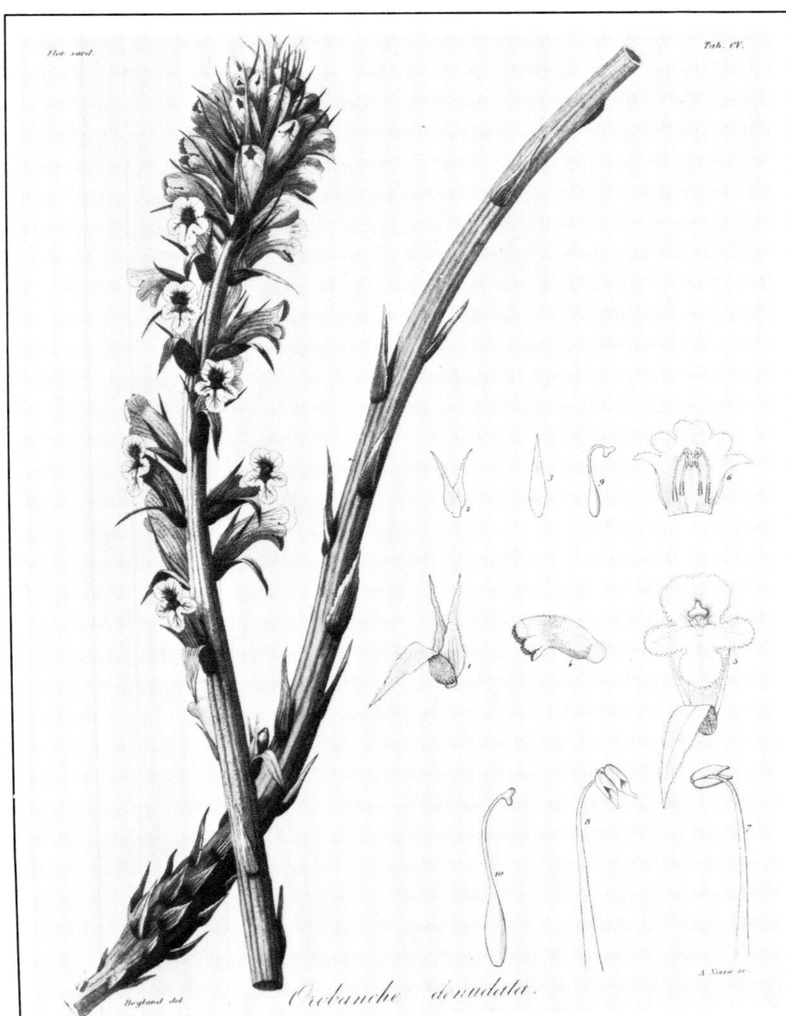

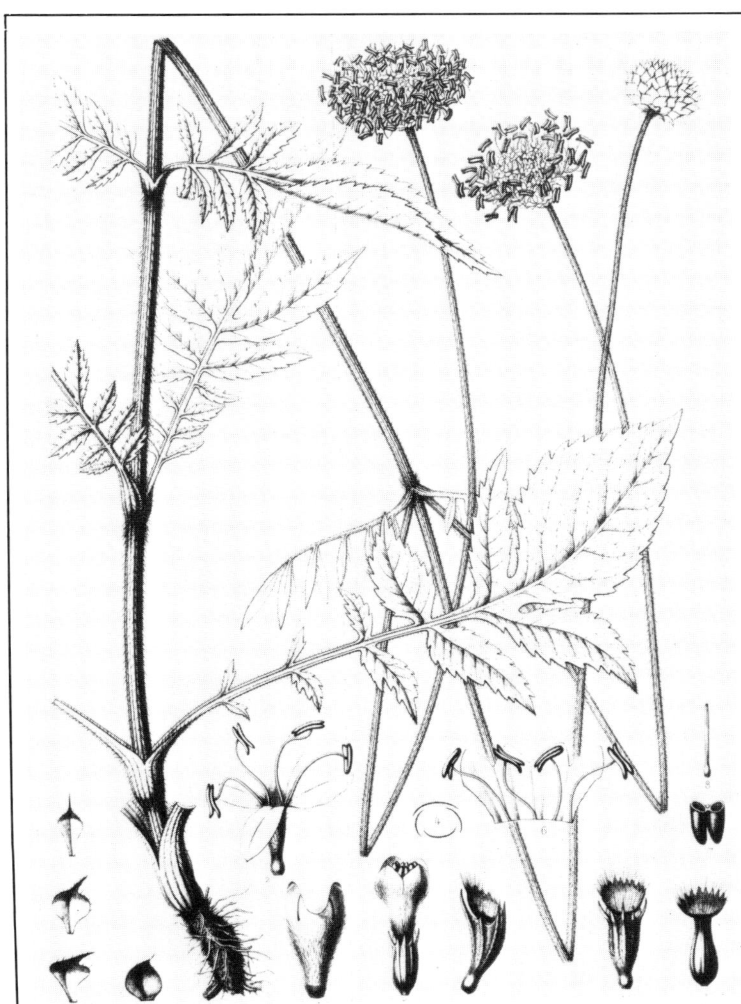

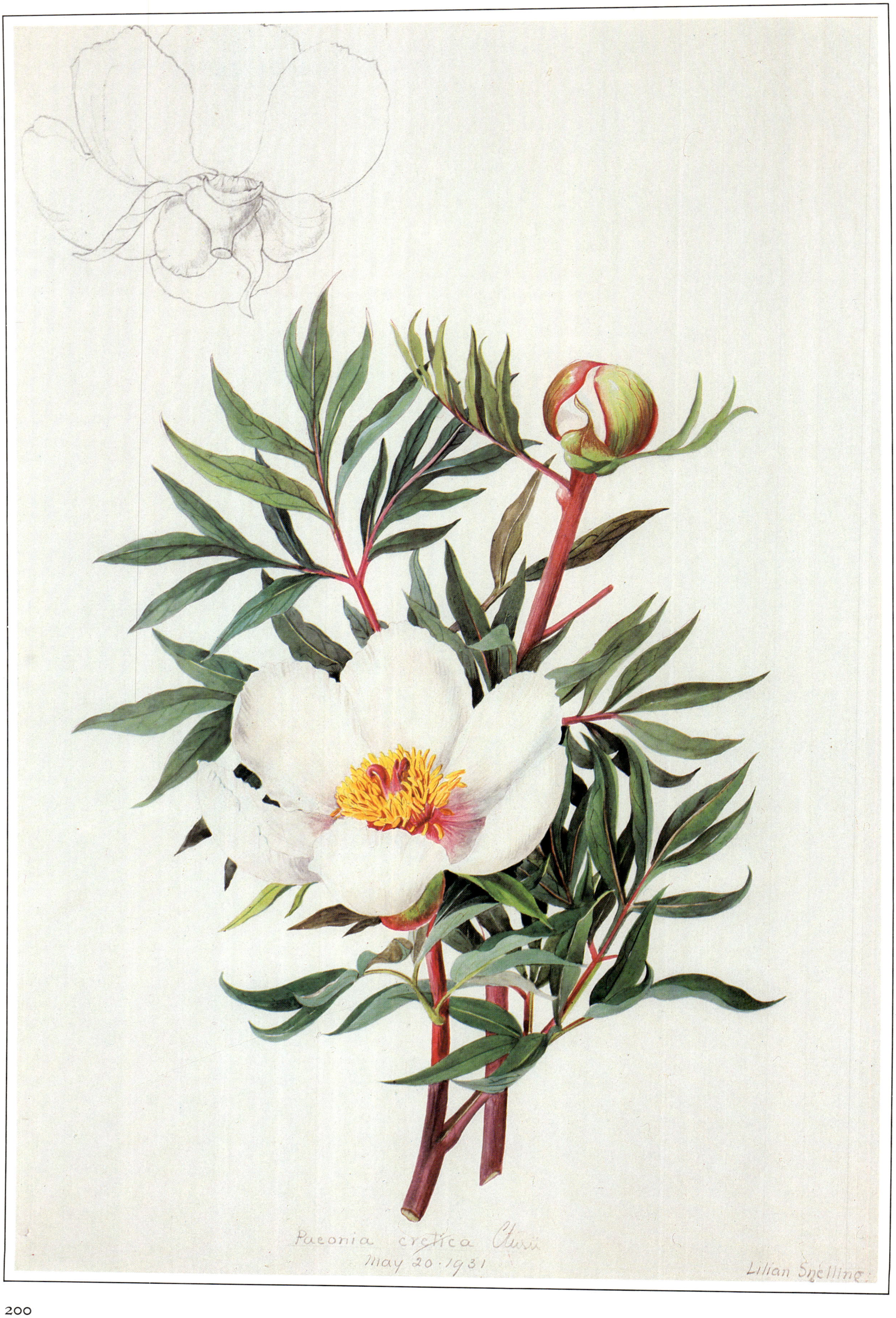
Paeonia cretica Clusii
May 20·1931
Lilian Snelling

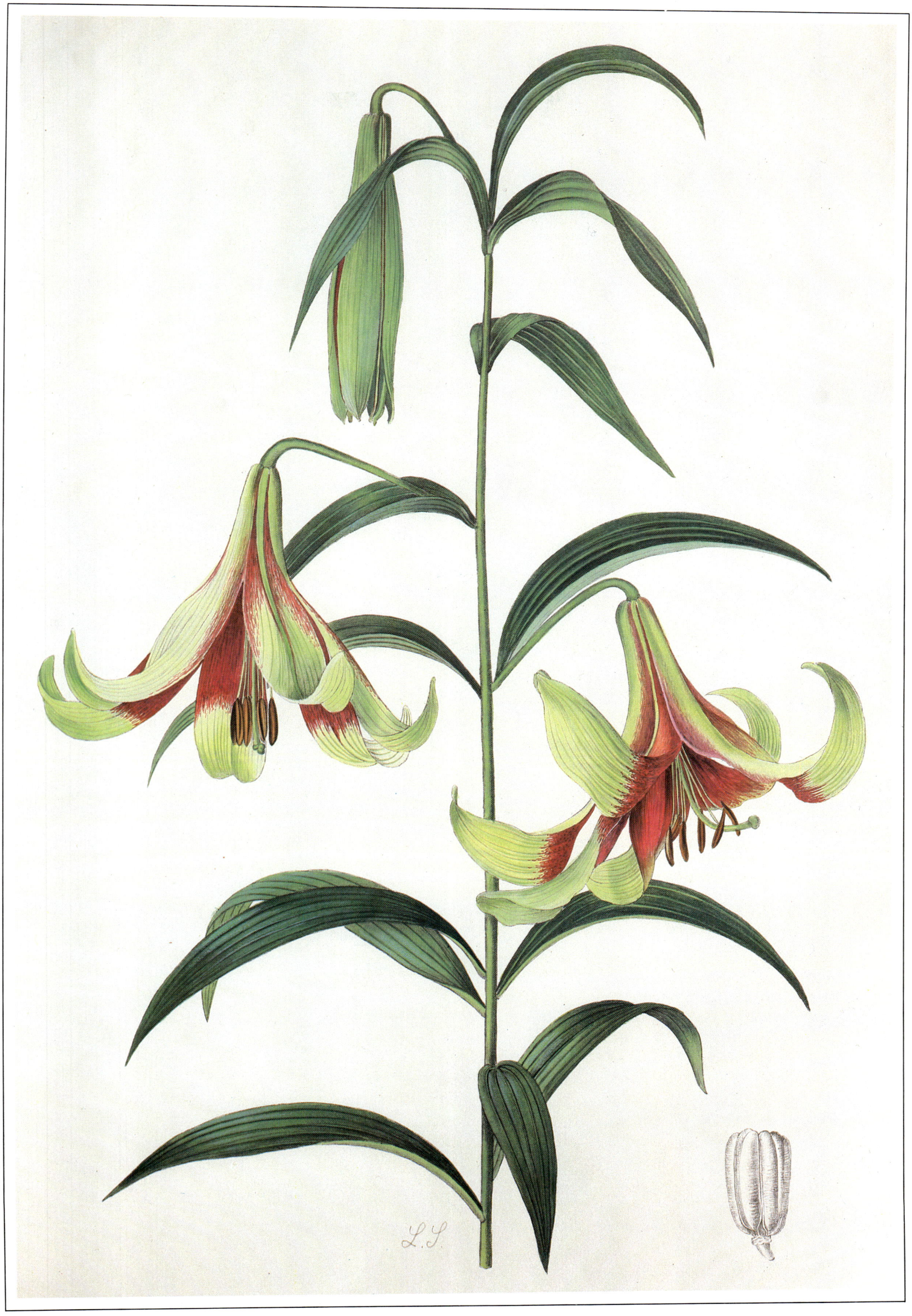
L. S.

to St Petersberg for his father to describe. He spent most of the time in Russian territory, but made one daring journey into China, evading the frontier guards and reaching Turfan, the first European to see the city since the early seventeenth century.

In England meanwhile, Sowerby's *English Botany* continued to be updated, with the last edition appearing from 1860 to 1886, and a final Volume 13 in 1892. Nothing rivalled it, although a small flora appeared which continued in use for the next century. *A Handbook of the British Flora* by George Bentham was first published in 1858. It was reprinted in 1865 with small wood engravings by W. H. Fitch alongside the text. The fifth edition was revised by Sir J. D. Hooker in 1887, and this 'Bentham and Hooker' remained the standard popular work on the British flora, with Fitch's illustrations in a separate volume, until 1952 and the appearance of the Clapham, Tutin and Warburg *Flora of the British Isles*.

In continental Europe, the nineteenth century saw the production of comprehensive national floras following the lead given by the *Flora Danica* (of which new parts continued to appear up to 1871) and by Curtis's *Flora Londinensis* and Sowerby's *English Botany*. The largest and finest of these floras, Reichenbach's *Icones Florae Germanicae et Helveticae*, was begun in 1837 and not completed until 1912. It contains 3744 plates, the earlier of which are engravings, the later lithographs, and it is still one of the most useful sources of illustrations of central European plants. Heinrich Gottlieb Ludwig Reichenbach (1793–1879) was one of the foremost German botanists of the century. From 1820, he was Inspector of the Royal Natural History Museum and Professor of Botany at Leipzig. The work was initiated in 1834 as illustrations of grasses and sedges, and continued from 1837 as an illustrated Flora of the whole of central Europe. Reichenbach drew nearly all his own illustrations, but a few were by Schnorr, the engraver of most of the plates. After 1852, Reichenbach's son, Heinrich Gustav Reichenbach became the main illustrator. All the plates are well drawn, less elegant but more naturalistic than similar French work, and those by the younger Reichenbach are, if anything, better than his father's. Heinrich Gustav Reichenbach worked in Hamburg Botanic Garden where he became a specialist in tropical orchids. The Flora was still unfinished when he died in 1889. Between 1869 and 1903 there was a break in publication, and the later parts with lithographic illustrations were by G. Beck von Mannagetta and F. G. Kohl.

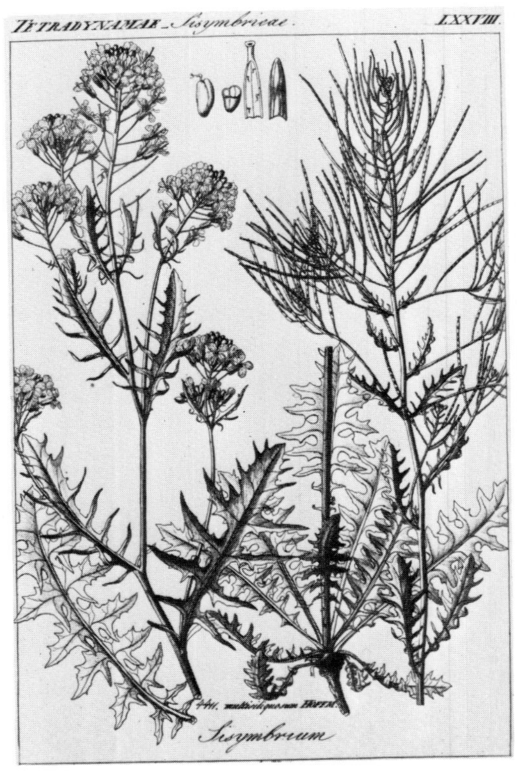

Sisymbrium austriacum. Lithograph from *Icones Florae Germanicae et Helveticae* (1837 onwards) by Ludwig Reichenbach.

Between 1864 and 1903, Alexis Jordan and Jules Fourreau produced a series of engravings of European plants under the title *Icones ad Floram Europeam . . .*; the illustrators included Fourreau himself and C. Delorme. Jordan named every variety he could recognise – such microspecies have come to be called 'Jordanons'. Not many are recognised today as distinct species, but the illustrations are good and still useful botanically and to anyone studying old garden flowers.

The first complete illustrated Flora of France was by the Abbé Hippolyte Coste, the *Flore Descriptive et Illustre de la France, de Corse et des Contrées Limitrophes*. It is still a standard work today. The three volumes were published in Paris between 1900 and 1906. The small, clear illustrations are placed opposite the admirably concise text; their form and style is similar to that of Fitch's illustrations for Bentham and Hooker.

Two illustrated Floras of the Riviera appeared in English to cater for the fashionable travellers and for those who wintered in the south of France for the sake of their health. The earlier, *A Contribution to the Flora of Mentone and to a Winter Flora of the Riviera, Including the Coast from Marseilles to Genoa* was by John Traherne Moggridge. It was published in parts between 1864 and 1871, with hand-coloured lithographs after Moggridge's drawings. The descriptions and ecological notes on the plants are most valuable. Wilfrid Blunt says that the lithographs 'give but a faint idea of the beauty and sureness of his line.' Moggridge died in France in 1874 at the early age of 32. A later and more comprehensive work is Clarence Bicknell's *Flowering Plants and Ferns of the Riviera and Neighbouring Mountains*, published in

Previous page:
Lilium nepalense. Hand-coloured lithograph by Lilian Snelling from the supplement to H. J. Elwes's *Monograph of the Genus Lilium*. The lithographs in the first supplementary volume (published 1934–40) are less heavy and more attractive than those by W. H. Fitch for the main part of the monograph. Lindley Library.

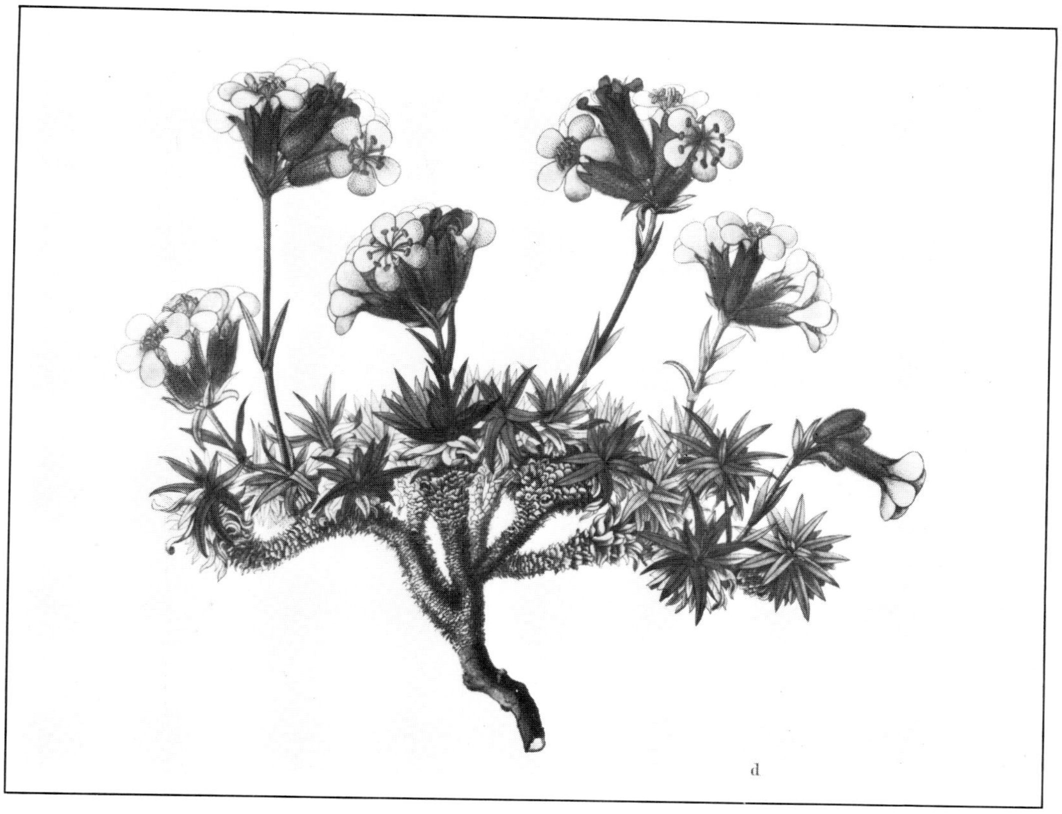

Saponaria caespitosa, an alpine soapwort from the central Pyrenees. Engraving by A. H. Payne after a drawing by the author from Heinrich Moritz Willkomm's *Icones et Descriptiones Plantarum Novarum . . . Hispaniae* (1852–62).

1885. It was arranged as a Flora, but the plates by Bicknell himself are reproduced rather unsatisfactorily by chromolithography and are not as attractive as Moggridge's, while the text is less interesting.

Although no Floras of Spain were produced by native Spaniards during the nineteenth century, two excellent works appeared on Spanish plants. The great Swiss botanist Edmund Boissier visited Andalucia in 1837. His *Voyage Botanique dans le Midi de l'Espagne* (1839–45) contains 181 fine engravings in line and stipple of the plants that he collected. The introduction gives something of the history of the production of the book. The artist, Jean-Christoph Heyland (1792–1866) was born in Frankfurt-am-Main, but spent nearly all his life in Geneva. He worked under the close supervision of Boissier, drawing mostly from dried specimens, but also using living plants raised from seed which Boissier had collected. In many of the illustrations the part which was designed to be hand coloured was worked in stipple only, while the part which was to remain uncoloured was engraved in line.

Boissier's main interest, though, was the flora of the Orient, but a German botanist, Heinrich Moritz Willkomm (1821–95) spent his whole life working on the flora of Spain. He published two beautifully illustrated works, with very clear engravings after his own drawings. The first was '*Icones et Descriptiones Plantarum Novarum, &c., Europae Austro-occidentalis; praecipue Hispaniae*' a quarto of which only two volumes were published, between 1852 and 1862. They contain 158 hand coloured engravings of the families Caryophyllaceae and the Cistaceae. His second work was also incomplete, a folio containing 183 hand-coloured lithographs, again after his own drawings, which appeared in parts between 1881 and 1892 and was entitled *Illustrationes Florae Hispaniae Insularumque Balearium*. The hand-coloured engravings of his earlier work are among the finest produced in the second half of the nineteenth century.

The best nineteenth-century work on Italian plants was on Sardinia, Guiseppe Moris's *Flora Sardoa*, published in Turin between 1837 and 1859, which has three volumes of good engravings after drawings by Magdalena Lisa. The book was not finished until 1904, when the volume on monocotyledons by Ugolino Martelli appeared, illustrated with rather good, delicately drawn lithographs. The first complete Italian Flora, by A. Fiori and G. Paoletti was published in Padua between 1895 and 1904, with a volume of illustrations by P. Brombin. It contains 4236 small figures of plants, after the style of Fitch in Bentham, or of Coste's *Flore de France*, but they give the impression of having been reduced more than was intended by the artist, as much of the detail is obscured.

Next page:
Iris polakii, opaque watercolour on paper, by Paul Furse, 1965, from a plant collected by the artist in the Iranian Azerbaijan in 1962. Lindley Library, Royal Horticultural Society, London.

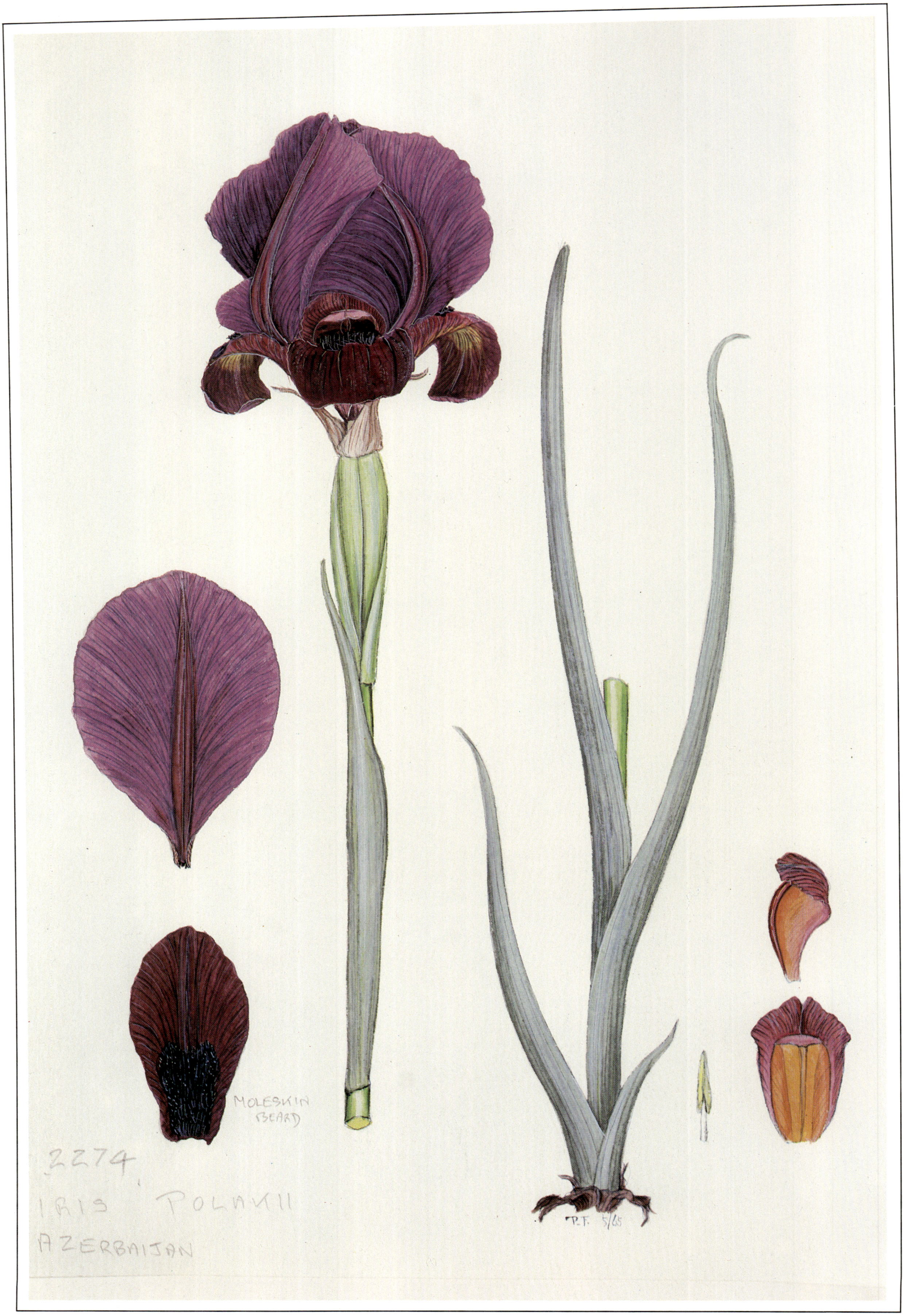
MOLESKIN BEARD
2274
IRIS POLAKII
AZERBAIJAN
P.F. 5/65

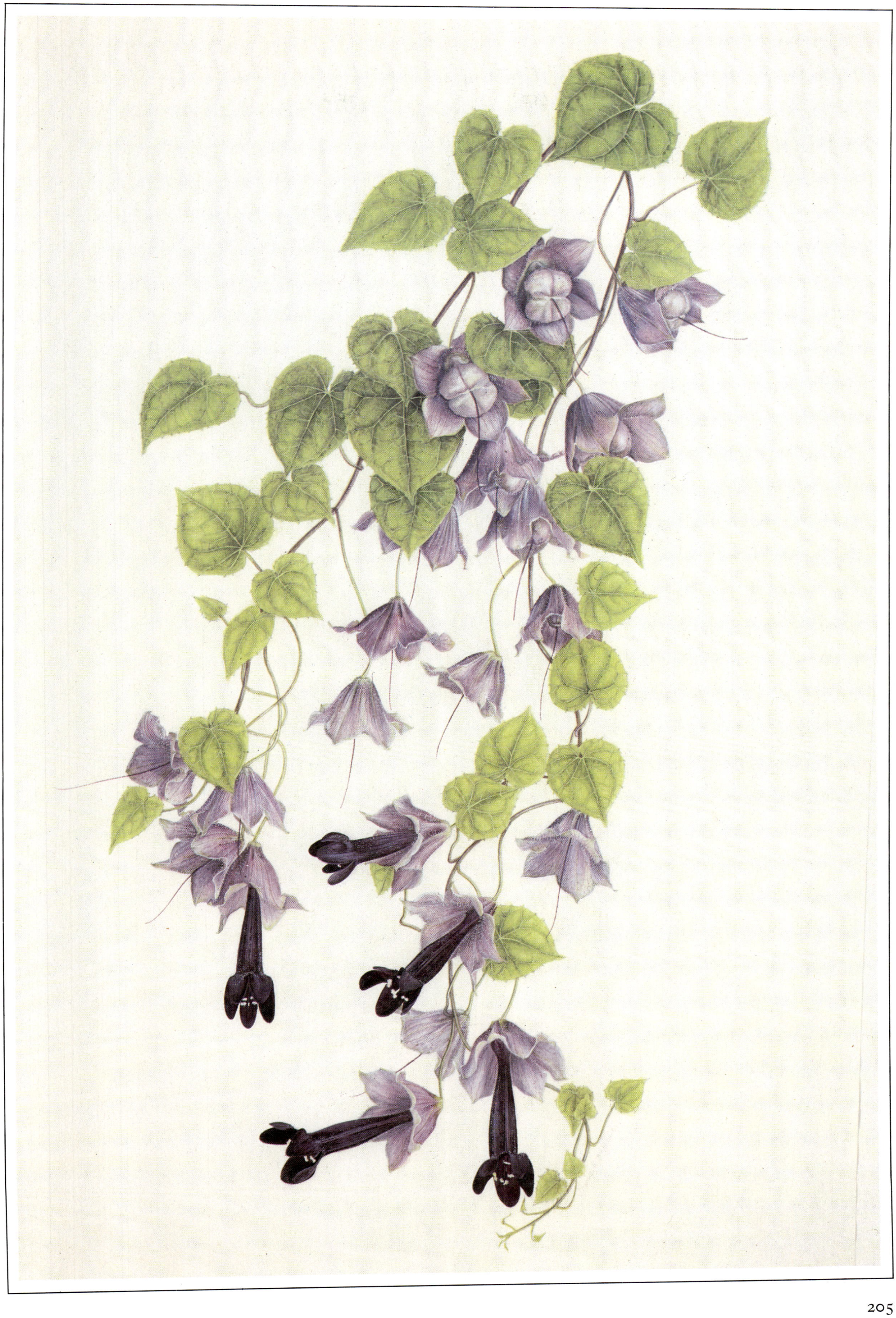

CHAPTER XIV

Monographs

Until the end of the eighteenth century, illustrated botanical books had concentrated on a particular geographical area, or had been florilegia, selections of interesting plants which had flowered in the gardens to which the artist had access. The nineteenth century saw the rise of the illustrated monograph on a single group, often a family or a genus. The most famous of them all is probably Redouté's *Les Roses* (1817–24), but this and, to a greater extent, *Les Liliacées* (1802–16) still have much of the florilegium in them, and do not attempt to be complete. Among the earliest attempts at comprehensive monographs were l'Heritier's *Geraniologia* (1791 or 1792), which remained unfinished, and his *Cornus* (1788).

One of the earliest of the true monographs, and still one of the most sumptuous, was Lambert's *Description of the Genus Pinus* (1803–24). Aylmer Bourke Lambert (1761–1842) was a keen amateur botanist and also, as the son of the Hon. Bridget Bourke, daughter and heiress of Viscount Mayo, a man of considerable private means. Though he never attained the eminence of his friends Sir Joseph Banks and Sir J. E. Smith, he built up a very important herbarium and library, including the Chilean collections of Ruiz and Pavon, and was one of the founder members of the Linnean Society of London. He is remembered chiefly for his *Pinus*, a splendid folio,

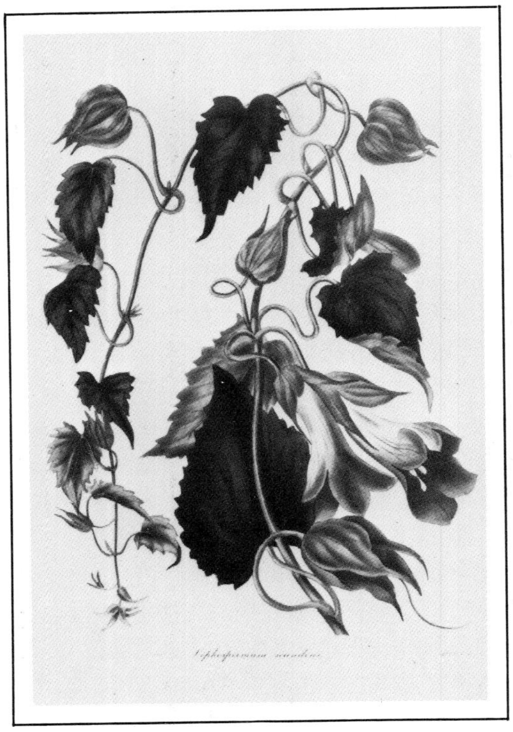

'*Lophospermum scandens*', now known as *Maurandia scandens*, from Mexico. Hand-coloured lithograph after a drawing by Mrs Bury from Benjamin Maund's *The Botanist* (Volume I, 1837).

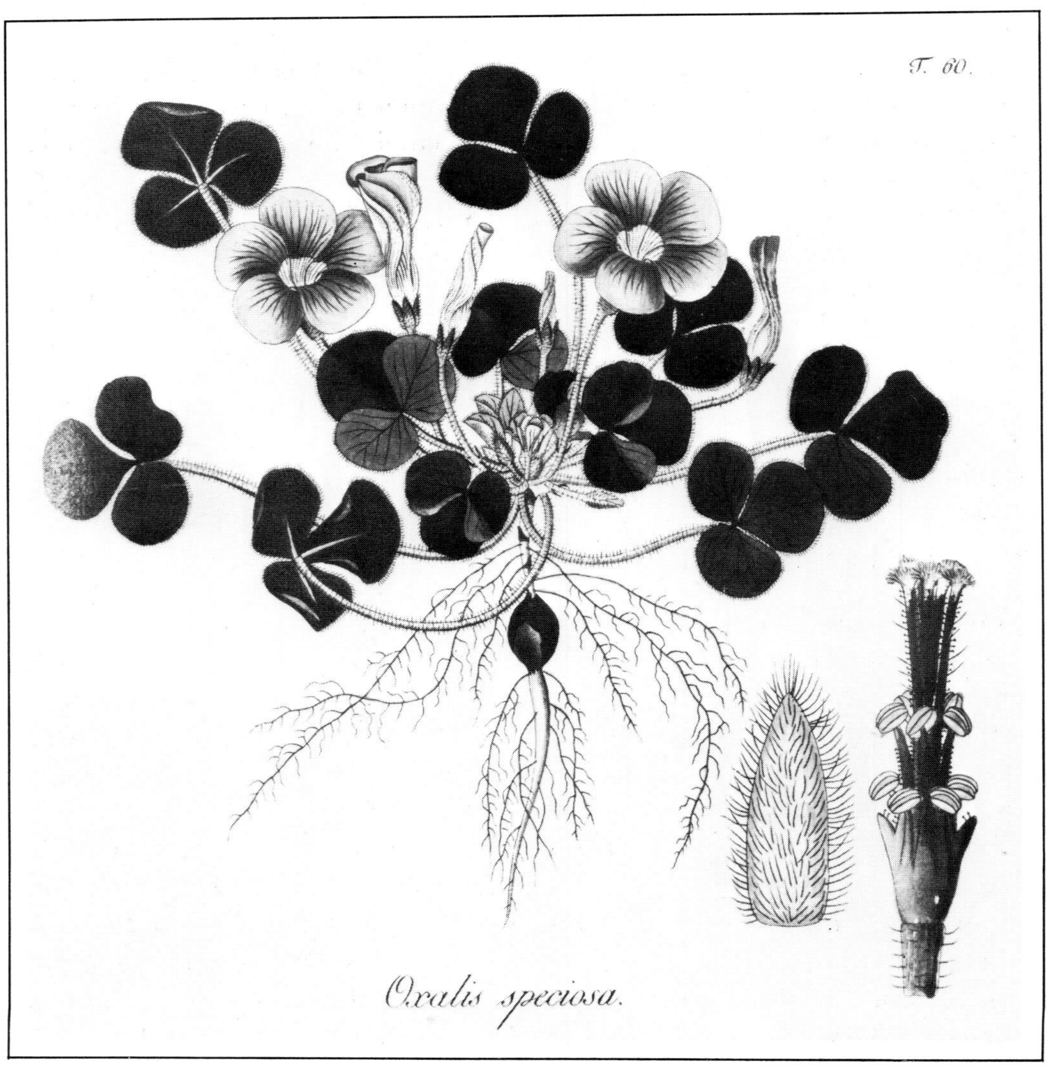

Left:
'*Oxalis speciosa*', now called *Oxalis variabilis*, from South Africa. Hand-coloured engraving from Nikolaus von Jacquin's monograph on *Oxalis* (1794).

Previous page:
Rhodochiton volubile, watercolour on paper, by Joanna Langhorne, around 1968. A member of the nightshade family, Solanaceae, this delicate climber was introduced into European gardens from the wet mountain forests of Mexico in 1829, but is still rarely seen. Collection of the artist.

Opposite page:
Top: the ivy-leaved geranium (*Pelargonium peltatum*), a commonly grown garden species that is a native of South Africa; hand-coloured engraving from Curtis's *Botanical Magazine* (Volume I, 1793). *Centre: Pelargonium praemorsum*, again from South Africa; hand-coloured engraving from *The Botanist's Repository* (1797–1812), edited by Henry C. Andrews. *Bottom:* a heath, *Erica massoni* (named after Francis Masson); hand-coloured engraving from *The Heathery, or a monograph of the genus Erica* (1804–06) by Henry C. Andrews.

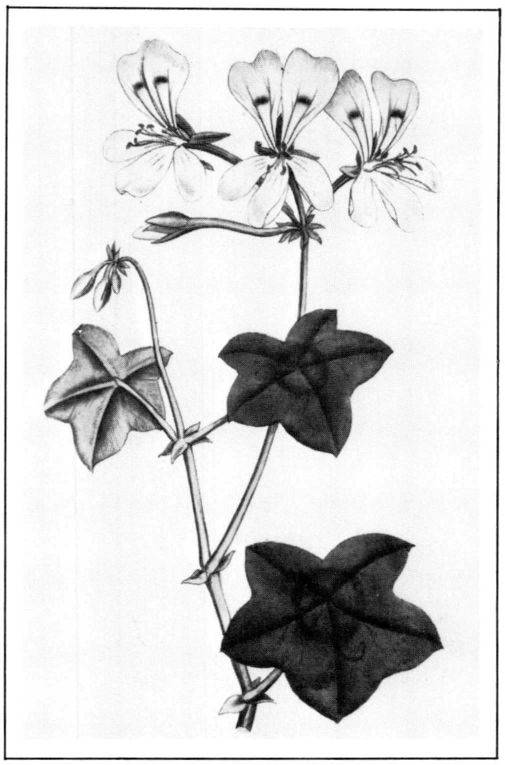

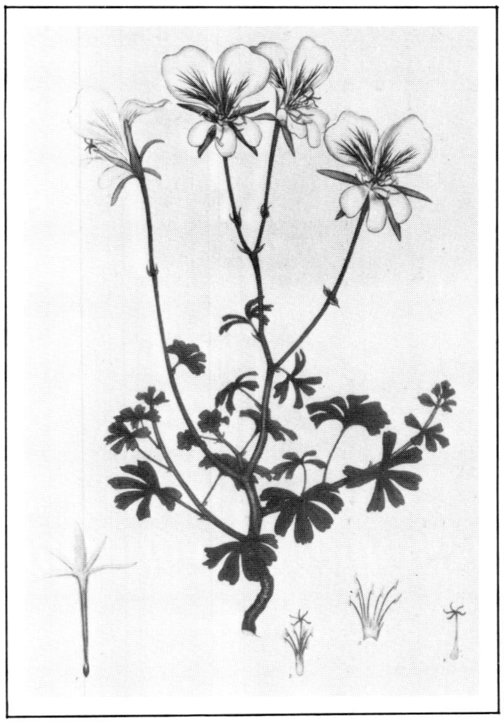

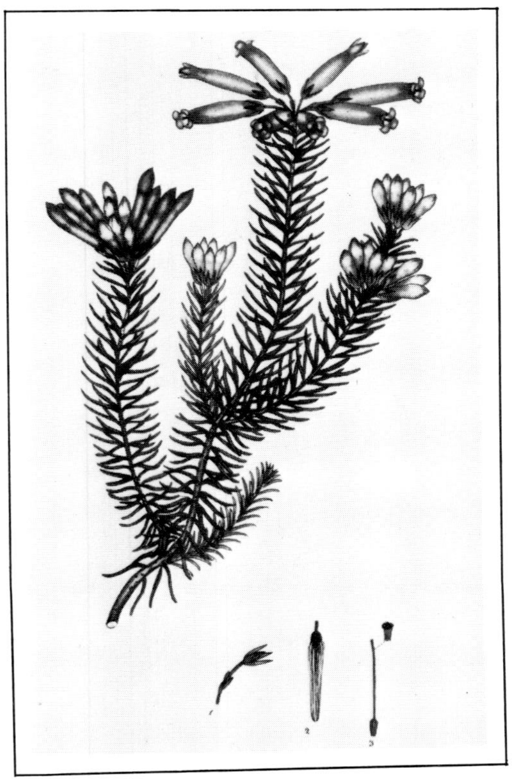

with 42 hand-coloured engravings after Ferdinand Bauer, with others by Francis Bauer and one each by Ehret, Sydney Parkinson and James Sowerby. It included most of the pines then known, and a second volume with an additional twelve plates by Sowerby, the Bauers and John Lindley appeared in 1824. The book's importance is shown by the appearance between 1828 and 1837 of a second edition in three volumes; a smaller, quarto edition appeared between 1837 and 1842. The plates are of the artistic and botanical that would be expected from the artists involved. Lambert's text includes descriptions of the species, some of which, such as *Pinus banksiana*, were being described for the first time.

A character very different from Lambert produced monographs at the same time. He was Henry C. Andrews, of whose life little is known except that he lived in Knightsbridge, then a rather racy village on the outskirts of London. His books were mostly small, cheap and popular, but they contained illustrations of a large number of new species, and the plates were clear and good, though lacking the finesse of Bauer or Redouté. Andrews was active from 1794 to 1830, and his earliest work was *The Botanist's Repository*, a journal similar in spirit and quality to Curtis's *Botanical Magazine*. Ten volumes appeared between 1797 and 1812. The success of this enterprise prompted Andrews to produce the four-volume *Coloured Engravings of Heaths* (1802), which was followed in 1804–06 by *The Heathery; or a Monograph of the Genus Erica*. Heaths were greatly in vogue at the time, with many exciting new species being introduced from South Africa. Most of Andrews's specimens came from the collection of the Marquis of Blandford at Whiteknights, Reading, or from the Hammersmith nursery of Lee and Kennedy (Andrews's father-in-law), whose catalogue listed 282 species and varieties. The *Erica* craze lasted until the middle of the nineteenth century, but today no more than a handful of Cape species are grown in Europe. Andrews both drew and engraved all his own plates. Floral details were carefully depicted, and the plates were printed in green ink before being hand-coloured. There was a concise botanical text including advice on growing heathers, together with Latin descriptions by Mr Wheeler, Demonstrator in Botany to the Society of Apothecaries at the Chelsea Physic Garden.

Andrews's second monograph, published in 1805, was devoted to another primarily South African genus which was having a great rush of popularity, *Geranium*. As for the heathers, Andrews insisted on working only from the living plant. He mentions the difficulties of classification caused by hybridisation of cultivated geraniums and complains of the division of the genus into *Geranium* and *Erodium* in l'Heritier's *Geraniologia*. If such generic divisions were generally adopted, he writes, the approach to botanical science 'would be so choked up with ill-shaped useless lumber, that, like a castle in a fairy tale, guarded by hideous dwarfs, none but the botanic Quixote would attempt investigation.' Nearly all the species that Andrews called *Geranium* are now placed in the genus *Pelargonium*. The standard of the plates is very high, often with three or more varieties of a species represented on a page. Andrews, who found widespread disagreement between botanists and gardeners when he looked for an author for the text of *The Botanist's Repository*, decided to write the text himself '. . . the author therefore thinks it much better to try his own strength, however weak, than to remain tottering between the support of two such unequal crutches'. In spite of this he did receive help from botanists for some of his later volumes. Andrews's third and least successful monograph was *Roses*, begun in 1805 and completed in 1828, in which the plates are stylised in drawing and garish in colouring.

Two minor but interesting monographs appeared in Paris in spite of the upheavals caused by the Revolution. In 1808, François Delaroche, who wrote the text for the fifth and sixth volumes of Redouté's *Les Liliacées*, published *Eryngiorum nec non Generis novi Alepideae Historia*, a monograph of the sea hollies. Fifty species are described, and the 32 elegantly engraved plates (some by Dien) are after drawings by Turpin, Poiteau and F. B.

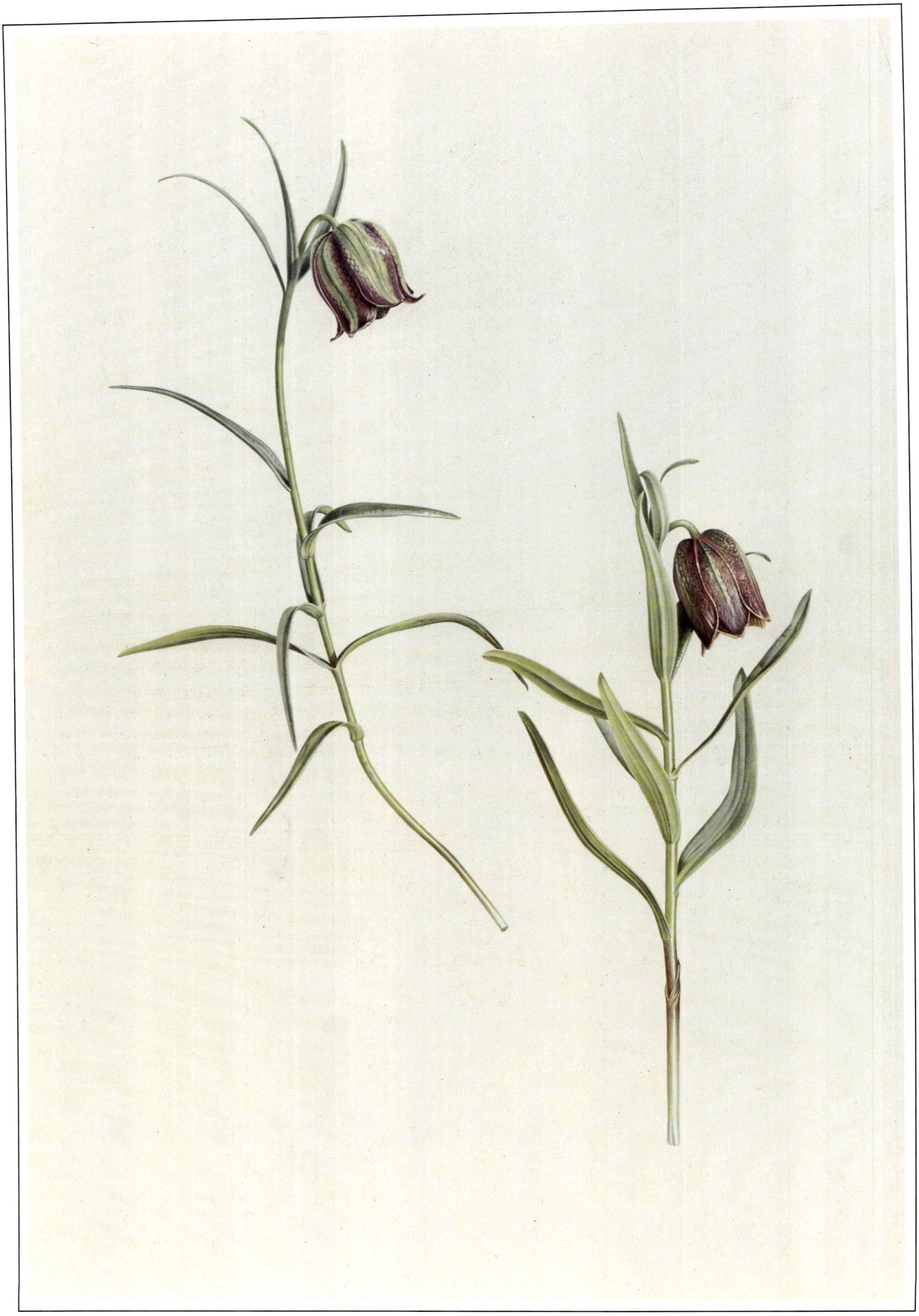

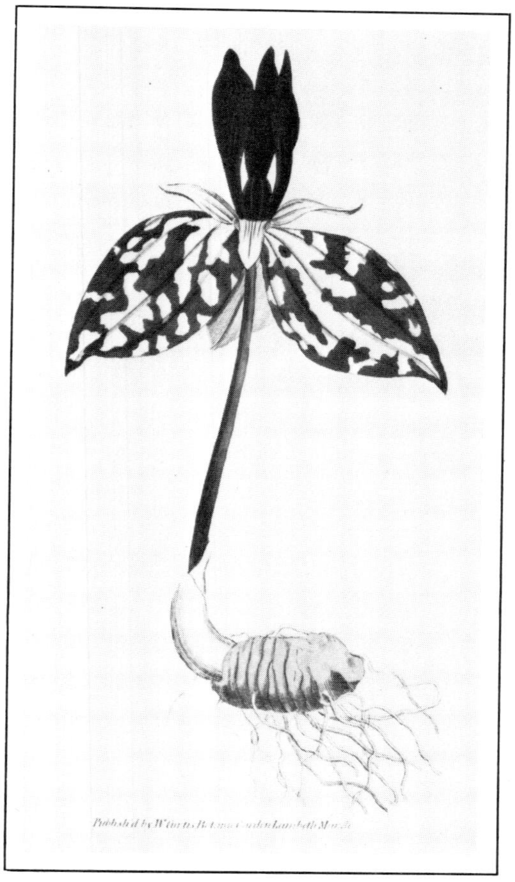

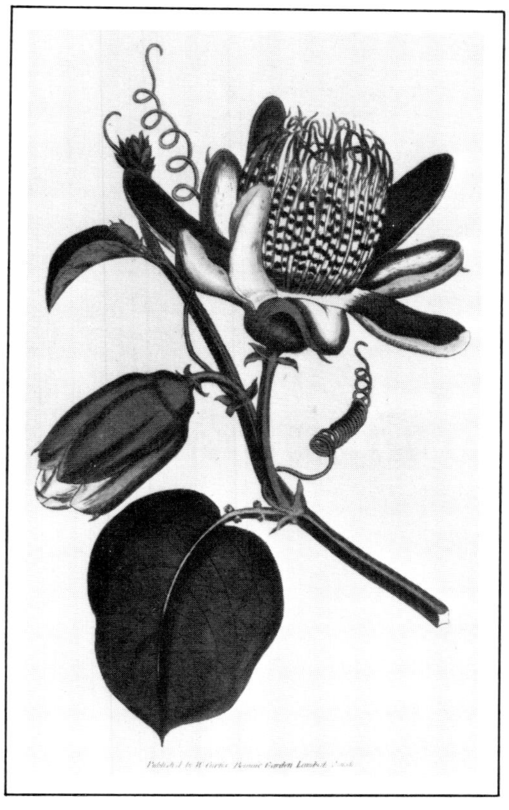

Two hand-coloured engravings from Curtis's *Botanical Magazine* (Volume II, 1796). *Top: Trillium sessile*, a native of woods in North America. *Above: Passiflora alata* from Peru.

Two fritillaries, *Fritillaria messanensis* and variety *atlantica* of the same species, found respectively in Sicily, Greece and Crete, and in the Atlas Mountains of Morocco. Painted in watercolour on vellum by Rory McEwen, 1977. Private collection.

Debalzac. Most copies are uncoloured but do justice to the decorative qualities of these unusual plants. A smaller monograph was an octavo on Madagascan orchids by Louis-Marie Aubert du Petit Thouars, who himself drew the 108 species from nature. The line engravings are careful and delicate. The book is dated 1804, but it did not appear until 1822.

The botanical tradition of Jacquin in Vienna was continued by Nicholas Thomas Host (1761–1834), who published two major works, a monograph on willows, *Salix* (1828) and an account of the grasses of Austria (1801–09). *Salix* has 105 delicately hand-coloured engravings, showing the species in flower and in mature leaf by J. Ibmayer (1801–30). The *Icones et Descriptiones Graminum Austriacorum* contains four hundred coloured engravings, also by Ibmayer, of sedges as well as grasses, and covers all the species then known in the Austrian Empire.

Another botanist working in Austria was Caspar von Sternberg, who produced three volumes on saxifrages entitled *Revisio Saxifragarum Iconibus Illustrata*. The first volume, containing 83 species, appeared in 1810, with supplements in 1822 and 1831, the last being published in Prague. In his introduction, Sternberg complains of the difficulties caused by the Napoleonic wars and of the problems of communication with gardeners in England, especially Adrian Hardy Haworth, who is best known for his work on succulents, but was also author of *Saxifragarum Enumeratio* (1821). Haworth divided *Saxifraga* into several genera, none of which were recognised by Sternberg. Small wonder that Andrews despaired of botanists, as Haworth was one of those who helped him with the text of *The Botanist's Repository*. The plates in Sternberg's monograph, the work of several artists, are variable in quality. The most notable of the artists was Jacob Sturm (1771–1848), the author of the small *Deutschlands Flora*, who came from a Nuremberg family of engravers, and engraved many of the plates, as well as drawing most of the later ones.

A unique book on grasses was the *Hortus Gramineus Woburnensis* by George Sinclair, who, for seventeen years was gardener to the sixth Duke of Bedford. The 'plates' consist of pressed specimens of grasses, their leaves painted a sometimes lurid green, pasted on the page. Red clover (*Trifolium pratense*) has been included, and in some copies its flower is painted red. Two editions appeared, in 1816 and 1825, but the second had engravings in place of the dried plants.

With the end of the Napoleonic wars in 1815, Europe became more prosperous, while the industrial revolution made some Englishmen extremely rich. Huge fortunes were made in industry and trade, and the rise in the price of corn caused by the war made the large landowners even wealthier. Huge and lavishly illustrated botanical books were published to take advantage of the new prosperity.

The perfect subject had first been seen by the Bohemian botanist Thomas Haenke in 1801 and subsequently noticed by many other travellers. Nevertheless, the Giant Waterlily was not named until 1837 when John Lindley christened it *Victoria regia* to honour the new Queen. It was collected by Sir Robert Schomburgk, a German by birth, who was travelling for the Royal Geographical Society up the River Berbice in British Guiana; the first plant to flower in Britain, at Chatsworth House in 1849, was grown from seeds sent to Sir W. J. Hooker, who gave a seedling to Joseph Paxton. With its amazingly engineered leaves, up to two yards in diameter, and its large flowers, it was a natural choice for a huge monograph. The first to appear was John Lindley's *Notice of Victoria Regia* (1837), with hand coloured lithographs by Gauci after drawings made in the wild by Schomburgk. This was a folio about 28 × 20 inches in size. Ten years later, Hooker himself produced a monograph with drawings by W. H. Fitch of plants growing at Syon House and at Kew. This is slightly larger, about 30 × 20 inches. Again the coloured plates are hand-coloured lithographs. An almost equally vast volume was produced in America in 1854, with chromolithographs by W. Sharp from specimens grown at Salem, Massachusetts. This showed details of the plant's life history. By this time, it was known that

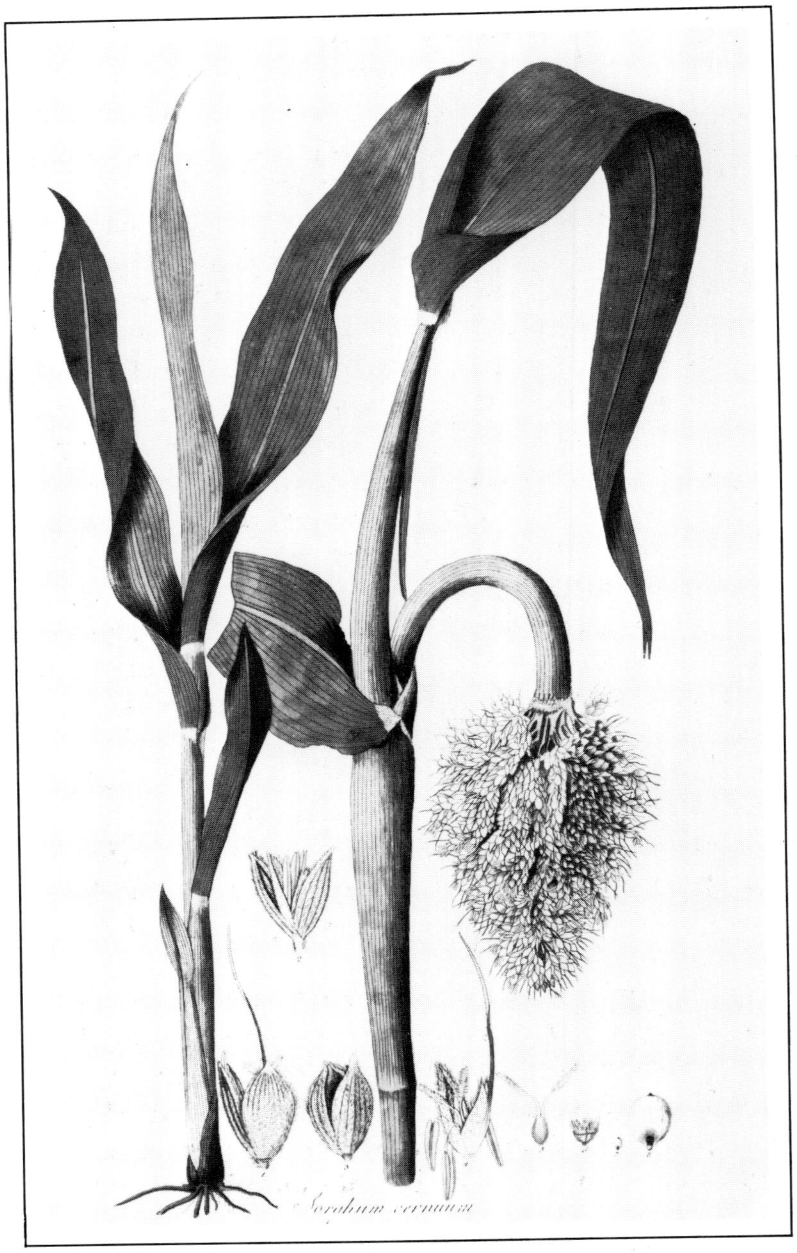

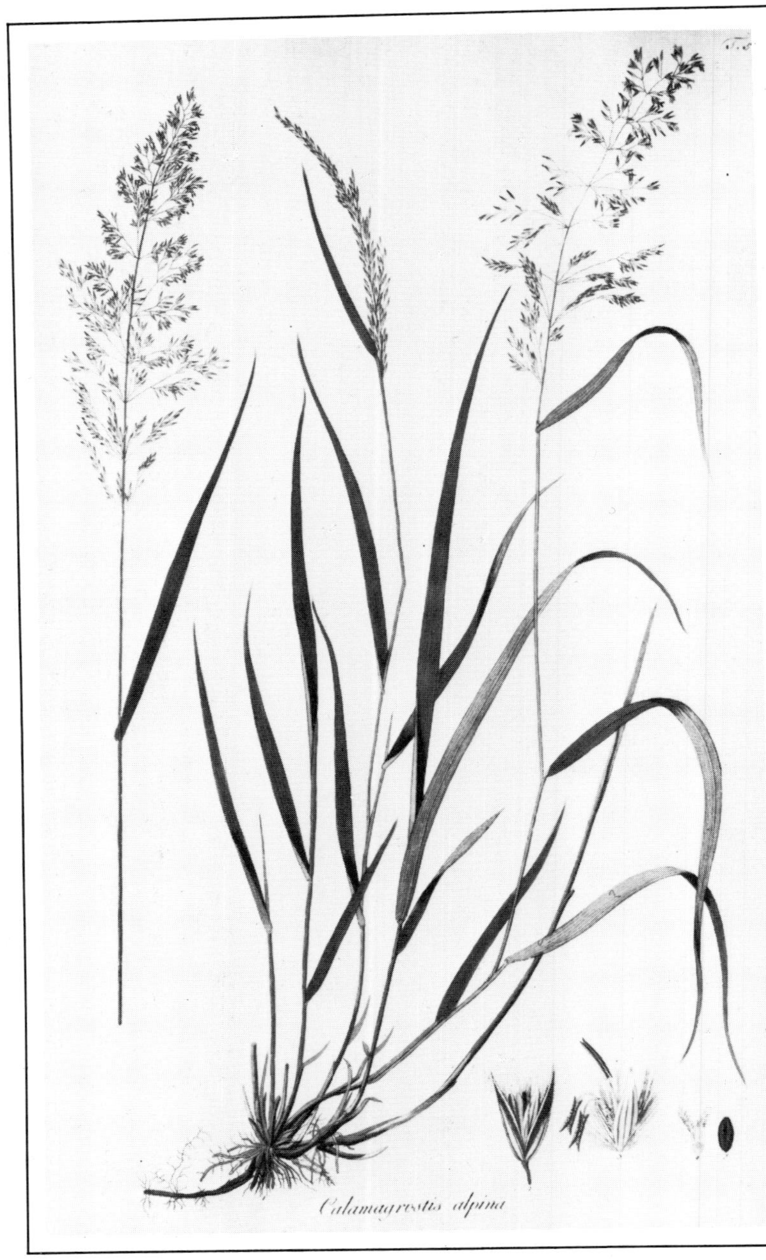

A tropical millet (*Sorghum bicolor*) and *Calamagrostis alpina.* Hand-coloured engravings by J. Ibmayer from *Icones et Descriptiones Graminum Austriacorum* (1801–09) by Nicolas Thomas Host.

Victoria regia was fairly common in slow rivers and backwaters of tropical South America east of the Andes.

South America also provided the subject matter for another enormous book, James Bateman's *Orchidaceae of Mexico and Guatemala* (1837–41), of which only 125 copies were printed. The work was dedicated to Queen Adelaide the widow of William IV. Apart from botanical plates, it contains vignettes and plates of native costume, jewelry and shells by Rudolph Ackermann and George Cruikshank. It was one of the first products of the 'Orchido-Mania, which now pervades all (and especially the Upper) classes,' and even extended to Windsor Castle itself.

James Bateman (1811–97) of Knypersley Hall in Staffordshire was a landowner with large private means and also a keen grower of orchids. Many of the plants were grown in his own greenhouses, or in other collections, such as those of the Duke of Devonshire or of the nurserymen Loddiges. Three collectors were responsible for most of these introductions: Skinner, Theodore Hartweg and Colley, who was commissioned by Bateman himself.

Skinner was a merchant from a well-known Scottish ecclesiastical family who lived in Guatemala, working in the indigo and cochineal trade. His early interest was in birds and insects; he did not turn his attention to botany until Bateman saw some specimens he had sent back to the Manchester Natural History Museum and wrote asking him to collect orchids, giving descriptions of some particularly desirable species. Skinner's first shipment were all new introductions, very carefully packed, and arrived in excellent condition. He continued to send regular consignments of orchids and introduced at least a hundred new species, including the now familiar *Odontoglossum grande* and *pulchellum* as well as *Lycaste skinneri* and *Cattleya*

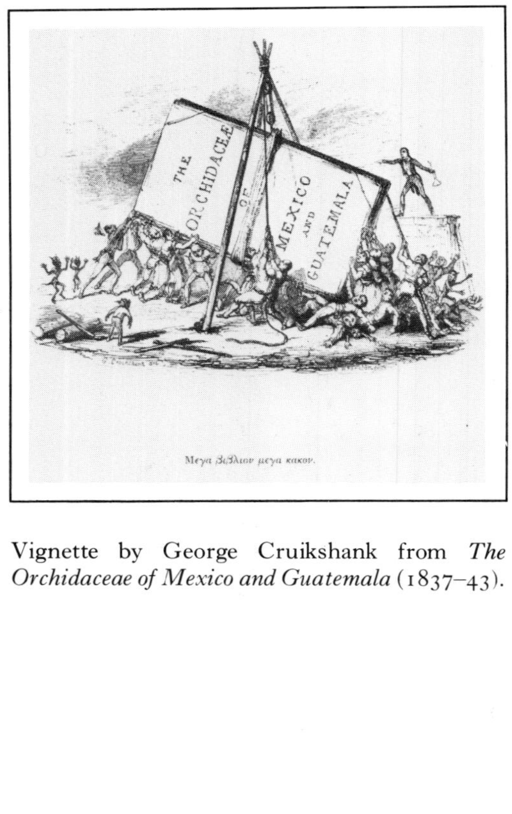

Vignette by George Cruikshank from *The Orchidaceae of Mexico and Guatemala* (1837–43).

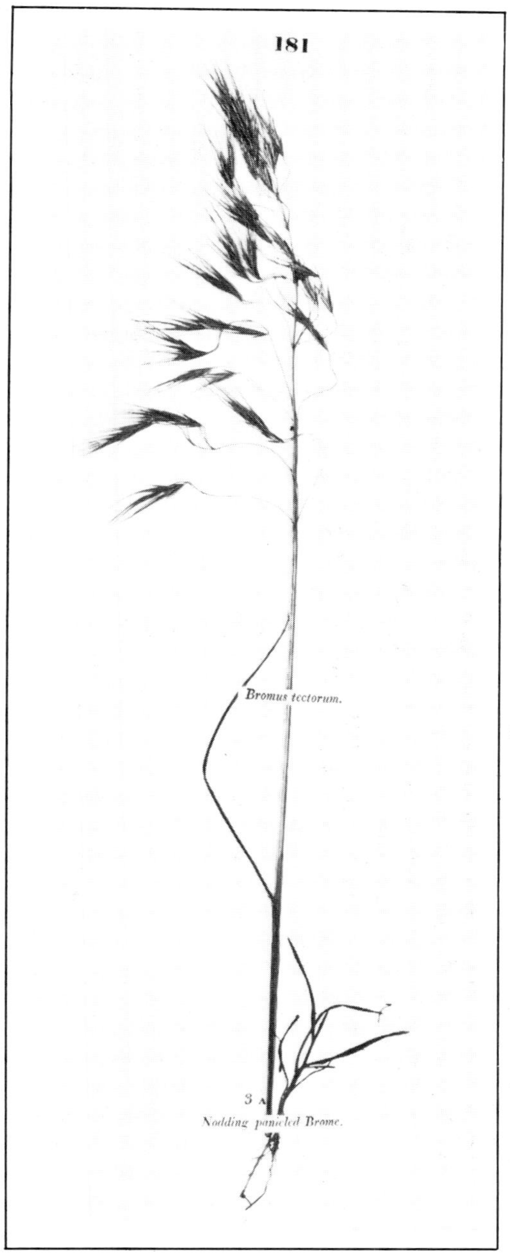

Anisantha tectorum, a pressed specimen from *Hortus Gramineus Woburnensis* (1816) by George Sinclair.

Ixias. Coloured lithograph from *The Ladies' Flower Garden of Ornamental Bulbous Plants* (1841) by Jane Wells Loudon.

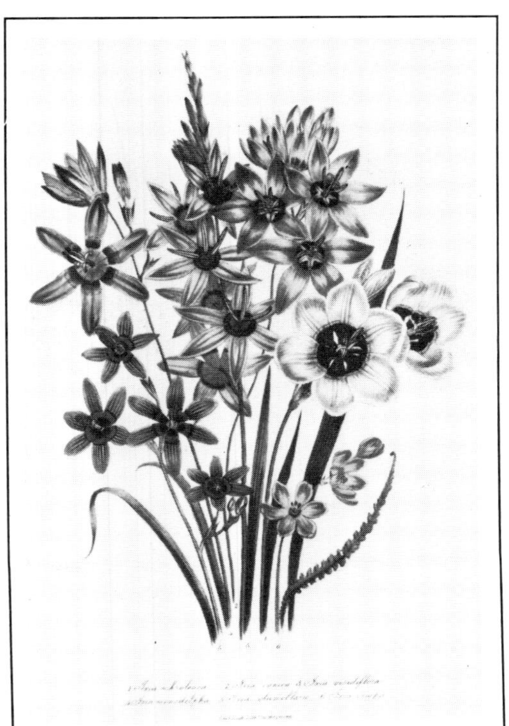

skinneri, both named in his honour. He was never to see the results of his collecting, as he died of yellow fever in Panama on his way home in 1867.

Theodore Hartweg, a German, was sent out to Central America in 1836 as a collector for the Horticultural Society, which was interested by the plants that were being sent back from Central America. He went to search particularly in the highlands of Mexico, from where he sent back both seeds and orchid plants. His seven-year stay was such a success that he was sent again from 1845 to 1848, this time to California as well as to Mexico. Many new plants were introduced by him, notably *Rhodochiton volubile* and many cool-growing *Odontoglossum* species, the subject of Bateman's *Monograph of Odontoglossum* (1864–74). This had lithographs by W. H. Fitch, as did Bateman's *A Second Century of Orchidaceous Plants* (1867).

The plates in the *Orchidaceae of Mexico and Guatemala* were mostly by Miss Drake and Mrs Withers; all were lithographed by M. Gauci. Seven of Mrs Withers's originals, which are in the Lindley Library of the Royal Horticultural Society, emphasise Gauci's great skill, and the high quality of the hand colouring – the original watercolours are very little superior to the published plates. Although Mrs Withers describes herself as 'Flower Painter to Queen Adelaide', little is known of her life, except that she taught flower painting. Some paintings of fruit by her, showing her great skill, were done for the Horticultural Society and are now in the Lindley Library: some of them were reproduced in the Society's *Transactions* and others in *The Pomological Magazine* as well as in Benjamin Maund's *The Botanist*. Miss Drake is also almost unknown, and little else of her work was published, except for some plates in Lindley's *Sertum Orchidaceum* and in *The Botanical Register*.

Other books devoted to orchids appeared at about the same time as Bateman's. John Lindley (1799–1865), the Secretary of the Horticultural Society, worked particularly on orchids; his *Sertum Orchidaceum* (1837–41) was dedicated to the Duke of Devonshire. The fifty hand-coloured plates lithographed by Gauci after Miss Drake and Schomburgk are inferior to Bateman's *Orchidaceae* in printing and much inferior in the quality of the colouring.

W. J. Hooker's *Century of Orchidaceous Plants* (1851) with lithographs by Fitch was the precursor of Bateman's second century; its hundred plates are reprinted from the *Botanical Magazine*. Orchidomania continued throughout the century; a later orchid book was Frederick Sander's *Reichenbachiana* (1888–94), illustrated with chromolithographs, mostly after Henry Moon. At the other end of the scale from the enormous orchid books, but in their own way no less successful, were the illustrated flower books aimed at gardeners, such as Mrs Loudon's *The Ladie's Flower Garden of Ornamental Bulbous Plants*, published in 1841 and containing 58 pages of illustrations lithographed after her own paintings, illustrating 305 species, many of them from South Africa. Jane Wells Webb (1807–58) had been left in poverty by the death of her father when she was seventeen and had to earn her living by writing. She began with a romantic work of science fiction called *The Mummy, a Tale of the Twenty-Second Century*, which was published in 1827. J. C. Loudon (1783–1843), a landscape gardener and publisher of horticultural and architectural books acquired a copy for review in one of his journals, and determined to make the acquaintance of the author, whom he presumed to be a man. They met in February 1830 and were married in September of the same year. After the success of her bulb book, Mrs Loudon produced a series similar in style on wild flowers, perennials, and ornamental greenhouse plants.

Orchidomania reached its peak from about 1830 to 1850; it was succeeded by Fernmania, which lasted until the end of the century and resulted in a flood of fern books, some of which were nature-printed. The first book to exploit the fern craze and show the beauties of nature-printing was Thomas Moore's folio, *The Ferns of Great Britain and Ireland*, edited by John Lindley and printed by Henry Bradbury, who was instrumental in introducing nature-printing to Great Britain. It appeared in 1855, and a smaller popular version appeared in 1859. Coloured inks were used for the printing: green for the fronds, brown for the roots and an appropriate colour for the stem. The results are especially beautiful for the fronds with their delicate toothing and veining, but the wiry fern roots also come out well. The stems are less good, as they tend to be crushed by the process and to appear thicker than they are.

W. J. Hooker had produced his first fern book much earlier, while he lived in Scotland and before he met Fitch. *Icones Filicum*, with illustrations by R. K. Greville appeared between 1827 and 1831. 250 of Greville's clear drawings showing good dissections of important reproductive parts were skilfully engraved by J. Swan, who made clever use of line shading. Both coloured and uncoloured copies were sold. Greville, a fine and very productive amateur botanist, was also the author and illustrator of *Scottish Cryptogamic Flora* (1823–28) and *Algae Britannicae* (1830). Hooker continued to publish fern books for the next thirty years: the *Genera Filicum* (1838–42), with drawings by Francis Bauer, the five volumes of *Species Filicum* (1846–64) and the more popular *Century of Ferns* (1854), *British Ferns* (1861) and *Garden Ferns* (1862), all with lithographs by W. H. Fitch.

Edward Joseph Lowe published the first of his many fern books in 1856 with the first of his eight-volume *Ferns, British and Exotic* and ending in 1895 with *Fern Growing; Fifty Years Experience in Crossing and Cultivation.* The plates of all these were colour printed and, useful though the books are, compare unfavourably with Hooker's works.

H. J. Elwes (1846–1922) of Colebourne in Gloucestershire was a traveller, sportsman, collector and gardener in the grand style. After being educated at Eton and spending five years in the Scots Guards, he never spent a complete year in England. In his youth he was an enthusiastic slaughterer of all sorts of game and a collector of birds and butterflies, but after his marriage in 1871 he became more interested in gardening and is said to have remarked to E. A. Bowles in later life, 'To think that I spent twenty of the best years of my life catching butterflies.' Between 1870 and 1914 he visited nearly every part of the world where hardy plants are likely to be found, with the exception of Australia and New Zealand. No less than 98 of the new plants that he collected and introduced were figured in the *Botanical Magazine*, including such now familiar garden plants as the giant snowdrop (*Galanthus elwesii*), the orchid *Pleione pricei*, and *Oxalis adenophylla*. His visit to Asia Minor in spring 1874 was particularly significant botanically:

Right:
Beech fern (*Phegopteris connectilis*). Nature print by Henry Bradbury from Thomas Moore's *The Ferns of Great Britain and Ireland* (1855), edited by John Lindley.
Far right:
Crocus banaticus, a beautiful and unusual species from Romania and southern Russia with many distinctive features including unequal petals and broad leaves. Hand-coloured lithograph by F. Huth after a drawing by the author from George Maw's *A Monograph of the Genus Crocus* (1886).

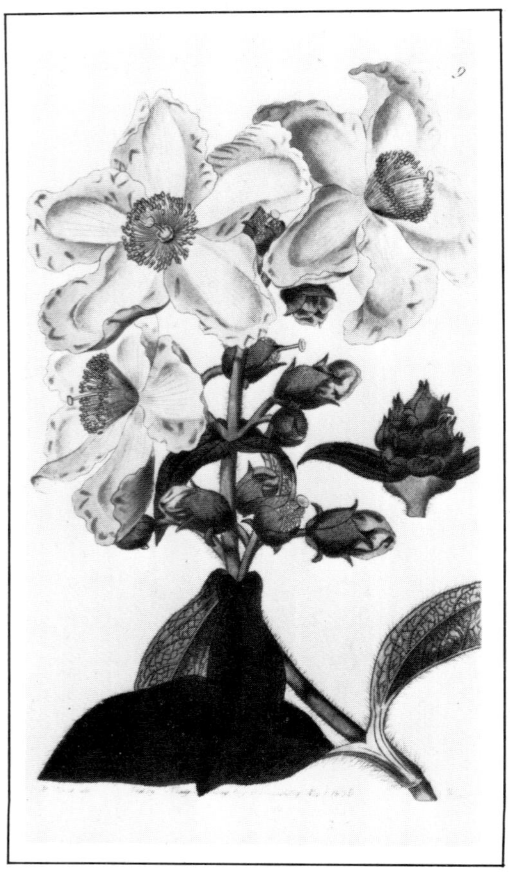

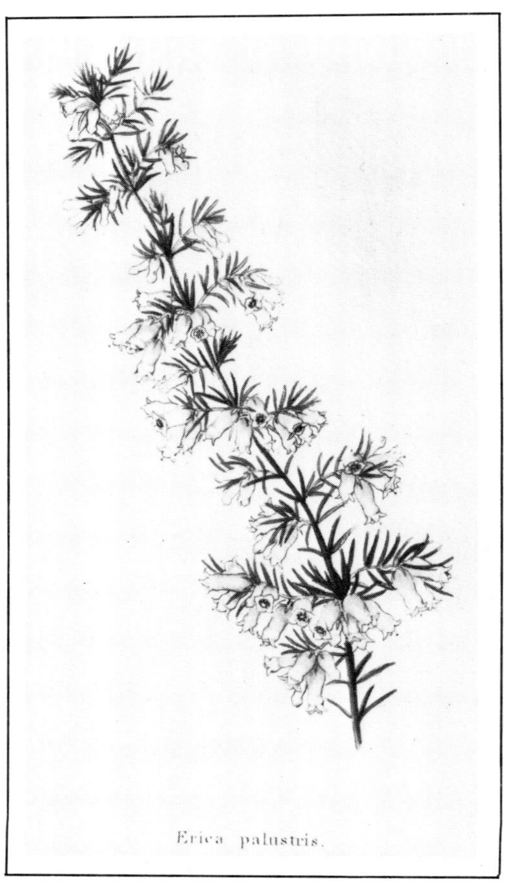

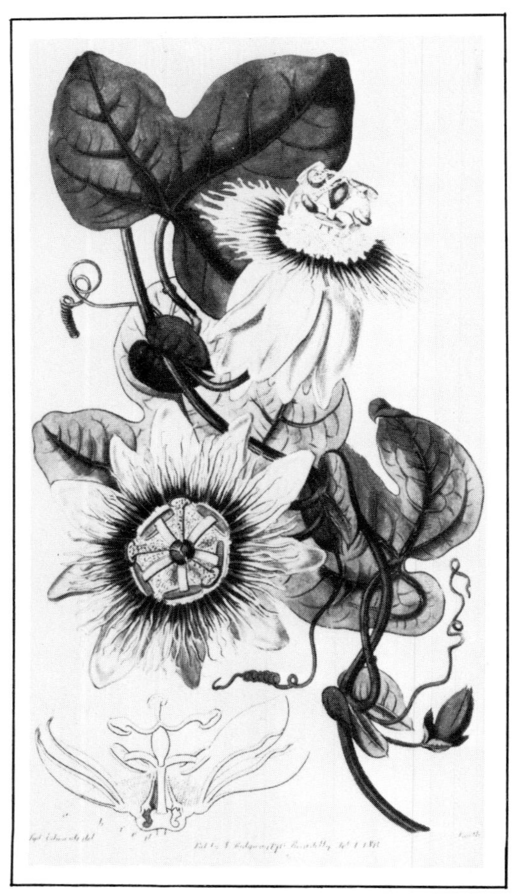

Far left:
A rockrose, *Cistus vaginatus*, from Tenerife; hand-coloured engraving by Watts after a drawing by M. Hart from *Cistineae* (1825–30) by Robert Sweet.
Left:
A heath, *Erica palustris*. Hand-coloured engraving by George Cooke from *The Botanical Cabinet* (1817–33) produced by Conrad Loddiges and Sons.
Below:
'*Passiflora glauca*'. Hand-coloured engraving by Smith after a drawing by Sydenham Edwards from Benjamin Maund's *The Botanical Register* (Volume I, 1815).

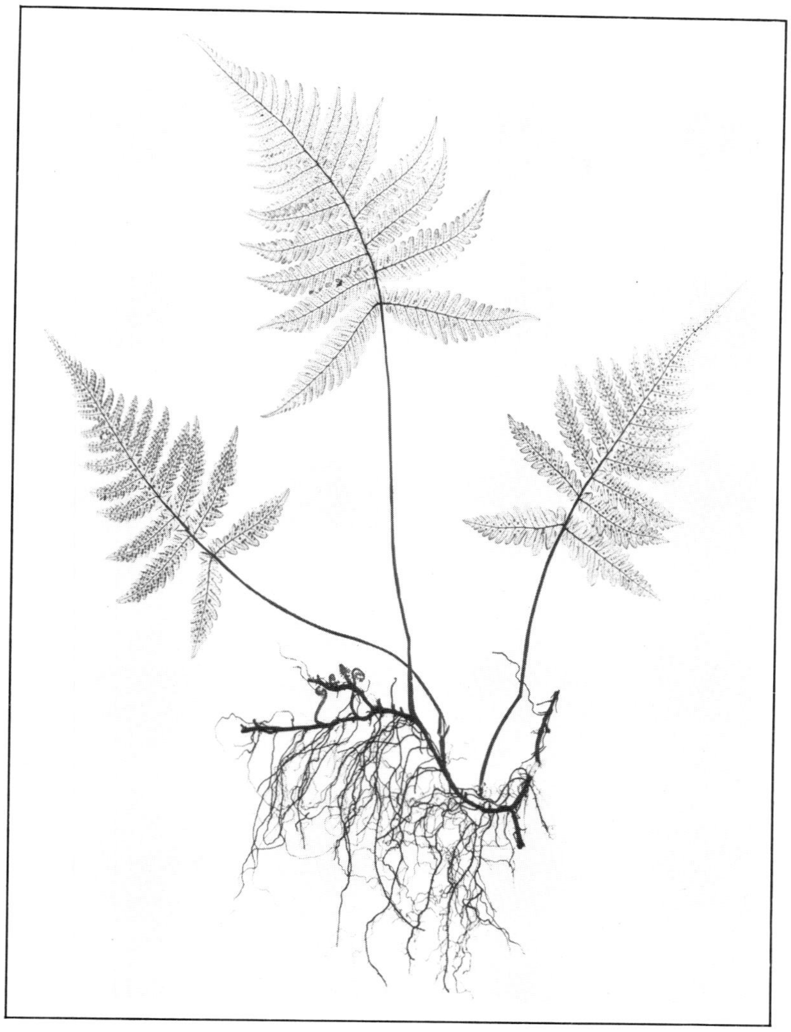

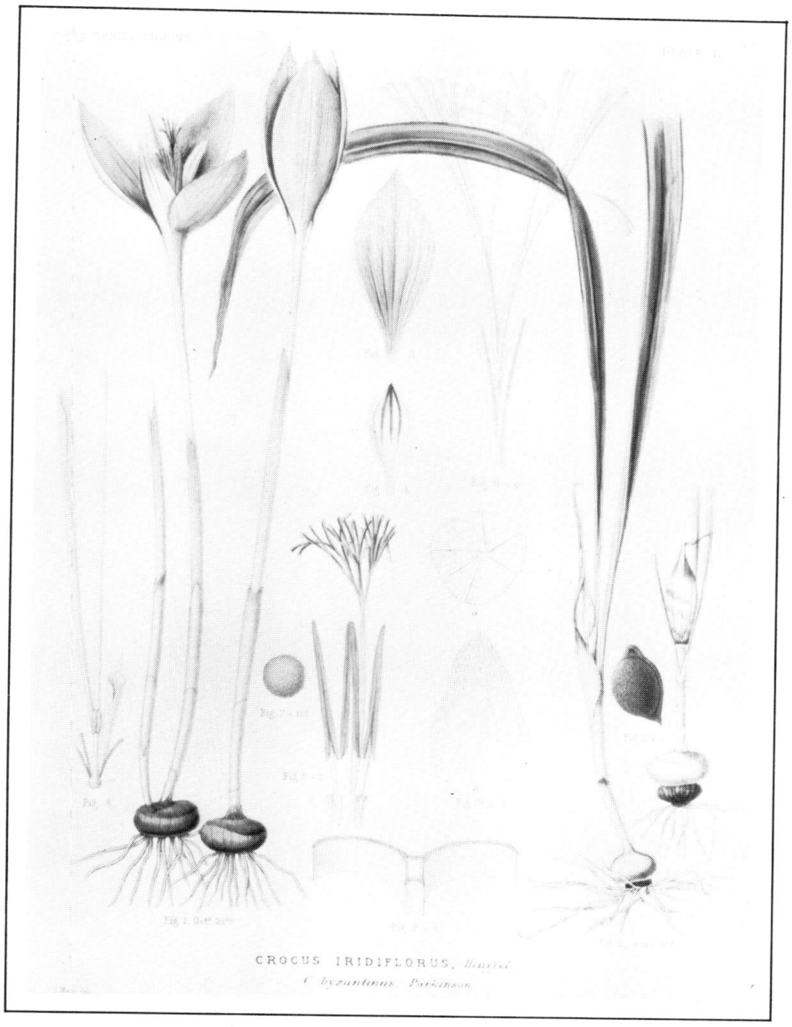

in spite of having most of his collection stolen, he introduced *Fritillaria elwesii* and several species of crocus and tulip, as well as the *Galanthus*. He also did some shooting and killed a particularly fine boar.

He had meanwhile acquired an interest in a tea estate in Darjeeling, and this gave him the chance of several visits into this botanically rich part of the Himalayas. It was here that he first became interested in lilies. He visited India in the winter of 1879–80 with his brother-in-law, F. O. Godman, and acquired bulbs of *Lilium wallichianum*. In 1888, again with Godman, and this time with his wife also, he visited Mexico, narrowly missing death when the train before theirs was destroyed by the collapse of a viaduct. 'We saw more dead men on this journey than ever before or since,' he wrote, describing how they had to walk past the piles of dead passengers and horses. This trip produced *Lilium humboldtii* among other species. In 1898, he visited Central Asia to shoot *Ovis ammon*, the legendary wild sheep; he also saw the rare Przewalski's horse, and brought back a species of lady's slipper orchid, *Cypripedium guttatum*. In 1901–02, he visited Chile, and collected *Nothofagus antarctica* and the Monkey Puzzle (*Araucaria araucana*) as well as a large number of butterflies. From Formosa, which he visited in 1912 with W. R. Price, he brought back *Lilium formosanum* variety *pricei*.

Finding that there was no good book on lilies to which he could refer, Elwes determined to write one himself. It was to be one of the finest monographs of all time. The first volume of his *Monograph of the Genus Lilium*, a folio with fifty hand-coloured lithographs by W. H. Fitch, appeared between 1877 and 1880; it included nearly all the lilies then known in cultivation. Elwes himself wrote the text, dismissing it in the introduction as 'not the work of a scientific botanist, but merely the result of a few years' horticultural study.' Fitch's plates are magnificent, with the flowers mostly drawn life size, but they lack subtlety and do not show the grace and poise which is such a characteristic of most *Lilium* species. By the beginning of the twentieth century, Elwes himself had introduced many new species, and others had been collected in China by such collectors as E. H. Wilson and George Forrest. Fitch started to plan a supplement, entrusting its production to A. S. Grove, son of Sir George Grove, who was a keen lily grower.

Aspidium lachense, drawn and lithographed by W. H. Fitch in *Species Filicium* (Volume IV, 1862) by W. J. Hooker.

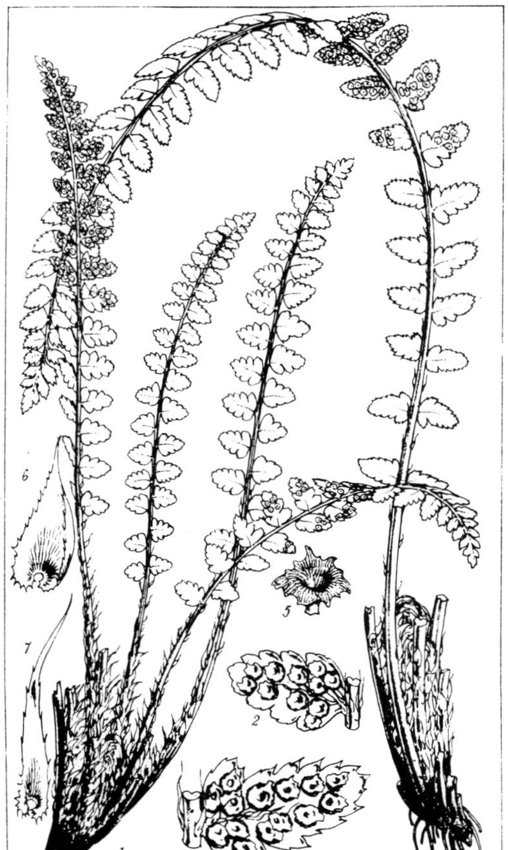

Grove in turn enlisted the help of a professional botanist, A. D. Cotton, then Keeper of the Herbarium at Kew. Lilian Snelling, who was the main artist for the *Botanical Magazine* from 1922 to 1952, was entrusted with the drawing of the plates, which have the grace which Fitch's illustrations lack. The first part of the supplement appeared in 1933, and other parts appeared regularly until Part VII in 1940. The cost of publication was generously underwritten by Dame Alice Godman, F. O. Godman's daughter. Since World War II, two more parts have appeared, the last in 1962, each with five illustrations by Margaret Stones and text by W. B. Turrill. Most were colour printed but forty copies were kept for hand colouring.

Apart from his interest in bulbs, which led to the *Monograph of the Genus Lilium*, Elwes also became an expert on trees, many species of which he planted on his Gloucestershire estate. This led him to produce his other great work, the seven volumes of *The Trees of Great Britain and Ireland* (1906–13). He collaborated with Augustine Henry who had returned from a lifetime working in China and lived, conveniently, at Kew. Native and introduced trees were included and each tree was studied 'from seed to timber'. Large and particularly fine specimens were listed, some of which can still be seen today. Elwes travelled all over the country looking at tree collections, and what he saw was added to his knowledge of many of the exotic species as they grew wild in their native lands. The two authors worked well together, each signing his own pieces of work. In order that difficult groups should not hold up production, the easier species were covered first. Details of leaves, flowers and seeds are illustrated with lithographs, and these are supplemented by excellent photographs of whole trees, many taken in the wild. Elwes undertook publication himself, travelling to Edinburgh to deal with the printers, with whom he established a good working relationship. His energy and boldness paid off, because the book made a profit – one of the very few large botanical books to have done so.

One of the best and most beautiful plant monographs ever written was *A Monograph of the Genus Crocus* (1886) by George Maw. The advertisement for the projected book quoted a testimonial from John Ruskin: 'Seven most exquisite drawings have come and are quite beyond criticism . . . I rejoice to see the *Croci* thus dealt with.' Maw had drawn the plates himself; they were reproduced by lithography and hand coloured. In almost every case the plates were drawn from living specimens, mostly grown by Maw and often also collected by him. Each plate shows a plant in flower and in leaf and has enlargements of seeds, styles, leaf cross-sections and corm tunics. The account of each species is introduced by a charming vignette of the habitat of the plant or of a city associated with it – the crags of the Taurus Mountains in Turkey, the Moorish palaces of Granada or the temples of Athens. Maw was a businessman, the head of a firm of decorative tile and pottery manufacturers, as well as an accomplished amateur geologist and botanist. He travelled widely, visiting the High Atlas Mountains with Sir J. D. Hooker and John Ball in 1871 and Asia Minor in 1877, and began studying crocuses in detail in 1875. His corms and specimens came from many sources, notably from British consuls who were then in all important cities in the region. Some, such as Mr A. Biliotti, the vice-consul in Trebizond, had crocuses named after them. Elwes also supplied him with many of his specimens. Maw's work was thorough and accurate and has only been superseded (and at the same time often confirmed) by recent studies at Kew in which details of the chromosomes, unknown in Maw's day, have been investigated.

Two more fine monographs with hand-coloured illustrations appeared in the last years of the nineteenth century: Frederick Janson Hanbury's *An Illustrated Monograph of the British Hieracea* published in eight parts between 1889 and 1898, and Alfred Fryer's *The Potamogetons (Pond Weeds) of the British Isles* published between 1898 and 1915. The monographs that followed them, such as Ellen Willmott's *The Genus Rosa* (1910) and W. R. Dykes's *The Genus Iris* (1913), were reproduced photomechanically and printed in colour.

Below:
Potamogeton × nitens (*P. gramineus × P. perfoliatus*), a hybrid pondweed that is found locally throughout the British Isles, drawn and lithographed by Robert Morgan in *The Potamogetons (Pondweeds) of the British Isles* (1898–1900) by Alfred Fryer and Arthur Bennett.
Bottom:
A hawkweed, *Hieracium sinuans*, part of a colour plate from *An Illustrated Monograph of the British Hieracia* (1889–98) by Frederick Hanbury.

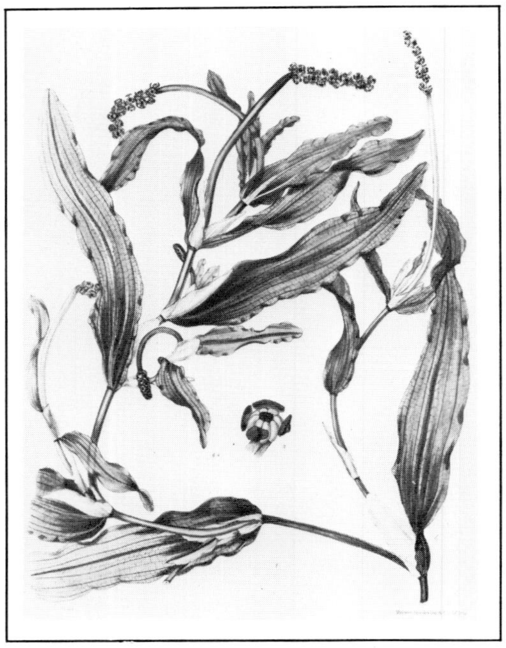

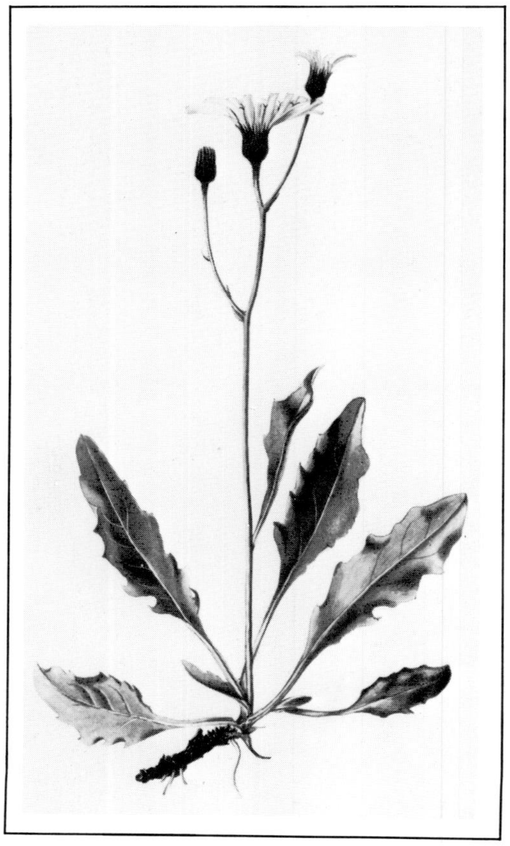

CHAPTER XV
Some Twentieth Century Illustrators

The influence of photomechanical reproduction processes on twentieth-century botanical illustration has been as profound as that of lithography in the nineteenth. No longer does the painter have to rely on the artistic talent of the lithographer or engraver to transform his pictures into images printed on the page – skill is still required to produce a good result, but it is a matter of technology rather than artistry. Modern high-speed printing has meant that in the twentieth century coloured books can be produced without difficulty in runs of thousands of copies. Hand colouring became a thing of the past, mainly because of the high cost of skilled labour, although it did not die out completely until the middle of the century – the plates in the *Botanical Magazine* were hand-coloured until 1948.

The first great colour-printed flower book of the twentieth century was *The Genus Rosa* (1900–14). Its compiler, Ellen Willmott, who was born in 1857, was one of the greatest figures in the late nineteenth and early twentieth century gardening. While her most famous contemporaries, William Robinson and Gertrude Jekyll, were primarily concerned with the design and planting of gardens, Miss Willmott concentrated on the plants themselves, introducing them from abroad, breeding new varieties and growing them in her gardens at Warley Place in Essex and Tresserve near Aix-les-Bains on the edge of the French Alps. Her father, who had made a modest fortune in property and on the stock market, bought Warley Place with 33 acres of land in 1875. Here Miss Willmott lived until her death in 1934. In her heyday, she employed 104 gardeners and grew no less than 100,000 different species and varieties of flower. The number of plants named either *willmottiana* or *warleyensis* after her or her garden are a monument to her importance as an introducer of new species. Her interest in Roses was greatly stimulated by E. H. Wilson's first expedition to China in 1903; and she was one of the rich amateurs who helped finance his collecting. Among the hundreds of new species he introduced was *Rosa willmottiae*.

Alfred Parsons RA, (1847–1920), whom Miss Willmott chose to illustrate her rose monograph, was a well-known painter and president of the Society of Painters in Watercolour. He was not a botanical illustrator – his only other published plant illustrations are eleven plates for *A Garden Flora*, which is an account of the trees and flowers grown in the gardens of Nymans, Sussex. His paintings are nevertheless very attractive, delicate and informal watercolour portraits of 130 individual species and varieties. The early colour printing is not altogether satisfactory, making the plates seem not so much delicate as faded – the greens are especially poor. The book was issued in twenty parts at one guinea each, and only a thousand copies were printed. Miss Willmott was both imperious and unbusinesslike and, driving her publishers, John Murray, nearly to distraction before the work was finally completed in 1914; it still stands unrivalled, both as an account of the species and as a source of illustrations of wild roses.

The notes on the history and cultivation of each species were written by Miss Willmott, and the botanical descriptions were by J. G. Baker, a professional botanist who worked at Kew Gardens. Miss Willmott continued her work on other genera, remaining at Warley until her death, although, because of the general economic state of the country and her former extravagance, she passed her later years in comparatively reduced circumstances.

Miss Willmott was also involved in another important monograph, *The Genus Iris* by W. R. Dykes. She was responsible for much of the preliminary work with Sir Michael Foster, a keen amateur grower of irises. On Sir Michael's death, his note books were passed to Miss Willmott, who collaborated on the book with W. R. Dykes, then a modern language master at Charterhouse School. The illustrations were by F. H. Round, who was drawing master at the school, and thus conveniently placed for receiving specimens of such evanescent flowers. Round's painting of all the flowers is good, but the leaves appear flatter even than they are in real life, and fade away rather unsatisfactorily at the bottom of the page in most plates. The book, published by the Cambridge University Press in 1913, contained 47 coloured plates by Round. Dykes left teaching in 1920 and was Secretary of the Royal Horticultural Society until his death at the age of 48 in a motor accident in 1925. After his death, his widow published his *Notes on Tulip Species* (1930), illustrated with 54 of her own paintings. Her drawing of the tulips is good, but the colour printing rather poor.

In many respects, the work of Miss Willmott and W. R. Dykes marked the end of the tradition of great flower books that had continued, more or less without a break, since the seventeenth century and depended quite heavily on the efforts of wealthy amateur botanists and gardeners, as well as on the patronage of usually royal or noble plant fanciers. Just as in the field of general publishing, World War I marked the start of a decline in the output of lavishly produced gift books – a decline that was hastened by the changed economic realities of the postwar world – so, too, did it mark the end of the spectacularly grand natural history book.

However, plant illustration survived, particularly in the world's botanic gardens and the museums, universities and learned societies with which they are associated. Many of the artists who have kept the tradition alive are represented among the 373 whose work is in the unique collection of the Hunt Botanical Library in Pittsburgh, a collection founded on that of Rachel McMasters Hunt (1882–1963).

The Royal Horticultural Society continued to be an important patron of botanical painting in England until the 1960s, through its management of the *Botanical Magazine*, which it took over in 1921 with the encouragement of H. J. Elwes, and through its publication of a series of botanical monographs of genera important to gardeners. The editorship of the *Botanical Magazine* remained at the Royal Botanic Gardens, Kew, where it had been since the days of Hooker and Fitch; it is at Kew that most of its artists have worked ever since.

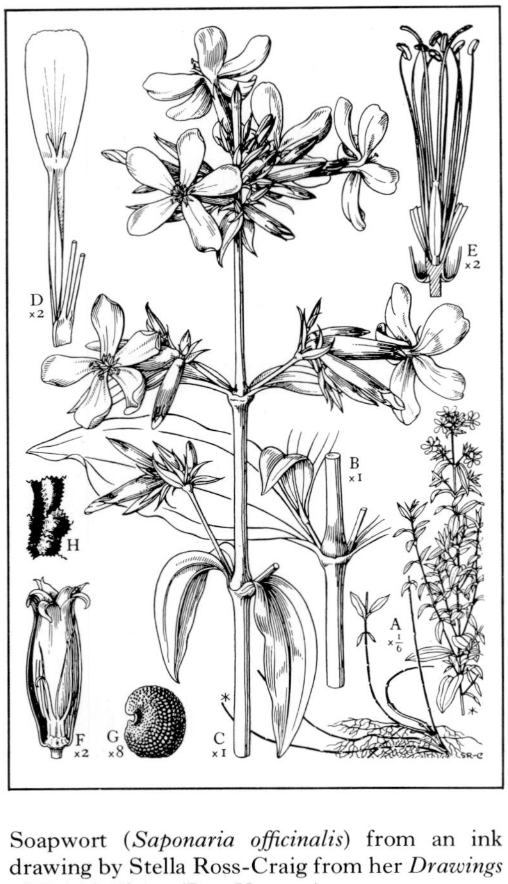

Soapwort (*Saponaria officinalis*) from an ink drawing by Stella Ross-Craig from her *Drawings of British Plants* (Part V, 1951).

During the period that the RHS managed the *Botanical Magazine*, Lilian Snelling (1879–1972) was its chief artist; she also contributed plates for other publications by the Society. The finest of these was *A Study of the Genus Paeonia* (1946) by F. C. Stern, an amateur gardener famous for his chalk garden at Highdown near Worthing in Sussex. The fifteen delicate coloured plates, mostly by Lilian Snelling, are well reproduced. The originals are in the Lindley Library, and one of them is reproduced here. Stern also wrote the smaller and more modest *Snowdrops and Snowflakes* (1956) illustrated with line drawings and coloured plates by E. A. Bowles and Margaret Stones.

Stella Ross Craig, born in 1906, worked all her life in London; she was on the staff of the Royal Botanic Gardens, Kew, from 1929 to 1960 and produced numerous plates for the *Botanical Magazine* from 1932 onwards. It is, however, for her black and white line drawings that she will be particularly remembered. Her finest work is the *Drawings of British Plants* which was published between 1948 and 1974. It contains over a thousand drawings of wild flowers from the British Isles, all of them absolutely clear, with no fussy details or stipple shading which might become blurred in the printing, but still giving a good idea of the character of the plant.

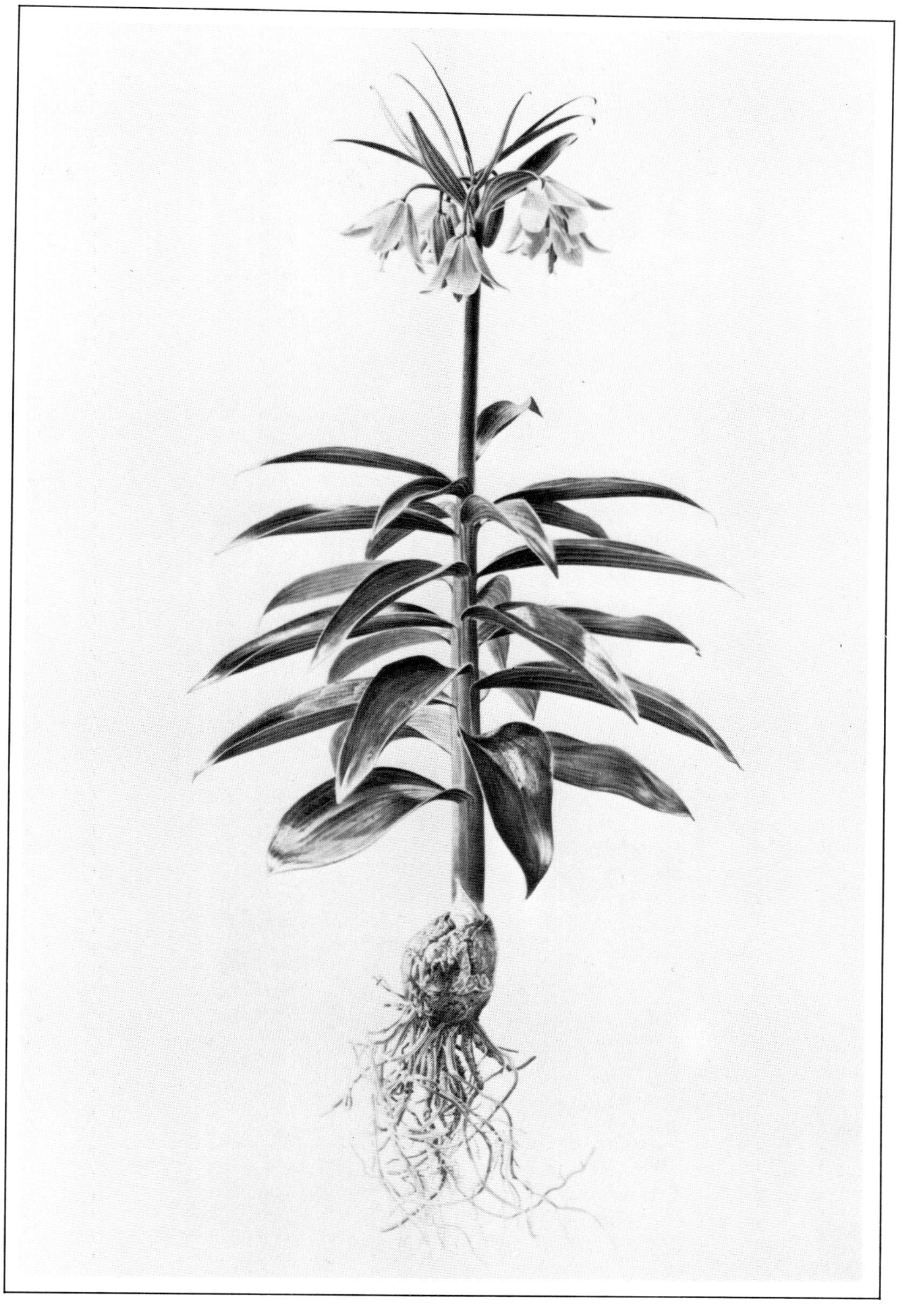

Fritillaria raddeana, watercolour on vellum, by Rory McEwen, 1978. From a plant collected by Paul Furse in northeastern Iran.

Margaret Stones is probably the foremost botanical illustrator living today. Born in 1921 in Victoria, Australia, she studied in the National Gallery Art School in Melbourne before coming to England in 1951. She worked at the British Museum (Natural History) and the Herbarium of the Jardin d'État in Brussels, as well as at the Royal Botanic Gardens, Kew. In 1955 she became artist to the *Botanical Magazine*, and since that time has produced most of the plates for it. She also illustrated supplements VIII and IX to Elwes's *Monograph of the Genus Lilium*, published in 1960 and 1962, and, most notably, produced the plates for *The Endemic Flora of Tasmania* (1967–78).

Mary Grierson was born in Bangor, North Wales, in 1912. She trained as an aerial photography interpreter in World War II and as a mapmaker before turning to flower painting. She studied under John Nash RA, who was primarily a painter but had illustrated R. Gathorne Hardy's *Wild Flowers in Britain* (1937), and his own *Poisonous Plants* (1947) and *English Garden Flowers* (1948). From 1960 to 1972, she was on the staff of the Royal Botanic Gardens, Kew, as a botanical artist and produced many plates for the *Botanical Magazine*, as well as numerous drawings for the *Kew Bulletin* and Hooker's *Icones Plantarum*. Her most recent published work has been in *Orchidaceae* by P. F. Hunt, a large folio published in London in 1974, in which her forty very good colour plates show the whole range of variation in this exceptionally diverse family. Others of her paintings have recently appeared in *The Plantsman.* She has also been particularly involved in paintings of endangered plants, and in making a series of paintings of tulip species for the Dutch bulb nursery of van Tubergen. Mary Grierson's main influence, however, has been on the next generation of young flower painters, through courses at Flatford Mill Field Centre, among them Joanna Langhorne, who was for some years on the staff at Kew.

The artists associated with Kew and their counterparts elsewhere in the world show that there is still a role in botany for the illustrator. Since the 35mm. camera came into widespread use, photography has replaced sketching as a method of recording plants found on expeditions. The artists tend to remain at home, painting plants grown in gardens or preserved in herbaria, sometimes using photographs as a guide to the habit or colours of the plant as it grows in the wild. Certainly for botanical details, such as flower dissections, photography is no rival for drawing.

Two factors have shaped botanical publishing in the years since World War II and particularly since 1960. At the popular end of the market, mainly in northern Europe, the advent of the international coedition has made it possible for popular flower books to be illustrated in colour to very high standards, as the investment can be recouped in a number of countries and languages. This development has been encouraged by a second factor: the emergence of photolithography as the dominant printing process for the production of illustrated books. The cost of colour has decreased greatly in relation to that of all the other elements that go into a book. Photolithography has also made possible the re-emergence of the splendidly produced monograph, although nowadays the primary appeal of the books as things of beauty rather than as botanical documents is emphasised by the fact that they are quite likely to be published by art galleries (like those of Margaret Mee on tropical South America) or by firms specialising in lavish limited editions, one of which was Wilfrid Blunt's *Tulips and Tulipomania* (1977), with illustrations by Rory McEwen. It is perhaps also relevant to mention here that the country which has produced the largest number of well-illustrated botanical works in the last two decades is South Africa, which has two great advantages: a uniquely rich native flora and the industries of gold and diamond mining to provide a clientele able to afford the books.

The talented amateur and the wealthy patron did not disappear completely from the world of botany at the start of World War I; both can be identified from the past two decades.

An amateur botanist who combined the talents of plant collector and traveller, photographer and flower painter was Paul Furse (1904–76). After entering the Navy at the age of thirteen and becoming inter-forces boxing champion, he had risen to the rank of Rear Admiral when he retired in 1959. Although he never received any artistic training, painting was in his blood. His great aunt was the redoubtable Marianne North, who travelled all over the tropics in the late nineteenth century painting flowers in their natural habitats; her highly coloured pictures – exotic trees against a backdrop of blue mountains, gingers in the steamy forests – can be seen in the gallery at Kew which she gave to the nation. Furse's grandmother also painted flowers, mostly bold and rhythmic watercolours of the alpines of Switzerland where she spent much of her life; these have never been published. His father, Charles Furse, was a fashionable Edwardian portrait painter. During early life, Paul Furse painted birds and landscapes as well as flowers, but later botanical painting became his chief pastime. Some of his early flower pieces are a carpet-like pattern of different flowers, and in the later botanical pieces many different forms of the same species often fill the page: a variation on this theme can be seen in the illustrations to *Collins Guide to Bulbs* where his paintings are to be seen alongside those of Paul Jones and Margaret Stones. *Fritillaria* became one of his favourite genera, and he did a series of paintings of these and lilies while he was in the navy, often painting on board his submarine. His style is very individual and very forceful, with much use of Chinese white to get highlights and the effect of bloom on the leaves of the bulbous plants he usually painted and to help imitate the fleshy texture that most of them have.

On his retirement from the navy, Furse started a series of plant-hunting expeditions to the Middle East: in 1960 to eastern Turkey with Patrick Synge, and in 1962, 1964 and 1966 to Iran and Afghanistan with his wife Polly, herself an accomplished landscape painter. The plant collecting did not leave much time for painting in the field, but Furse made sketches and copious notes, took photographs and brought back bulbs and seeds which were grown at the Royal Horticultural Society's garden at Wisley and at Kew. The flowers were painted when they flowered in cultivation, and a large number of these paintings are preserved in the Lindley Library of the RHS and at Kew, where they form a useful supplement to the herbarium specimens. Not many of Paul Furse's paintings have been published, but some can be seen in the RHS Lily Yearbooks, some in Synge's *Lilies* (1980), and in the Alpine Garden Society's *Guide to Fritillaries*, which, although not written by him, is based largely on his work.

The elements which traditionally went into really great flower books were reunited in *The Endemic Flora of Tasmania*. The combination of a rich, knowledgeable and enthusiastic sponsor, a scientific botanist, a fine painter with the ability to produce a sustained output of excellent work and a printer who works to the highest standards. The sponsor and instigator of the work was Lord Talbot de Malahide, an English diplomat and Irish baron. Born in 1912, Milo Talbot had an early career typical of the English aristocracy. After Winchester and Trinity College, Cambridge, he entered the diplomatic service in 1937 and served in Turkey and Lebanon and as ambassador in Laos in 1955–56. Meanwhile he had inherited Malahide Castle near Dublin and the family estates near Fingal, Tasmania. The garden at Malahide castle was ideal for growing the hardier Antipodean plants, having little frost but not as much rain as the gardens on the west coast of Ireland, and Lord Talbot soon built up a garden famous for its rare and interesting southern hemisphere plants. He commissioned a series of paintings of Tasmanian plants, and it was from these that the idea for the Flora developed. Lord Talbot died in 1973 before the book was completed, but the responsibility for the project was taken over by his sister, Rose Talbot, who has over-seen the completion of the work. Publication was in six parts between 1967 and 1978, and the 155 coloured plates by Margaret Stones illustrate 254 species, all endemic to Tasmania. The main text was written by W. M. Curtis, Reader in Botany at Hobart University, and the notes on cultivation were written by Lord Talbot. Margaret Stones visited Tasmania for some weeks in 1962, but most of the plates were drawn in England from living plants, either collected by Lord Talbot as seeds and grown at Kew or Malahide, or else were sent by air from Tasmania, where many amateur and professional botanists had been set to search for them. The originals for the plates are now kept in Launceston, Tasmania. The plates are very reminiscent of some of the Australian work of Ferdinand Bauer, not just in their subject matter but also in the beautiful drawings of botanical details in most of the plates. Even minute sedges and Centrolepidaceae have been made into elegant and colourful plates.

Enlightened individual sponsorship of the sort that produced *The Endemic Flora of Tasmania* is inevitably going to remain very unusual in the contemporary world. On the other hand, the bulk of the illustration undertaken for commercial publishers will tend to be of a functional nature, for inclusion in popular floras. Yet work of the highest artistic and botanical quality continues to be done by illustrators working mainly in scholarly surroundings. It is difficult to see how any great new illustrators will manage to emerge from the herbarium or the botanic garden to reach a public in the way that, say, Redouté or James Sowerby or Crispin de Passe did in the past.

References

ANDERSON, Frank J. *An Illustrated History of the Herbals.* New York, 1977.
ARBER, Agnes *Herbals. Their Origin and Evolution. A Chapter in the History of Botany.* Cambridge, 1912, reprinted 1938, 1953.
BIBLIOTHEQUE ROYALE ALBERT 1e *Vélins du Muséum. Peintures sur vélin de la collection du Muséum National d'Histoire Naturelle de Paris.* Brussels, 1977.
BLUNT, Wilfrid *The Art of Botanical Illustration.* London, 1950.
BLUNT, Wilfrid, and RAPHAEL, Sandra *The Illustrated Herbal.* London, 1979.
CAVE, Roderick, and WAKEMAN, Geoffrey *Typographica Naturalis.* Wymondham, 1967.
COATS, Alice M. *The Book of Flowers.* London, 1973.
COATS, Alice M. *The Quest for Plants.* London, 1969.
COATS, Alice M. *The Treasury of Flowers.* London, 1975.
COX, E. H. M. *Plant Hunting in China.* London, 1945.
DANIELS, Gilbert S. *Artists from the Royal Botanic Gardens, Kew.* Pittsburgh, 1974.
DESMOND, Ray *Dictionary of British and Irish Botanists and Horticulturists including Plant Collectors and Botanical Artists.* London, 1977.
DUNTHORNE, Gordon *Flower and Fruit Prints of the 18th and early 19th centuries.* London, 1970.
FREEMAN, R. B. *British Natural History Books 1495–1900. A Handlist.* London, 1980.
GRIGSON, Geoffrey (Ed.) *Thornton's Temple of Flora.* London, 1951, 1972.
HADFIELD, Miles, HARLING, Robert, and HIGHTON, Leonie. *British Gardeners. A Biographical Dictionary.* London, 1980.
HATTON, Richard G. *Handbook of Plant and Floral Ornament from Early Herbals.* London, 1960. Originally published in 1909 as *The Craftsman's Plant Book.*
HENREY, Blanche *British Botanical and Horticultural Literature before 1800.* Vol. I 16th century and 17th century. Vol. II 18th century. Vol. III 19th century, bibliography.
HUNT INSTITUTE FOR BOTANICAL DOCUMENTATION *A Catalogue of Redoutéana.* Pittsburg, 1963.
HUNT INSTITUTE FOR BOTANICAL DOCUMENTATION *The Hunt Botanical Catalogue.* Vol. I Printed Books 1477–1700; compiled by Jane Quimby. Pittsburgh, 1958. Vol. II: Part i, Printed Books 1701–1800; Part ii, Introduction to Printed Books 1701–1800; compiled by Allan Stephenson. Pittsburgh, 1961.
HUTTON, Paul, and SMITH, Lawrence *Flowers in Art from East and West.* London, 1979.
JACKSON, Benjamin Daydon *Guide to the Literature of Botany.* London, 1881, reprinted Koenigstein, 1974.
LEGER, C. *Redouté et son Temps.* Paris, 1945
LEVI D'ANCONA, Mirella *The Garden of the Renaissance.* Florence, 1977.
LEMMON, Kenneth *The Golden Age of Plant Hunters.* London, 1968.
NISSEN, Claus *Die Botanische Buchillustration.* Stuttgart, 1966.
PIERPOINT MORGAN LIBRARY *Flowers in Books and Drawings c940–1840.* New York, 1980.
PRITZEL, G. A. *Thesaurus Literaturae Botanicae.* Leipzig, 1872.
RICE EISENDRATH, Erna *Portraits of Plants. A Limited Study of the Icones in : Ann. Missouri Bot. Gard. 48, 1961, 291–327.*
SITWELL, Sacheverell, and BLUNT, Wilfrid *Great Flower Books 1700–1900.* London, 1956.
STAPF, Otto *Index Londiniensis. To illustrations of Flowering Plants, Ferns and Fern Allies,* Vols I–VI. Oxford, 1929–31. Supp. compiled by W: C. Worsdell. Oxford, 1941.
SYNGE, P. M. *R. H. S. Dictionary of Gardening.* Oxford, 1956.
TREVIRANUS, L. C. *Die Anwendung des Holzschnittes zur bildlichen Darstellung von Pflanzen.* Leipzig, 1855.
WHITTLE, Tyler *The Plant Hunters. 3,450 years of searching for Green Treasure.* London, 1970.

Index

The index concentrates on book titles and proper names mentioned in the book, particularly on authors, artists and engravers. Page numbers in italics indicate references in picture captions.